CAUCASIAN BATTLEFIELDS

PLATE I

THE FRONTIER OF TURKEY AND THE U.S.S.R., 1945
RAVINE OF THE ARPA-ÇAY

CAUCASIAN BATTLEFIELDS

A HISTORY OF THE WARS ON THE TURCO-CAUCASIAN BORDER 1828-1921

BY

W. E. D. ALLEN

AND THE LATE

PAUL MURATOFF

THE BATTERY PRESS
Nashville

In Association With
ARTICLES OF WAR, LTD
Skokie

Reprinted with permission
The Battery Press, Inc.
P. O. Box 198885
Nashville, Tennessee 37219
Ninth in The Battery Press European War Series
1999
ISBN: 0-89839-296-9
Printed in the United States of America

THE BOOK IS DEDICATED
BY THE SURVIVING AUTHOR
TO THE MEMORY OF OLD FRIENDS OF
THREE GENERATIONS

★

JOHN BADDELEY
CHARLES HAGBERG WRIGHT   ARTHUR HINKS
noble Victorians, lovers of learning who lived out their span in useful scholarship

ARNOLD WILSON   STANLEY CASSON
moderns in the Elizabethan style, killed on active service in their sixth decade

HARRY DE VILLEOUTREY
MARK PILKINGTON   PETER FARNHAM
companions in desert places, lovers of strange lore in the little time they had for learning

★

BY THE SAME AUTHORS

BY W. E. D. ALLEN:

A History of the Georgian People (1932)

The Ukraine: A History (1940)

Guerrilla War in Abyssinia (1943)

BY THE LATE PAUL MURATOFF AND W. E. D. ALLEN:

The Russian Campaigns, 1941–45 (in two volumes, 1943 and 1946)

CONTENTS

BOOK I

THE RUSSIAN CONQUEST OF CAUCASIA

BOOK II

THE CAUCASIAN THEATRE DURING THE CRIMEAN WAR

BOOK III

THE CAUCASIAN CAMPAIGN OF 1877–8

BOOK IV

THE CAUCASIAN CAMPAIGNS OF WORLD WAR I, 1914–18

LIST OF MAPS

The maps are based on the old Russian sheets of Caucasia printed by the Caucasian section of the Imperial General Staff, 10 versts: 1 inch; on the British sheets, 1:200,000 and 1:210,000; and on the more modern Turkish sheets, 1:800,000 (*Harita Genel Direktörlüğü*). Details have been added from maps published in works cited in the Bibliography.

For reasons of economy, it has proved necessary to reduce the number of maps originally prepared by the late P. P. Muratoff, and to make one map serve the purpose of different operations over the same area. For this reason symbols illustrating troop movements have in some cases been eliminated. On certain maps lack of space has made it necessary to omit a few place-names mentioned in the related text. A system of cross-references to alternative maps is intended to go some way to remedy their omission. In all cases it should be emphasized that the maps are approximative, and the reader who wishes to consider the topography of the area in detail is referred to the series of sheets cited above which are available in the Map Room of the Royal Geographical Society.

Mr Victor Nehring, of the Staff of the Royal Geographical Society, has been responsible for making new drawings from the original drafts by the late P. P. Muratoff.

The following maps will found on a sheet in the rear cover pocket: 1,14, 30,38.

LIST OF PLATES

PREFACE

THE joint author of this book, Paul Paulovich Muratoff, has not lived to see the work in print. He died on 5 October 1950. Paul Paulovich was responsible for the collation of the Russian sources and for the preparation of the original drafts of the maps. His constant counsel and his meticulous attention to detail were invaluable in the preparation of the book, and the lack of his supervision of the work in its final stages may be marked by errors which he would not have allowed to pass. For such errors as exist in the text and maps, I, as his surviving collaborator, must be held responsible.

The genesis of the book was over thirty years ago when, as a boy, I was fascinated by Baddeley's *Russian Conquest of the Caucasus*, by Lake's story of the defence of Kars, and Norman's *Campaign of 1877*. In later years travels in Caucasia and Turkey—and the encouragement of John Baddeley during the last decade of his life—inspired me with the ambition to contribute a companion and sequel to the *Conquest* and to bring the story down to the end of the First World War.

Much material accumulated over the years and some of it was published in the form of articles in the *Journal* of the Royal Geographical Society and in the *Army Quarterly*. With the approach of the Germans to the northern Caucasus in the summer of 1942 the subject seemed to acquire more than academic interest. At that time I was in England after the East African campaign and, during some enforced leisure, I determined to revert to my old subject. I was fortunate in securing the collaboration of Paul Paulovich Muratoff, himself a lover of military history, an officer of the old Imperial Russian Army and veteran of Augustovo, who had spent some time in the western Caucasus and the Crimea.

The subsequent writing of the book illustrates the minor vicissitudes which are the common experience of our time.

The two authors had infrequent opportunities of meeting. My own library of reference books was inaccessible, and we were dependent on the Caucasian collections—built up in great part by Baddeley—at the London Library and the Royal Geographical Society. The staffs of those two institutions pursued their normal avocations through the worst days of the war, and without their constant help research in London would have proved impracticable. Although neighbouring buildings were evacuated, and the London Library itself was badly damaged, Baddeley's friend and contemporary, Sir Charles Hagberg Wright, had continued to attend the library until his death in 1941, and the ageless Mr Cox still greeted the inquirer with omniscient kindness. The Royal Geographical Society also was hit, but the magnificent Hinks, sleeping often on the premises, was always there with his vigorous cheerful welcome. Hinks had passed his seventieth year; Hagberg Wright was over eighty; Mr Cox, now, is nearing ninety. Certainly, the fabulous Victorians set the style.

Through the hazards of the London blitzes, Paul Paulovich was working on the Russian sources. My wife, in spare evenings, typed out the raw material which followed me to my obscure pursuits. During the autumn of 1942, I remember working on the Sarikamiş battle in quarters in a condemned school on the Yorkshire moors. Baratov's Persian campaign kept a wintry course about the north Atlantic, and I was in the Kiği-Oğnot battles when we touched the West African coast.

Through the Beirut summer of 1943, my duties left me little time, but some evenings I turned to track the Turks from Erzincan to Kars and on the strange campaign which brought them to Baku. When my wife and I arrived in Mosul in early summer of 1944, the book was almost complete in draft, and a temporary sinecure at an average temperature of 108° in the shade gave us the chance to visit some of the ground of the 1914–17 campaigns—Sulaimaniya and Ruwandiz and the Kurdish alps as far as Amadiya. From the levies' camp in Sir Amadiya I could observe the ranges of Hakâri just beyond the Turkish frontier. With Peter Farnham, beloved of his levies and deeply read in the lore of the Iraq border, we concocted all kinds of plans. By the favour of the Turkish authorities I visited Hakâri in the summer of 1945, coming from Erzurum and Kars, along the Murat-su, seeing the battlefields of 1915–16 and rounding the lovely shores of Van. Peter Farnham had been killed in a jeep accident in Mosul some months before. Such is the luck of the world.

From time to time during the last quarter of a century, I have had opportunities of talking to veterans of the campaigns of 1914–18. They include the late General Baratov, the late General Kâzim Karabekir, His Excellency Rauf Orbay and General Kvinitadze. To these and to some score of officers and other ranks who fought on the Turkish or Russian sides in the campaigns of 1914–18, I am indebted for the authentic colour of epic days.

General Nuri Yamut, Chief of the Turkish General Staff, has shown great kindness in sending me books which were otherwise inaccessible.

It is not possible to name the several hundred persons—including numerous competent and hospitable officers and officials and many others casually met on journeys in Turkey—who helped directly or indirectly to complete the book. I may be allowed, perhaps, to recall with particular regard Esat Öner, the portly and jovial *vali* of Kars, whose question *Vouloir est-ce que c'est pouvoir?* will always remain in my memory as a tart comment on literary aspirations; Muvaffak Uyanik, former Director of Education in Diyarbekir, archaeologist and alpinist; and Colonel Nazmi Sevgen, with his store of unusual learning.

In Turkey, I was lucky to serve under three ambassadors, the late Sir Maurice Peterson, Sir David Kelly and Sir Noel Charles, all kindly and forbearing and ready to tolerate my interpretation of my duties as ambulatory rather than sedentary. I am in their debt for many happy days.

To Mr R. J. L. Kingsford and the Syndics of the University Press I am grateful for steady encouragement in a work which they can hardly regard as a profitable publishing venture.

My wife's sturdy toughness on horseback and by truck and jeep and *kayik* on some journeys which men of the country hesitate to take may be attributed to her love of adventure. Her devoted perseverance in typing and re-typing the MS. and her patience during the long hours spent on writing and talking about the book are other aspects of her remarkable gift for friendship.

W. E. D. ALLEN

WHITECHURCH, CO. WATERFORD
8 *September* 1952

THE SPELLING OF PLACE-NAMES

THE standardization of the spelling of place-names has presented formidable difficulties, notably for a frontier region where so many languages overlap. For example, since some of the principal reference sources have been Russian, it has been necessary to anglicize the Russian forms of Turkish place-names originally transliterated from the Arabic into the Cyrillic alphabet.

The abolition of the Arabic characters and the adoption of a Latin alphabet in Turkey (1928) has tended to accentuate rather than simplify the problem, because the numerous Turkish place-names in Caucasia and Persia continue to be transcribed from the Arabic alphabet according to the English or Russian systems.

Since the present work is a military history and not a study in philology, it has been decided to adopt a rough-and-ready solution which will doubtless provoke the criticism of purists—to whom the surviving author readily offers his apology.

The modern Turkish system used on the series of staff maps (*Harita Genel Direktörlüğü*, 1936) and on the excellent physical maps of Turkey issued by the Turkish Ministry of Education has been used for all place-names within the modern boundaries of the Turkish Republic. Exception has been made in the case of a few historical names of Georgian or Armenian origin (e.g. Ardanuch, Shavsheti). A simplification of the Turkish genitive termination has also been adopted (see below). In order to ease understanding of Turkish geographical terms, the older system of hyphening has been retained (e.g. *Yalnizçamdağlari* is written *Yalnizçam-dağ*). An attempt has also been made to translate Turkish, Russian and Georgian place-names, particularly when these may be taken as descriptive of local features: e.g. *Yalnizçam-dağ* (T.) = Lone Pine Mountain; *Gory Mokriya* (R.) = Wet Mountains; *Taşli-çay* (T.) = Stony River; *Tskhenis-dzqali* (G.) = Horse's River. One of the most striking characteristics of oriental place-names is their simplicity of conception. In this respect, as the expression of the straightforward peasant mind, they may be compared to the toponomy created in Australia or Canada during the last century or even to old Celtic toponomy. (What can be more 'highland' in flavour than 'Gap of the Sword', rendered in Turkish as *Kiliç-gedik*, and quite incomprehensible in the anglicized transcription of the Russian transformation, *Kluch-gyaduk*?)

A further difficulty in the rendering of names arises from the numerous changes in nomenclature and spelling adopted during the last twenty years by the authorities in Turkey and also in the Caucasian republics. Erzerum has become Erzurum, Erivan has become Erevan or Yerevan. Names familiar in the old wars have disappeared from the map. The historic *Soğanli-dağ* (Onion Mountains) have become *Pasinlersiradağlari* (Pasinler chain of mountains); *Dram-dağ* is now marked *Şeytan-dağ*; *Karakilise* = *Karaköse*, etc. In these cases the names made familiar in the chapters on the earlier wars have been retained, but the new names are indicated in brackets in the Index.

The traditional English forms of well-known names have been retained instead of new forms or the native variant now used on modern maps, but in the case of lesser known names the newer variant has been adopted. For instance, *Trebizond* is used for *Trabzon*, *Tiflis* for *Tbilisi*, *Batum* for *Batomi*, *Sukhum* for *Tskhumi* or *Suhumi*; but *Kutaisi* has been adopted for the older *Kutais*, *Giresun* for *Kerasunt*. Alternative forms are indicated in the Index. Soviet forms adopted after 1922 in Caucasia are also given in the Index, e.g. *Leninakan* for *Aleksandropol* (T. *Gümrü*), *Kirovabad* for *Elizavetpol* (T. *Gence* = *Ganja*).

Where translations of topographical names are not given in the text or footnotes, they are given in brackets in the Index (T. = Turkish, R. = Russian, G. = Georgian, A. = Armenian, C. = Cherkess, P. = Persian).

TURKISH ALPHABET

There are twenty-nine letters in the latinized Turkish alphabet. The majority are pronounced as in English. The following Turkish letters differ, however, from their English equivalents:

TURKISH	ENGLISH	REMARKS
c	j	English *j*, e.g. *Cemal* (Jemal), *Erzincan* (Erzinjan). But *j* has been retained for place-names outside the present Turkish frontier even when such names are clearly of Turkish origin, e.g. *Balajari*, *Delijan*, *Ganja*.
ç	ch	As in *church*, e.g. *çay* (chay); *Çakmak* (Chakmak). But *ch* has been retained for rendering names outside the Turkish frontier except where such names consist of common Turkish words, e.g. *Gökçay*, *Türkmençay*.
g	g	Pronounced *gy* before *â*, *û* and soft vowels. When written *ğ* it is generally, but not always, mute; e.g. *Boğaz* pronounced Bo-az; *Soğanli*, pronounced So-anli.

TURKISH	ENGLISH	REMARKS
h	h	*kh* is not used for Arabic خ: *Harput*, not Kharput; *Harşit*, not Kharşit. But *kh* has been retained for rendering خ outside the Turkish border; e.g. *Khoy*, *Shemakha*, and for the traditional *khan* in its honorific form.
i and ı	i	No distinction has been made between Turkish *i* and the slurred *ı*. Strictly, the capital letter also carries a dot: *İstanbul*.
j	j	Equivalent to French *j*, this letter is rarely used, and chiefly in words of French and Persian origin, e.g. *Manjil*.
k	k	Pronounced *ky* before *â*, *û* and soft vowels. The letter *q* is not used for Arabic ق: *Kars*, *kara*, *kizil*, not qara, qizil.
ö	ö	As in German.
s	s	As in English; *ss* is not used.
ş	sh	*Yeşil* = Yeshil.
ü	ü	As in German.
y	y, i	Used for *i* in the second half of diphthongs, *Köprüköy*.

Turkish grammatical forms in toponymics. The Turkish plural is formed by adding to the noun *-lar*, or *-ler*, according to whether the vowels of the stem are hard or soft. The genitive case is expressed by adding the suffixes *-i*, *-ı*, *-u*, *-ü*, or *-si*, *-sı*, *-su*, *-sü*, to the object possessed: *Kophanlari* = 'hans' (resthouses) of Kop; *Torosdağlari* = mountains of Taurus; *Çildirgölü* = lake of Çildir. The suffix, however, is often omitted, both in speaking and writing. The equivalent in English would be to speak of 'Çildir Lake' or 'Lake Çildir' (*Çildirgöl*) instead of 'lake of Çildir' (*Çildirgölü*). Further, the genitive suffix is never added to a substantive when qualified by an adjective or by a substantive used adjectivally: *Karadeniz*, Black Sea; *Tavşan-dağ*, Hare Mountain.

As the use of the Turkish plural and genitive suffixes in geographical names is extremely complicated and by no means standardized, it has been abandoned for practical purposes in this work. Thus, 'Lone Pine Mountains' strictly should be written *Yalnizçamdağlari* or even *Yalnizçamsiradağlari*, lit. 'chain of the mountains of the solitary pine'. However, it has been rendered *Yalnizçam-dağ*, which, at worst, is the equivalent of 'Lone Pine Mountain'.

RUSSIAN ALPHABET

RUSSIAN	ENGLISH	REMARKS	RUSSIAN	ENGLISH	REMARKS
а	a		п	p	
б	b		р	r	
в	v		с	s	
г	g	As in *get*	т	t	
д	d		у	u	Pronounced as 'oo' in *boot*
е or ѣ	e	See Note 1, below	ф or ѳ	f	
ё	ё	Pronounced as 'yo' in *yonder*	х	kh	
ж	zh	Pronounced as 's' in *pleasure*	ц	ts	
			ч	ch	
з	z		ш	sh	
и or і	i		щ	shch	
й	y	See Note 2, below	ы	y	See Note 2, below
к	k		э	e	
л	l		ю	yu	
м	m		я	ya	
н	n		ь		Soft sound; see Note 3, below
о	o				

NOTE 1. The Russian е is pronounced 'ye' (as in *yes*), and wherever a Russian е follows a vowel, it has been transliterated as *ye*. However, if the е is initial (as in *Ermolov*) or follows a consonant, it has been rendered as *e* (though this is not strictly correct phonetically). It follows that whenever *ye* comes after a consonant, that consonant is soft in Russian, e.g. *Muravyev* (see also Note 3 below).

NOTE 2. The English letter *y* has been used for both the Russian й and the Russian ы (which is nearly unpronounceable in English). However, as ы never follows a vowel, while й never follows a consonant, no confusion can arise.

In the terminations ій (ий in modern spelling) and ый the final й has not been transliterated, ій having been rendered as *i* and ый as *y*. Therefore, whenever, *ii* occurs, the two letters should be sounded separately.

A slight difficulty has arisen in the transliteration of such terminations as ыя, which, if rendered strictly in accordance with the system, would be *yya*; this is so formidable a combination that *iya*, though not strictly correct, has been used in its stead.

NOTE 3. *Soft Consonants*. Whenever a soft consonant is followed by a vowel, the vowel is preceded by a *y* in the English rendering (see Note 1 above).

TRANSLITERATION OF GEORGIAN NAMES

In the transliteration of Georgian names the general principles adopted by the editors of *Georgica* have been followed (see articles by A. Gugushvili, *Georgica*, vol. 1, no. 1, 'The Georgian Alphabet'; and nos. 2 and 3, 'Classification of Georgian Sounds'). For simplicity diacritical marks have been omitted (e.g. *Kartli* for K'art'li, *Mtiuleti* for Mt'iulet'i).

TRANSLITERATION OF ARABIC, PERSIAN AND ARMENIAN NAMES

In general the principles laid down by the Permanent Commission on Geographical Names have been followed. Again, diacritical marks have been omitted. Except in commonly accepted forms (e.g. *Iraq*) the rendering of the Arabic Qaf ق by *q* has been abandoned in favour of the Turkish *k* with a view to avoiding anomalies in the rendering of numerous place-names in Persia and Iraq containing Turkish elements (e.g. *Kizil Arvat*, *Khanikin*).

ACKNOWLEDGEMENT

For the final co-ordination of the transliteration of Turkish and Russian names and for the preparation of the Index I am indebted to my secretary, Mr Anthony Mango. His painstaking interest and ordered method have proved a necessary discipline to my own impatience, to which must be attributed the surviving errors.

W. E. D. A.

Across the Caucasian peninsula and its home basin, the Caspian Sea, Russia is in immediate contact with the whole mass of Muslim Asia. From the Caucasian peninsula Russia may reach out everywhere where it may prove necessary....For Russia the Caucasian peninsula is a *bridge* thrown from the Russian shore into the heart of the Asian continent, a *wall* shielding Central Asia from inimical influences and an *advance bastion* protecting both seas—the Black and the Caspian. Occupation of this country constituted a prime state necessity.

FADEYEV, *Sochineniya*, I (St Petersburg, 1889), p. 10;
cited by KHACHAPURIDZE, *Istoriya Gruzii*
(Tiflis, 1949), p. 22

Thus the past is on top of us and with us all the time; and there is only one way of mastering it even remotely in any one sector: by knowing how these things have come to be, which helps to understand their nature, character, and their correlation, or lack of correlation, to the present realities of life.

Professor L. B. NAMIER, *Avenues of History*
(London, 1952), p. 2

BOOK I

THE RUSSIAN CONQUEST OF CAUCASIA

CHAPTER I

THE HISTORICAL GEOGRAPHY OF CAUCASIA

MAP I

THROUGHOUT history Caucasia has been a borderland. The main chain of the Caucasus Mountains, stretching from the Taman peninsula on the Azov Sea to the Apsheron peninsula on the Caspian, is the natural limit of the mountain zone of the Middle East comprising the Armenian and Iranian highlands—themselves a westerly extension of the Himalayan system.

In the north the slopes of the main chain of the Caucasus descend to the north Caucasian steppe. This steppe in its western part is a projection of the grasslands which are fed by the rivers flowing to the Black Sea. The Kuban, like the Don, belongs essentially to the Pontic river system. The low hills of the Stavropol ridge running north to Manych form the divide between the western and eastern steppes. Eastward the north Caucasian steppe disappears in the sandy flats stretching to the Volga and forming part of the old bed of the Caspian Sea, an extension of the arid zone of the Aralo-Caspian depression. This steppe is typical of the desiccated lands which edge the 'great heartland'[1] of inner Asia. Appropriately enough, this north Caucasian steppe is inhabited by the nomad Mongol-speaking Kalmyks.

The main chain of the Caucasus Mountains consists of a series of parallel ridges. These ridges are linked by necks or saddles which give access from north to south of the main chain. The connecting ridges often form wide upland glens, sometimes at a great elevation. Typical examples are upper Svaneti and the Tush and Khevsur glens in north-eastern Georgia. Here communities have lived in almost complete isolation during all the known period of history.[2]

The granitic backbone of the Caucasus chain runs as a single wall right across the Caucasian isthmus from sea to sea and separates the basins of the rivers of the northern slope from those of the southern. It thus represents the transverse water divide of Caucasia. The overlapping parallel ranges are all shorter in length than the main chain, but the principal parallel range of Bokovoy (R. 'Flanking') is higher

[1] The expression was first used by H. J. Mackinder, *Democratic Ideals and Reality* (1919), p. 96.

[2] Strabo and other classical writers mention the Svans.

than the main chain and gives rise to some of the most notable peaks of the Caucasian system, namely, mounts Elbrus (5630 m.), Koshtan-tau (5198 m.), Kazbek (5043 m.) and Adai-khokh (4410 m.).[1] The peaks of the Bokovoy generally exceed in height those of the main ridge: Ushba (4697 m.), Tetnuld (4858 m.) and Shkara (5182 m.).

The western Caucasus, under the influence of the moist climate of the Black Sea basin, is heavily forested and the snow-line is lower than in the eastern mountains, which come under the desiccating influence of the Aralo-Caspian depression.[2]

There are three principal lines of access over the main chain from north to south. The Coast Road (*Beregovoye Shossé*) is an artificial route which has been constructed during the last half-century.[3] A motor-road and a railway have been built by the Russians along the line of the Black Sea coast from Novorossiysk to Kutaisi in Georgia.[4]

The second line of access from north to south is the Georgian Military Road (*Voyennaya Gruzinskaya Doroga*) from Vladikavkaz (Ordzhonikidze) to Tiflis (Tbilisi). It runs up the valley of the Terek through the gorge of Daryal, which is as celebrated in the history and legend of the Middle East as the famous Cilician Gates through the Taurus. The road crosses the watershed by the *col* of Krestovy (now known under the Georgian form *Juari*, 'Cross') and descends the valley of the Aragvi to its junction with the Kura some twenty miles above Tiflis. The Romans and the Persians, and later the Georgian kings, maintained fortifications and a garrison at the Daryal defile (the 'Caspian' or 'Iberian' Gates), but the severities of the route (so well described in Lermontov's *Hero of Our Times*) seem to have made it impractical as a main line of invasion for barbarian or medieval armies. It was only at the end of the eighteenth century that the Georgian Military Road became a regular highway when General Todleben, the conqueror of Berlin during the Seven Years War, crossed the pass with two battalions and four guns; the Georgians, who had a fort at the defile, were his allies.

The third route follows the sandy foreshore of the Caspian between the foothills of the Dağistan mountains and the sea. The narrowest

[1] *Tau* (tor) and *khokh* (hoch) are Indo-European (Ossetian) toponyms.

[2] A profile of the western Caucasus from Anapa to Elbrus is to be found in *Sbor. Mat.* XXIII, opp. p. 16. Profiles of the flanking ridges, showing geological structure, are in Javakhishvili, *Geography of Georgia* (G.).

[3] During the Russo-Turkish War of 1877–8 a track existed which was used by Shelkovnikov's column. (The late P. P. Muratoff, as a student engineer, worked on the construction of the new route. W.E.D.A.)

[4] One section of the railway was incomplete (1942).

point between the mountains and the sea is at Derbent,[1] where the gap is about six miles in width. This coastal route has always been the historic highway along which armies have passed the Caucasus. The Sassanid kings of Persia were accustomed to maintain a garrison on the so-called Caucasian Wall, but neither the Persians nor their successors, the Arabs of the eastern caliphate, could permanently check the incursions of the northern nomad peoples. The Khazars made many invasions of Transcaucasia and Persian Azerbaijan along the route by the Caspian coast. Derbent constituted the most important frontier fortress covering the wealthy Caspian provinces of the caliphate; it could be reinforced from Barda'a (P. *Partav*), the metropolis of eastern Transcaucasia in the triangle formed by the rivers Kura and Terter. But in the tenth century the Russians (Warangs) actually surprised and sacked Barda'a by an amphibious operation directed across the Caspian from the lower Volga.

In the thirteenth century the Mongols, coming from Persian Azerbaijan, moved north by the Derbent gap against the Russian dukedoms; and, during the wars between the Iranian Il-khans and the Mongols of the Golden Horde settled on the Volga, the Caspian foreshore was the main passage-way for the contending armies. When, in the eighteenth century, Peter the Great undertook the first Transcaucasian expedition of modern Russian arms he repeated from the lower Volga the amphibious strategy of his Warangian predecessors.

The route along the Caspian shore has always had a prime importance which derives from the strategic unity of the Caspian area with the two great estuaries of the Volga and the Kura-Aras falling to the oblong inland sea.[2]

There are other passes over the Caucasus, the best-known of which are the Mamison (Imeretian Military Road) connecting the upper valley of the Ardon, a tributary of the Terek, with the valley of the Rion, and the Klukhor Pass (2816 m.), giving access from the valley

[1] P. *Dar-band*; A. *Bab-al-abwab* ('Gate of Gates'); also called the 'Albanian', but *not* 'Caspian' Gate. For the Caucasian Wall at Derbent, see Baddeley, *Rugged Flanks*, I, pp. 155 ff.; for description and plan, see R. von Erckert, *Der Kaukasus und seine Völker* (Leipzig, 1887), pp. 216 ff.

[2] See further M. N. Gersevanov, 'Istoricheski ocherk torgovykh putey soobshcheniya v drevnem Zakavkazye', in *Sbor. Sved.* I; and an interesting article by A. N. Dyachkov-Tarasov, 'Neizvestny drevni torgovy put' iz Khorezma v Vizantiyu cherez Kavkaz', in *N.V.* XXVIII; *ibid.* XVIII, see V. N. Khudadov, 'Napravleniye putey soobshcheniya i tsentry Zakavkazya'. On the early Warangian raids across the Caspian there is a rare work by B. Dorn: *Kaspi: o pokhodakh drevnikh Russkikh v Tabaristan s dopolnitelnymi svedeniyami o drugikh nabegakh ikh na priberezhya Kaspiyskago morya*, St Petersburg, 1875.

of the Kuban to that of the Kodor. There are some seventy other tracks and paths across the main chain, some of which are only suitable for pack transport and others for men marching in single file. In many cases they are blocked by snow except for two or three months a year.[1]

In Transcaucasia, south of the main chain, the same diagonal division into western and eastern areas is reproduced. The Little Caucasus (Suram or Meshkian Mountains), thrown off at right angles from the main chain, forms a ridge 1525–1830 m. in height linking the Great Caucasus with the Armenian highlands. To the west of the Suram Mountains the Rion basin is an emerged part of the bed of the Black Sea and belongs to the climatic and natural world of the Pontus. East of the Suram Mountains, the rivers Kura and Aras (Araks), coming down from the high plateau of Armenia, flow through the arid steppe of Azerbaijan which forms, as does the Kalmyk steppe to the north of the main chain, a part of the Aralo-Caspian depression.

The ranges forming the 'peripheral rim'[2] of the Armenian plateau run parallel with the main chain of the Caucasus in a general direction south-east to north-west. These ranges are in fact a continuation of the Elburz chain which constitutes the connecting fold between the central Asian complex of ranges and the Caucasian and Armenian systems. The Armenian 'rim' is continued in the Pontic Alps which build the southern containing wall of the Black Sea basin. Strategically, the ranges between Persian Elburz and the Pontic Alps represent a second mountain wall, parallel with the main chain of the Caucasus, covering the approaches from the Eurasian plain to the Iranian plateau and the Mesopotamian lowland. These ranges form extensive upland plateaux, averaging levels of 2135–2440 m. (7000–8000 ft.): the 'lake district' round the sources of the Kura; the Kars plateau; Zangezur and Karabağ to the north of the Aras; and Karadağ to the south of the river. Rising above the level of the plateaux are isolated massifs like Alagöz (4095 m.) and Ararat (5165 m.).

This peripheral rim of the Armenian plateau is cut through by the valleys of the Kura and the Aras, flowing into the Caspian, and the valley of the Çoruh falling to the Black Sea. The rim is ribbed also by a series of ranges running in a general direction west to north-east.

[1] Baddeley, *Rugged Flanks*, I, pp. 125 *et seq.*, notes two practicable routes for pack animals in the area between the Mamison and Daryal passes. East of the Georgian Military Road horsemen can by-pass the Daryal gorge by the difficult route up the R. Assa and over the Arkhotis Pass (*ibid.* pp. 168 *et seq.*).

[2] The expression is from Lynch, *Armenia: Travels and Studies* (2 vols. 1901).

The valleys of the Kura, Aras and Çoruh form natural alleys of access between, but not across, these ranges, thus:

Yeşil-Irmak—Kelkit-Irmak—Çoruh,
Upper Euphrates (Firat)—Aras—(Kura),
Eastern Euphrates (Murat-su).

A limited number of 'saddles'—giving rise often to the upper valley of an affluent of one of the main rivers—allows access from one diagonal to another. The best-known are those across the Pontic Alps from Trebizond on the Black Sea to Erzurum, which lies behind the dividing ridge between the Euphrates and the Aras. The Zigana and Vavuk passes (both over 1830 m.) link Trebizond with Bayburt in the valley of the Çoruh. The Kop-dağ saddle carries the road over the divide between the Çoruh and the Kara-su, the northern arm of the Euphrates. Other saddles link the lower Çoruh with the upper Kura; the Oltu-çay affluent of the Çoruh with the Kars-çay and Arpa-çay affluents of the Aras, and these again with the upper Kura; and the valley of the Aras with that of the eastern Euphrates (Murat-su).

War in the borderland between the Caucasus and the mountain ranges covering the approaches to the Iranian and Anatolian plateaux has essentially the character of mountain war: battles for saddles and defiles, followed by attacks on fortified localities commanding main alleys of communication along river valleys. Transport and supply difficulties impose the use of relatively small forces accustomed to the rigours of climate in the Armenian highlands. The experience of former wars shows that when the commanding saddles have been lost fortified localities can sustain long sieges owing to the difficulty of transporting heavy artillery and the supplies necessary to the support of a siege force (Kars, 1855; Erzurum, 1877–8). In the recurrent Russo-Turkish wars fortified localities have succumbed to famine (Kars, 1855) or to surprise storm (Kars, 1877; Erzurum, 1916).

A notable lesson of the successive Russo-Turkish wars over the same terrain and under similar climatic conditions is the progressive capacity of man to endure and overcome hardships imposed by natural conditions (compare the operations against Erzurum in the winters of 1877–8 and 1915–16).

The Alps as a barrier between the peoples of the north European plain and the Mediterranean basin may in some ways be compared to the main chain of the Caucasus which barred the descent of the Eurasiatic nomads into the old civilized lands of the Middle East. Again, as Savoy and the Lombardy plain constituted the historica battleground between Mediterranean peoples in conflict, so the Armenian highlands and the valleys of the Rion and the Kura which form the trough-like depression of Transcaucasia were the natural arena of the successive 'world powers' of the Middle East. The historic frontier between the East Roman empire, dominating the Black Sea basin and the Anatolian plateau, and the successive Iranian empires (Parthian and Sassanid) lay along the ridge of the Suram. Each side sought to control the *glacis* sloping down beyond *their* side of the Suram 'wall'.[1] In the sixth and seventh centuries A.D., Lazika and Iberia (western and eastern Georgia on either side of the Suram) were the scene of 'world war' between the rival imperial powers of Byzantium and Sassanian Persia.[2] And in the third decade of the seventh century the Emperor Heraclius, by his campaign in Transcaucasia, prepared the way for his epic passage of the Zagros passes and his descent on the Sassanian capital in Mesopotamia.

The natural line of division between a power controlling the Ponto-Anatolian area and the Caspian-Iranian region has always been the Suram ridge and the main chain of the Zagros running in a general direction south-south-east. Between the fortress of Akhaltzikhe (G., 'New castle') at the head of the Kura gorges and Mount Ararat, the centre of the Armenian 'knot',[3] the Akhalkalakis-dzqali (G., 'Newtown water'), the Arpa-çay (T., 'Barley river') and the middle course of the Aras have offered convenient boundaries, leaving to cover Anatolian territory the Arsiani-Soğanli and Ağri-dağ chains and, as bastions protecting eastern Transcaucasia, the Somkhetian mountains (R. *Gory Mokriya*, 'Wet mountains'), the Alagöz massif and the great complex of Shah-dağ and Karabağ east and south-east of Lake Sevan. This natural division was recognized in the Turco-Persian partition of Caucasia in 1636.

During the recurrent Turco-Persian wars of the sixteenth–eighteenth centuries, the fortress of Kars was always the advance base at the head of the great Anatolian trunk road, Sivas-Erzincan-Erzurum. The importance of Kars lay in the fact that it covered the

[1] The terms *amier* and *imier*, 'this side' and 'that side', constantly recur in the Georgian Annals. Cf. Wakhusht, *Description géographique*, p. 336, n. 1.

[2] For the first two Lazic wars see Procopius, analysed in Brosset, *Histoire de la Géorgie: Additions et Eclaircissements*, Add. IV, 'Sur le Royaume de Lazique'. The campaigns of Heraclius have been studied by N. H. Baynes in *Eng. Hist. Rev.* XIX (1904), 'The first campaign of Heraclius against Persia'; and in the *United Services Magazine*, XLVI–XLVII (1912–14), 'A critical study of the military operations of the Emperor Heraclius'. See also *B.S.O.A.S.*, 1944, XI (2), 'Roman and Byzantine campaigns in Atropatene', by V. Minorsky.

[3] The expression is from Dudley Stamp, *Asia: An Economic and Regional Geography*.

twin Turkish fortress at Akhaltzikhe and made possible a rapid Turkish advance both down the Kura gorges to Gori and along the affluents flowing to the middle valley of that river. Such line of advance at once turned the line of the Suram and threatened Tiflis. Tiflis covered all the middle and lower Kura and was the key to eastern Transcaucasia as far as the Caspian. Kars was the key to Tiflis and hence has been described as the 'key to Transcaucasia'.[1]

On the Persian side the fortress of Erevan, at the foot of the great complex of mountains round Lake Sevan (A. *Sevan*; T. *Gök-çay*; 'Blue water'), commanded the valley of the Aras and stood on the flank of any advance from Kars down the affluents of the Kura against Tiflis. From Erevan an advance across the passes of the Ağri-dağ could, first, cut communications between Van and Erzurum and, secondly, threaten the Lake Van area and the route through Bitlis to Diyarbekir and the upper Tigris. Thus the main barrier of the Zagros and the difficult pass of Kotur covering Van might be turned.

In general terms the line through Suram, Akhalkalakis-dzqali, Arpa-çay, Aras and Zagros remained the frontier between the two Middle Eastern powers—Ottoman Turkey and Persia—until the end of the eighteenth century. Both Mussulman powers pursued a policy of indirect rule along the Caucasian borderland. On the whole the local potentates in eastern Transcaucasia, the kings of Kartli and Kakheti (east Georgia) and various Armenian meliks and Tartar khans, depended on the Persians; the western potentates, the Cherkess chieftains, the Georgian princes in Imereti, Mingrelia and Guria, the Muslim Georgian paşas of Akhaltzikhe and the Muslim-Georgian *dere-beyler* ('lords of the valley') in Acaristan and Lazistan, owed fealty to the sultan. The Georgian Muslim princes were in fact among the most determined opponents of the Russians during the early wars of the nineteenth century.

The Kura-Aras estuary, with the rich winter grazing of the Mugan steppe and the adjoining summer pastures in the highlands of Kara-bağ and the Meskhian lake district, were perfect bases for the mounted armies of the Asiatic nomads, and from here the Mongols advanced to the conquest of the Volga valley and the Eurasian plain. In later centuries the Russians used the other great Caspian estuary of the Volga as the base for their advance into the Middle-Eastern lands and (in part) for their conquest of the nomad homelands in central Asia.

The development of Russian history has been largely influenced by

[1] By the delegates of the Transcaucasian Seym at the Conference of Trebizond (March 1918) (cf. *Dok. Mat.*).

the existence of the two marine basins lying one on either side of the Caucasian isthmus—the Black and the Caspian Seas. Greek civilization, radiating round the Pontic basin, spread up the rivers flowing into it and coloured the whole background and life of the first Russian state on the Dniepr (Kiev, tenth–twelfth centuries). Similarly, oriental influences (Arabic, Jewish and Turkish) spread across the Caspian Sea and up the Volga. The early Volga states, Khazar, Kipçak and Tartar (Golden Horde), were essentially oriental in character. Muscovite Russia, which emerged in the fifteenth and sixteenth centuries in the area between the upper Don and the middle Volga, was a synthesis of the two old and passing worlds of Orthodox Byzantium and the Tartar-Mongols who had themselves developed and declined under Islamic Persian and even Chinese influences. Muscovy, a state and a civilization, arising out of the riverlands of the Eurasiatic plain, was an entirely new historical phenomenon, which was destined to modify and transform neighbouring and older societies.

Before they became a 'European' power on the Black Sea and the Baltic, the Muscovites mastered the great Eurasian rivers: northern rivers like the Pechora flowing to the Arctic; and the Volga pouring into the land-locked Caspian beyond which lay the inmost lands of the Asiatic peoples.

Chief instrument of this Russian advance from the region of the upper Don and the middle Volga to the Caspian Sea, the Caucasus and the central Asian steppes, were the Cossacks, communities of men with many purely anarchic tendencies but serving military and colonial ends—a peculiar and original manifestation of the Russian genius.

By the end of the sixteenth century the Russians had finally set their power on the Volga and defeated the great Turkish attempt to fight off their progress to the south (1569–70). During the following century the Don Cossacks were challenging Turkish control of the Azov Sea, and on the Caspian shore the Terek (Terski) Cossacks had set up forts along the Terek and the Sunja where they came into contact and conflict with the numerous and formidable Muslim tribes of Chechnia and Dağistan.

In 1722 a state of civil war in Persia encouraged the Emperor Peter I to attempt the conquest of the Caspian coasts. The first regular army trained on modern European lines to operate in Caucasia was transported by sea from the mouth of the Volga to the Terek; and Peter, with no great difficulty, occupied all the country south of Derbent as far as the khanate of Kuba (north of Baku). Russian troops were also landed on the southern coast of the Caspian in the

Persian provinces of Ghilan and Mazandaran. Unfavourable weather conditions on the Caspian and the outbreak of epidemics checked further progress; and the death of Peter, followed by changes in Russian policy, and later the rise of a powerful military dictatorship in Persia under Nadir Shah, delayed Russian expansion round the shores of the Caspian for half a century. The campaign of 1722, however, had demonstrated the practicability of a relatively rapid advance along the Caspian flank of the main chain of the Caucasus.

During the second quarter of the eighteenth century the foreign advisers of the Empress Anne, Münnich and Lacy, developed the system of Cossack advance and colonization into an imperial tactic and made these semi-anarchic communities into the instruments of a calculated and long-sighted policy. The Cossack 'lines'—'stations' (*stanitsy*) or military settlements linked by forts—were extended. Thus there was the Orenburg line over against the steppe Kazaks beyond the Urals; the Ukrainian line between the Dniepr and the Donetz protecting the newly settled lands of the south from incursions by the Tartars of the Crimea; and the Caucasian line built first along the Terek and connecting the Terek with the lower Don. At the same time, in Caucasia, the Russians increased their influence with the Georgian kings in Kartli and Imereti, and they made a useful alliance with the Kabardan princes—Cherkess nobles ruling a mixed population of Cherkesses and Tartars and controlling the northern slopes of the main chain between the upper waters of the Kuban and the Terek. The Kabardan alliance allowed the Russians to isolate the Cherkess communities along the Kuban and the coast of the Black Sea, who were under Turkish influence, from the tribes of Dağistan who remained more or less within the Persian sphere.[1]

To understand the political strategy which formed the background

[1] *B.S.E.*, vol. on U.S.S.R. (1948 ed.), has some useful maps illustrating the process of Russian expansion (between p. 351 and p. 418). See also E. D. Felitsin's military-historical maps of N.W. and N.E. Caucasia published in Potto's *Istoricheski ocherk Kavkazskikh voyn ot ikh nachala do prisoyedineniya Gruzii* (Tiflis, 1899). Recent studies covering aspects of the Russian approach to Caucasia are: N. A. Smirnov, *Rossiya i Turtsiya v XVI–XVII vv.* (2 vols., Moscow, 1946); A. A. Novoselski, *Borba Moskovskogo gosudarstva s Tatarami v XVII veke* (Moscow, 1948); B. H. Sumner, *Peter the Great and the Ottoman Empire* (Oxford, 1949). Dr Halil Inalcik has utilized Turkish sources in a short but brilliant study on 'The origin of the Ottoman-Russian rivalry and the Don-Volga canal (1569)', published in English and Turkish in vol. XLVI of the *Belleten* of the Turkish Historical Association (*Türk Tarih Kurumu*), 1948. Cf. also P. A. Sadikov, 'Pokhod Tatar i Turok na Astrakhan v 1569 g.', in *Istoricheskiye Zapiski*, XXII (1947), pp. 132–66; and E. N. Kusheva, 'Politika russkogo gosudarstva na severnom Kavkaze v 1552–53 gg.', *ibid.* XXXIV (1950), pp. 236–87.

of military operations in the Caucasus, it is necessary to focus the different peoples of Caucasia within the framework of Middle Eastern history.

The *Cherkesses* (T. *Çerkes*) or Circassians represent a stock formerly much more widely distributed. Cherkess place-names are to be found in many parts of the Ukraine in forms such as Psiol (C. *pse*, 'water') and Kremenchug. The Cherkesses remained an important element in the population of the Crimea until the eighteenth century, and they occupied most of the north-western Caucasus and the basin of the Kuban until the Russian conquest in the middle of the nineteenth century. Old Russian documents frequently refer to the Cossacks as Cherkesses, and there was clearly a substantial substratum of Cherkess blood in the very mixed population of the northern Caucasus and the Black Sea coastlands as far west as the Dniepr.[1] But this is only one aspect of the remarkable dispersion of the Cherkesses. In the classic world and down to the Middle Ages, the slave market was the normal mechanism for maintaining the labour supply of civilized countries. Hence it was an instrument for imposing a continual process of redistribution of population. As slaves and soldiers the Cherkesses were celebrated in the Byzantine and Islamic worlds, and they finally rose to fame in the Mamluk corps which dominated Egypt from the later Middle Ages until the generation of Napoleon. The Mamluk corps was, in fact, a sort of foreign legion which was recruited largely from the western Caucasus and the Don steppe.[2]

The *Ossetians* (R. *Ossetiny*) who occupy the central Caucasus on both sides of the watershed are anthropologically very mixed, but speak a language which is Indo-European in structure. They possibly represent the survival of elements from the Germanic wanderings in the south Russian steppe during the first centuries A.D. The Ossetians were on the whole friendly to the Russian power, and their presence on both sides of the watershed and along the line of the Georgian Military Road favoured the relatively bloodless expansion of Russian influence south of the centre of the chain.[3]

[1] Compare Klyuchevski's work showing that there is a noticeable Finnic ethnic element in the population of the Volga basin and north Russia.

[2] It was the Mamluks who first defeated the Mongols at the Battle of Ain Jalut (1260), one of the decisive battles of history. The Mamluks still maintained themselves in the Sudan after Mehmet Ali's massacre of the Cairo garrison, and one of the principal reasons for the Egyptian viceroy's conquest of the Sudan was his fear that the Mamluks might re-form and succeed in building a rival state on the upper Nile. The last remnants of the Mamluks scattered to Darfur and beyond to Wadai, and I am told that their descendants are still to be met with occasionally in those parts.

[3] The Ossetians call themselves Os or As (also, 'Iron'). The form Ossete derives

The varied tribes of Dağistan (T., 'Mountain land') are of mixed origin. The *Chechens*, inhabiting the beech forests which cover the northern slopes of the eastern Caucasus between the valleys of the Terek and the Sunja, are remotely related to the other autochthonous Caucasian peoples. The democratic basis of their tribal life—almost anarchic in character in contrast to the hierarchic organization of the Cherkesses—made them formidable opponents of the Cossacks, whose social background they may be said in some ways to have shared.

The *Andis* and *Avars* of Dağistan proper may be descended from fugitive elements from some of the older peoples of Asia Minor and Transcaucasia; the Avars, in particular, have been identified with the classical Albanians who formerly inhabited the lower valleys of the Kura and Aras. Among the tribes of Dağistan a land-owning class, mostly descended from Tartar intruders, constituted a conservative element which was opposed to the popular leadership of Shamyl during the first half of the nineteenth century and which was often inclined to take sides with the imperial authorities and to seek governmental support for their privileges.[1]

In Dağistan, and, indeed, throughout Muslim Transcaucasia, the *Arabs* cannot be ignored as a creative factor which has influenced the cultural mould of modern populations. The Arab caliphate, which in essence was half Iranian, ruled the eastern Caucasus from the seventh to the eleventh century A.D., for about the same length of time as the Romans ruled in Britain. Arabic continued to be the language of religion and culture in Dağistan even after the Revolution of 1917.

The *Georgians* are the most numerous and at the same time the most civilized single group in Transcaucasia.[2] This peculiarly gifted

from the Georgian locational termination, Os-eti (Os-land). The Os gave their name to the As-ov Sea and, under their medieval name of Alans, to the Daryal gorge (P. *Dar-i-Alan*, 'Gate of the Alans'). Much material has been published on the Ossetians, notably in *Sbor. Mat.*; Baddeley, *Rugged Flanks*, summarizes material to 1938. See also Chursin, *Osetiny: Etnograficheski Ocherk* (Tiflis, 1925), and Melikset-bekov, *Yugo-Osetiya* (Tiflis, 1925). See also articles under 'Allan' by Bartold in *E.I.* and by Zeki Velidi Togan in *I.A.* Togan believes that the As-Allan people were occupying the steppe round the northern and eastern shores of the Caspian until the early Middle Ages, and in his *Turkistan Tarihi* (Istanbul, 1929) he puts forward the view that there may have been a submerged Allan element in the ethnic make-up of the (Oğuz) Turkomans. Vernadski, *Ancient Russia* (Yale, 1943), attaches great importance to the As-Allan element in early Russian history.

[1] For a learned and charming account of the Chechens and peoples of Dağistan at the close of the last century, see Baddeley, *Rugged Flanks*.

[2] At the beginning of the nineteenth century the Georgians did not number more than 700,000. The Cherkesses, greatly reduced by the plague and later by migrations, probably numbered about 600,000; the tribes of the eastern Caucasus

people who produced a brilliant medieval culture have inherited an original literature and a rich art.

The *Western Georgians* speak the distinctive Mingrelian dialect of Georgian and they are generally described as belonging to the Svano-Colchian group. This group comprises the *Svans* of the southern slope of the main chain (upper valleys of the rivers Kodor and Ingur), the *Mingrelians* and *Imeretians* of the Rion basin, and the *Acars* and *Lazes* of the Pontic Alps. The Svano-Colchians include descendants of the most primitive populations of the Caucasian isthmus. There is a long-headed element along the Black Sea coast (notably among the *Abkhaz*, a group distinct both from the Svano-Colchians and the Cherkesses), and negrito types have been observed.

The *Eastern Georgians* (*Kartlians*, *Kakhians* and *Meskhians*) represent an early movement into the Kura valley from Asia Minor which took place between the seventh and fifth centuries B.C. Place-names in Pontus and Cappadocia indicate that the Kartlo-Meskhian group at one period occupied a much wider area than they do to-day.

There may be remote racial connexions between the Georgians and elements among the *Kurds*.[1] The Kurds are certainly a very mixed people. The Kurmanci language is classified as Indo-European, but Basil Nikitin and others have shown that it contains many non-European elements. Nordic types are not uncommon among the Kurds. At the same time the Zazas of the Dersim represent an autochthonous element which is not purely Kurdish and is derived from an older stratum. The Kurdish tribes are pastoralist, and they range the mountains of the Anti-Taurus and the Taurus as far north as the upper Euphrates and the Aras and south to the Zagros. The age-old antagonism between the nomad and the farmer brought them into

were about half a million at the time of the Murid wars (cf. Baddeley, *Conquest*, p. xxix). The Georgians now number over 3,000,000, while the number of Armenians has risen from under a million to 2,000,000. For the Georgians, see Allen, *History of the Georgian People* (London, 1932), and section by same author in *The Baltic and Caucasian States* (ed. John Buchan); also arts. 'Gruziya' in *B.S.E.*, and 'Tiflis' by V. Minorsky in *E.I.*

[1] The classic *Kardukhoi* (Kard-ukh-oi) of Xenophon is composed of the Georgian nominal root *Kart* with Armenian and Greek plural suffixes added. The non-Indo-European elements among the Kurds represent the survivors of the pre-Indo-European and non-Semitic peoples who formerly occupied all Caucasia and Asia Minor and the country as far south as the Persian Gulf. For an excellent bibliography on the Kurds, see V. Minorsky's article in *E.I.* *Georgica* contains numerous articles by C. F. Lehmann Haupt, M. Tsereteli, A. Safrastian and others on east Anatolian and Caucasian origins to 1937. More recent views are summarized by Prince Cyril Toumanoff in 'Mediaeval Georgian Historical Literature', *Traditio*, I (New York, 1943). Reference should also be made to numerous articles in the fifty-odd volumes of the Turkish Historical Association.

natural conflict with the settled Armenian population which was scattered over all the country between the Mediterranean Taurus and the Kura.

The *Armenians* (who call themselves *Hai-kh*) represent late intrusions into Asia Minor which can be correlated with the movement of the Meskhians and Iberians into the Kura valley. The Armenian language, again, is classified as Indo-European, but Professor Nikolay Marr and other students of the language have shown the extent to which it is penetrated with pre-Indo-European elements.[1]

Physically there are no very marked differences between the various peoples of Asia Minor and the Caucasian isthmus.

> That peculiar type which is native to these parts, so-called autochthonous—'out of the ground'—is quite permanent, always resurgent, mastering the bodies of new masters. The strong-boned physique, the broad square head with its thick growth of wavy hair and beard, the wide dark eyes, the sallow skin, are bred to these mountain countries from the Ice Age. This type is called Armenoid, or more appropriately Alpine; it has spread to Europe in prehistoric times along the mountain belt as far as the Pyrenees, and over Iran to the Pamirs; and it has filtered down through Palestine to Egypt, surviving as all types, mostly in those parts which were likest in condition to its homeland. The peasantry of western Asia, whether they be called Georgian, Armenian or Turkish, are of this autochthonous stock—out of the ground.[2]

Appropriately, as the arid valleys and plateaux of eastern Transcaucasia form naturally a westerly extension of the Aralo-Caspian depression, they are inhabited by a *Turkish* or *Tartar* population (the two names are really interchangeable), akin in speech and way of life to the Turco-Tartars of central Asia. The Turks of Azerbaijan (Azeris) use a form of Turkish which is related to the dialects of Anatolia, the Crimea and the Turkish-speaking districts of Persia—including southern Azerbaijan, Hamadan and Khorasan. This form of Turkish derives originally from the Turkoman tribes of the Transcaspian steppe and is distinct from the *eastern* (Cagatay) Turkish spoken by the Özbeks and (with variations) by the Kazaks and Kirgiz

[1] The most recent general work on Armenian history is by René Grousset, *Histoire de l'Arménie* (Paris, 1949). See also *B.S.E.*, art. 'Armeniya'; and article on Armenia by A. Safrastian in *E.B.* (14th ed.).

[2] Quoted from W. E. D. Allen, *History of the Georgian People*, p. 21. For the anthropology of Caucasia, see art. by A. Javakhishvili, 'The Caucasian race', in *Georgica*, I (1936), pp. 2–3. For recent views on Anatolian anthropology see articles in *Belleten* of Turkish Historical Association, and Bayan Afet Inan, *L'Anatolie, le pays de la 'race' turque: recherches sur les caractères anthropologiques de la Turquie* (Geneva, 1941).

of central Asia. The linguistic influence of the Turkomans (Oğuz) from beyond the Caspian has been strong among all the western Turks. From the Turkomans stemmed the Selcuks, the Osmanlis and other later clans like the Black and White Sheep (Kara and Ak Koyunlus), the Afshars and the Kacars, who have dominated the political and military life of Iran and Anatolia during the past nine hundred years.

The Azeri Turks adhere to the Shi'a heresy of Persia (by contrast to the Sunni Osmanlis and the Sunni tribes of Dağistan), and Azerbaijan was, in fact, the homeland of the Shi'a renaissance and of the great Sufi family of Safawi shahs who were themselves partly of Turkish origin and used Turkish speech.

Culturally eastern Transcaucasia has been very much under Iranian influences since the remote period when the Iranian Sassanids were holding Derbent against the northern nomads. Some Turkish elements penetrated there soon after the era of the Hijra.[1] One of the Arab geographers of the ninth century A.D. describes the typically Turkish physical type and compares the harsh dialect of Azerbaijan with that of Khurasan. It is worth noting that, while the Volga and Crimea Tartars have passed increasingly under the influence of their Russian neighbours and the Anatolian Turks feel the attraction of the Mediterranean world, the cultural background of the Caucasian Turks, like that of the Turks of central Asia, is very largely Iranian. The Turanian and Iranian worlds have for two thousand years, probably for very much longer, been subject to mutual blending and to all kinds of linguistic and cultural cross-currents; in recent centuries, while neighbouring Turkish groups have been attracted by the ancient civilizations of Iran, Turks have formed a substantial proportion of the population of the Persian empire, and sturdy Turkish clans like the Shahsevens[2] of Karadağ have always been the core of the armed forces of successive Persian dynasties—the latter themselves often of Turkish blood.[3]

[1] The infiltration of Turco-Mongol elements into Anatolia and Caucasia began in very early times. See Prof. A. Zeki Velidi Togan, *Umumi Türk Tarihine Giriş* (General Survey of Turkish History) (Istanbul, 1946), I.

[2] Lit. 'Lovers of the Shah'.

[3] Nizami (1140/41–*c.* 1198), one of the many Azeris who have enriched Persian culture, was a native of Ganja. He flourished in the same half-century as the celebrated Georgian poets, Shota Rustaveli and Sargis Tmogveli, who were themselves both under the influence of the Persian literary tradition. Fuzuli (?–*c.* 1556), one of the most moving of Turkish poets, used the Turki of Azerbaijan and Persian with equal facility, as did his near contemporary the first Safawi Shah Ismail. There is no comprehensive work on the Turks of Azerbaijan. The articles

Both Turks and Persians were imperial peoples, but the Turks proved more vital as a military power than the Persians. In the fifteenth century the Osmanli Turks had revived the classic style of 'world power' in the Middle East; and the history of the development of the Russian empire in Asia is that of conflict for control of the basins of the Black Sea and of the Caspian between this classic style of world power and the new world power which had arisen in the Eurasian plain. Modern Russian imperial power in its new expression as a union of soviet republics represents a blending of Slavic, Turkic and Caucasic elements.

The Caucasus, which became towards the middle of the eighteenth century the scene of the epic struggle between the old, Middle Eastern, and new, Eurasian, powers, was a borderland in which the most diverse social forms were in process of growth and decay. Here came in contact the ancient imperialisms of Turkey and Persia and the new half-European imperialism of Romanov Russia; anarchic communities of Cossacks and groups of Great Russian serfs and even German dissenters settled in villages in recently occupied lands; the hierarchic clans of the Cherkesses and the democratic predatory septs of the Chechens; the fanatical tribes of Dağistan and the feudal Muslim lords of Meskhia and Acaristan; the rustic principalities of western Georgia and the mosaic of miniature city states which had been gathered together into the kingdom of Irakli II of Kartli and Kakheti.[1]

In the great struggle between the Russians and the Turks which ranged all round the shores of the Black Sea from the Balkans to the Caucasus during the years 1769–74, the Turks tried to loosen the hold which the Russians had already gained over the passes through the central Caucasus by their alliance with the Kabardans. But their efforts to detach the Kabardans from Russia and their promotion of a general movement of the Cherkesses and the Chechens against the

in *E.I.* under 'Shirwan', 'Adharbaidjan' and 'Azeri' merit revision. The articles in *I.A.* under 'Azerbaycan' (A. Z. V. Togan) and 'Azeri' (M. F. Köprülü) are more serious. See also art. under 'Azerbaidzhanskaya S.S.R.' in *B.S.E.* and numerous articles in *Izvestiya obshchestva obsledovaniya i izucheniya Azerbaidzhana* (Baku, 1926–). For topography, see *E.I.* arts. under 'Shekki' (Nukha), 'Kubba' (Kuba) and 'Sheh-sewan' (Shahsevens), by V. Minorsky, and *idem*, 'Transcaucasica', in *Journal Asiatique* (1930), pp. 41–111.

[1] Tiflis, Erevan, Ganja and Nahçivan were included in the dominions of Irakli. The towns along the Kura and the Aras and places like Derbent, Baku, Sheki and Shemakhi, with their independent khans, had, like Tabriz and Akhaltzikhe with their virtually independent governors and their powerful merchants' corporations, many of the characteristics of city states in contrast to the rural principalities of west Georgia and the tribal areas in the mountains. Cf. Allen, *History*, chap. XVII.

Caucasian line failed in the event. The Russians made their first passage of the Daryal defile (through which they did not delay to build a carriage road), and an interesting campaign followed, during the course of which Russo-Georgian forces crossed the Suram, occupied Kutaisi, the capital of the kings of Imereti, and laid siege to the Turkish fortress at Poti. Another detachment operated in the Kura defile in the direction of the Turkish stronghold at Akhaltzikhe (T. *Ahiska*).[1]

Following the Treaty of Küçük Kaynarca in 1774, Kabarda and Osseti passed definitely under Russian control. The Georgian Military Road was put in good condition and protected by fortified posts. In 1783 the east Georgian kingdoms of Kartli and Kakheti became associated by treaty with the Russian crown; at the same time a successful expedition against the tribes of Dağistan led to the temporary occupation of Derbent. The Caucasian line was further consolidated. A new line was traced from Rostov-on-Don to Mozdok on the Terek—covered by ten forts and by numerous *stanitsy*, populated by Cossacks and by specially transferred serfs. Suvorov was ordered to create the Kuban line from the mouth of the river to *stanitsa* Kavkazskaya on the Caucasian line. Suvorov's fortified posts prevented the Cherkesses from crossing the Kuban and joining forces with the nomad Nogays and Kalmyks of the steppe. Many thousands of these latter were forced to migrate beyond the Volga, and Russian peasants were settled over the steppe between the Don and the Kuban.

During the Russo-Turkish War of 1787–91 the Turks made a final effort to challenge the Russian position in the northern Caucasus. The Turkish fortress of Anapa controlling the mouth of the Kuban and the Straits of Kerch was the scene of severe fighting in successive years, and it was only stormed by the Russians in the last year of the war. Batal Paşa, who marched from the coast to the region of the upper Kuban with a great horde of Turks and Cherkesses, was defeated by a relatively inferior Russian force near the site which used to bear his name.[2]

As a result of the two Russo-Turkish wars of the last half of the eighteenth century, the Russians were firmly established along the

[1] Cf. *Georgica*, IV, V, 'The Battle of Aspindza', by J. Gogebashvili.

[2] Batalpashinsk, now renamed Cherkessk. Cf. Baddeley, *Conquest*, and for a detailed summary of these campaigns, Potto, *Istoricheski ocherk*; Dubrovin, *Istoriya voyn...na Kavkaze*, and Butkov, *Materialy dlya novoy istorii Kavkaza ot 1722 do 1803 gg.*, 3 vols., St Petersburg, 1869 (Vol. III contains a valuable documentation of printed and MSS. sources).

river lines north of the main chain, while their influence and partial control had been extended over Kabarda and Osseti and the Georgian principalities of Kakheti, Kartli, Imereti and Mingrelia to the south of the main chain. They were thus astride the watershed with control of the ways over by the Mamison and Daryal passes. The Turks retained their influence over the Mussulman Cherkesses and Abkhazians and held the forts which gave access to the western mountains at Poti, Anakliya and Sukhum.

In eastern Caucasia, a revival of the military power of Persia under Aga Muhammad Khan (founder of the Turkoman Kacar dynasty which survived into the third decade of the twentieth century) resulted in an attempt to restore the ancient Persian hegemony over the valley of the middle Kura. In 1795 the Persians took and sacked Tiflis and occupied most of Kartli and Kakheti, but in the following year combined land and sea operations from Astrakhan placed the whole of eastern Transcaucasia easily in Russian hands. Derbent was occupied; then the khanates of Kuba and Baku and later those of Shirvan (Shemakhi), Sheki (Nukha) and Karabağ (Shusha). The capture of the last gave the Russians a first hold on the peripheral rim of the Armenian plateau. Ganja,[1] the principal centre of Islam in Transcaucasia, was taken. The Russians reached the Aras river, and the way to Erevan and Tabriz lay open to them. The campaign, conducted with small forces, revealed to the Russians the real weakness of the two Mussulman powers in the Middle East, and brilliant perspectives seemed to open up just as Empress Catherine II died. Her successor, Paul I, was opposed to a continuance of his mother's policies everywhere and ordered a general withdrawal to the Caucasian line; but renewed attacks on Kakheti from Dağistan imposed the annexation of Georgia to the Russian crown in 1801.

The following decade saw further conflict with both Turkey and Persia, complicated by the repercussions of the Napoleonic wars.

In 1812 the Treaty of Bucharest with Turkey left the Russians in control of all the Rion basin to the west of the Suram, and also of Abkhazia, along the south-west slope of the main chain falling to the Black Sea. On the Mingrelian coast the Turks only retained the fort of Poti; but farther north the strong fortress of Anapa had been restored to them. They still held Akhaltzikhe and Atskhur, strongholds which dominated the gorges of the Kura and threatened the route over the Suram between Kutaisi and Gori. So long, in fact, as

[1] Renamed by the Russians Elizavetpol; now Kirovabad.

they failed to control the Black Sea coast between Anapa and the River Bzyb, and while the Turks remained masters of the upper Kura, the Russian hold on western Transcaucasia was incomplete and very insecure.

In 1813 the Treaty of Gulistan with Persia confirmed the Russians in possession of the Caspian coast as far as the khanate of Lankuran. But the Persians still held the khanates of Erevan and Nahçivan, whence they could make a rapid advance up the valleys of the Akstafa and Borchalu rivers against Tiflis and the middle valley of the Kura. And, as in the west the mountain block of Circassia remained unconquered, so in the east the formidable massif of Dağistan, inhabited by even more warlike and certainly more fanatical tribes, resisted all attempts at Russian pacification.

Dağistan was, in fact, destined to remain a major military problem of the Russian empire for the following fifty years.

In 1821 the ambitions of the Persian Valihad, Abbas Mirza, were diverted against the Turks by Russian influences, anxious to defeat the contemporary British policy of consolidating the interests of the two Muslim powers in the Caucasian area. The Persians for a short time occupied Bayazit and Kars. British diplomacy was, however, successful in restoring the *status quo*.

In 1826 rumours of the death of Alexander I and of internal disorders in Russia induced the restless Persian prince, who was sadly lacking in perspective, to make an effort to reverse the Treaty of Gulistan. The Muslim populations of Lankuran and Karabağ revolted and Ganja opened its gates to Abbas Mirza. The Russians were at first disturbed by the prospect of a holy war; but the Persians at Ganja were easily defeated, and in the spring of 1827 General Paskevich undertook a campaign of final conquest in eastern Transcaucasia. In June the Russians opened the siege of Erevan, while a field force, advancing along the valley of the Aras, occupied Echmiadzin, the holy city of the Armenians and the seat of their patriarch, and took possession of the khanate of Nahçivan with the border fort of Abbasabad which covered the road to Tabriz.

Abbas Mirza manoeuvred ineffectually with a large force (about 30,000, mostly irregular horse). In September Erevan fell, and on 15 October the Russians entered Tabriz. Peace negotiations failed to mature, and in January 1828 the Russians took Dilman and Urmia and Ardabil. The road to Teheran lay open to them.

On 18 February 1828 the Treaty of Turkmençay finally confirmed the Russians in the possession of the khanates of Erevan and Nahçivan, north of the Aras, and of the khanate of Talysh to the south of the embouchure of the Kura-Aras into the Caspian Sea.

The Russians had thus consolidated their possession of the whole of eastern Transcaucasia. Their strategy, military and political, since 1722 had, in fact, constituted a sustained turning movement of the main chain of the Caucasus. The mountains themselves remained unconquered; on the west from the Taman peninsula as far as the sources of the Kuban and the Ingur; on the east from the upper Terek to the foothills of Dağistan overlooking the Caspian. But the conquest by the Russians of the Caspian foreshore from the mouth of the Terek along by Derbent to Baku, combined with their control of the waters of the Caspian Sea, had made possible the subjugation of the plains of the Kura and Aras and of the mountain massifs beyond (Somkheti, Shah-dağ and Karabağ) before they had secured their grip on the main chain. Political alliances with the Kabardans, the Ossetians and the Georgians had allowed the Russians access through the Daryal Pass to the valley of the middle Kura, and the friendship of the Imeretians and Mingrelians had given them the Suram ridge and the easy conquest of the Rion basin. The Cherkesses remained unconquered, but they were now contained on the south as well as on the north and cut off from contact, except by sea, with their Turkish allies.

The Treaty of Turkmençay had definitely eliminated the Persians as a factor in Transcaucasian politics and had cut off the warlike tribes of Dağistan from all direct contact with the Islamic (and outer) world. Iran had been the dominant cultural and political influence in Caucasia for two thousand years. In future Caucasian Islam was to look to the Turks alone for support. The Turks who, at the beginning of the eighteenth century, had been the masters of the mouths of all the rivers flowing into the Black Sea, who in 1711 had forced the great Peter to a capitulation on the Prut, and who as late as 1735 had successfully sustained a defensive war against the allied armies of Austria and Russia, were now fighting a losing battle to hold the passes giving access to their inner homeland in Anatolia. During the century following Peter's Caspian campaign of 1722, the main chain of the Caucasus mountains had ceased to be the age-old barrier protecting the Middle Eastern lands against invasion from the north. Caucasia had become the base from which Russian power, political and military, was to be directed westward across Anatolia towards the Mediterranean, southward across Persia towards the Indian Ocean, and eastward across the Caspian into the heart of Asia.

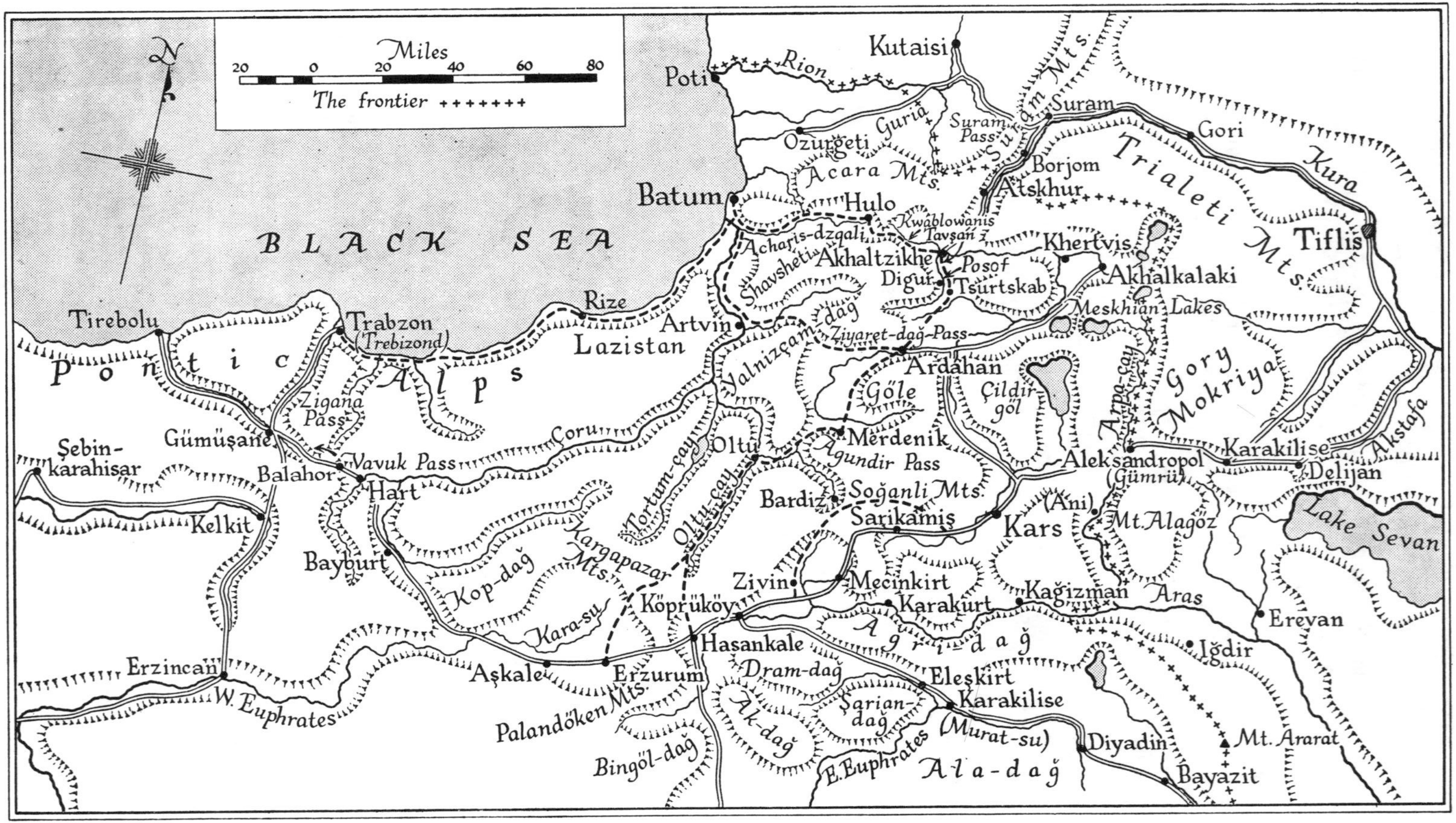

Map 2. The campaign of 1828–9

CHAPTER II

THE RUSSO-TURKISH CAMPAIGN OF 1828–9

MAPS 2, 3; cf. also 4

THE causes and objectives of the Russo-Turkish war of 1828–9 were mainly European. The Greek insurrection and the battle of Navarino gave the Russians the opportunity of settling accounts with Turkey without, for once, having to watch the prospect of intervention by the western powers.

In Asia, Russian objectives were limited, firstly, to the expulsion of the Turks from the Caucasian coast including the fortresses of Anapa, Poti and Batum; and, secondly, to the conquest of the paşalik of Akhaltzikhe and the establishment of a satisfactory military frontier.

The Turkish frontier in 1828 ran from Poti along the lower course of the Rion and over the Suram Mountains to a point in the Kura defile between Borjom and Atskhur. Thence through the region of the Meskhian lakes to the long ravine of the Arpa-çay. In theory the paşaliks of Asia Minor could raise about 200,000 regular and irregular troops, but the military resources of Sultan Mahmut II had been weakened by the long Greek war and as a result of the massacre of the janissaries, who had not yet been adequately replaced by the new *nizam* (regular) units in process of formation. It was the age of the great paşas in Turkey, and the eastern marches were largely controlled by the paşa of Akhaltzikhe,[1] who was master of all the upper Kura country as far as Oltu, and by the Himshiashvili (T. *Himşioğullar*) clan, Muslim Georgian beys, who ruled the mountains between the Çoruh and the Gurian lowlands. The forces which these Muslim lords could put into the field were in fact more formidable than those of the Turkish regular army.

The Turks held a strategic frontier which was almost perfect. The fortress of Batum, backed by steep wooded hills, covered the coast and was one of a line of fortified posts, extending from Trebizond to Anapa. The citadel and town of Akhaltzikhe with its warlike population defended the valley of the Kura at the entry to the Borjom defile, where the strong point of Atskhur was also in Turkish hands. To the

[1] Until 1744, the paşalik of Akhaltzikhe had been hereditary in the family of Jaqeli—Muslim Georgians, themselves descended from the ancient atabegs of the medieval Georgian kingdom. Cf. Brosset, *Histoire de la Géorgie*, II (1), pp. 199 ff., 'Histoire de Samtzkhé ou Saathabago'.

south of the Çildir-Akbaba massif, Kars covered the main route across the Arpa-çay from Aleksandropol and Erevan. Southward again of the massif of Alaca-dağ, the steep ravine of the Aras was impassable to large bodies of troops, and the line of the Ağri-dağ protected the right of the Turkish forces deployed along the frontier.

But the Turkish strength was more apparent than real. Their troops were scattered in inadequate detachments. There were garrisons of 2000 men each at Batum and Poti, 5000 at Anapa, 6000 nizams and 4000 irregulars held Akhaltzikhe, with 1000 men forward at each of the forts of Atskhur and Akhalkalaki. Some 30,000 nizams and 10,000 irregulars were concentrating at Erzurum and Kars. On the right flank was a garrison of 2000 in the strong hill-fortress of Bayazit, and the paşa of Muş was collecting another 5000 men for the defence of Van. The Turks had plenty of guns but their fire-power was feeble, and some of the fortress pieces had served since the days when Sultan Murat IV was reorganizing the frontier in the second quarter of the seventeenth century. The walls and moats of Kars and Akhaltzikhe had been designed at the beginning of the eighteenth century, and Kars had been unable to withstand a Persian attack a few years before.

The shadow of the Ottoman eighteenth century hung over the crumbling walls and neglected cannon of the frontier towns. The gaudy figures of the janissaries, slovenly and insolent, had only just departed, and the blue tunics and red fezzes of the *gavur* sultan's nizam troops were a strange innovation. Strings of young Circassians and kidnapped Gurian peasants were still collected in the alleys of Akhaltzikhe and marched off to Batum and Trebizond for shipment to the slave markets of Istanbul and Alexandria.[1]

The Russian army, tempered in the Napoleonic wars and officered by men with twenty-five years' experience of Caucasian campaigning, was in fine condition. A tradition of uninterrupted victories against oriental peoples gave a remarkable *élan* to the pupils of Ermolov and the comrades of Kotlarevski. Service in the Caucasian army could satisfy the most ambitious and the most desperate men in Russia, and an epic literature was already growing up round the Caucasian regiments.

The Russian general, Paskevich, had at his disposal 51 battalions of regular infantry, 11 squadrons of regular cavalry, 17 regiments of

[1] In 1828, Akhaltzikhe, the principal entrepôt of the Caucasian slave trade, was the most important city in eastern Turkey after Erzurum and Trebizond. Its population was estimated by Gamba at 40,000, mostly Turks or Georgian Muslims with a few thousand Christian Armenians.

Kuban and Terek Cossacks and 154 guns, rather less than 60,000 men, of whom about one-quarter were detached for internal security duties in the northern Caucasus. There remained available for battle action 36 battalions, 8 regular squadrons, 13 Cossack regiments and 112 guns, disposed as follows:

Khoy-Julfa (observing Persia)	General Pankratiev	6 bns, 2 regts Cossacks, 16 guns
Erevan (observing Bayazit)	General Chavchavadze	5 bns, 1 regt Cossacks, 10 guns
Borjom (covering Kura defile)	General Popov	2 bns, 2 regts Cossacks, 4 guns
Kutaisi (operating against Guria)	General Hesse	8 bns, 1 regt Cossacks, 14 guns
Gümrü (Aleksandropol)[1] (striking force)	General Paskevich	15 bns, 8 sqns, 7 regts Cossacks, 68 guns

The campaign opened in May with the surrender of the Turkish garrison at Anapa to a combined operation of the Russian fleet[2] and troops from the Caucasian line. This fortunate event relieved Paskevich of anxiety with regard to his rear where a concerted movement of the Cherkesses with Turkish support might have been expected (as in the war of 1787–91).

Paskevich directed his main operation against Kars. A successful assault on the fortress would cut off the Turks, defending the paşalik of Akhaltzikhe, from the main Turkish base at Erzurum, the centre for concentrating supplies and reinforcements from Asia Minor and Trebizond.

On 14 June the small Russian army crossed the frontier river Arpa-çay; and Paskevich took station to the south of Kars where he was able to flank any relieving force advancing from Erzurum. On 17 June, the seraskier Köse Mehmet[3] did, in fact, order an advance from Erzurum over the Soğanli-dağ, but on that day the Russians took up a position on the Kars-çay only four miles from the fortress and athwart the main Erzurum road. Paskevich was still without siege artillery, but he established his field guns on the Şorah heights to the west of the Kars-çay from which he could drop shells into the town over the defending walls.

[1] T. *Gümürü*, renamed Aleksandropol in honour of Emperor Alexander I.

[2] Under the Scots Admiral A. S. Greig (see *D.N.B.*). For history of Anapa, see *Kav. Sbor.* III (1), p. 176, n. 1.

[3] Köse Mehmet is the 'Kiossa' of Monteith and the Russian authors. *Köse* means 'beardless'—a most unusual condition in a paşa of this period—and Köse Mehmet may have been a eunuch like his near contemporary, Ağa Muhammad Khan of Persia.

Following a bombardment, the Russian infantry took the suburbs to the east of the Kars-çay, and on the morning of 23 June they captured the wall which divided the town from the fortress itself. The Turkish commander then surrendered with 2000 nizams and 150 guns; but the rest of the garrison succeeded in withdrawing. Russian losses amounted to less than 400 men.

Köse Mehmet had already crossed the Soğanli ridge and he now halted to await the Russians. Paskevich sent a strong advance-guard along the Erzurum road; he intended, however, to operate not against Erzurum but against Akhaltzikhe.

The great upland plain of Kars, the scene of the approaching campaign, lies at an altitude of between 1830 and 2440 m., and, within the perimeter of its surrounding ranges, occupies an area of some 7000 square miles. The rolling downlands, with their rich herbage feeding the noted herds of Göle and Çildir, afford ideal conditions for the movement of cavalry during the late summer and autumn months. Equally it is difficult to defend this great plain by forts or fixed positions. The historic fortified positions have always been sited at the entries and exits: on the north-west at Ardanuch and Ardahan commanding the routes over the Yalnizçam-dağ to the valley of the Çoruh; on the north-east at Akhaltzikhe where the Kura enters the long gorge through the mountains to Kartli; on the east at Kars, blocking the gap between the massif of Çildir-dağ (Akbaba-dağ) to the north and the Alaca-dağ massif to the south.

Since classical times the population of this borderland has been mixed. Until the Ottoman conquest in the fifteenth century the Georgians held a line of fortified towns and castles running west to north-east from Tortum and Oltu to Ardahan and Akhaltzikhe. Kars and Ani on the Arpa-çay were city states successively ruled by princes of Armenian, Arab, Kurdish, Georgian and Turkish blood. But the rich downland, stretching between the few towns with their commanding citadels, early became the grazing grounds of nomad Turkish tribes (*tümenler*), who remained a vigorous fighting population dangerous to their settled neighbours.[1]

Running in a general direction west to north-east, the long level ridge of Yalnizçam-dağ separates the Kars plain from the valley of the

[1] See Prof. A. Caferoğlu, *Materials Relative to the Dialects of our Eastern Provinces: Dialects of the Provinces of Kars, Erzurum and Çoruh* (T.) (Burhaneddin Basimevi, Istanbul, 1942). Prof. Caferoğlu finds villages of the Karakoyunlu, Dünbüllü, Kacar, Afşar, Bayat and other well-known Turkish tribes in the Kars province.

Çoruh and forms the watershed between streams flowing northward to the Çoruh and the Black Sea and eastward to the Kura.[1]

Rising from numerous sources in the beautiful ravines and marshes at the foothills of the Göle massif—itself the axis of the ranges of Yalnizçam and Soğanli—the Kura, after a first loop westward, follows the general north-easterly direction of the Yalnizçam-dağ. After passing through Ardahan with its sinister old citadel, its shabby streets and fine stone bridge over the river, the Kura enters a precipitous ravine which separates the northern spurs of Çildir-dağ from the massif of Dokuz-pinar ('Nine springs'), an easterly buttress of the Yalnizçam-dağ. After the intake of the Akhalkalakis-dzqali (G., 'Newtown water') at Khertvis, the Kura loops west until, below Akhaltzikhe, it receives the united streams of Posof-çay and Kwablowanis-dzqali and turns north-north-east through the long defile of Borjom.

The principal route from Akhaltzikhe ran north of the Kura to Akhalkalaki. It then turned west over a saddle between Çildir-dağ and the Dokuz-pinar to Ardahan; from here it forked south-east to Kars, north-west by west over the Ziyaret-dağ to Ardanuch and the valley of the Çoruh, and south-west over the cols of Göle to the upper valley of the Oltu-çay and Erzurum.

A movement from Kars on Akhaltzikhe by way of Ardahan might not have been without danger even for a well-equipped Russian force, much superior to its opponents in discipline and fighting power. In the Soğanli-dağ, the seraskier occupied a flanking position in relation to Ardahan, while the numerous and elusive Turkish pastoralists, scattered over the vast downlands, might have been expected to rally to his support. Paskevich took the bold decision of leading his army across the upland tracks from Kars to Akhalkalaki by the high col across the Akbaba-dağ overlooking the eastern shore of the majestic lake of Çildir.[2]

The preparations for the Akhaltzikhe expedition took about three

[1] The name Yalnizçam-dağlar (T., 'Lone pine mountains') indicates the character of this barren range. The old Georgian name for the whole range was Arsiani, now used on Turkish maps to describe a single peak south-west of Posof. The Georgian name, *Qwa Qrile* ('Rockstrewn'), for a part of the Arsiani is as descriptive as *Yalnizçam*.

[2] This beautiful lake, swept often by great storms, is the haunt in summer of numerous flocks of pelican (*Pelecanus crispus*). Its position on an international frontier has made it difficult of access to travellers. In 1021, a battle was fought near the lake between the Byzantine emperor, Basil Bulgaroktonos, and the Georgian king, George I. In 1578, Lala Mustafa Paşa, the conqueror of Cyprus, defeated the Persian army of Tokmak Khan in the same region.

weeks. General Bergmann with 6 battalions and strong artillery support was left to hold Kars, and, on 17 July, Paskevich with the rest of his forces began the difficult march over the Akbaba uplands. In summer the pasture of these uplands, which attracts the flocks of the nomads from all the country between the Aras and the Black Sea, is in fine condition, gay with flowers; and the long animal convoys of the Russians could live on the country.

Paskevich needed five days to cover the eighty-odd miles between Kars and the Akhalkalakis-dzqali. On 25 July he attacked the fort of Akhalkalaki and met a desperate resistance from the garrison of 1000 Laz irregulars, of whom only 300 survived to surrender. Khertvis, at the junction of the Akhalkalakis-dzqali with the Kura, was taken without a shot. In the meantime news arrived that in the north-west General Hesse had taken Poti with 2000 prisoners and 40 guns (15 July). In the first days of August Paskevich approached Akhaltzikhe.

A dangerous situation was developing. Paskevich had relied too much on the slow movements of the Turks with their large forces of irregulars, who marched often with thousands of livestock and great crowds of women and camp followers.[1] Informed of Paskevich's expedition against Akhaltzikhe, Köse Mehmet moved his forces from the Soğanli-dağ to Ardahan and further across the saddle between the Yalnizçam-dağ and the Dokuz-pinar massif in the general direction of the Posof-çay. About 35,000 Turks were approaching Akhaltzikhe from the south almost simultaneously with the advance of Paskevich from the east.

Two-thirds of the Turkish forces were irregulars and they had with them only 14 guns. But Paskevich, who had 5000 infantry, 3000 cavalry and 30–40 guns (a force numerically smaller than the garrison of Akhaltzikhe alone), doubted whether it were feasible to continue the operation. He received, however, at this moment reinforcements of 2500 men from Tiflis and, counting also on the co-operation of Popov's force in the Borjom defile (4 battalions and 2 regiments of Cossacks), he decided to give battle before Akhaltzikhe.

Köse Mehmet had dispersed his forces in four main camps on both sides of the Posof-çay, for he was governed by the necessity of considering water supply and grazing for his great numbers of troops, camp followers and animals. The Turks remained passive, and they

[1] In this respect the Turkish army of 120 years ago, prior to reforms of Mahmud II and his Prussian military mission (see von Moltke, *The Russians in Bulgaria and Rumelia in 1828 and 1829*), may be compared to the type of Abyssinian army which fought the campaign of 1935–6 and which took the field as irregulars in 1941. Cf. W. E. D. Allen, *Guerrilla War in Abyssinia* (London, 1943).

failed either to prevent the Russians from occupying the Tavşan (T., 'Hare') heights lying between Akhaltzikhe and the junction of the Posof-çay with the Kura, or to stop the junction of Popov (coming up the Kura defile and by-passing the Turkish fort at Atskhur) with Paskevich.

On the morning of 7 August Paskevich, leaving part of his force to guard the Tavşan camp, carried out a bold manoeuvre round the fortress with 5 battalions of infantry, all his cavalry and most of his guns. A threat on the Turkish left flank, covering the Ardahan road and their principal supply dump, induced the seraskier to weaken his right flank which received the full force of the Russian attack, although the advance was itself enfiladed from the guns of the fortress.

The Turkish position was taken by storm, and the seraskier was able with difficulty to retire into the fortress with 5000 nizams. Another 6000 regulars were dispersed or killed; with the Cossacks in pursuit, the Laz and Kurdish irregulars scattered all over the countryside, and enormous booty in animals and supplies fell into the hands of the Russians.

Following the bombardment of Akhaltzikhe on 13 and 14 August, the Russians took the city and fortress by storm during desperate street fighting on the 15th and 16th. On the latter day the seraskier finally capitulated in the citadel, having been granted the right to withdraw with 4000 of his nizam troops. Total Russian losses between 7 and 16 August amounted to about 1000 men.

Immediately after the storming of Akhaltzikhe, Paskevich sent a column to take Atskhur which surrendered without a fight. The direct route from Tiflis up the Kura valley via the Borjom defile to Akhaltzikhe and the upper Kura was thus opened to the Russians.

Another column under General Muravyev advanced on Ardahan, but it was found that this place had already been occupied without resistance by a force coming from Kars.

Meanwhile, on the extreme left flank of the Russian line, General Chavchavadze, operating from Erevan southward across the Aras, had marched over the Ağri-dağ and taken the Turkish fortress of Bayazit, thus cutting the main route of communication between Erzurum and Tabriz. Chavchavadze had then marched westward through Diyadin and occupied the fort of Toprak-kale (Eleşkirt) in the rich plain through which run the first affluents of the Murat-su, the eastern branch of the Euphrates. Westward the Cossacks had reached points within sixty miles of Erzurum; southward the way lay open to Muş and Bitlis and the whole region of the upper Euphrates.

With the approach of winter Paskevich decided to suspend active operations. Troops were suffering from fatigue and epidemics, particularly those who had been operating in Guria. Garrisons of 3000 men each were left to hold Akhaltzikhe and Kars (Bebutov and Bergmann). Hesse went into winter quarters at Kutaisi and Chavchavadze concentrated at Bayazit. Paskevich withdrew the rest of his forces into Georgia.

The results of the campaign of 1828 had surpassed all expectation. Troops engaged in the field had not exceeded 18,000 infantry and 7000 horse, with about 100 guns. With the conquest of Kars and the fortresses of the paşalik of Akhaltzikhe all Russian objectives had been secured and, with the exception of the coastal region and the Çoruh valley, an ideal frontier had been attained. All at a cost of some 2500 killed and wounded and an equivalent number in sick and dead from epidemics.

In the north the capture of Anapa and an outbreak of plague among the Cherkesses gave security on the Black Sea and the right flank of the Caucasian line. The capture of Poti made possible henceforth direct sea communication between the Crimea and the Azov ports and Transcaucasia.

In the eastern mountains, the news of Paskevich's victories and reports of his favourable treatment of the Muslim population had impressed the mountaineers of Chechnia and Dağistan, while neither the Persians nor these tribes, who were influenced by Persian policy, were likely to move in view of Chavchavadze's success in the Eleşkirt valley.

During the year 1828, Russian progress on the Balkan front had not been remarkable. They had suffered severe losses in battle and from epidemics and were held before the fortresses of the quadrilateral, Ruschuk, Silistria, Shumla, Varna (Varna alone had fallen). Sultan Mahmut II was determined to continue the war, and he carried through numerous reforms in the administration and army organization.

On the Caucasian front a new seraskier, Salih Paşa, was sent to take over from Köse Mehmet; meantime Hakki Paşa of Sivas was appointed his deputy. The paşas of Trebizond and Muş were ordered to recruit forces locally and to participate in operations. Only 10,000 men survived from the army of the dismissed seraskier, but his successor planned to raise by the spring a new army of 60,000 men with 70 guns in the region of Erzurum. The Turks, however, remained dependent for immediate action on the irregular forces which the powerful dere-beys of Lazistan and Acaristan might choose to raise. The most active and influential of these was Ahmet Bey Himşioğlu of Hulo, who with his brothers ruled over the whole of upland Acaristan. Paskevich had opened negotiations with him, but the Turks won their man with the promise of the paşalik of Akhaltzikhe, and Ahmet Bey, secure in his snowbound highlands, made preparations for an invasion of the valley of the Kura.

The Turks were less successful with the Kurds. The great Kurdish chieftains resented the reforming policy of Mahmut II and scented his hostility to feudal rights. They were already on the edge of the series of risings against the new Ottoman state which were to trouble Kurdistan from Upper Iraq to the Taurus during the thirties and forties. At the same time, through the aid of Mussulman officers in his service, Paskevich was establishing friendly relations with some of the Kurdish tribal chieftains. As a result, the paşa of Muş was unable to recruit any large force of Kurdish cavalry and he was more or less immobilized.

During the winter Paskevich visited St Petersburg and impressed the Emperor Nicholas I with the possibilities of a strong offensive in Asia Minor. Paskevich proposed, first, the conquest of Erzurum and the Armenian highlands; secondly, a combined operation against Trebizond with the support of the Russian fleet (then dominant in the Black Sea as a result of the sinking of the Turkish fleet at Navarino); and thirdly, with Trebizond as base, a move on Sivas, possibly before the autumn. Paskevich regarded Sivas as the nodal point of all Anatolian communications, threatening Istanbul's connexions with Armenia, Syria and Iraq.

Paskevich was probably correct in his general appreciation, although Kayseri, rather than Sivas, may properly be regarded as the nodal point of Anatolian communications. But Paskevich's thesis that Turkey might be conquered more easily through Asia Minor than across the Balkans was confirmed less than five years later when the Egyptian army of Ibrahim Paşa, supported by a fleet cruising along the southern coast of Anatolia, passed the Taurus and advanced victoriously on Konya and Kütahya. The Danube, with its system of fortresses, as well as the main chain of the Balkans, easily reinforced and supplied from the centre of the empire on the Bosporus, presented formidable obstacles to an advance from the north. On the other hand, the Turks could only with difficulty and tardily supply the Caucasian front; and once the Armenian highlands were conquered and armies (which could not be replaced in time) dispersed, the highway across the Anatolian plateau lay open to an invader. Such an advance, however, presupposed command of the Black Sea by the Russians, with combined operations along the coast and supplies available by sea for each stage of the advance. Only in 1828–9 did the Russians have command of the Black Sea; in the three subsequent wars it was in the hands of the Turks and their allies. Successful operations against a defender of Anatolia are, in fact, to a considerable extent dependent on sea-power.

Paskevich's plans were based on three conditions:

(*a*) The arrival of adequate reinforcements: his force available for active operations had been reduced, in November 1828, to 13,000 infantry, 5000 cavalry and 68 guns.

(*b*) An understanding with the Kurdish tribal chieftains of the Armenian highlands and the upper Euphrates valley.

(*c*) An absolute guarantee of continued peace with Persia.

Recruits numbering 20,000 had been allotted to the Caucasian army, but these could not reach the Caucasus, undergo training and be available in the line before July. Paskevich was reduced to withdrawing isolated units from the Caucasian line and to organizing Georgian and Mussulman militia.

Relations with Persia were complicated by the counter-intrigues of Turkish official and unofficial agents. The Russian envoy, the writer Griboyedov, was killed in a riotous attack on the Russian legation in Teheran.[1] But Paskevich managed to influence Abbas Mirza by large gifts and frequent threats to his provincial capital at Tabriz and secured his continued neutrality.

The Turks took the offensive as early as February 1829 when Ahmet Bey of Hulo crossed the main ridge of the Yalnizçam-dağ deep in winter snow with about 12,000 Lazes and Acars and 3000 Turks with 6 guns. He reached Diğur (on modern maps, Posof) at the junction of the Kwablowani and Posof streams and sent forward his brother to observe the Borjom defile. Bebutov was holding Akhaltzikhe with only 3 battalions and 4 field guns; the Turkish fortifications were in ruins, and the Russians were forced to retire into the citadel leaving the Armenian part of the population to be slaughtered by the triumphant Lazes. The situation of the Russians remained dangerous for

[1] Alexander, *Travels*, I, chap. IX, gives an entertaining account of the visit of a Persian embassy to the Russian court following the death of Griboyedov. Cf. also I. K. Enikolopov, *A. S. Griboyedov v Gruzii i Persii: istoriko-biograficheski ocherk* (Zakniga, 1929). Griboyedov was the brother of Countess Paskevich. In December 1825 he had been arrested as a Decabrist at Ekaterinodar on the Caucasian line by the Viceroy Ermolov in person. (See *Arest Griboyedova* in E. T. Veidenbaum's *Kavkazskiye etyudy* (Tiflis, 1901), pp. 261–7.)

some days, but the garrison was eventually relieved by the advance of Colonel Burtsov, with 2 battalions, 1 regiment of Cossacks and 6 guns, up the Borjom defile, and they made a successful sortie, captured Ahmet's artillery, and raised the siege.[1]

Simultaneously with the operation against Akhaltzikhe, the Turks had made an advance from Batum with about 7000 irregulars but they were defeated by a small force (mostly Gurian militia) under General Hesse, who drove them back and occupied Kobuleti.

At the end of April, Ahmet Bey again took the offensive in the region of the Posof-çay, but he was defeated by Burtsov at Tsurtskab, a village to the south of Diğur. The numerical weakness of the Russians, the wide dispersion of their forces in small detachments, and the large numbers of irregulars available to the Turks, still encouraged the latter to persist in the offensive against Akhaltzikhe.

At the end of April, Salih Paşa had concentrated about 50,000 nizams and irregulars between Erzurum and Hasankale and in the region of the Soğanli-dağ, and he hastened to send Hakki Paşa to the support of the Laz forces along the Posof-çay. The Turks passed through Ardahan (where there was no Russian garrison) and engaged Burtsov's small force in fierce fighting at Tsurtskab.

Muravyev (3 battalions, 1 regiment of dragoons and 1 regiment of Muslim volunteer horse) now advanced on Ardahan along the road from Kars and threatened the Turkish rear. On 2 June the Turks were again defeated at Tsurtskab and their force of nearly 8000 dispersed into the Yalnizçam Mountains. Paskevich, who himself arrived at Ardahan at this juncture, ordered the place to be garrisoned and fortified. The abortive Turkish operation had delayed the opening of the Russian offensive by two or three weeks. But only now, at the beginning of June, did weather conditions really allow of large-scale operations.

Paskevich had collected in the region of Kars a striking force which, after deduction of troops in garrisons and on lines of communication, amounted to 18 battalions (12,000), 12 regiments of Cossacks and mounted irregulars (6000) and 70 guns.

The Turks had available in the Soğanli-dağ area about 50,000 men of whom 30,000 were regulars (nizams).

The Soğanli[2] Mountains bar all the roads leading from the east and

[1] In *Kav. Sbor.* I, pp. 1–122, 'Vospominaniya iz Kavkazskoy stariny', V. Andreyev gives an interesting personal account of the Akhaltzikhe affair.

[2] In modern Turkish the name, Pasinler-sira-dağ, has been substituted for the historic Soğanli-dağ.

the north-east to Erzurum. The River Aras, turning east below Köprüköy, cuts a long defile past the southern wall of the Soğanli. South of this defile rises the massif of Dram-dağ,[1] which divides the Pasin valley from that of Eleşkirt (upper Murat-su). The passage of troops coming from Bayazit through the Eleşkirt valley and over the Dram-dağ by the Kara-derbent defile and the col of Velibaba was practicable. But an operation from Bayazit against Erzurum was

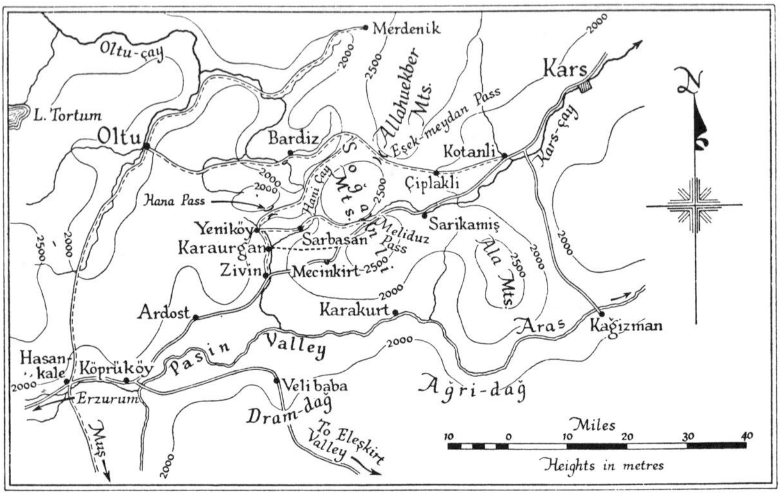

Map 3. The Soğanli-dağ

ruled out by Paskevich, since Bayazit was too distant from his principal bases at Aleksandropol and Erevan, and the line of advance was exposed to counter-attack from the south from Muş and Van.

The most direct operational approach to Erzurum was along the main caravan road from Kars, but this route passed over the most difficult part of the Soğanli-dağ. Here the buttresses of the principal ridge extend far to the north-east and form the eastern rim of the upland basin in which Kars lies. The main road from Kars to Erzurum followed the upper course of the Kars-çay, and at a distance of forty miles from the fortress entered the mountains west of the village of Sarikamiş ('Yellow reeds'). Five miles beyond Sarikamiş the road climbed the stiff fir-covered slopes of the Soğanli-dağ to the watershed at the Meliduz col (2440 m.), from where it gradually descended to the village of Mecinkirt (eighteen miles from Sarikamiş), which is situated on an upland shelf slightly inclined to the south.

[1] The name, Dram-dağ, has disappeared from modern maps; it constituted the western shoulder of the Ağri-dağ.

From Mecinkirt begins the descent along the southern slopes of the Soğanli. These southern slopes are by no means so precipitous as the northern and they incline over a distance of forty miles to the village of Köprüköy where there is a fine old stone bridge over the Aras. From Köprüköy through Hasankale thirty miles of good road led to the capital of Turkish Armenia.

There was another road across the Soğanli which was not in those days regarded as so practicable. Twenty-five miles out from Kars, it forked north-west from the first road at Kotanli, passed Çiplakli and approached the northern flank of the Soğanli ridge which here falls abruptly to the ravine of the Bardiz-çay, an affluent of the Oltu. In the valley lies the village of Bardiz, but the road, without touching Bardiz, mounted to the main plateau of the Soğanli and then, after some miles, began a gradual descent, following the line of the Zivin-çay flowing north to south through Yeniköy ('New village'), Karaurgan and Zivin. All these villages lie in the wide ravine formed by the stream. From the pass over Bardiz to the village of Zivin is about twenty miles. There was a track, little used, over the high plateau from the col of Meliduz to Zivin (fifteen miles), and another track from Sarikamiş direct to Yeniköy.[1] After Zivin, the main road descended the southern slopes of the Soğanli by easy gradients over the thirty-five miles to the Köprüköy bridge across the Aras.

At the beginning of June 1829 about 20,000 troops (of whom half were nizams) under Hakki Paşa, held a strongly fortified position on the Meliduz col covered by 24 guns. A detachment 2000 strong, under Osman Paşa, watched the Yeniköy-Zivin track. The principal force under the seraskier Salih Paşa, 20,000 strong with 20 guns, had been concentrated at Zivin by 15 June. But liaison had not been established with Hakki Paşa, and the two groups were separated by the deep ravine of the Hani-çay and fifteen miles of difficult country.

By this date (15 June) not only the advance-guard but the main body of the Russians had already approached the high plateau of the Soğanli without meeting with any resistance.

At the beginning of June Paskevich had received information as to the disposition of the Turkish forces. He formed the conclusion that movement by the direct road to Kars and attack on the frontal position of Hakki Paşa would allow the Turks to bring up reinforcements. He decided, therefore, to use Burtsov's detachment only (2000 men) for

[1] This latter track, after 1877, was converted to a road which became part of the route from Erzurum to Kars; and the older road between Mecinkirt and Meliduz declined to a secondary status.

a diversion against the position of Hakki at Meliduz. At the same time he directed a force of 5000 under Muravyev, including strong cavalry support, to march by the western track through the Eşekmeydan ('Donkey market') Pass on Yeniköy and Karaurgan. The third and main column under Pankratiev (with Paskevich) was to follow the mountain path which went up from Sarikamiş to meet the Karaurgan track at Yeniköy. Once on the Soğanli plateau, it was intended that the main column should cross the Hani-çay ravine at Sarbasan and then turn east to outflank the position of Hakki Paşa at Meliduz. This manoeuvre of Pankratiev's column would be covered from the Turks concentrated at Zivin by Muravyev's advance on Zivin from Yeniköy and Karaurgan. After the defeat of Hakki Paşa, Paskevich intended to concentrate all his forces against the seraskier at Zivin. In such wise the Russian commander planned two consecutive blows against two Turkish groups which had failed to establish satisfactory liaison over a stretch of mountainous country. The first conditions of success lay in mobility and surprise.

The Russian offensive movement began on the evening of 13 June and continued through the night. On the morning of the 14th, Burtsov's advance-guard came in contact with the Turks on the Meliduz position. By the evening of that day, Paskevich's main striking force under Pankratiev, after a hard climb of twenty-six miles with guns over a difficult and continuously ascending path, reached the high plateau and established, at the same time, contact with Burtsov's detachment.

On the following day (15 June), Muravyev's Cossacks and irregulars supported by two infantry battalions, suddenly attacked Osman Paşa's detachment in his camp over Bardiz. The Turks were dispersed or captured. The paşa himself, finding that he was now cut off from Hakki Paşa's camp by Russian forces, fled along the Zivin road.

During the 15th and 16th the Russian troops were concentrating on the plateau. Hakki was now aware of the general direction of the Russian advance and awaited their attack. At the same time he sent to the seraskier to hasten the movement of reinforcements from Zivin along the road through Mecinkirt to Meliduz.

Information which came to Paskevich now suggested a daring change in his operational plan. Burtsov had reported that the Turkish position at Meliduz was very strong and well fortified. At the same time, Muravyev's patrols, passing through Yeniköy and Karaurgan, had located the big Turkish camp at Zivin, while local inhabitants reported that the greater part of the seraskier's troops were still on the march from Hasankale and Köprüköy to Zivin. Paskevich came to the conclusion that it would be wiser to strike first against the Zivin group which had not yet completed its concentration; therefore, on 17 June, he issued new orders: Muravyev was to open an attack in the direction Karaurgan-Zivin with the support of the greater part of Pankratiev's force; Pankratiev himself was to cross the Hani-çay ravine and cut the communications of Hakki Paşa with Karaurgan; while Burtsov was to continue to demonstrate before the Meliduz position.

On 18 June, Muravyev's advance forces came in contact with the Turks to the north of Karaurgan. On the same day the seraskier arrived at the Zivin camp where he had now about 15,000 nizams and 5000 irregular horse. His appreciation of the situation convinced him that only Russian patrols were operating in front of Karaurgan and that the main attack was still to be directed against Hakki Paşa at Meliduz. He took the initiative and ordered his horse to attack the Russians descending from the north to Karaurgan.

The Turkish cavalry ran into the whole of Muravyev's column (4 battalions, 1 regiment of Cossacks and 1 regiment of Muslim irregulars with 20 guns) and they were thrown back in disorder. Meantime the column led by Paskevich himself (7 battalions, 3 dragoon and Cossack regiments and 24 guns) came down the Sarbasan heights and occupied Karaurgan from the east.

In front of Meliduz, Hakki Paşa counter-attacked Burtsov with 5000 horse. Burtsov had only 3 battalions, 1 Cossack regiment and 1 regiment of Muslim irregular horse, but he stood firm and threw back the Turks to their main position. He received some timely help from Pankratiev who, with a small detachment, crossed the ravine to the south of Sarbasan and reached the path from Karaurgan to Meliduz.

After the defeat of his cavalry, the seraskier prepared to defend his position on the Zivin heights, and awaited the arrival of reinforcements from the south (some 20,000 men were strung out between Erzurum and Zivin). He hoped to give battle the next day, after having concentrated his forces and established liaison with Hakki at Meliduz. But, during the afternoon of the 19th, Paskevich continued the offensive movement from Karaurgan towards Zivin.

Muravyev crossed the Hani ravine to the west of the main track and, after ascending the more accessible part of the Zivin Heights, turned completely the Turkish left flank. Paskevich's column marched quickly down the main valley and occupied Zivin village. By the evening the whole of the Zivin camp was in the hands of the Russians; all artillery and stores and great trains of pack animals were taken without resistance. The Turks fled precipitately along the Erzurum road and, during the night, the approaching reinforcements were infected with the panic. The large forces at the disposal of the seraskier poured in disorder into the Pasin valley. He himself, with his staff, galloped to Hasankale.

Without losing a moment, Paskevich, during the night, prepared the second part of his operation, hoping that the enthusiasm of victory among his troops would overcome the fatigue of the last week's marching and fighting. Only a few cavalry detachments pursued the Turks to the south, while, under cover of darkness, almost all the Russian forces were brought up for an attack on Hakki Paşa's position early in the morning.

At 7 a.m. Paskevich had concentrated for the attack about 9000 infantry, 5000 cavalry and 56 guns: on the left Burtsov and Pankratiev; in the centre, along the path Karaurgan-Meliduz, Paskevich himself; on the right, along the path Zivin-Meliduz, Muravyev. On the extreme right the cavalry of General Osten-Sacken were directed to cut the Turkish line of retreat from Mecinkirt. Some Turkish prisoners were sent to Hakki Paşa to inform him of the seraskier's defeat, but he decided to resist. However, his troops were already demoralized; their resistance was weak and before midday Hakki surrendered. His delay had, in the meantime, given the bulk of his troops the time to disperse and the Russians took only 1500 prisoners (with 20 guns). The failure of Osten-Sacken to cut the flight of Hakki's army was strongly censured by Paskevich.

The Soğanli operation of 13–20 June 1829 has been described in some detail because this brilliant manoeuvre of Paskevich remained the classic model for all the subsequent campaigns along the Caucasian border. The Russian losses were astonishingly small, not more than 300 men. The Turks lost about 3000 killed and wounded (including Osman Paşa's detachment). These losses were slight for an army of a nominal strength exceeding 60,000, but all the Turkish artillery was lost to the Russians and, more important, the morale of the troops was thoroughly shaken. Even the nizams began to desert, and the seraskier's army thawed like snow in the spring.

The campaign of 1829 was in essentials decided by a single blow. This amazingly easy Russian success must be explained by the defects of the Turkish military organization of the period, which was in a phase of transition following the reforms only recently initiated by

Sultan Mahmut II. The Turks were destined to face a further series of defeats in Asia Minor during the Egyptian campaigns of the thirties. Turkish troops could fight with courage and they were ready to show their traditional stubbornness in defence. But their artillery was badly instructed, their cavalry undisciplined, and their infantry incapable of manoeuvre. The officers were not properly trained, and the staff were frequently appointed because of the favour of individual paşas or of the court. However, the results which Paskevich attained with so small a force and with such insignificant losses testify to the perfection of his strategy and to the soundness of his tactics applied to the special conditions of the Caucasian border.

Extremely difficult conditions of terrain were overcome. Since he had to take no fortresses, Paskevich avoided frontal attacks and preferred manoeuvre, relying on the factors of mobility and surprise. By his ascent of the Soğanli plateau, Paskevich had boldly left his bases far behind, and his only source of supply became the train of 3000 carts following the army. During the march to Karaurgan and the fighting of 19–20 June, this train remained on the plateau defended only by a very small detachment. During these critical days Paskevich compared his columns to ships which had left the coast and were navigating in open sea. But the risks which he took gave him the initiative and remarkable mobility in action. Further, it is clear that only highly trained and well exercised troops could have sustained the fatigue of eight days of continuous fighting and marching in difficult mountain country. The sole period of relative *détente* which troops enjoyed was during the days of concentration and redistribution of forces on the plateau (15–16 June). The example of Paskevich was forgotten in the campaigns of 1853–6 and 1877–8 and only recovered and followed by Yudenich in 1915–16.

Following their defeat, the Turks had little chance of further successful resistance in front of Erzurum. On 23 June Paskevich arrived at Köprüköy, where the bulk of his forces was concentrated and where he received the elders of the Kurdish tribes from Ağri-dağ and Bingöl. On 24 June the Turks abandoned Hasankale, a fort defended by 30 guns, and on 25 June the seraskier surrendered Erzurum with the remnants of his army and 150 fortress guns. The vast stores which had been concentrated there made it possible for Paskevich to continue his campaign without moving up supplies over the long road from Kars and Aleksandropol.

Paskevich now faced the problem of a deeper invasion of Anatolia with the limited forces at his disposal. The numerous Armenian population of the eastern vilayets welcomed the Russian victory as their own, and Paskevich, by an intelligent policy, was successful in conciliating a part of the Muslim population.[1] The four regiments of Muslim horse recruited among the Kabardans, Circassians and Tartars, and even from Chechnia and Dağistan, proved invaluable for internal security purposes, and Paskevich appointed a Muslim aristocrat of Cherkess origin, General Bekovich-Cherkasski, governor of Erzurum, with an advisory council partly recruited from Turkish officials and notables. The Kurdish chiefs to the south of the Aras were not unfriendly, and through them Paskevich had succeeded in neutralizing the paşa of Muş at the beginning of the campaign. The paşa of Van, however, had proved more formidable, and, at the time of the Turkish operation against Akhaltzikhe, he had invaded the Eleşkirt valley with some nizam units, 6 guns and several thousand Kurdish cavalry, and laid siege to the small Russian garrison of Bayazit (6 June). Fighting continued over a period of three weeks and the Turks penetrated into the town. Russian losses were relatively heavy (about 400). After the occupation of Erzurum, Paskevich sent a small force to the relief of Bayazit, and the troops of the paşa of Van were dispersed.

Prospects of continued Turkish defence in the eastern vilayets—while new armies were formed in central Anatolia—were now entirely dependent on the organization of irregular resistance. The relatively small numbers of the Russians, scattered over a great extent of country, favoured the development of an irregular strategy by the defence. Further, the attack on Bayazit and the earlier operations of the Acars against Akhaltzikhe had demonstrated the possibilities of irregular action. Wide mountainous areas, inhabited by warlike Muslim tribes, lay on both sides of the long line of communications between Kars and Erzurum. While the Kurdish chieftains to the south of the Aras were a doubtful element who intrigued impartially with Paskevich's agents and with the paşas of Van and Muş, the Muslim dere-beys of Lazistan and Acaristan, Georgian by blood and speech, were, like the Slav Muslim landowners of Bosnia, fanatically anti-Christian, stoutly conservative, and attached by many personal interests to the court in Istanbul. At the same time the Turks were in

[1] 'He was everywhere received with loud acclamations, in which the Kurdish Beys joined with all their hearts, saying: "Take Erzeroum, and you will be willingly joined by all our tribes".... The Kurds collected great quantities of sheep, cattle and other plunder, which materially assisted the army; and General Paskevich's skill in handling the different tribes of his complicated government was equal to his military talents.' Monteith, *Kars and Erzeroum.*

a position to supply the Muslims of the Pontic Alps with plenty of guns and ammunition.

Paskevich intended to secure Trebizond as an essential sea base for further operations against Sivas and central Anatolia. The co-operation of the Russian Black Sea fleet was essential to the success of this operation, but the fleet (11 ships of the line, 8 frigates, 12 corvettes) was at that moment fully engaged in the decisive operations off the coast of Bulgaria (Varna and Sizeboli), the result of which was soon to terminate the war.

At the beginning of July Paskevich sent Burtsov, with 2 battalions, 6 guns and a detachment of Cossacks and Muslim irregular horse, to occupy the ancient town of Bayburt in the valley of the upper Çoruh on the main road from Erzurum to Trebizond. The difficult col of Kop-dağ (2691 m.) on the watershed between the Kara-su and the Çoruh was successfully passed, and the eighty-mile march to Bayburt was made without resistance. When, however, Burtsov attacked a concentration of Laz tribesmen on the wide alm of Hart-ovasi to the north of Bayburt, the Russians were repulsed with a loss of 300 men and 18 officers (including the gallant Burtsov himself). Paskevich decided that it was vital to redeem this defeat if the spread of the irregular movement everywhere were to be checked. He marched with all his available force (6000 infantry and cavalry) to Bayburt.

Meantime from Trebizond Osman Paşa (who had proclaimed himself seraskier following the surrender of Salih), collecting a force of some 6000 nizams, moved through Gümüşane and took up a position at Balahor, covering the col of Vavuk (1973 m.). Paskevich rightly appreciated the dispersion of his opponent's forces and struck two sharp blows before they could combine, first, against the Laz levies at Hart (about 6000) and, secondly, against Osman Paşa at Balahor. Both operations were successful: the Laz were scattered, and Osman was defeated with the loss of his guns and animal trains. He retired into the valley of the Kelkit. The Russians occupied Gümüşane and, following along the Kelkit valley, their patrols came within sight of Şebinkarahisar.

The advance of the Russian forces in Europe across the Balkans convinced Paskevich that the war was approaching its end and that no further decision could be obtained in Anatolia. The Turkish regular forces had been practically eliminated, but caution was imposed on Paskevich by his own lack of numbers, and the initiative, as winter approached, really remained with the guerrillas. The existence of considerable irregular forces in the mountainous belt of the Pontic

Alps and Acaristan protected the two Turkish ports of Batum and Trebizond and prevented Paskevich from establishing direct sea communication between the Russian Black Sea ports and the army in Asia Minor.

In the middle of August, General Osten-Sacken crossed the eastern ridge of the Yalnizçam[1] from Akhaltzikhe and invaded upland Acaristan, while at the same time General Hesse undertook a forward movement in Guria. Osten-Sacken occupied Hulo, the capital of Ahmet Bey, in the upper valley of the Acharis-dzqali; but with no further news from Hesse, he abstained from a march down the valley of the Acharis-dzqali against Batum and withdrew to Akhaltzikhe before the autumn rains.

At the beginning of September, Hesse, with a force of 2500 Russians and 3000 Gurians, with 7 guns, had moved from Ozurgeti through Kobuleti on Batum. The Turks had prepared a strong position on the heights of Tzikhis-dziri covered by the River Kintrishi, and here Hesse was repulsed, on 16 September, with a loss of over 600 men.

The action at Tzikhis-dziri took place several days after the conclusion of an armistice at Adrianople, but news of this event had not yet reached the Caucasian army. At the beginning of September, Paskevich, in anticipation of the end of the war, had already begun to send back troops to winter quarters in Georgia. This movement was interpreted as a sign of weakness throughout the occupied territories. The paşa of Van immediately showed activity, and his Kurdish irregulars occupied Hinis and appeared at Aşkale on the road from Erzurum to Bayburt. At the same time, the paşa of Trebizond, although already informed of the armistice, moved against Bayburt and was joined by large numbers of Laz. On 28 September Paskevich defeated him in the last action of the war of 1828–9.

The terms of the Treaty of Adrianople, governed by wider issues of European politics, were moderate. On the Caucasian frontier, the Russians only retained the strong places of Akhaltzikhe, Akhalkalaki and Atskhur—all of which might be deemed to cover the Borjom defile.

The whole of the paşaliks of Erzurum, Bayazit and Kars were restored to the Turks with the south-west part of the paşalik of Akhaltzikhe including Ardahan and the sources of the Kura. In Guria, not only Batum but also Kobuleti remained to the Turks, and the frontier here was marked by the Fort of St Nicholas on the north bank of the Notanebi stream.

[1] In Old Georgian, *Arsiani*; Arsiyan, on modern Turkish maps.

Anapa, from which the Cherkesses had, till the war, been supplied with arms by the Turks, and Poti, an inferior port which gave, however, facilities for direct communication between the Crimea and Transcaucasia, passed to the Russians, and the Turks abandoned all claims to suzerainty in Circassia.

This latter was, in many ways, the most important condition of the Treaty of Adrianople; for by it international sanction was given to the Russian position in the Caucasus, and the Turks retained no pretensions to interfere with the measures of pacification which were to become the main task of the Russian army during the following quarter of a century. The evacuation by the Russians of the eastern vilayets led to a certain redistribution of population, for many thousands of Armenians who had shown open sympathy to the Russians, followed the army when it withdrew and were settled in the newly incorporated regions of Erevan, Akhalkalaki and Akhaltzikhe.[1]

The Russian campaigns of 1828 and 1829 may be regarded as a military and political reconnaissance of eastern Asia Minor. The defence of the ways to Erzurum and further into central Anatolia was provided, not by the fortress of Kars, but by the natural bastion of the Soğanli-dağ ridge. In relation to Asia Minor, the significance of both Kars and Akhaltzikhe was that of advance positions, the defence of which might delay invasion by a whole campaigning season. The Akhaltzikhe area had been shown to have great defensive possibilities, and with the loss of the control of the exit from the Borjom defile, the Turks were to experience great difficulty in future campaigns in holding the Kars plain since the fortress of Kars blocked only one of the two lines of invasion. Paskevich himself fully appreciated the significance of the acquisition of Akhaltzikhe.

In the campaign of 1829 the Russians had become acquainted with the ways leading over the Soğanli-dağ, and Paskevich, during the withdrawal of his army, did not neglect to have a reconnaissance made of the track from Erzurum to Ardahan which went through Nariman to Penek and along the Oltu-çay, then along the ridge of the Yalniz-çam-dağ into the valley of the upper Kura. The Oltu-çay route was to prove of importance during the campaigns of 1877 and 1914.

Paskevich had also clearly grasped the strategic importance of the Bayazit region as a possible base for an offensive operation against Erzurum along the Eleşkirt valley. Through this fertile valley, rich in supplies, the Soğanli position could be turned from the south. The

[1] Erevan province, later the core of S.S.R. Armenia, had at this period a majority of Turkish Muslims.

Eleşkirt route, however, could itself be turned from the south from the direction of Van-Bitlis-Muş, and an advance from Bayazit on Erzurum would impose covering operations in the Lake Van area (such as were undertaken in 1914–15).

The final phases of the campaign of 1829 indicated that the first stage of an offensive against Sivas, directed to the conquest of central Anatolia, required the control of the triangle Bayburt-Trebizond-Şebinkarahisar. To realize this objective, combined operations between the Caucasian army and a Russian fleet in the Black Sea were necessary.

The complete domination of the Black Sea by the Russian fleet was, after 1829, the first condition preliminary to further Russian expansion in the Middle East. The creation of a strong Black Sea fleet became a prime military aim of the government of Nicholas I. At the same time it became clear that the security of Asia Minor could only be ensured by the limitation of the strength of the Russian Black Sea fleet and of Russian naval bases on the Black Sea. In the war of 1853–6 the Turks, with the support of the western maritime powers (Britain, France and Sardinia) achieved this object. In the early seventies, the renewal of Russian plans to construct a Black Sea fleet was one of the causes contributory to the outbreak of the war of 1877–8. On this occasion, the Turks retained naval superiority in the Black Sea. Again in 1914, the arrival of the *Goeben* and *Breslau* in the Bosporus was an important insurance for the Turks in Anatolia and was one of the factors which gave the Committee of Union and Progress the confidence to declare war on Russia.

The campaigns of 1828–9 also indicated the potential importance of guerrilla action where regular forces, relatively weak in numbers, are operating over a wide expanse of territory. Had the Kurds to the south of the Aras proved as active as the Laz and Acars, Paskevich might have found himself in a difficult position after the fall of Erzurum. As it was, the hardest fighting in the 1829 campaign had been against the Acars round Akhaltzikhe and against the Laz in the upper Çoruh valley. The Kurds remained a doubtful factor, and Paskevich deserved credit for the skill of his policy towards them. Their neutrality gave Paskevich all the advantages which Napier enjoyed nearly forty years later in Abyssinia, when a passive but potentially dangerous mountain population allowed a relatively small force to advance to Magdala in the heart of the Ethiopian highlands. In neither case were the regular forces really the masters of the country over which they were advancing. As guerrillas, liable to operate against either of two combatant armies, the Kurds, like the Abyssinians, have remained a factor of local importance down to the present time.[1]

[1] In the first quarter of the sixteenth century it had been Kurdish defection to the Turks, organized by a Kurd, the astute historian and statesman, Idrisi of Bitlis, which had enabled Sultan Selim I to expel the garrisons of Shah Ismail and push back the frontier of Persia from the Euphrates to the Zagros. The Kurds continued to be a factor in the Turco-Persian frontier wars of the three succeeding centuries. The great *ağas* were relatively content with the loose feudal relationship with the Porte which had been introduced on the advice of Idrisi. It was only when the nineteenth-century *étatisme* of Mahmut II disturbed the hereditary rights of the *ağas* that the Kurds became a source of danger on the eastern frontier of Turkey. See Bibliographical and Supplementary Notes, pp. 536–7.

CHAPTER III

RUSSIAN PACIFICATION OF THE MAIN CHAIN OF THE CAUCASUS. DEVELOPMENT OF BRITISH INTEREST IN CAUCASIA 1830–52

MAP I

THE occupation by the Russians of the whole of the Transcaucasian trough (Rion and Kura-Aras basins) and of long sections of the peripheral rim of the Armenian highland zone, while the mountain tribes of the main chain of the Caucasus still remained unsubdued, presented the Russian government with a major problem of internal security—only comparable to the contemporary problem which was developing for the British on the North-West Frontier of India.

At the same time, the appearance of a Russian army in front of Constantinople in 1829 and also the remarkably easy victories of Paskevich on the borders of Persia and Turkey had promoted no little nervousness in England. This tendency to uneasiness was subsequently aggravated, in 1833, by the arrival of a Russian fleet in the Bosporus to support Sultan Mahmut II against the pretensions of his rebellious vassal, Mehmet Ali Paşa of Egypt. The terms of the treaty concluded at Hunkâr Iskelesi between the Russians and the Turks seemed to indicate that the Russians aimed at establishing a permanent protectorate over the whole of the Ottoman empire. The international situation was complicated by the support which the government of King Louis Philippe was according to Mehmet Ali, and, in the outcome, the British combined with the Russians to check the threatened ascendancy of the Egyptian viceroy. Nevertheless, from the early thirties many serious persons in England were convinced that the Russians were planning a comprehensive penetration of Asia Minor directed towards the acquisition of ports on the Mediterranean and the Persian Gulf. The Foreign Office refrained from initiating an openly anti-Russian policy in the East, but opposition to the progress of Russian expansion became a tradition with successive British ambassadors accredited to the Sublime Porte (Ponsonby, 1833–41, and Stratford Canning, 1842–56). This attitude did not fail to produce certain embarrassments for the Russians in the Black Sea and notably in Caucasia.

The Turks were well informed in regard to all the intricacies of politics among the mountain tribes, and they were able to interest influential Englishmen in the possibilities inherent in supporting the tribes in their resistance to the establishment of Russian rule. News of the sharp character of the struggle against the Russians in Circassia, as also in Chechnia and Dağistan, reached Istanbul in 1832. Already in 1834, Ponsonby, with what Melbourne described as his ridiculous russophobia, was proposing that aid should be given to the Cherkesses and even that British ships should be sent to the Caucasian coast. His reports to London, in fact, gave a false impression of what was happening in Caucasia, and the Turks themselves, despite their knowledge in detail, arrived at no intelligent appreciation of the events which were taking place in the decade 1830–40. On the other hand, the Russian government, which was ready to develop plans for a systematic liquidation of Cherkess resistance after the conclusion of the war of 1828–9, was completely taken by surprise by the formidable character of the insurrection which swept the eastern Caucasus during 1830–2.

Paskevich, during his brief viceroyalty in Caucasia, had only time to stabilize the situation in Ossetia, where orderly conditions were essential to the security of the Georgian Military Road. His plans for the pacification of Circassia were incomplete when he was ordered to take command against the Polish insurrection in 1831. Along the western coast Russian posts were being pushed forward gradually from Anapa round the magnificent Bay of Sujuk. A port was under construction at Novorossiysk and, farther to the south, at Gelenjik, a new naval and military base was in process of organization. Paskevich had intended, first, to unite Gelenjik with the valley of the Kuban by a new fortified line across the mountains. Later, he proposed to construct a coastal line from Gelenjik to Sukhum in order to prevent all intercourse by sea between the Cherkesses and the Turks. The unexpected movement in the eastern Caucasus made it necessary to postpone the completion of this plan for several years.

As early as the Russo-Turkish war of 1769–74, the Muslim mountaineers of Chechnia had responded to the call to *gazavat* ('holy war') preached by the mysterious adventurer Sheikh Mansur (who was perhaps an Italian renegade in Turkish pay).[1] During the following half-century both Turkish and Persian agents were active in the mountains, but neither had had any serious influence on the development of the fanatical puritan movement of the Murids[2] (which

[1] Baddeley, *Conquest*, p. 48; cf. N. Smirnov in *Voprosy Istorii*, x, pp. 193–9.

[2] *Murid* (T., properly *mürit*) = disciple, follower; A. V. Moran, *Türkçe-Ingilizce Sözlük*.

was in many ways comparable to the contemporary Wahabi movement in Arabia). In Muridism mystic elements were blended with social motives. The Murids constituted a kind of monastic and military Muslim order in which all were equal and all were obliged to obey the spiritual leader (*imam*). At the same time the Murids were relieved from all obligations to the village elders and landowners (*bekler*); their most sacred duty and object in life was to die in battle against the infidel (*gawur*).

In 1830, the first imam, Kazi Mullah, gathered several hundred Murids at the *aul*[1] of Gimri and raised the standard of gazavat. He counted, not without reason, on the immediate support of many thousands of Avars and Chechens, smouldering with anger at the punitive expeditions of the Russian General Ermolov during the years preceding the war of 1828–9, and excited by the fermenting propaganda of Turkish emissaries who had been working in the mountains during the recent war. The imam destroyed several isolated Russian detachments and appeared, with rapidly growing forces, in front of the forts along the left of the Caucasian line. He crossed the Sunja and the Terek, burned a number of Cossack *stanitsy* and raided the town of Kizlyar. The Avars, at the same time, threatened Derbent, and the other Lesghi tribes began to raid into the valleys of Kakheti, while large bands of Chechens surrounded Vladikavkaz.

The Russians were taken completely by surprise, and their effort to suppress the gazavat lasted three years and involved the concentration of considerable forces. Finally, a column of 10,000 Russian infantry and Cossacks with artillery, under General Velyaminov, took Gimri, the first stronghold of the Murids, and Kazi Mullah himself was killed. Russian losses during the three years' operations were not less than 3000 killed and wounded.

The Russians had, however, failed to understand the real force of the Murids, and the gazavat soon flamed up under a new and even more formidable imam, Shamyl. The strength of the Murids lay in their conception of war as an end in itself and as a vehicle for self-purification and self-sacrifice. The imam was striving to attain neither material ends nor the political independence of the Dağistanis and Chechens. In their conceptions the Murids differed essentially from the Cherkesses upon whose minds religion lay lightly enough and whose chieftains were inspired by pretensions to political independence under Turkish protection, and by the desire to save their communities and lands from Cossack usurpation.

[1] Mountain village.

Shamyl knew how to excite the enthusiasm and fanaticism of the Murids; but his puritanism and ruthless insistence on obedience and sacrifice tended to alienate the villagers who were not among the elect, but who were expected to suffer the exactions of the Murids and the reprisals of the Russians, while his radicalism roused the antagonism of the powerful *beks* of Dağistan, many of whom, driven out of their properties by the Murids, sought the protection of the Russians. Gradually, the movement, born in Dağistan, began to find a greater strength in Chechnia, where a certain natural primitive communism favoured the egalitarian tenets of the imam.

In Istanbul Shamyl's movement was incorrectly appreciated. The Turks saw in it a response to their own political propaganda, and they failed to understand the mystical aspects of the Murid heresy. In British circles the movement was taken to be a national and political struggle against the Russian empire, and as such comparable to the various anti-imperial movements in different parts of Europe and the Near East which were at that time stimulating the sympathies and enthusiasms of English Liberals. Among those private individuals who espoused the cause of the Cherkesses it was hoped that it might be possible to co-ordinate the two movements in the eastern and western mountains. The travellers, Longworth and Bell, made more than one excursion to the Cherkess coast where they established relations with the leaders of the tribes and (somewhat indiscreetly) held out hopes of the material help and diplomatic support of the British government. The smuggling of arms and ammunition along the coast from Trebizond was organized, and some influential Turks, with British turcophils like David Urquhart, set up in Istanbul a Cherkess political centre at the head of which was a certain Sefer Bey who had been educated in Russia. Under considerable misapprehension as to local conditions, Turkish mullahs were sent to preach the gazavat among the Cherkesses.[1]

In 1837, the leaders of the Cherkess tribes, impressed probably by a recent visit of Longworth and Bell, sent an ultimatum to the Russian authorities on the Kuban. They demanded cessation of work on the new fortified line which was under construction from Olginsk bridge-head on the Kuban to Gelenjik and which, when complete, would enclose all the north-western part of the Cherkess lands.

In the meantime, the Emperor Nicholas I had decided to take the strongest measures to bring the Caucasian war to an end. During the

[1] David Urquhart was for some time First Secretary at the Istanbul embassy (cf. *D.N.B.*). For biography of Sefer Bey, see *Kav. Sbor.* III (1), pp. 176–7.

autumn of 1837, the emperor made a tour of his Caucasian dominions; he landed at Gelenjik and then at Redutkale, whence he proceeded to Kutaisi, Tiflis and Erevan. An invitation was sent to Shamyl to appear at the emperor's court in Tiflis where he was promised a pardon. After some hesitation, the imam refused the prospect of a favourable personal settlement; the whole conception of gazavat imposed a continuing struggle. For the following spring, General Count Grabbe received the emperor's order to take Shamyl's stronghold at Ahulgo and to put an end to the Murids.

Comprehensive operations had already been initiated in the western Caucasus; these followed the plan which had been prepared by Paskevich seven years before. In June 1837, 3000 men with 16 guns were landed at Adler at the mouth of the River Mzymta. At the same time the Russians gradually pushed forward their posts along the coast to the south of Gelenjik.

During April and May 1838 strong Russian detachments were landed at Sochi and Tuapse where, after some fighting with the Cherkesses, the construction of forts was begun. In the following year a major action took place between Gelenjik and Tuapse where, along the valleys of the Shahe and the Psesuape, were many flourishing Cherkess villages. A whole Russian infantry brigade, supported by marines and Cossacks, was engaged at the mouth of the Shahe and suffered substantial losses. However, the building of the fort at Golovinsk proceeded, as well as that of a number of other forts along the coast from Gelenjik to Adler. The Black Sea coast line seemed to have been created; but further events were to prove that it was easier to build this line than to maintain it.

Meanwhile in the eastern Caucasus, Grabbe had met with initial success. After a siege of eighty days he had taken Ahulgo, Shamyl's stronghold in the gorge of the Andi-köysu. But Grabbe had over 3000 casualties out of a total force of 8500, and the imam himself made good his escape. He took refuge in the inaccessible forests of Ichkeria in the southern part of Chechnia, which is divided from Dağistan by the high ridge to the north of the Andi-köysu. In the following spring (1840) Shamyl, encouraged by news from the western Caucasus, was already renewing his raids.

The winter of 1839–40 had been a hard one for the Cherkesses. During the fighting of the two previous years many of their thriving settlements had been laid waste, and thousands of their cattle had been driven behind the Caucasian line by the Cossacks. All important points along the coast were in the hands of the Russians and communications with Turkey were interrupted. Cherkess lands were now confined between the Kuban, the line from that river to Gelenjik, and another along the Laba.

But their extremity united the tribes as never before; with the spring of 1840 a general insurrection flared up which soon assumed serious proportions. Formidable forces, some of them 10,000 or 12,000 strong, began to attack the Russian 'lines' and the fortified posts along the coast. Forts Lazarev, Velyaminov, Mikhaylovsk and others were stormed and their garrisons massacred. But after the first surprise Russian resistance hardened. An attack on Golovinsk was beaten off. The Cherkesses now united their forces to attack the fort of Abinsk, the key point on the new line covering the way over from the lower Kuban to Gelenjik. The small Russian garrison of four infantry companies, some Cossacks and twelve guns put up a desperate and successful resistance. In the final attack across the deep ditch covering the main redoubt of the fort, the Cherkesses left 685 dead. This repulse, and the arrival of Russian naval units to protect the coastal posts, damped the ardour of the Cherkesses. In the western mountains the situation became more or less stabilized. Guerrilla war continued; some Russian forts were evacuated, while others were strengthened and more heavily garrisoned. The Cherkesses had lost confidence in the promises made by individuals professing to represent the Turkish or British governments, and their disappointment had an important bearing on the course of events in Caucasia during the campaigns of 1853–6. The Russians, on their side, were unable to complete the pacification of the western Caucasus, first, because of the continued successes of Shamyl in the eastern Caucasus and, later, because of the deterioration of the international situation in the Black Sea.

Between the years 1840 and 1842 the Russians lost about 5000 men in numerous encounters with the Murids. The casualties of Grabbe's costly expedition against Ahulgo brought these losses up to nearly 9000. In 1843 Shamyl again took the initiative and transferred his raids from Chechnia to Dağistan. He was joined by thousands of new recruits, including the famous Haci Murat (immortalized in Tolstoy's story of that name). Russian losses rose during the year to 12,000, and two score of Russian guns fell into the hands of the Murids who had plenty of trained deserters from the Russian forts to man them. All Chechnia was in a state of insurrection, while in Dağistan the Russians were forced to evacuate many important points. The Murid movement now inspired serious anxiety in St Petersburg, and the

Emperor Nicholas ordered the whole of the 5th Army Corps to the Caucasus. The operations against the Cherkesses were suspended, and not less than 30,000 men with a numerous artillery were concentrated against the terrible imam. This concentration was the maximum which the limited communications of the period would allow. In the spring of 1845 the command was entrusted to Prince Vorontsov.

Vorontsov well understood the necessity for a systematic conquest of eastern Caucasia; the lines must be gradually advanced; forts, roads and bridges built; forests cut down and the mountaineers deprived of access to grazing areas. The conception of a long siege of the natural Caucasian fortress had been outlined by General Velyaminov fifteen years earlier, in the days of Ermolov; later, Petersburg, impressed by the easy successes of Paskevich against the Turks, had considered the long siege unnecessary, and the lightning raids of the Murids had been countered by the clumsy expeditions against the successive strongholds of the imams. The emperor continued to be impressed with the possibility of terminating the mountain war by one blow aimed at the capture of the person of the imam, and it was he who overrode Vorontsov's judgement in deciding the campaign of 1845.

In the late spring of that year, two columns, composed of 18,000 infantry and Cossacks with 50 guns, under the personal command of Vorontsov, after a difficult march penetrated to the upper ravine of the Andi-köysu. The imam again withdrew into the forests of Ichkeria and concentrated his forces in the region of the auls of Veden and Dargo. Already embarrassed for supplies, Vorontsov decided to follow the imam with rather more than half his force. Having crossed the wild mountains to the north of the Andi-köysu, Vorontsov entered the district of Dargo where he found that the imam still avoided battle, allowing the Russians to occupy the aul of Dargo.

Without having brought his enemy to battle, Vorontsov now tried to return to his base at Grozny through the Chechnian beech forests where the Murid sharpshooters awaited the Russians behind every tree. Horses and baggage were abandoned and the numbers of the wounded grew. Vorontsov left Dargo on 13 July, and it took him a whole week to cover the thirty miles through the forests to Grozny. He lost 4000 men, including three general officers and 200 other officers.[1]

[1] The Dargo affair is described in Vorontsov's report in *Akty*, v, Docs. 383–4 (with maps). See also Baddeley, *Conquest*, chap. XXIV for a detailed account in English based on *Akty*.

Shamyl, with inflated prestige and growing numbers, appeared in the region of Vladikavkaz and threatened the Georgian Military Road. His mounted bands raided Kabarda, and a direct contact with the Cherkesses seemed likely.

But in General Freitag, commanding on the Sunja, and in the Cossack General Sleptsov, the Russians had men who could rise to the emergency. Shamyl's bold movement lost its original impetus; the Kabardans, traditionally friendly to the Russians and dominated by their feudal chieftains, proved lukewarm, and the Cherkesses beyond the upper Kuban failed to respond to the exhortations of the imam. Shamyl feared for his communications with Chechnia, and when he turned back he narrowly escaped being surrounded at the crossing of the Terek.[1]

The climax of the Murid movement proved to have been reached and passed in the years 1845–6, but the fact was not observable at the time.

[1] See N. Gorchakov, 'Vtorzheniye Shamilya v Kabardu v 1846 godu', in *Kav. Sbor.* IV, pp. 19–38.

M. F. Fedorov has left a delightful personal account of the life of an officer in the western Caucasus in 'Pokhodniya zapiski na Kavkaze s 1835 po 1842 god', in *Kav. Sbor.* III, pp. 1–219. A poet of some talent, he cultivated the society of the numerous Decabrist exiles who were posted to the Caucasus. Less sophisticated, but equally interesting, are 'Souvenirs d'un officier ayant servi dans le Caucase (1835–1838)', included by M. F. Brosset in vol. II of his *Collection d'historiens arméniens*, pp. 485–696.

Much useful information on the western Caucasus is given by A. Yurov in 'Tri goda na Kavkaze, 1837–1839', *Kav. Sbor.* VIII, pp. 1–240, and IX, pp. 1–155. See also *ibid.* II, pp. 1–74, 'Ocherk polozheniya voyennykh del na Kavkaze s nachala 1838 do kontsa 1842 goda'; and I. Miroslavski, 'Vzryv Mikhaylovskago ukrepleniya v 1840 godu', *ibid.* IV, pp. 1–18. For later years, see I. Drozdov, 'Obzor voyennykh deystviy na zapadnom Kavkaze s 1848 po 1856 god', *ibid.* X, pp. 497–584, and XI, pp. 465–98. The containing operation of 1851 is described by Gen. Evdokimov in 'Ekspeditsiya 1851-go goda na pravom flange Kavkazskoy linii', *ibid.* VIII, pp. 307–34.

BOOK II

THE CAUCASIAN THEATRE DURING THE CRIMEAN WAR

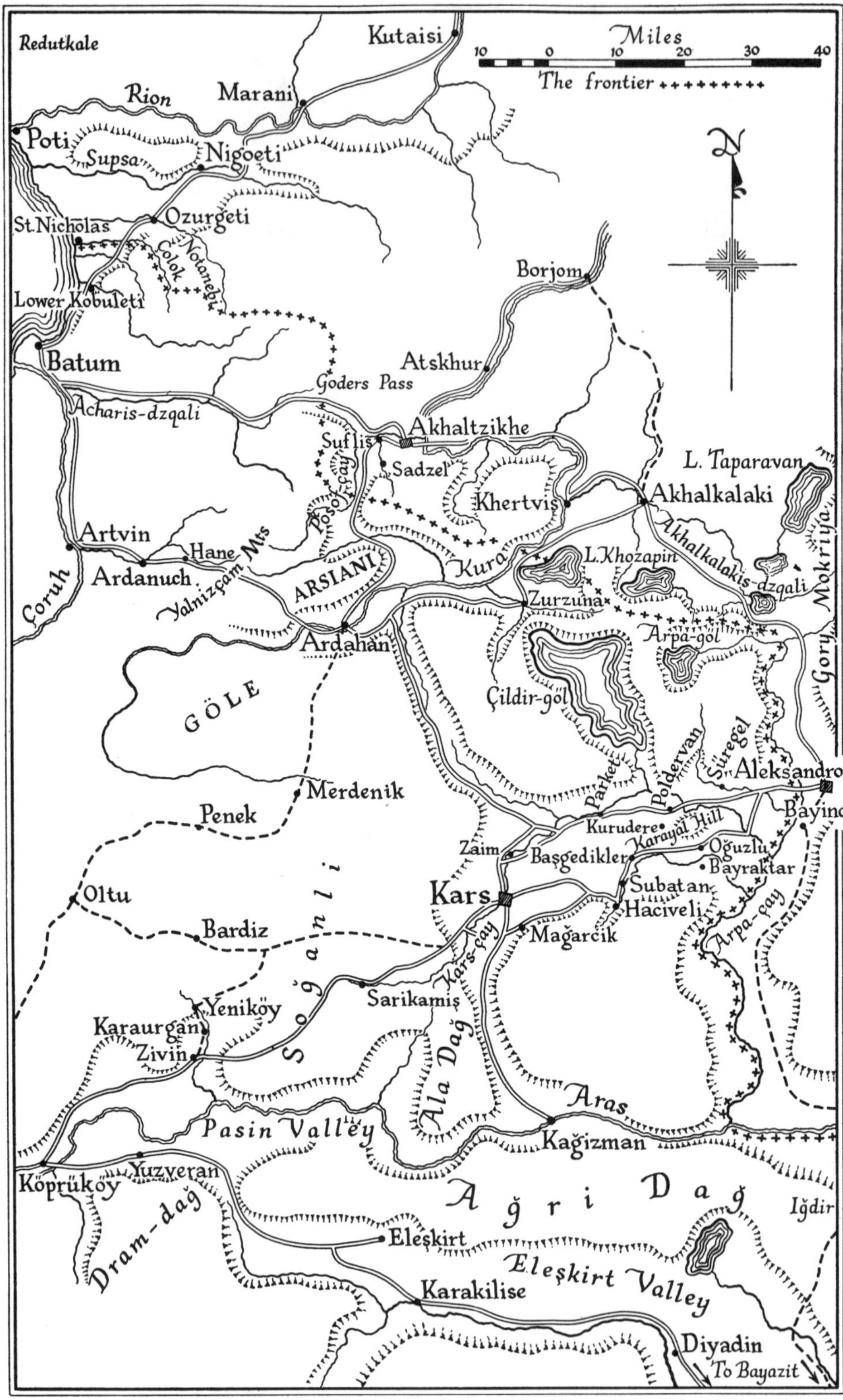

Map 4. The Russo-Turkish frontier, 1853–6

CHAPTER IV

FRONTIER OPERATIONS, AUTUMN 1853

MAP 4

DURING the revolutionary crisis of 1848–9 in Europe, Russian armed intervention to preserve the Austro-Hungarian monarchy had been inspired by the dislike of the Emperor Nicholas I for all movements directed against established governments. But the nascent nationalism of the Slav groups in Austria-Hungary had been stirred by the appearance of the Russian army in the middle Danube valley and by the imposing manifestation of Slav power. The insurrections which soon afterwards broke out in the neighbouring Turkish provinces of Bosnia and Hercegovina and the war in Montenegro indicated a spread and an aggravation of the ferment among the Slavs in eastern Europe.

The Emperor Nicholas, in spite of his disapproval of all disruptive movements, did not ignore the realities of the situation, and his policy now—in contrast to that initiated by the Treaty of Hunkâr Iskelesi at the beginning of his reign—was directed to securing a peaceful partition, by agreement among the principal European powers, of the territories of the Ottoman empire.

During the same decade, in England, alarm at the expansion of Russian influence both in the Near and Middle East and in central Europe (exemplified by Russian intervention against the Revolution of 1848) was tending to stimulate a more decisive policy in the eastern Mediterranean. This policy found support in Paris where the new emperor, Napoleon III, was anxious to popularize his recent *coup d'état* by an attempt to revive the glories of Napoleonic France.

The Turkish court and imperial bureaucracy, which was on the wave of the modernist movement initiated by the late Sultan Mahmut II, saw in the combination of international circumstances and ambitions an opportunity of securing the support of the western powers to stabilize the position of the Ottoman empire in the Balkans and the Near East and to put an effective check to the dangerous Russian effort directed to attaining naval control of the Black Sea.

When, in the summer of 1853, following the failure of the Menshikov mission, the Russians occupied the Rumanian principalities, the Turks, confident of the support of the western powers, adopted the bold policy of declaring war on Russia, and their troops entered Wallachia where they secured an initial success at Oltenitsa (1 November). An offensive on the Caucasian front was at the same time undertaken by the Turks, and some heavy fighting had taken place before the Russian naval attack on a Turkish squadron at Sinop (Sinope) (30 November) provoked the entry of Great Britain and France into the war.

Since the disastrous experiences of the campaigns against the Russians and the Egyptians in the twenties and thirties, the imperial Ottoman army had been in process of transformation from an Asiatic army of eighteenth-century type (such as took the field against Paskevich) into something approaching a European army of the post-Napoleonic period. During the preceding two decades scores of European officers had been engaged in training cadres. Among these were some distinguished soldiers including the Prussians, Colonel (later Field-Marshal) von Moltke and Colonel von Kutschkowski, the British colonel, Fenwick Williams, R.E., that gallant English adventurer of the Hungarian War, Colonel Guyon, and the Polish general, Chrzanowski.[1] At the same time British naval officers were engaged in the reorganization of the Ottoman navy.

Already by 1837 the Turkish army had an effective strength of 100 nizam (regular) battalions, more or less trained, six regiments of *süvari* (regular cavalry), three regiments of field artillery, and forty battalions of *redifs* (infantry of the reserve). In 1842 conscription (for all Muslim subjects of the sultan) on the Prussian model was introduced, and by 1850 the number of nizam battalions had been doubled, and that of redif battalions quadrupled. Although the roll of an average battalion did not exceed 500–600 men, the strength of the Turkish regular army now approached 250,000 men, which might be supplemented by about 150,000 men of irregular formations, the majority of whom were available for recruitment in Asiatic Turkey. It had been decided to strengthen the field artillery by the creation of six regiments of seventy-two guns each, but this increase had not been attained by 1853.

The proposed organization in six army corps met with difficulties owing to the lack of suitable senior officers for staff appointments. Some excellent officers were to be found in the artillery and personnel

[1] For Williams and Guyon (Hurşit Paşa) see *D.N.B.* For von Moltke's mission see his *Briefe* and R. Wagner, *Moltke und Mühlbach zusammen unter den Halbmonde* (Berlin, 1893). The activities of Chrzanowski and other Poles in Turkey are described in a rare little book by T. Gasztowitt, *La Pologne et l'Islam* (Paris, 1907).

had been strengthened by the arrival of numerous *émigré* Hungarians after 1848–9, but the weakness of the Turkish army continued to lie in the command and particularly in the high command. The Turkish army remained incapable of efficient manoeuvre and therefore unsuitable to carry out any offensive operation. Its strength lay in the defensive—as was proved in the fighting along the Danube.

In Asia, the Turks had undertaken the fortification on modern lines of Trebizond, Erzurum and Batum; much attention had been paid to Kars, which, under the supervision of Colonel Williams, had been converted into a fortified camp defended by modern works. Plans had also been made for the fortification of Ardahan, but these were not complete when war broke out.

The Turkish navy in the Black Sea was inferior in tonnage and weight of fire and also in training to that of the Russians.

During the summer of 1853 the Turks gradually concentrated considerable forces along the Caucasian frontier. Armenia was the territory of IV Army Corps (which existed only on paper). In effect, Turkish forces in October 1853 consisted of several detachments, each 15,000–20,000 strong, distributed respectively in the districts of Erzurum, Kars, Ardahan and Batum.

Politically, the Turks were concerned to force the issue before the winter in order to take advantage of existing trends of policy and opinion in Britain and France, while the Russians were inclined to play for time and to avoid military action for as long as possible. Hostilities, when they broke out, therefore took the form of a Turkish offensive and of a relatively passive defence of the frontier by the Russians.

The attention of the Russians was focused on the Danube where the bulk of the Turkish forces were concentrating, where Allied intervention was expected, and where the attitude of the Austro-Hungarian government gave cause for concern. The Caucasian army was therefore maintained at the lowest strength consistent with the requirements of defence.

Troops in Caucasia were dispersed over a vast territory and their duties and problems were numerous. In the western Caucasus it was necessary to defend the Kuban and Laba lines and the coastal posts against attack by the Cherkesses, and at the same time to anticipate possible Turkish or Allied landings. In the eastern Caucasus it was important to maintain the garrisons in Chechnia and Dağistan and to protect Kakheti against the raids of the Murids who might be expected to attempt collaboration with the Turks.

In the autumn of 1853, there were five infantry divisions in Caucasia (grenadiers, and 13th, 19th, 20th and 21st), of which the 19th and part of the strength of the 20th and 21st were retained in the mountains to support the Cossacks along the lines and in Dağistan. Units representing another division were dispersed throughout the Muslim districts of eastern Transcaucasia and along the Persian border. Prince Vorontsov therefore had rather less than two divisions, supported by ten Cossack regiments, some dragoons and irregular formations of Georgians, Imeretians and Gurians, with some Muslim mounted militia—in all about 23,000 men—with which to defend the Transcaucasian frontier. He may be said to have somewhat exaggerated his difficulties when he declared to the emperor that only a few battalions were ready to meet the Turks.

The situation in the western Caucasus remained more satisfactory than might have been anticipated. In August Shamyl gathered some 10,000 mountaineers and raided into Kakheti across the main ridge of the Caucasus. But General Prince Argutinski crossed the mountains from the valley of the Samur and threatened the rear of the imam, who hastened to retire without having effected any serious action. The importance of the raid was exaggerated by the Turks and increased the hopes which they placed on Shamyl. At the same time it influenced the disposition of Russian forces and impelled the cautious Vorontsov to retain large reserves in Kakheti and the region of Tiflis in spite of the news of Turkish concentrations along the frontier.[1]

The defences of the frontier were in poor condition. Along the Black Sea, opposite the Fort of St Nicholas[2] and to the north of the Rion at Redutkale, there was one battalion and some Cossacks. The old Turkish fortifications at Atskhur, Akhaltzikhe and Akhalkalaki remained as they had been in 1829. The new fortifications round Aleksandropol had not been completed.

The real defence—wisely in view of the known Turkish weakness in manoeuvre—was left to field forces. In fact, the Russians did not expect an early attack by the Turks, and at the beginning of November the principal group on the frontier, at Aleksandropol, had not completed its concentration.

The distribution of the Russian units available for operations along the frontier was as shown on p. 61.

[1] S. Porembski, 'Vtorzheniye Shamilya v Dzharo-Belakanski Okrug v 1853 g.', *Kav. Sbor.* XI, pp. 499–524.

[2] Fort St Nicholas, north of the frontier river Çolok, was held by two companies of infantry.

Force	Dist. H.Q. and Commander	Strength
Gurian	Ozurgeti Gen. Prince Andronikov	Litovski Regt (13th Inf. Div.) 1 battn Kurinski Regt (20th Inf. Div.) 1 Cossack regt Gurian militia 2 field batteries (4500 men with 16 guns)
Borjom	Akhaltzikhe Gen. Prince Andronikov	Vilnenski Regt (13th Inf. Div.) 1 Cossack regt Gurian and Imeretian militia 1 field battery
	Atskhur, Borjom, Suram, Akhalkalaki	Brestski Regt (13th Inf. Div.) Belostokski Regt (13th Inf. Div.) 3 field batteries
	Kutaisi	General reserve
Aleksandropol	Gen. Prince Bebutov	Erivanski Grenadiers Gruzinski Grenadiers 1 battn Caucasian Rifles 1 sapper battn Nizhegorodski Dragoon Regt 2 Cossack regts 3 field batteries (10,000 men and 24 guns)
Erevan	Erevan	2 battns Mingrelski Grenadiers 1 battn Shirvanski Regt (21st Inf. Div.) 1 Cossack regt 1 field battery (3000 men and 8 guns)
Tiflis	Tiflis	5 battns from 20th and 21st Inf. Div. 3 field batteries

On the night of 27–28 October, Ahmet Paşa, in command at Batum, took the Russian post at Fort St Nicholas by surprise and made the garrison prisoners. A Russian battalion with two guns, sent from Ozurgeti (now Makharadze), arrived too late and encountered the Turks in considerable strength (about 5000). During the next few days, the Turks, reinforced to 15,000, made the passage of the frontier river, the Çolok; the Gurian field force thereupon withdrew from Ozurgeti and retired to Marani on the Rion in order to cover the roads to Kutaisi. On orders from Tiflis, the garrison of Redutkale was also evacuated, and the sea communications of Transcaucasia with the Crimea via Poti were abandoned without a fight. The Turks, however, in view of the approaching rains, did not press their advance beyond Ozurgeti. At the end of the month the annihilation of the Turkish squadron at Sinop by Admiral Nakhimov indicated that the hasty Russian evacuation of the coast had been unnecessary.

Ali Paşa, commanding at Ardahan, proved less successful. With about 18,000 men, half of whom were Laz and other irregular formations, he advanced at the beginning of November in the direction of Akhaltzikhe and the Borjom defile. Attacks by the irregulars on Atskhur and Akhalkalaki were repulsed. On 13 November, Ali began to concentrate the bulk of his forces round the village of Suflis on the right bank of the Posof-çay within three miles of Akhaltzikhe. He brought the town under fire, but preferred to await the arrival of further reinforcements before attacking. Prince Andronikov, who had recently arrived in Akhaltzikhe, himself decided to attack, although his forces were less than half those of the Turks.

Ali Paşa was occupying a strong position on the heights marked by the villages of Suflis and Sadzel. Andronikov intended to make a frontal attack and at the same time to turn the Turkish left by a flanking movement up the left bank of the Posof-çay, but when he observed the Turks massing to meet the flank attack, he suddenly—in the tradition of Paskevich—reversed his plan and drove home the frontal attack. The Vilnenski (Vilna) Regiment stormed the height on which Sadzel stood; the Turkish defenders, crowded behind the low walls of the gardens of Suflis, were then caught between the converging attacks of the Vilnenski men from Sadzel and the Belostok and Brest regiments which moved up on the Russian right after fording the breast-high waters of the Posof-çay. The Turks began to withdraw in good order, but their rearguard was broken and all their guns taken, and the retreat became a rout as the Cossacks and mounted Imeretians rode down the disordered infantry. Ali Paşa lost 4000 in killed, wounded and prisoners against a Russian loss of 400, while the Laz and other irregulars (*başibozuks*)[1] scattered over the countryside.

The simultaneous Turkish offensive from Kars against Aleksandropol also ended in disaster. At the beginning of November, Abdi Paşa, with 30,000 men and 40 guns, slowly marched from Kars in the direction of Aleksandropol where the Russian concentration was known to be incomplete. The Turks made their main camp round the Armenian village of Süregel, fifteen miles from Aleksandropol on the Turkish side of the Arpa-çay, and their patrols crossed the deep ravine of the river and penetrated into Russian territory. Abdi Paşa then moved the bulk of his artillery to the heights above the village of Bayindir on the Russian side of the Arpa-çay commanding the road from Aleksandropol to Erevan.

On 10 November Prince Bebutov, still unaware that hostilities had

[1] *Başibozuk*, a term which became familiar in nineteenth-century newspapers, means literally 'broken-head', and has a pejorative sense as applied by professional soldiers to irregular levies.

begun, sent a strong reconnaissance force consisting of the bulk of the field troops at Aleksandropol, under Colonel Prince Orbeliani, in the direction of Bayindir. Descending the ravine which runs down to the Arpa-çay, Orbeliani was surprised by the fire of the Turkish batteries from the heights above Bayindir. The Russian infantry deployed and showed great composure for several hours under the concentrated fire of the forty Turkish guns. But it was impossible to find stations for the Russian artillery; Orbeliani hesitated to attack the strong Turkish position, while he could not retire without exposing himself to the attack of the numerous Turkish cavalry. The Russian commander was fortunate only in the lack of initiative shown by Abdi Paşa. As evening drew on, Bebutov moved out from Aleksandropol with all the field troops remaining there—three battalions and some Nizhegorodski dragoons—and, marching along the Arpa-çay, seemed to threaten the communications of the Turks with the right bank. This demonstration was sufficient to stultify the action of the cautious Abdi who limbered up his guns and began to retire in the direction of Süregel. Orbeliani, in his turn, withdrew to Aleksandropol, with losses amounting to 1000 men, about 20 per cent of his force.

Bebutov had been lucky in extricating the ineffectual Orbeliani from a very dangerous situation, and Abdi Paşa had missed the opportunity of destroying the principal Russian field force in Transcaucasia at one blow in the first week of the campaign. Never was the inadequacy of the Turkish high command at this period more dramatically demonstrated.

Some reinforcements for Caucasia were slowly arriving from Russia, elements of the 13th Infantry Division by sea, and the 18th by land, involving, in the days before the development of the Russian railway system, long route marches during which effectives, under adverse conditions, were often reduced by 50%. A small part of these reinforcements reached Bebutov at the end of November and, after receiving news of Andronikov's victory at Suflis, he decided to begin an offensive movement. The same news discouraged Abdi Paşa, who withdrew from Süregel in the direction of Kars. He took up a strong position covering the Aleksandropol-Kars road on the heights marked by the village of Başgedikler overlooking the valley of the Kars-çay from the south. Abdi had forty-eight guns in position, of which twenty were concentrated in one big battery on a hill (Oğuzlu) covering the right flank of the Turks. There were 36,000 men in the Başgedikler camp, including 20,000 regular infantry and one brigade of cavalry; the rest were başibozuks and Kurds of doubtful value either in the defence of a fixed position or in a battle of manoeuvre. When on 1 December it was reported that the Russians were advancing from Süregel, the cautious Abdi was in Kars and the boisterous and self-confident Ahmet Paşa, who was a critic of his chief's handling of the Bayindir action, was in command. Ahmet, with great numerical preponderance, decided to attack the Russians without delay. A typical encounter battle developed which favoured the seasoned Caucasian infantry and was beyond the manoeuvre capacity of the young and half-trained Turkish regular army.

The foreign officers who were advising Ahmet designed a movement whereby 5000 nizams and 5000 horse were to turn the Russian right flank in depth, while 10,000 başibozuks and Kurds overwhelmed the left flank. Bebutov anticipated the outflanking tactic, and while his reserves held the disorderly Turkish attacks, he directed a main assault against the twenty-gun battery on Oğuzlu hill. The first attack by the Gruzinski (Georgian) Grenadiers was repulsed with heavy losses, including Orbeliani and most senior officers killed. A second attack, more ably led by Prince Bagration, commanding the Erivanski Grenadiers, captured the battery; after some fierce bayonet fighting among the stone houses of Oğuzlu village the nizams broke. The attack on the Russian left by the Kurds and başibozuks failed to develop after the loss of Oğuzlu hill; on the right Prince (Yason) Chavchavadze, with six squadrons of Nizhegorodski Dragoons and some mounted Georgian militia, had fought off heavy attacks by eight battalions of nizams and the Turkish regular cavalry brigade; finally, a counter-attack by the dragoons and a regiment of Terek Cossacks routed the Turkish cavalry and captured four guns.

The whole of the Turkish army was now in retreat towards Kars; their losses in the Başgedikler battle were heavy—half their artillery and all their supply train, with 6000 men killed, wounded or prisoners. The Russian losses amounted to 1300 men, which included most of the strength of the gallant and unfortunate Georgian Grenadier Regiment.

On the day before the Başgedikler battle the naval action off Sinop (Sinope) took place, which provoked the entry of Britain and France into the war. At the beginning of November a Turkish proposal to send a strong squadron to the Caucasian coast had met with the objections of the British ambassador; but about the middle of the month a squadron of inferior strength (seven frigates, three corvettes and three armed steamers) was dispatched to Sinop, and the Russians contended that it was designed to protect the transport of reinforcements to Trebizond and Batum. On 30 November, Admiral Nakhimov's squadron (six sail of line and two frigates) entered the port of Sinop and, after a hot cannonade, burned or sank all the Turkish ships with the exception of one steamer which escaped.[1] The Russian losses were about 200, but large numbers of seamen and soldiers aboard the Turkish ships 'drank the sherbet of martyrdom'.

For the time the two victories of Başgedikler and Sinop effectively secured the Russian position in Caucasia, but the action at Sinop had a devastating effect on public opinion in Britain and France, and the entry of those two powers into the war on 3 January 1854 at once threatened all Russian prospects in the Black Sea. The outlook in Caucasia was regarded by the highest authorities on the spot with particular foreboding.

[1] The Russian ships were armed with General Paixhan's shell guns, patented as early as 1824. 'It was the effect of Russian shells on the Turkish ships at Sinope that roused the naval mind' to the necessity of armour; see Publications of the Navy Records Society, vol. LXXXIII, *Russian War, 1854; Baltic and Black Sea*, Captain Dewar's introduction to Part 2, pp. 205, 211.

Namik Kemal, a celebrated Turkish writer of the nineteenth century, was the author of a drama, *Akif Bey*, which has as background the battle of Sinop. The scene is laid among the people of Çürüksu.

CHAPTER V

OPERATIONS IN GURIA AND KAKHETI. BATTLE OF KURUDERE. SUMMER–AUTUMN 1854

MAP 4

THE viceroy of the Caucasus, Prince Vorontsov, regarded the arrival of the Franco-British fleet in the Black Sea as a formidable danger, not only to the Caucasian coast but, in general, to Russian rule south of the main chain. Tiflis insistently demanded reinforcements; in the early spring three dragoon regiments and six regiments of Don Cossacks arrived; but with the deterioration of the position on the Danube and the threatened invasion of Russia, it became obvious that no further troops could be expected. Harassed and depressed the viceroy fell into a decline and left Tiflis; General Read[1] acted in his stead. This officer was frankly pessimistic. He anticipated the landing of a Franco-British expeditionary force and a Persian attack along the Aras. In a report to the emperor he advocated the evacuation of all the Muslim territories in eastern Caucasia—including Derbent and the hardly won positions in Dağistan—and the temporary sacrifice of Georgia, Imereti, Mingrelia and Guria. He proposed to limit the Russian defence to holding the Caucasian line north of the main chain, along the rivers Kuban, Sunja and Sulak.

The Emperor Nicholas I fiercely rejected Read's defeatist proposals and he was replaced as head of the military administration in Caucasia by General Prince Baryatinski (the future captor of Shamyl). The officers commanding on the frontier certainly did not share the fears of Tiflis, and circumstances were to witness that the emperor was correct in his appreciation of the situation.

Political fears proved to be exaggerated, and events illustrated the extent to which policy and strategy can lack foundation when based on the assumption of internal movements within an enemy country—even when the government of that country itself anticipates such movements.

[1] A veteran of the Napoleonic wars, General N. A. Read met his end at the Battle of the Chernaya (R. *Chernaya Ryechka*). Despite his name, *R.B.S.* makes no mention of his descent from one of the many Scottish officers who served Russia in the sixteenth century. But the fact that he was a noble of the Smolensk government—where a number of Scots officers were endowed with estates—points to a Scots origin for the general's family.

Varied difficulties might develop in a territory where a numerous population composed of mutually antagonistic communities were subject to the attraction of different foreign influences and where the occupying forces, scattered in relatively small detachments over a wide area, did not exceed 100,000 regular troops and Cossacks. But a determined policy, a bold front and vigorous action proved equal to the situation.

The possibility of widespread attacks by the Cherkesses—with Allied support from the sea—and the likelihood of an invasion of Kakheti by Shamyl were real dangers, and there was no absolute certainty that the 'peaceful' Muslim population of eastern Transcaucasia would remain loyal if they were disturbed by the news of Allied and Turkish successes or Persian intervention.

The very possibility of Turkish invasion and Muslim attacks gave the Russian administration the enthusiastic support of the Christian Georgians and Armenians. The Russians had had little trouble with the Georgians since the Kakhetian revolt of 1812, and none with the Armenians. Some Georgian officers had been infected by the Decabrist movement and there had been minor insurrections, notably in Guria, which had been really agrarian in character. During the war of 1828–9 the Georgians (with the exception of the Gurians who had been directly threatened by the Turks) had shown reluctance to join the colours, and Paskevich had had indeed more success in recruiting Muslims for his two excellent regiments of mounted irregulars. During the last quarter of a century, however, the Georgian landed families had come much under the influence of Russian life (particularly military), and many Georgian officers, or officers of mixed Georgian-Armenian origin, held commands in the imperial army—notably Andronikov, Bebutov, Orbeliani, Bagration, Chavchavadze and Eristov. At this period there was no serious anti-Russian sentiment in Georgia, either among the gentry or among the peasantry; the development of Georgian nationalism later in the century was a phenomenon of a new *bourgeois* environment.[1]

[1] The peasant revolts in Mingrelia and south Ossetia (1819–20) and in Guria (1841) were agrarian rather than nationalist. Russian troops intervened to protect the local landowners; see *Istoriya S.S.R.* II, pp. 266–7, 277–8; and G. K. Khachapuridze, *Guriyskoye vosstaniye v 1841 g.* (Tiflis, 1931). For the efforts of Ermolov and succeeding viceroys to encourage a military and intellectual *élite* among the Caucasian nobility, see *R.B.S.* under J. K. Bakikhanov, V. I. Bebutov, A. G. Chavchavadze and N. D. Eristov (Eristavi). See also Khachapuridze, 'K voprosu o kulturnykh svyazyakh Rossii i Gruzii v pervuyu polovinu XIX veka', in *Voprosy Istorii*, 1946, Nos. 5–6, pp. 76–89. Yet there was a romantical legitimist ferment among the Georgian aristocracy in favour of the Bagratids as late as 1832. See the

Moved by fear of the Turks and of their Muslim neighbours, the peasantry were not far behind the gentry in their loyalty to the Russian connexion. The appeal of Prince Vorontsov, a popular viceroy, was met with enthusiasm. Four regular battalions of the line were recruited in Imereti and eastern Georgia, while in Kartli and Kakheti, threatened by the dreaded Murids of Shamyl, ten *druzhiny* (lit. 'brotherhoods', broadly 'commandos') were formed of 1000 men each. In Imereti 5000 militia were raised; 3000 in Guria—traditionally excellent marksmen—about 2000 in Mingrelia. Some of these formations, particularly the Gurian and Imeretian *druzhiny*, were to show fine fighting qualities, and they constituted a useful addition of 20,000 men to the Russian forces in Caucasia.

The hopes which the Turks placed in the mountaineers were soon dissipated: there proved to be little community of interest between the Murids and the Cherkesses, and the latter were themselves divided by inter-tribal rivalries. When, in the spring of 1854, the Allied fleet appeared in the Black Sea and the Russians proceeded to evacuate all the coastal forts from Anapa southward,[1] the moment seemed to have arrived for action by the Cherkesses. But Shamyl's *naib* or emissary to the Cherkesses, Muhammad Amin, arrived in Istanbul to complain of his difficulties.[2] The chief of the Cherkess Committee set up in Istanbul, Sefer Bey, an *émigré* who had at one period served in a Russian cavalry regiment, was sent to Sukhum with another *émigré* who called himself Behçet Paşa. Behçet remained in Sukhum to negotiate with the Abkhaz, while Sefer proceeded to Tuapse. Here Sefer found little enthusiasm among the important Shapsugh tribe, but, on proceeding to Anapa, he met with a rather better reception from the Natukhai. At Sukhum, Behçet found the ruling family of Abkhazia, the Sharvashidzes, divided; the Christian princes adhered to the Russian connexion, but Iskander (Alexander), a Muslim, was prepared to deal with the Turks and asked for the recognition of his independence as ruling prince and for the inclusion in his territory of the adjoining Mingrelian district of Samurzakan.

Prince Alexander's territorial ambitions made more difficult negotiations with the neighbouring Dadianis of Mingrelia, the last

account of the arrest, examination and pardon of Prince A. G. Chavchavadze in *R.B.S.* under that name. Also cf. *Akty*, VIII, 'Gruzinski tsarski dom', correspondence between Nesselrode and Baron Rosen, Docs. 134–60 (1832–3). The elderly pretender, Alexander Bagration, was in Tabriz in Oct. 1833. For his personality and adventures during the Kakhetian rising of 1812, see Monteith, *Kars and Erzeroum*, pp. 72–8.

[1] See *Russian War, 1854*, Navy Records Soc. Publ. vol. LXXXIII, nos, 16, 17, 18, 44.

[2] For biography of Muhammad Amin, see *Kav. Kal.* XVI (1861), p. 86.

among the Georgian princely families which had maintained a certain independence under the Russian régime. A French tutor at the country-house court of Zugdidi had a certain influence with the regent, Princess Dadiani;[1] but the lady preferred caution, for the Mingrelian peasantry had suffered for centuries from the depredations of the Abkhaz and the Turks. The various conversations languished, while all the coast from Anapa to Redutkale became a no-man's land; small Russian contingents remained at the mouth of the Kodor and at other points near the coast.[2]

The Turks still adhered to their plans for a Caucasian offensive and made proposals to the Allies for landings along the coast and for a threefold offensive from Batum, Ardahan and Kars against Kutaisi, Akhaltzikhe and Aleksandropol—with a general rising of the Muslim population and the occupation of Tiflis as the ultimate objectives.

The Allies received the Turkish proposals with little real interest. The British objectives were limited to the destruction of the Russian naval position in the Black Sea and to the stabilization of the situation of the Ottoman empire in the Balkans and Asia Minor. Palmerston, perhaps, alone envisaged the wider perspectives of a general European offensive against the Russian power, involving the exploitation of potential national movements within the Russian dominions; but Prussian and Austrian sensitiveness over the Polish question prevented the further development of this theme.

Louis Napoleon sought a romantic *révanche* for his uncle's debacle in 1812 and, more immediately, a reversal of the serious diplomatic defeat of the Orléans monarchy during the Near Eastern crisis of 1841–2. He looked for the easy successes which an alliance with British sea-power might be expected to give.

The destruction of the Russian Black Sea fleet became the first aim of the Allies; all other objectives seemed to be secondary. But the destruction of the Russian fleet involved the capture of its strongly fortified base at Sevastopol, which it would have been dangerous to attack only from the sea. The necessity for a land attack on Sevastopol was realized, but such a land attack imposed a concentration of transport which alone excluded any ambitious operations off the Caucasian coast. This problem of transport occupied all the attention and all the efforts of the Allied governments; temporary bases were

[1] The princess was a daughter of A. G. Chavchavadze.

[2] Curious details of Turkish and Allied relations with the Cherkess tribes are given by Major Osman Bey in *Kav. Sbor.* II, pp. 143–214, 'Vospominaniya 1855 g.' A son of the former grand vizier, Kibrizli Mehmet Paşa, Osman Bey had served on the staffs of Mustafa Zarif and Selim before being taken prisoner by the Russians.

created on Turkish territory at the Bulgarian Black Sea ports of Varna and Kavarna, and it became important to hold the better part of the Turkish army in the fortresses of the quadrilateral in order to protect these bases from Russian attacks through the Dobrogea or from across the Danube.

British and French observers had visited the Caucasian coast, but they had been little impressed either by the prospects of effective action by the Cherkesses or by the condition of the Turkish forces in Armenia.

It is interesting to speculate on the course of events had the main Allied action been directed towards the expulsion of the Russians from Caucasia. The Allies, with their command of the sea, would have had superior transport facilities, and they might have cut Russian communications through Rostov with the northern Caucasus; but the Russians would still have been able to transport reinforcements by the long Volga-Caspian water route to Derbent and Baku. In spite of the gloomy prognostications of General Read, it is likely that the Russians would have succeeded in checking any really formidable movement on the part of the Murids, who were now in definite decline; at the same time the Allies could not have obtained any decision by landings on the Circassian coast, and had they made a descent (as they actually did in 1855) on the swampy Mingrelian shore, they would have become involved in difficult fighting so soon as the Russians had concentrated along the Suram Mountains.

A conquest of Caucasia, if achieved, could not have been consolidated unless the Allies had been prepared to maintain a permanent military protectorate there to confront the reviving and resentful power of Russia. The Turks, with their empire already showing signs of breaking up into its component nationalities, were certainly not capable of sustaining the duties and difficulties of a protecting power over the varied peoples of Caucasia. And a western power could only have undertaken the protectorate and defence against a Russian *révanche* on the basis of secure lines of communication which must have involved, *inter alia*, occupation of the Straits and a veiled protectorate, also, over the Ottoman empire with control of its armed forces.

To a sea-power like Britain the prospective occupation and protection of Caucasia presented problems comparable in magnitude with those of India and Egypt, with the difference that a position in either India or Egypt could be developed primarily from the sea, while, in Caucasia, an occupying sea-power would be constantly threatened by an advantageously situated land-power, and sea communications would remain dependent on many varying contingencies in international politics. The same problem was to present itself, in modern form, in 1919–20, and was again to be decided negatively for the same political and strategic reasons.

Allied officers who had visited Trebizond, Erzurum and Kars had remained sceptical as to the prospects of the Turkish armies on the Caucasian frontier. The troops had come through a hard winter; they were poorly clad and badly fed; typhus and other cold and dirt diseases were rife in the cantonments of the Kars and Erzurum highlands, while at Batum a virulent form of malaria had ravaged half the troops stationed there. Nevertheless, new contingents were constantly arriving both by sea and by the overland routes. The corps of Selim Paşa, between Batum and Ozurgeti, amounted to 40,000 men, half of whom were regular formations. At Kars the new seraskier, Mustafa Zarif Paşa, commanded a force of some 40,000 regulars and 20,000 irregulars; and there was a further reserve of 20,000 at Erzurum. At Bayazit, gathered for an attack against Erevan, were 8000 regulars and about 10,000 Kurds under another Selim Paşa. The Ardahan region was held only by irregular contingents of Laz and Kurds. Excluding fortress pieces, many of which were of antiquated type, the Turkish artillery had less than half the weight of that available to the Russians. Under all the known conditions, it might appear to an impartial observer that Turkish offensive ambitions were out of proportion to the force available.

Turkish plans for a general offensive were known to the Russian commanders in the field who, encouraged by the victories of the campaign of 1853, determined to meet the Turkish advance with a counter-offensive. Disposable forces were limited by the requirements of internal security.

It will be observed from the following order of battle that the whole of the 19th Infantry Division and parts of the 20th and 21st Infantry Divisions with all the Kuban and Terek and some of the Don Cossack regiments, were held on the Caucasian and Lesghian lines with a view to meeting the threat of attack by the Cherkesses and Lesghians. These units amounted to approximately one-half of the total of Russian armed forces available in Caucasia.[1]

[1] The ratio may be compared with the one-third of the Italian armed forces held on internal security duties during the Abyssinian campaign of 1941, and it indicates the value of the *threat* of internal revolt within the territory of an enemy power; this *threat* is often more valuable, from the aspect of immobilizing enemy troops, than is action which fails to explode to the extent anticipated by the enemy.

The Russian order of battle was:

Force	H.Q. and commander	Strength
Guria	Kutaisi	Litovski Regt (13th Inf. Div.)
	Andronikov	Brestski Regt (13th Inf. Div.)
		2 battns Kurinski Regt (20th Inf. Div.)
		1 Cossack regt
		Gurian militia
		Imeretian militia
		1 field battery, 1 mountain battery
		(approx. 9000 men and 10 guns)
Akhaltzikhe	Gen. Kovalevski	Vilnenski Regt (13th Inf. Div.)
		Belostokski Regt (13th Inf. Div.)
		1 Georgian line battn
		2 Cossack regts
		Georgian militia
		3 field batteries
		(approx. 9000 men and 24 guns)
Aleksandropol	Bebutov	Gruzinski Grenadiers
		Erivanski Grenadiers
		Tulski Regt (18th Inf. Div.)
		Belevski Regt (18th Inf. Div.)
(2 battns in garrison at Akhalkalaki)		Ryazhski Regt (18th Inf. Div.)
		Tverski Dragoons, Novorossiyski Dragoons, Nizhegorodski Dragoons (These 3 regts constituted the Dragoon Bde)
		2 Cossack regts
		2 regts Muslim irregular horse
		6 field batteries
		2½ mountain batteries
		(approx. 20,000 men and 68 guns)
Erevan	Gen. Baron Wrangel	2 battns Shirvanski Regt (21st Inf. Div.)
		1 battn Mingrelski Regt (21st Inf. Div.)
		1 battn Tifliski Regt (21st Inf. Div.)
		1 Cossack regt
		1 regt Muslim irregular horse
		1½ field batteries
		(approx. 2000 men and 12 guns)
General reserve	Tiflis	Ryazanski Regt (18th Inf. Div.)
	Baryatinski	Several battns of 20th Inf. Div.
		2 batteries
	Lesghian line (Kakheti)	Parts of 21st Inf. Div.
	Persian border	4 sqns Nizhegorodski Dragoons
		Several Cossack regts
Caucasian line	Northern Caucasus	19th Inf. Div.
		Part of 20th Inf. Div.
		Kuban Cossack regts
		Terek Cossack regts
Black Sea coast	Between Sukhum and Redutkale	Don Cossack regts
		Some line battns

The campaign of 1854 began rather late in the spring, only at the beginning of June, and, at first, in the lowlands of the west where climatic conditions were most favourable. The absence of important

Turkish forces in the Ardahan region made it possible to strengthen the Guria force with detachments from Akhaltzikhe. During the first week in June, Andronikov sent forward a strong reconnaissance under Colonel Prince Eristov (two infantry battalions, 1000 Gurian militia, 300 Cossacks and 4 guns) from Marani to Nigoeti along the road to Ozurgeti. On 8 June, Eristov was attacked, but not surprised, by a force of 8000 Laz irregulars under Hasan Bey. After a sharp action the attack was beaten off with a loss to the Turks of 2000 men and 2 guns. On 15 June Andronikov brought his main force up to Ozurgeti, and Selim Paşa retired across the frontier stream of the Çolok.

The Turks entrenched a position covering the road from Ozurgeti to Kobuleti, facing the bridge over the Çolok with their right flank to the road and their left covered by dense forest. They did not find it necessary to destroy the bridge, although Laz irregulars watched the thick undergrowth along the river bank. Thirteen Turkish guns were established on rising ground, and their field of fire lay across a ravine which gave the Russians some dead ground for their advance.

On the morning of 15 June the Turks found that the Russians were already across the Çolok which they had passed by the bridge and by fords which had been found for them by the Gurians. Selim Paşa allowed Andronikov to deploy his columns without interference. On his right, facing the Turkish left in the forest, Andronikov placed General Maidel with two Kurinski and two Litovski battalions; on his left, by the bridge, General Brunner with the two remaining Litovski battalions and two Brest battalions. In reserve were the other two Brest battalions, one Belostok battalion and the Cossacks with the mounted Gurians and Imeretians. Andronikov's modest artillery (one field and one mountain battery) proved adequate to fight down the fire of the Turkish guns.

Andronikov's main attack was directed against the Turkish left, where the two Kurinski battalions were sent forward in a frontal attack while the Litovskis turned the position through the forest. But the gallantry of the celebrated Kurinskis failed to capture the stubbornly defended Turkish trenches and they were thrown back with heavy losses. At this difficult moment Andronikov acted with promptness in throwing in all his reserve, while two of Brunner's battalions were withdrawn from the Russian left to strengthen the flanking movement through the forest.

In the second assault the infantry again reached the Turkish trenches, and the Turks finally broke before the wild charge of the Gurian and Imeretian horse. Selim Paşa, with the loss of 4000 men and all his guns, retired on Batum. The Russian losses of 1500 men were heavy among the Kurinski men and the Georgian militia, who had fought with great gallantry in their first action.

Andronikov did not even occupy Kobuleti (since he had no instructions to do so), but withdrew behind the Çolok with his defensive task complete. Operations in Guria were suspended for the rest of the year.

The decisive operations on the extreme Russian right flank were soon followed by a useful success on the left. At the end of June, the Erevan force, under General Baron Wrangel, which had been brought up to a strength of about 5000 men with 12 guns, advanced across the Turkish frontier from Iğdir, where the northern slopes of the Ağri-dağ come down to the flat plain of the Aras valley. At the beginning of July, Wrangel had information that the Turks (estimated at 9000 infantry and 7000 irregulars with only 4 guns) had occupied the pass of Çengel (2037 m.) on the main track from Bayazit to Iğdir over the Ağri-dağ. After a difficult march Wrangel attacked the Turkish position in the defile between the massifs of Çengel and Pamuk-dağ and drove them out with a loss of 2500 in killed and prisoners. A part of the Turkish force scattered into the desolate highlands of the Ağri-dağ and most of the Kurdish irregulars dispersed to their villages. The Russian losses in this affair were less than 400 men, and on 19 July Wrangel was able to occupy Bayazit where he found substantial reserves of supplies and fodder. (Cf. Map 13.)

During June and July the danger of a Murid diversion against the Russian rear was also dissipated. The imam, with rather weaker forces than in the previous year, made an attempt to invade Kakheti. He descended into the Alazani valley but failed to capture any of the Russian posts along the Lesghian line. With a small mobile mounted column (four squadrons of Nizhegorodski Dragoons, Cossacks and Kakhetian *druzhiny*, 2500 men in all), Prince David Chavchavadze met the Murid horde, 8000 strong, at Shildi on 16 July and dispersed them.[1] The autumn and winter passed without further incident along the Lesghian line.

[1] Shamyl withdrew over the main chain, but he had time to send a band to burn Chavchavadze's castle of Tsinondali, where the ladies of the family were kidnapped and removed into captivity. The intervention of Colonel Fenwick Williams (then in Kars) was sought, and after some months the ladies were exchanged against Shamyl's favourite son who was held as a hostage in Russia. See Prince Baratov, 'Nashestviye skopishch Shamilya na Kakhetiyu v 1854 g.', *Kav. Sbor.* I, pp. 237–68; and *ibid.* XI, pp. 525–72, V. Soltan, 'Ocherk voyennykh deystviy v Dagestane v 1854 g.' See also Baddeley, *Conquest*, p. 449 and n. 3. Sir Oliver Wardrop, who was at Tsinondali thirty years later, gives a lively account

A decision in the campaign of 1854 was to be found neither on the flanks nor in the rear but on the main front between Kars and Aleksandropol, where the principal forces of the combatants were concentrated for action. Prince Bebutov had available about 20,000 troops (13,000 infantry, 3000 regular cavalry and 4000 mounted irregulars) with 68 guns. The eight grenadier battalions and the sixteen squadrons of the dragoon brigade were among the best troops in the Russian army and far superior in quality to any troops which the Turks could bring against them. On the other hand, the army of Mustafa Zarif Paşa, including the 6000 fortress troops in Kars, were numerically three times as strong as the Russians. The seraskier could put on to the field of battle about fifty battalions of regular infantry (35,000 men), four *süvari* regiments and 14,000 mounted irregulars. He had 80 field pieces. While the irregulars were mostly Kurds, accustomed to the conditions of the country, the nizam formations were the pick of the Anatolian and Syrian recruits, some of them relatively well trained. A number of British, French, Polish and Hungarian officers of European experience were attached to the seraskier's headquarters staff, including the Anglo-Hungarian Guyon and Colonel Kmety, both officers who had gained reputations in the field during the campaigns in Hungary in 1848–9.

At the beginning of May the Turkish army was concentrated round Kars with advance forces at Haciveli on the Kuru-dere,[1] a westerly affluent of the Arpa-çay which curves north and east of the massif of Alaca-dağ.

Bebutov, preoccupied with questions of supply for men and animals did not move out of Aleksandropol until the middle of June when the late spring grass on the Kars plateau gave promise of excellent grazing for his horses. The Russian camp was established to the south of the Kars-çay between Poldervan (on the Aleksandropol-Kars road) and the village of Kurudere. On the extreme left of the Russian position was a solitary eminence, Karayal, which served as an excellent observation post. Not far from Karayal was Oğuzlu village and the battlefield of the previous year.

The Russians remained in observation in their camp between 2 July

of the episode (*Kingdom of Georgia*, London, 1888, pp. 103–7). According to Wardrop, Jemaleddin, Shamyl's son, 'was a most amiable man; had become perfectly Russian in his way of life, and spoke Russian, French and German fluently.... It was with deep regret that he left civilization to return to the wild life of his native mountains, and in 1858 he died of a broken heart.'

[1] During the campaign of 1877 the Kuru-dere ('Dry Valley') was known as Mavrik-çay.

and 4 August. During the same period the Turks were proceeding with a very leisurely deployment in the direction of the Russian frontier. By the middle of July they had moved the bulk of their forces to the neighbourhood of Haciveli, and their tents were scattered across the plateau which gradually rises to the slopes of Alaca-dağ. Their front extended eight or ten miles from the Kuru stream on the right, to the two heights called Great and Little Yahni on the left. (Cf. Map 16.)

The two armies remained in observation for a period of three weeks without any action except a few affairs between patrols; both commanders felt some doubts about the outcome of a pitched battle and awaited the approach of autumn to retire to their respective bases. Neither Tiflis nor Istanbul, however, was satisfied with continued inactivity; and finally the seraskier's foreign officers induced him to try the issue of battle.

On the night of 3–4 August, Russian patrols reported the movement of Turkish baggage trains in the direction of Kars. Interpreting this as the beginning of a Turkish withdrawal, and wishing to put an end to the criticism of his inactivity, Bebutov decided to pursue the retreating enemy. On the night of 4–5 August Russian trains were despatched in the direction of Aleksandropol, and at dawn on the 5th the Russians began a general advance. They soon encountered the advancing Turkish army and at Kurudere, as at Başgedikler a year before, an encounter battle was fought.

The English soldier of fortune, Guyon, who had prepared the plan of battle at Başgedikler, was again responsible for the general plan; he again attempted to make use of the numerical superiority of the Turks to overwhelm both the Russian flanks. Guyon proposed first to take the isolated Russian observation post on the Karayal hill by a surprise attack of başibozuks at dawn. Here four battalions of Istanbul Rifles were to entrench themselves, supported by the advance of the sixteen Syrian battalions of Kerim Paşa with two *süvari* regiments and thirty-two guns. It was intended that the intrepid Kerim Paşa should attract the attention of the main Russian force, while Colonel Kmety, moving to the left with sixteen Anatolian battalions, two *süvari* regiments and eighteen guns, was to attack the Russian right across the low ground which lay between Karayal and the village of Kurudere. At the same time the Kurdish irregular cavalry was to ride round the Russian right flank in the neighbourhood of the village of Poldervan. Mustafa Zarif held twelve battalions and thirty guns in reserve.

The weakness of Guyon's plan was that the Turkish forces were directed to operate in three widely separated groups: before Kmety could come into action, Kerim Paşa had to sustain the whole weight of the Russian resistance, while the Russian superiority in the quality of their cavalry and horsed artillery might allow time for Bebutov to engage each attacking force in turn.

Bebutov, aware from an early hour of the morning that a major battle was impending, hastily began to deploy his troops; he made, however, the error, curious in a commander, of assuming that the Karayal observation post had been reinforced when it had, in fact, been overlooked. The başibozuks therefore took the hill by a sudden attack, and the Istanbul riflemen entrenched themselves upon it, taking up an enfilading position in relation to the Russian line. Bebutov found the position dangerous and ordered General Belyavski's brigade (Belevski Infantry Regiment, two battalions of the Tula Regiment and one battery, with two Tula battalions in reserve) to counter-attack. At the same time four squadrons of Novorossiysk Dragoons engaged the başibozuks on the extreme left.

To the right of Belyavski, keeping liaison with the main infantry force, was a cavalry group consisting of the Tver Dragoons, six squadrons of Nizhegorodski Dragoons, Cossacks, Muslim irregular horse and one horsed battery. To the right, again, of the cavalry the grenadier brigade, covering the main artillery park of five batteries, was hastily deploying along the northern edge of the low ground in front of Kurudere village. Bebutov was under the impression that the grenadiers were facing the Turkish centre, but in reality they had in front of them only the left wing of Kerim Paşa's group. Kmety was already beginning his movement to the left against the Russian right, which at that moment was only covered by six squadrons of Novorossiysk Dragoons, some Cossacks and Muslim irregulars and one mounted battery. Two battalions of the Ryazhski Regiment with units of Cossacks and Muslim horse and one battery were all the reserve available.

Fierce fighting soon developed round the Karayal Hill. The attack of the Istanbul rifle battalions forced Belyavski to swing his left flank back while, at the same time, the Syrians, supported by two batteries, assailed his exposed right. Belyavski now had to abandon his task of taking the hill, for he was hard put to it to maintain his position. The Tver Dragoons, by a brilliant charge, broke the attack of the Syrian infantry and overran the two Turkish batteries, but the Turks were able to recover eight of their guns.

Kerim Paşa brought up reinforcements, and the Istanbul Rifles descended from Karayal round the Russian left, while the *süvari* boldly charged a square formed by one of the Belevski battalions. On Belyavski's right, the cavalry (Nizhegorodski Dragoons and Don Cossacks supported by a mounted battery firing grape-shot) renewed the attack on the Syrian infantry. The Syrians fought well and, at one moment, took four of the Russian guns.

The Nizhegorodski Dragoons suffered heavily, losing half their men and twenty-three out of thirty-two officers. But their desperate charges gave the infantry time to recover, and Belyavski himself rallied the Caucasian riflemen and some mixed companies of the Russian regiments to a bayonet charge. The Syrians had taken heavy punishment and they now began to retire to the south. At the same time the Istanbul Rifles started to withdraw from Karayal, pursued in the direction of Oğuzlu by the Novorossiysk Dragoons. As the regulars broke, the başibozuks and Kurds dispersed. It was still only eight o'clock in the morning, but the fighting on their left flank had now turned definitely in favour of the Russians.

As soon as the position on the left seemed sufficiently stable, Bebutov undertook an attack on the right which he intended to be decisive. At about 8 a.m. along the northern end of the low ground in front of Kurudere, the grenadiers deployed in two lines supported by the fire of five batteries. Mustafa Zarif threw in the whole of his reserves in infantry and guns to support the threatened left flank of Kerim Paşa. But the Russian artillery soon began to silence the Turkish guns, while the grenadiers attacked in dense columns. They encountered a stubborn resistance from the Syrian infantry, and one battalion of Georgian Grenadiers, in resisting a counter-attack, lost two-thirds of its strength. Suddenly the Turkish reserve battalions, which had been brought last into the battle, recoiled and then broke in disorder. The Turkish gunners began to limber up, while one *süvari* regiment charged the advancing Russian infantry, who were forced to form a square. Bebutov sent in the troops of his own escort to check the Turkish horse.

In the meantime, the Turkish left, under Colonel Kmety, was continuing its now useless flanking movement. Early in the morning, swarms of Kurds and başibozuks appeared along the Kars-çay in the direction of Poldervan. Bebutov strengthened his right with a Cossack regiment and a brigade of Muslim horse. It was sufficient reinforcement to hold the irregulars, but after 8 a.m., when Kmety began to attack with his strong force of Anatolian infantry, Bebutov sent in the Tver Dragoons and what remained of his reserve (two battalions of the Ryazhski Regiment and one battery) under General Bagovut. Heavy fighting started about 9 a.m. The Russian cavalry easily dispersed the Kurds and başibozuks, then broke the *süvari* brigade and captured its mounted battery. The Anatolian infantry attacked in waves, but they were held by the two Ryazhski battalions and their flanks enveloped by the Tver and Novorossiysk Dragoons. After 10 a.m. Bebutov, now secure on his left, reinforced the right with two Tula battalions and one battery. At the same time, informed of the defeat of the Turkish right flank and centre, Kmety began to draw off his men. The Russian cavalry proved too exhausted to pursue.

The bulk of the Turkish army retired in three separate groups on Kars in good order; they were able to save most of their artillery. The losses of the seraskier were considerable: not less than 8000 killed and wounded, with 2000 prisoners and 16 guns. About 10,000 irregulars deserted to their villages after the defeat. Russian losses were sensible: 3000, representing 15% of the troops engaged.

Guyon's plan of battle was certainly a logical one, but he ignored the half-trained condition of the troops at his disposal. On the other hand, Bebutov, in spite of the costly neglect to occupy Karayal in strength, had once more demonstrated his great fighting experience, his capacity for acute observation and his understanding of quickly moving events. The Russian commander always felt the nerve of the battle and moved his troops just when necessary and without losing time. He skilfully took advantage of the disposition of the Russian forces within a semicircle formed by the three distinct Turkish groups. The battle was made up of three separate consecutive actions, in each of which the Turks were defeated. The Turkish nizams, particularly the Syrian battalions, fought with great courage; the fire of the new Turkish rifle battalions was excellent; the new regular cavalry units attacked with boldness and a spirit of sacrifice; and the guns were served competently but with a certain lack of mobility. It was clear to the Russians that the newly organized Turkish army was not yet capable of a war of manoeuvre but might be a serious enemy behind fortifications.

With his modest forces reduced by casualties, Bebutov was not prepared to undertake an operation against Kars and, with the approach of autumn, he retired across the frontier to Aleksandropol.

The victory of Kurudere had immediate repercussions in relieving Russian embarrassments both in the mountains and along the Persian border. Shamyl was impressed by the failure of his friends to secure any successes against the Russians and, not only during the winter but also during the campaigning season of 1855, he continued to maintain a cautious and expectant attitude without undertaking any serious move either in Chechnia or in Kakheti.

Negotiations for an agreement with the Persians, which had been dragged out without finality since the summer of 1853, were now rapidly concluded. A 'secret' convention was signed under which the Persians confirmed their neutrality and the Russians waived the Persian indemnity which had been outstanding since the Treaty of Turkmençay.[1]

[1] For Russo-Persian relations prior to and during the Crimean War, see *Akty*, x, Docs. 667–718 and *ibid.* xi, Docs. 486–623. Cf. also Sykes, *History of Persia* (London, 1921), pp. 346–7. Britain actually became involved in war with Persia in 1856.

CHAPTER VI

SIEGE OF KARS. OPERATIONS IN ELEŞKIRT REGION. TURKISH INVASION OF WESTERN GEORGIA, 1855–6

MAPS 4, 5, 6

DURING the early autumn of 1854 the attention of the Allied high command was wholly concentrated on operations in the Crimea where a rapid decision was expected. In November, however, the necessity for a long siege of Sevastopol became clear. The likelihood of the war continuing through the summer of 1855 imposed a reorganization of the Turkish army in Asia Minor and the military mission of Colonel (now Major-General) Fenwick Williams, R.E., was accordingly strengthened. At the same time the seraskier, Mustafa Zarif, was dismissed, and Şükrü Paşa, an officer equally inefficient and more obstinate, was appointed in his stead.

The disorganization of the Turkish army on the Caucasian frontier after the defeats of 1853–4 presented Fenwick Williams and his staff with an impossible task in view of the resources and facilities available. The sanitary services, deplorable enough in the Allied camps in the Crimea, were, in Asia Minor, veritably non-existent; the Turkish strength had, in fact, suffered more from incompetent organization than from the action of the Russians. In May 1854 Turkish effective strength on the Caucasian frontier had been about 120,000. In May 1855 it had fallen to less than 70,000. Losses during the campaign of 1854 in killed, wounded and prisoners had not exceeded 15,000, but during the winter epidemics were allowed to carry off twice that number; desertions during the same period were estimated at 20,000. Replacements did not amount to more than 10,000.

Williams came to the conclusion that it was not possible to organize a field army for an offensive during 1855. He decided to limit his efforts to the strengthening of the fortified camp at Kars with the object of presenting a serious obstacle to a Russian advance into Anatolia.

The Kars garrison, reorganized out of the remnants of the fine Anatolian and Syrian battalions which had survived the battle of

Kurudere and the epidemics of the winter months, in May 1855 amounted to sixteen battalions of nizams, three battalions of Istanbul Rifles, one brigade (eight battalions) of redifs, one *süvari* regiment and eighty field pieces besides fortress guns. In addition to these 15,000 regulars, there were contingents of Laz and Kurd irregulars and armed inhabitants of the town of Kars, amounting to another 9000.

A reserve, concentrated at Erzurum, did not exceed 5000 men, and there were a further 5000 scattered in small detachments round Ardahan, Bardiz and Oltu. An independent corps under Veli Paşa, composed of about 7000 regular infantry and an equal number of irregulars, occupied the Eleşkirt valley on the line of approach from Bayazit to Erzurum. In the Batum area and at various points along the coast, Selim Paşa commanded not more than 15,000 men.

The Turkish forces were clearly inadequate, and their usefulness was further reduced by bad organization of lines of communication and difficulties of supply. The two strongest groups, at Kars and Eleşkirt, were separated by considerable distances from their principal base at Erzurum.

The Russian government, satisfied with the results of the Caucasian operations of 1854, had under consideration plans of rather wider scope for the campaigning season of 1855. While the Franco-British fleet completely dominated the Black Sea and Allied armies were established in the Crimea, and while, at the same time, the Austro-Hungarian government was adopting an attitude which was becoming increasingly inimical, it was impractical to undertake any major operations against the Turkish dominion in the Balkans. The Emperor Nicholas I therefore contemplated the possibility of striking a serious blow against the Turks in Asia, a blow which might have the effect at a later date of diverting Franco-British forces from action in European Russia. The Caucasian troops were not numerous and it was difficult to reinforce them, but their high standard of training and their enthusiasm made them a formidable weapon. General Muravyev, who had held a command under Paskevich, was appointed viceroy in succession to Vorontsov. He brought to the Caucasus a cool and experienced head and a character naturally optimistic but tempered by prudence.

At the beginning of 1855, Muravyev undertook a regrouping of forces in the Caucasus. He decided to reinforce the field groups by eight battalions, three regiments of Cossacks and three batteries taken from the Caucasian and Lesghian lines; at the same time a number of changes were made in the commands.

PLATE II

A. THE CAMP OF ÖMER PAŞA AT ZUGDIDI, NOVEMBER 1855

B. THE CITADEL OF KARS IN 1877

The order of battle of the Russian army in Caucasia at the beginning of May 1855 was approximately as follows:

Group	Strength	Remarks
Guria	Brestski Regt (4 battns) 1 *plastun*[1] battn (Cossack Foot) Gurian militia 8 guns	H.Q. Ozurgeti O.C. Gen. Brunner
Mingrelia	Litovski Regt (4 battns) 2 Black Sea line battns 1 Georgian line battn Imeretian militia 2 Don Cossack regts 8 guns	H.Q. Marani O.C. Gen. Prince Bagration-Mukhranski
Reserve	2 battns Kurinski Regt 3 Black Sea line battns 1 Georgian line battn Imeretian militia 16 guns	In station at Marani at the junction of the Rion and the Tskhenis-dzqali Total strength in west Georgia 16,000 men and 32 guns
Kakheti	4 battns Tiflisski Grenadier Regt 7 other battns 4 sqns Nizhegorodski Dragoons 3 Don Cossack regts Georgian mounted militia	Concentrated in two groups, at Qwareli on the upper Alazani, and Zakatali; with Cossacks and dragoons and 1 horse battery forming mobile column[2]
Elizavetpol	Mingrelski Grenadiers (3 battns)[2]	—
Tiflis	About 6000 infantry	Group in general reserve
Akhaltzikhe	3 regts from 13th Inf. Div. 1 regt from 18th Inf. Div. 1 battn Georgian militia 1 Don Cossack regt 1 regt Muslim mounted irregulars 2 field batteries	Area included Akhalkalaki O.C. Gen. Kovalevski Total about 16,000 infantry, 7000 cavalry, 72 guns
Aleksandropol	Grenadier Brigade 3 regts 18th Inf. Div. Rifle battns Sapper battns Dragoon brigade 5 regts Cossacks 2 regts Muslim mounted irregulars 9 field batteries	O.C. Gen. Count Muravyev Total about 16,000 infantry, 7000 cavalry, 72 guns
Erevan	1 composite inf. regt. 1 Cossack regt 1 regt Muslim irregular horse 1 regt Kurd irregulars 1 field battery	O.C. Gen. Suslov Total about 3000 infantry, 2000 cavalry, 8 guns

[1] For the explanation of the word *plastun*, see Baddeley, *Rugged Flanks*, II, p. 44.

[2] A part or the whole of these units was subsequently withdrawn from security duties and used to reinforce the field forces.

Neither Shamyl in Dağistan nor the Turks of Ardahan with the neighbouring Laz showed any signs of activity, and as a result the Kakheti and Akhaltzikhe forces and the general reserve round Tiflis were gradually weakened in favour of the field army.

As Muravyev intended to move simultaneously all three of the field forces, these must be regarded as having an operational unity representing a strength of 18,000 infantry, 10,000 cavalry and 94 guns. This strength was later reinforced from the interior by 11,000 infantry and 32 guns, making a total field strength of 29,000 infantry, 10,000 cavalry and 126 guns. Having in view the limited communications of the period and the restrictions on supply imposed by the presence of the Allied fleet in the Black Sea, it was probably the maximum strength which could be maintained effectively in the field, and it was certainly sufficient for an offensive operation in the direction of Kars and Erzurum.

Muravyev realized the defects of the Turkish army opposed to him. At the same time the situation in Asia Minor was very different from that which had given the Russians their easy victories in 1828–9. The Turkish troops, although weakened by epidemics, were well armed, and many units were by no means badly trained. The incapacity of the Turkish high command was tempered by the presence of numerous foreign officers, who had been successful, at least, in converting Kars into a powerful modern fortress of the period. Muravyev was well aware of the competence with which the fortifications of Kars had been prepared. He on his side had no siege artillery available; he decided therefore to use his great superiority in the mounted arm to establish a mobile blockade. He showed, in fact, great ingenuity in carrying through the successful blockade—and ultimate reduction—of a first-class fortress largely by the strategic use of his superiority in cavalry. Muravyev was informed as to the resources of the garrison of Kars, for intelligence was not difficult in a town where most of the Armenian section of the population were hostile to the Turkish occupying force; and he calculated that, if an effective blockade could be established, the Kars army would be approaching a critical situation in September and that their position would become desperate by November. Events proved that his calculation was remarkably exact.

Muravyev's tactics comprised the observation of the fortress of Kars by strong infantry detachments stationed at points where they could check sorties by the garrison and at the same time provide, if necessary, support to the cavalry who were dispersed over a wide area. The cavalry sweep was directed to the destruction of all Turkish supply depots within a radius of a hundred miles round the fortress.

The strategy of Muravyev, reasonable and practical, was based on the exploitation of the defects in Turkish organization and on the distribution of the Turkish forces which, to the Russian commander-in-chief, appeared to be very faulty. The geographical position of Kars made it, in terms of nineteenth-century warfare, an excellent advance base for an offensive against Transcaucasia; but in relation to the defence of Asia Minor its situation had a rather negative value. Muravyev, with his experience of Paskevich's operation against Erzurum, realized that the real defensive position covering Erzurum was to be found in the region of the Soğanli-dağ. He held the view that if he had had available one more infantry division, he could have left 15,000 men to mask Kars and could have undertaken an offensive against Erzurum, leaving the isolated army at Kars to surrender later as the result of lack of supplies. As it was, Muravyev was of the opinion that the Turks might well have left only half their force in Kars and, with the balance, might have effected a junction on the Soğanli-dağ with Veli Paşa's force from Eleşkirt. Such a combination, with the limited forces at the disposal of Muravyev, might have placed the Russians, in June, in a difficult position; for their existing strength was, in the view of Muravyev, inadequate to contain Kars and at the same time to carry out a field operation against a Turkish army concentrated between Kars and Erzurum. Throughout the summer and autumn of 1855 Muravyev continued to expect the dispatch of relieving forces via Trebizond and Erzurum, and his appreciation of the situation justified the view held by Fenwick Williams and underlines the error of the alternative scheme for a Transcaucasian offensive adumbrated by the Turkish commander-in-chief, Ömer Paşa.

The problem which confronted the Turkish and Allied commanders in Asia Minor in 1855 was the preservation of a Turkish army on the Caucasian frontier and the defence of Erzurum and the ways leading into Anatolia until such time as a decision had been secured in the Crimea. The concentration on the *offensive* base at Kars and the neglect of the wider problems of Anatolian defence, which implied the preservation of a Turkish field army, proved in many ways unfortunate.

The retention of the bulk of the Turkish army in Kars was, in fact, the result of circumstances rather than of any false strategic conceptions. The Kars army was, after the defeat at Kurudere and the epidemics of the winter, so weak that it was practically immobilized. Williams judged the Kars troops to be incapable of any major movement. His maximal objective became the retention of the Russian army outside the fortress and the prevention of any advance on Erzurum. He felt that if an early decision could be attained at Sevastopol, time might remain to send reinforcements and supplies via Trebizond and Erzurum. In general he regarded the Turkish armies as incapable of manoeuvre in the field and was very sceptical with regard to Ömer Paşa's projects for a diversion in western Georgia.

This diversion remained the favourite topic of discussion in Turkish official circles during the spring of 1855. The Allies had already vetoed plans for a diversion in western Georgia before the Turkish defeats of the previous year. The plan, however, was now pressed by the Turkish commander-in-chief, Ömer Paşa. This ambitious soldier was a Croat by origin; he enjoyed a military reputation which had been won largely in operations against the dissident subject races of the empire, and he was surrounded by Polish and Hungarian *émigré* officers who were, in fact, soldiers of fortune with strong anti-Russian emotions. Ömer Paşa was commanding the Turkish army in the Crimea which was stationed in two groups, at Eupatoria and Balaklava, and which had been kept in observation with little opportunity for distinguished action. Ömer's circle at Allied headquarters and at Istanbul now began to insist that the bulk of the Turkish troops in the Crimea should be transferred to the Caucasian coast. It was proposed to unite with them reinforcements from the Balkans and from Tunisia, and Ömer himself envisaged an advance on Kutaisi, and subsequently on Tiflis, which might threaten a union with the Murids descending into Kakheti and the isolation of all Russian forces on the Turkish frontier.

The objections of the Allied command became even more emphatic when, on 18 June, the attempt to storm Sevastopol ended in disaster. Discussions of the plan, however, continued in Istanbul, and the Turks won the qualified support of the British ambassador, although General Fenwick Williams, by letter, continued to protest.

In the middle of July, Ömer, pressing his plans at Allied general headquarters in the Crimea, offered to replace troops removed from the Crimea by a further corps of 20,000 which had been completing training in Bulgaria under the supervision of the French General Vivien. It was difficult to object to this proposal, but both Marshal Pelissier and General Simpson maintained their refusal to support the general conception of a Georgian diversion. On 17 July, Ömer Paşa left for Istanbul; in the same ship travelled Lieutenant-Colonel Sissaud with the mission of dissuading the Turkish government from the undertaking. Discussions continued throughout the month of August until, finally, the British government withdrew their objections; Allied general headquarters received certain assurances, and preparations for the realization of Ömer's plan were undertaken. In the meantime three months of the campaigning season in Caucasia had passed and the position of the Turkish forces had further deteriorated.

On 24 May, Muravyev's army had crossed the Arpa-çay and established a camp on the same ground as the previous year, at Poldervan and Kurudere. Patrols occupied the two Yahni hills six miles from the fortress of Kars. On 3 June, Muravyev was joined by General Kovalevski and part of the Akhaltzikhe force. His strength now amounted to about 20,000 infantry and 7000 cavalry with eighty guns. On 6 June, with the object of effecting the complete blockade of Kars, Muravyev moved his main camp to the village of Mağarcik, five miles south-south-east of Kars, commanding the roads to Erzurum and Kağizman.

On 16 June Muravyev, leaving a force of 11,000 infantry and cavalry with forty guns in observation of the fortress and another 5000 men on line of communication to Aleksandropol, marched with the rest of his army in the direction of the Soğanli-dağ. He met with no resistance and destroyed the considerable reserves of Turkish supplies accumulated at Bardiz, Yeniköy, Karaurgan and Zivin.

Muravyev now decided to undertake an operation against Veli Paşa's force in the Eleşkirt valley. This force amounted to about 7000 regulars with an equal force of irregulars and twenty guns. At Bayazit, Suslov was in observation, but, in spite of Muravyev's order, he had hesitated to advance in view of his own limited numbers. In mid-June, Veli, fearing that the movement by Muravyev across the Soğanli might cut him off from Erzurum, evacuated the Eleşkirt valley without pressure from Suslov and, crossing the Dram-dağ, took up a position near Köprüköy covering the historic bridge over the Aras.

While Suslov now moved cautiously into the Eleşkirt valley, Muravyev sent forward Prince Dondukov-Korsakov with all the cavalry to intercept Veli in the Pasin valley. This officer was ordered to establish liaison with Suslov and then to turn swiftly west with the object of cutting off the Turks from Hasankale (Pasinler), and their line of retreat on Erzurum. At dawn on 21 June, Dondukov-Korsakov reached the northern bank of the Aras where he found Suslov on the other side of the river at Yuzveran. He now, however, found himself under the orders of this hesitant officer who continued to watch Köprüköy, considering his force insufficient to initiate an attack on the Turks. Suslov's caution defeated Muravyev's plan. On the night of the 21st, Veli, fearing the movement which was taking place across the Soğanli, withdrew from Köprüköy on Hasankale. In the morning he abandoned even this fortified position and retired on to the Deve-boyun ('Camel's neck') ridge covering Erzurum. The same day Dondukov-Korsakov's cavalry occupied Hasankale.[1] Muravyev, after censuring Suslov, sent that officer back to the Eleşkirt valley and himself returned to his camp before Kars.

Panic reigned in Erzurum, and Veli Paşa, an incompetent officer with no control over his half-trained and demoralized troops, was making preparations to fall back to Bayburt on the road to Trebizond.

Muravyev was aware that it was practicable to occupy Erzurum probably without a serious action. He deliberately refrained from doing so. He considered that the forces which he had left in front of Kars (about 15,000) were inadequate to contain the garrison, and his experience of the campaign of 1829 discouraged him from attempting to occupy Erzurum and advance to the Çoruh valley with the relatively small forces at his disposal. He feared to divide his strength and he had little confidence (probably rightly after his experience with Suslov) in the capacities of his subordinate commanders. Muravyev was obsessed also with the possibility of strong Allied forces being sent to Trebizond, and there is no doubt that a Russian occupation of Erzurum would have diverted to Trebizond Ömer Paşa's army which was then preparing for the descent on Mingrelia.

Muravyev was correct in his practical appreciation, but he undoubtedly showed lack of political imagination. The moral effect of the capture of Erzurum would have been important, and it would have done much to set off the impending loss of Sevastopol by the Russians. It is possible also that the news of the fall of Erzurum might have precipitated the capitulation of Kars where the garrison was already suffering severely from shortage of supplies. In war, political and psychological issues are inseparable from military. Before Erzurum, Muravyev abandoned the chance of an easy triumph because he overlooked the psychological value of such a chance, but three months later political and psychological factors forced him to undertake the costly assault on Kars.

Muravyev continued to believe that the Turks would send rein-

[1] A. Dondukov-Korsakov gives his account of this operation in *Kav. Sbor.* I, pp. 289–368, 'Vospominaniya o kampanii 1855 g. v Aziatskoy Turtsii'.

forcements via Trebizond to Erzurum; he considered any move for the relief of Kars via Kutaisi, the Kura defile and Akhaltzikhe as quite impracticable, and this view was, in fact, shared by Fenwick Williams. About 20 September, however, Muravyev received definite information that 8000 Turks had arrived at Batum, and that their commander, Ömer Paşa, had proceeded to Redutkale and Sukhum where, evidently, further debarcations were expected. It became clear to the Russian commander-in-chief that the Turks were preparing a serious diversion in western Transcaucasia with a threat to Kutaisi and, perhaps, to Tiflis. On 23 September Kars citadel fired a salute and exultant cheers were heard from the walls. It soon became known in the Russian camp that the Turks were celebrating the fall of Sevastopol—the news of which had reached them before the Russians. Muravyev called together his military council, and it was unanimously decided to storm Kars and to concentrate all available forces for this purpose.

This decision constituted a reversal of the whole of Muravyev's plan of campaign, for he had decided to await the results of his mobile blockade which he had anticipated would bring about the capitulation of the fortress before the end of November. He had information that the garrison would be at the end of its resources after another six or eight weeks, and that, in the meantime, a cholera epidemic was seriously weakening its strength. He estimated that the morale of the garrison would now be raised by the news of the fall of Sevastopol and the landing of Ömer Paşa in Georgia, and that the hands of the British officers who were opposing capitulation would be strengthened proportionately. At the same time he realized that his own troops were likely to be depressed by the recent news and that they would at the same time resent a withdrawal without a fight. He understood also that the fall of Sevastopol must have shaken public opinion in Russia and that a retreat now from before Kars would be a new and cruel blow to the confidence of the country. As before Erzurum he had allowed military principles to overrule political considerations, so now he ignored obvious military objections in favour of a desperate solution which was imposed by the political and general strategical situation.

The fortress of Kars in 1855 had become a real stronghold in comparison to what it had been in Paskevich's time. The citadel had in the new conditions of war lost its importance. The defence was based on a series of forts and redoubts which surrounded the fortified camp into which the army of Mustafa Zarif had retired after the defeat at Kurudere. Under the supervision first of Guyon (recalled to Istanbul after Kurudere) and later of Fenwick Williams, the strong Karadağ heights along the right bank of the Kars-çay to the north of the citadel had been heavily fortified. These heights fall almost vertically to the river, and they descend very steeply to the north and east, commanding all the approaches from the Russian frontier.

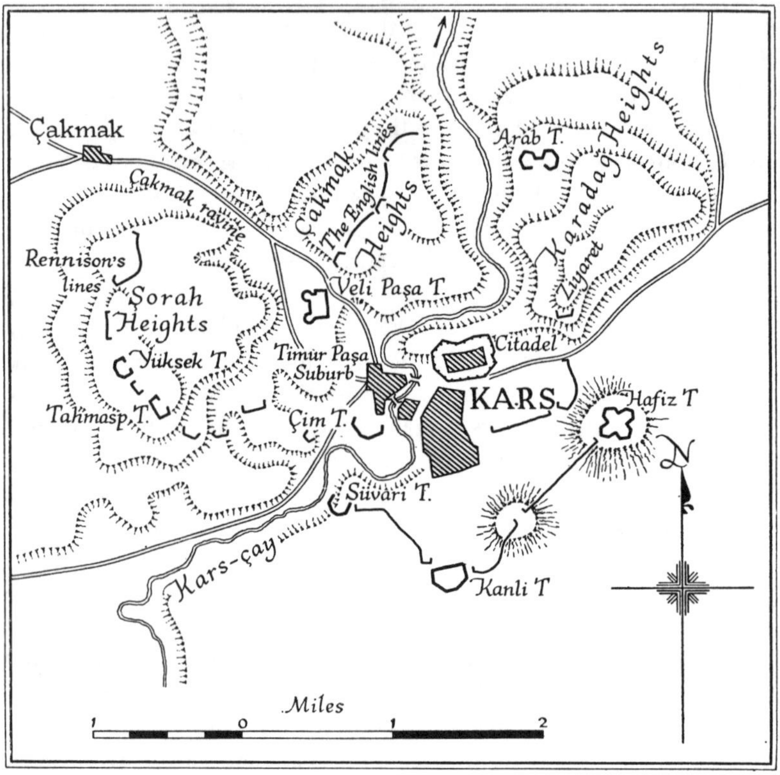

Map 5. The siege of Kars, 1855

A strong fort, Arab Tabia, had been built on the northern ridge of Karadağ above the river and another fort, Ziyaret Tabia, commanded the approach to the heights from the east.

The Karadağ heights were more accessible from the south. The Turks had therefore created to the south of Karadağ and to the south-east of the Orta-kapi suburb of the town a system of forts, redoubts and trenches to cover the approach over the flat ground. The left face of the defensive line stretched from Karadağ to the new fort of Hafiz Tabia; the central face from Hafiz Tabia to the stronger and newest fort of Kanli Tabia. The ground between the two forts was defended by trenches and two redoubts.

From Kanli Tabia the right face of the fortified area turned north-west and reached the bank of the Kars-çay, at the fort of Süvari Tabia, which had a field of fire over the river. The forts were all solidly built and armed with modern heavy guns; the trench system between them covered a distance of over four miles. Muravyev had no siege artillery and therefore he did not propose to attempt an attack on Karadağ and the three forts on the plain. His attention was directed to the left (western) bank of the Kars-çay where the Armenian suburb of Timur Paşa lay between the river and the flank of Çakmak-dağ. Artillery on Çakmak-dağ could at once dominate the town and the forts in the plain.

The slopes of Çakmak-dağ gradually inclined to the north-west to a ravine near a village of the same name. From the village a road ran along the ravine to Timur Paşa and the principal bridge over the Kars-çay. To defend this area the Turks, with their British advisers, had constructed the new and strong fort of Veli Paşa Tabia to the west of the Armenian suburb. Another fort, Çim Tabia, neither so strong nor so well situated, had been built on the left bank of the river to the south of Timur Paşa.

Not satisfied with the defence of the Çakmak ravine, which could be covered only by the guns of Arab and Veli Paşa Tabias, the British sapper officers had devised a fortified line, strengthened by three redoubts, to the north-west of Veli Paşa Tabia; this position was called by the Turks the 'English lines'.

To the south again of the Çakmak position the Şorah ridge, dominating the Kars-çay southward from the suburb of Timur Paşa, had also been fortified. Here the forts of Tahmasp Tabia and Yüksek Tabia could bring a cross-fire to bear on the relatively weak gap by Çim Tabia. Between Yüksek Tabia and the bluff with which the Şorah ridge ends, a line of trenches had been dug which the British called 'Rennison's lines'. The Şorah position was, in fact, an advance defended locality, and Muravyev decided to attack this part of the Kars fortifications since he saw that the undertaking might be carried through without siege artillery and even without serious artillery preparation.

British and Turkish officers in Kars had not failed to observe the Russian interest in the Şorah position, and when Muravyev launched his attack he had not the advantage of surprise. He was unable to make use of his great superiority in cavalry, but he had at his disposal a considerable infantry force (about 23,000) and ninety-eight guns, of which only forty were brought into action to support the storming parties. Muravyev was in fact relying desperately on the factor of surprise (which he did not enjoy) and on the weight of the bayonet attack of his veteran Caucasian battalions. The action was timed to begin at dawn on 29 September, and all preliminary movements had to be completed in the dark. As a result there was considerable muddle and some mistakes in direction with consequent delay. The attack on the Şorah heights began when dawn was already bright, and what had been planned as a night attack developed into an assault in broad daylight. The operation had been badly timed, and the factors on which the hope of success had been based (surprise, darkness and swift co-ordinated attack) were already absent.

The Turks, at the same time, aware that an attack was likely, had been able to concentrate picked troops in the threatened lines: the Istanbul Rifle battalions and the reorganized remnants of the Syrians who had done so bravely at Kurudere. Colonel Kmety, Lieutenant Teesdale[1] and old Kerim Paşa (the only Turkish general officer in Kars of courage and capacity) were present with the troops.

The Russian right column, the grenadier brigade under General Maidel, ascending the steep slopes in front of Tahmasp Tabia stormed to the attack at 5 a.m. (instead of 4 a.m. as planned). In spite of heavy grape-shot and rifle fire, the Erivanski Grenadiers took with the bayonet the line of trenches between Tahmasp Tabia and the Kars-çay. But the Mingrelskis, who had to capture the redoubt itself, were unable to make the escalade (since they had not been provided with ladders) and they were driven back with heavy losses. Other units which had managed to penetrate to the ground between Tahmasp and Veli Paşa were caught by the fire of the latter fort and also suffered severe casualties.

While it had not been possible to develop the initial success of Maidel's column on the right, Kovalevski's troops on the left had met with definite failure. In the darkness Kovalevski led his men against the centre of Rennison's lines instead of against the north-west end of the Şorah position, as had been intended. The Vilnenski Regiment came into a terrible semicircle of fire, and in a few minutes the Russians had suffered grave casualties. The Belevski Regiment, attacking to the right of the Vilnenski men, also had heavy losses. The remnants of both regiments withdrew to the ravine whence their attack had

[1] Teesdale, who was only twenty-two, was awarded the V.C. and the C.B. for his services in the defence of Kars (see *D.N.B.*).

started. The left column was in no position to renew the assault. Kovalevski had been killed.

The intermediate column consisting of the Tulski and Ryazhski Regiments under General Prince Gagarin, destined to support both Kovalevski and Maidel, had been given no very clear orders. Two of the Tula battalions were involved in the withdrawal of the Belevskis on their left, while the rest of the column after a partially successful attack on Yüksek Tabia were counter-attacked by the Turkish infantry (under Teesdale) and joined in the retreat of Maidel's men.

Dawn was past, and Muravyev was aware that his surprise had failed, but he persisted in attacks on Tahmasp Tabia. Meanwhile the onslaught of the Belostokski Regiment (General Bazin), supported by the dragoons and Cossacks of General Baklanov, penetrated the 'English lines' on the Çakmak heights whence the Turks withdrew in disorder to the shelter of Veli Paşa Tabia. But Muravyev had already exhausted the available reserves in the attacks on Tahmasp Tabia, and Bazin could not develop his success. Threatened by a Turkish counter-attack Bazin withdrew with only moderate losses. A final attack by two battalions of the reserve under Colonel Kauffman (later to be the conqueror of Turkistan) penetrated to the rear of the Şorah position where the infantry came under the fire of the guns of Veli Paşa Tabia and were forced to withdraw into cover of the Çakmak ravine. This final incident demonstrated the importance which the ravine might have had in a Russian attack, but its significance had not been realized by Muravyev's staff.

At 11 a.m. the Russian commander broke off the action and ordered a withdrawal to the positions occupied before the attack.

The Russian losses were imposing: Muravyev had lost 60% of all infantry taking part in the engagement—8000 killed and wounded. Turkish losses were light, about 1500. On the evening of 29 September Muravyev had only twelve to thirteen battalions and his cavalry fit for action, and the balance of forces at Kars had shifted definitely in favour of the Turks.

Both the method and the execution of Muravyev's assault are open to criticism. He shared with the Allied command at Sevastopol the error of having used his troops in the old-fashioned column of assault against the modern rifles with which a part of the defenders (the Istanbul battalions) were armed. Had the Russian attack succeeded, the capture of the Şorah heights would not have made the defences of Kars untenable. Muravyev was aware of the limited advantages which victory would give him, but he would have been satisfied to be able to bring his field batteries into action against the town and its suburbs and against the road and bridge which connected the fortress with the south-western forts. Under these conditions it is likely that Veli Paşa Tabia could not have continued to resist for more than a few days. Muravyev's intention had been to accelerate the natural course of events in Kars by disorganization of the defence rather than to effect an immediate capture of the fortress.

The Turks had no horses (these had all died or been slaughtered during the period of blockade), and they were incapable of making a serious sortie and of manoeuvring against the Russians in the field. Kmety alone favoured a sortie. The British officers were convinced that his terrible losses would compel Muravyev to raise the siege and retire to Aleksandropol.

Muravyev had been wrong, not perhaps in his decision to attack, but in his dispositions and his inability to correct them in the field. He showed, however, courage and judgement in continuing the siege. His pertinacity had a decisive influence now on the course of the Caucasian campaign and on the position in which the Russians were to find themselves at the coming peace conference. Muravyev knew well the psychology of the Ottoman commanders who were inclined always to await events rather than to create them. Every day which passed served further to discourage and weaken the defenders of Kars and to make an effective sortie less likely. At the same time Muravyev's knowledge of Ottoman command helped him not to exaggerate the possibilities of a Turkish move from Erzurum or of a rapid advance by Ömer Paşa in Mingrelia. Veli Paşa, like most of the Turkish commanders of the time, was no soldier but a bureaucratic careerist, and he had already proved a timid and incompetent commander. Muravyev had met the ambitious Ömer Paşa some years before, and he had formed the opinion that this officer would be reluctant to risk his career by any hazardous operation in western Georgia; he rightly guessed that Ömer, in fact, was intent on a military parade in Mingrelia which might compensate him for the lack of opportunities which he had had in the Crimea. As it was, the season was already too advanced to allow of the Mingrelian operation reaching a decisive phase, in spite of the ineptitude shown by the Russian commander, Bagration-Mukhranski. Muravyev estimated that, even should Ömer reach Kutaisi before the onset of the heavy rains (mid-November), the state of the garrison of Kars would force them to capitulate in time to allow the Russian commander to transfer the bulk of his force to the Suram Mountains, covering the middle valley of the Kura and Tiflis. Meantime, in view of the quiet conditions in eastern Transcaucasia and of the failure of Shamyl to appear in the field, Muravyev was able to allocate reinforcements between his own forces in front of Kars and Bagration-Mukhranski in Mingrelia. During the month of October, the scarred troops in front of Kars were recuperated in the excellent new camp which Muravyev had built at Vladikars where a plenitude of food and forage contrasted with the conditions of destitution in the beleaguered fortress.

On 6 November, finally, Teesdale appeared with a flag of truce at the Russian advance posts and asked for an interview on the following day between the Russian commander and General Fenwick Williams. Two days later the formal capitulation of Kars was completed, with all guns delivered in good condition and 24,000 prisoners, of whom 4000 were sick or wounded. The British and Turkish officers retained their swords, and a provision was made for the protection of the Hungarian and Polish officers in the Turkish service (Kmety and Kolman, in fact, made their escape to Erzurum before the capitulation). Muravyev's calculations, as to the capacity of Kars to resist without relief or further supplies, had proved remarkably correct.

The planning of Ömer Paşa's expedition to the Caucasian coast had envisaged the concentration of a force of 45,000 men: of them 15,000 already comprised the force at Batum; 20,000 were to be transported from Bulgaria; and 10,000 picked troops, including some battalions of the Istanbul (Sultan's) Rifles were to be transferred from the Crimea. Although not more than 30,000–35,000 troops were concentrated by the end of September, the transportation was carried out relatively efficiently. Of the 15,000 horses estimated by British officers as necessary for lines of communication, Ömer had not been able to collect more than 5000; he had no cavalry except the so-called Ottoman Cossacks, less than 1000 strong, recruited from Polish refugees and the descendants of Cossack fugitives who had settled at the mouths of the Danube.[1] Ömer hoped, rather optimistically, to find plenty of irregular cavalry among the Cherkesses and Abkhaz.

[1] For the romantic history of the Ottoman Cossacks, see W. E. D. Allen, *The Ukraine: A History*, p. 259, note on the Zaporogians after the destruction of the *Sech*. Fugitives to Turkey after the destruction of the *Sech* in 1775, these Cossacks settled at the mouths of the Danube on Turkish territory. Following the Russo-Turkish War of 1828–9, the remnants of their descendants migrated to Anatolia. Some of them were to be found during the nineteenth century pursuing their hereditary calling of fishermen on the small islands of Beyşehir-göl in the Isaurian Taurus (see Murray's *Guide to Asia Minor*, where they are erroneously described as the descendants of prisoners of war of 1806–12). The islands on the lake of

Similar political considerations induced Ömer to make his base at Sukhum rather than at Redutkale. The distance from Sukhum to Kutaisi is 110 miles, and the way lay through forest country and across numerous streams; roads were scarce and undeveloped and bridges virtually non-existent. From Redutkale to Kutaisi the distance is only seventy miles along an excellent post-road which passed, however, for a part of the way along a causeway with marshy ground on either side. Redutkale again presented difficulties as a base owing to the lack of dry ground for camps. At the same time Ömer Paşa was largely guided in his choice of a base by the exaggerated hopes which he placed on the alliance of the Cherkesses and Abkhaz. In the event, Shamyl's *naib*, Muhammad Amin, proved to have little influence among the Cherkess chieftains, who, with the approach of winter, preferred to await events. Shamyl, also, was not prepared to undertake serious operations so late in the year.[1] In Abkhazia, the alliance of Iskander Sharvashidze proved of little use beyond allowing the Turks undisturbed occupation of Sukhum; and not more than 400 Abkhazian irregular horse appeared in the field, where their depredations against the Mingrelian peasantry were to prove an embarrassment.

Bagration, who expected a simultaneous Turkish advance from Sukhum, Redutkale and Batum, adopted an extremely cautious plan. He did not propose to offer battle before the Turks reached the Ingur; he kept the four battalions of the Litovski Regiment, with some Cossacks and four guns, in observation along that river while he concentrated the bulk of his forces at Zugdidi and behind the Tskhenis-dzqali ('Horse's river'). At the same time he gave orders to Brunner to evacuate Guria and to bring his men back behind the angle made by the junction of the Tskhenis-dzqali with the Rion.

The lack of initiative of Bagration in allowing the Turkish concentration without interference promised no good, particularly as the Turkish forces available were twice as strong as the Russian group to the north of the Rion.

Beyşehir are now deserted, but some of the Cossacks are still to be found in the village of Kazak-köy near Akşehir. They speak Russian and have a church where they follow the practice of the Old Believers.

These Cossacks were called 'Nekrasovtsy' after their first leader, Nekrasov. There is a full study of their history in the rare *Izvestiya obshchestva lyubiteley izucheniya Kubanskoy oblasti*, Part 2 (1900), pp. 1–74, by P. P. Korolenko, the well-known antiquary and historian of the Kuban. (For his biography, see *ibid.* Part 6 (1913), pp. 3–14.)

[1] For events in Dağistan see V. Soltan, 'Obzor sobytiy v Dagestane v 1855 i 1856 gg.', *Kav. Sbor.* XII, pp. 479–530.

Ömer Paşa had to reserve substantial forces to protect his coastal bases (about 10,000 men), since these might well have been attacked by a Russian commander of initiative who could thus have delayed the deployment of the offensive. Ömer had actually available as a striking force at the beginning of October four brigades, each of eight battalions of nizams, and one rifle battalion; twenty-seven field and ten mountain guns.

Several scores of western officers—mostly Hungarians and Poles—were available to improve the command. The chief of staff, Ferhat Paşa, was a Hungarian (Stein); Colonel Simmons represented the Allied high command, while Lieutenant J. A. Ballard and other British officers, who had been responsible for training the rifle battalions and the gunners, were present in the field.

On 8 October the news of the Russian repulse before Kars reached Sukhum, and Ömer Paşa decided that it was a propitious moment to open his campaign. On the 15th his troops, advancing along the coast, occupied Ochemchiri and Anaklia, the latter a small town at the mouth of the Ingur. His main body advanced through Samurzakan in the direction of the same river. The wooded country of Samurzakan was, in those days, thinly populated and poor in resources. Transport was inadequate, but all supplies had to be carried by road. There were only two roads available of which one followed the coast to Anaklia, while the other—and the worse—ran through the hills and crossed the Ingur at the village of Rukhi, twenty-five miles above its mouth and five miles from the town of Zugdidi (G., 'Great oak'). The latter, described as the main Sukhum road, was only a track through the forest, and Ömer's troops took twenty days to cover the fifty miles between Sukhum and the right bank of the Ingur, which they reached on 4 November. The average speed of the advance was therefore not more than two to two and a half miles per day—and this in spite of the fact that the weather was unusually dry for the time of the year.

The dispositions of Bagration for the defence of the Ingur line were very faulty; in fact, his actions may serve as a classical example of how a river line must not be defended.

The Ingur, like most of the Caucasian rivers which descend from the mountain to the Colchian plain, is broad but shallow and, in many places, divided by islands and banks into several channels. Usually only one of these channels, often the most narrow, is fairly deep and may constitute a serious obstacle since the speed of the current is frequently very rapid. But in dry weather (and the last fortnight of October and the first of November were, in 1855, proving exceptionally dry) fords may be found almost everywhere. These fords change capriciously, but the local men know them. Under such conditions it was practically impossible to prevent the Turks from crossing the Ingur somewhere between Rukhi and the mouth, and the obvious strategy of Bagration was to rely on an active defence: to keep a strong mobile force available at one central point ready to strike at the Turks when a part of their army was already across the river. Bagration

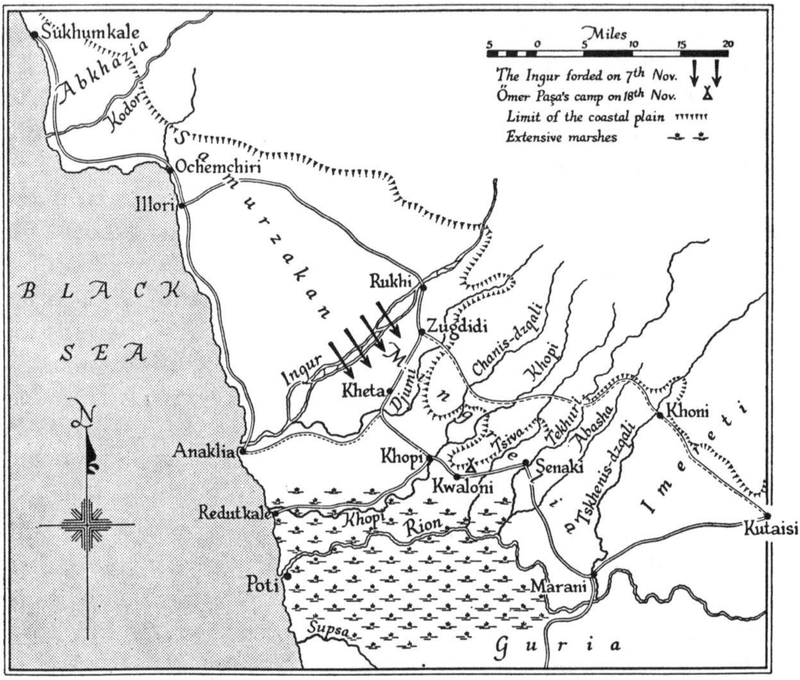

Map 6. The Ingur campaign, 1855

chose exactly the opposite course: he distributed his strength in small detachments along the Ingur with each body virtually isolated from its neighbour.

The principal detachment (2½ battalions, some Cossacks and Mingrelian militia: about 3000 men with four guns) lay at Rukhi. Six miles farther down the river another battalion was in observation of a ford and, a mile and a half lower 1½ battalions with four guns kept watch at a third ford. Two miles farther to the south other possible fords were held by half a battalion with Cossacks and two guns. Another 1½ battalions were stationed along the lower Ingur where they could give warning of any movement from Anaklia. Thus two-thirds of the Russian force were dispersed; as a central reserve Bagration retained two Kurinski battalions, two guns and the Imeretian mounted militia (total, about 2000).

Ömer Paşa spent two days on the right bank of the Ingur, and on the morning of 7 November he attacked. His plan was appropriate to the faulty disposition of the Russian forces. With one brigade he made a feint in the direction of the Rukhi crossing, where the opening of a lively artillery cannonade attracted Bagration himself.

Ömer, meantime, had sent his three remaining brigades against the more southerly fords; the thick undergrowth allowed the Turkish riflemen to approach the river bank without being observed. Turkish artillery engaged the Russian four-gun battery at the third ford to the south, while a whole Turkish brigade crossed at an undefended ford a little farther up the river and took the guns in the rear. At the fourth ford, where only half a battalion and the Cossacks were stationed, another Turkish brigade crossed the river in spite of the resistance of these troops. At the upper ford the guns were already taken and the two infantry battalions (having lost all their officers) were in retreat. Too late the excellent Kurinski battalions and the Mingrelian horse were brought into action. There was confused fighting in the woods and, as evening approached, Ömer Paşa broke off the action. Bagration ordered a general retreat on Kheta. He had lost the day—with three guns and 600 casualties. Turkish losses did not exceed 400, including one British officer killed.[1]

Bagration had given his adversary an easy victory, but Ömer, who had thus succeeded in enhancing his prestige both at Istanbul and at Allied general headquarters, did not appear keen on the military exploitation of his success, and he seemed almost indifferent to the fate of Kars. He spent two days on the left bank of the Ingur and only occupied Zugdidi (the capital of Mingrelia, whence the princess-regent had already fled) on 10 November. In the agreeable surroundings of Zugdidi, Ömer stayed five days, thus losing the last week of the favourable dry weather. Contact with the Russians was lost.

The conduct of Bagration, however, was almost equally strange. He abandoned his original intention of fighting delaying actions on the successive river lines of the Tsiva and Tekhuri and did not stop until he had crossed the Tskhenis-dzqali. Bagration had thus abandoned the whole of Mingrelia while Ömer remained inactive at Zugdidi. At the same time he ordered Brunner to retire to Marani

[1] Laurence Oliphant's account from the English side may be compared with an anonymous description in *Kav. Sbor.* v, pp. 335–43, 'Ingurski boy' (with plan).

(although the Turks had made no effort to advance from Batum), thus abandoning the unfortunate and loyal population of Guria to the mercies of the başibozuks.

On 18 November Ömer attained the River Tsiva, taking eleven days to cover the twenty miles from the Ingur. Here the autumnal rains caught him. He now wisely changed his principal base from Sukhum to Redutkale, only fifteen miles by a good road from his camp on the Tsiva. On 2 December Ömer's advance troops occupied Senaki on the Tekhuri and his patrols reached the right bank of the Tskhenis-dzqali.

The rains were now becoming incessant, and the Turkish camp on the Tsiva was transformed into a bog. Ömer's troops were without bread and his animals without forage. By the second week in December the Turkish army was submerged in the Mingrelian mud, familiar enough to the Caucasian army. In these circumstances, and with the news that Kars had capitulated, Ömer ordered a withdrawal, first behind the Tekhuri, later behind the Tsiva.

If the Caucasian expedition of Ömer Paşa did not end now in complete catastrophe, it was due to the ineptitude of his adversary, Prince Bagration-Mukhranski. By 18 November the latter had concentrated on a strong position behind the Tskhenis-dzqali the remains of his own force, together with Brunner's troops from Guria and recent reinforcements which had arrived from Tiflis—in all, twenty battalions and the mounted militia of Guria, Imeretia and eastern Georgia (the Mingrelians had deserted, following on the evacuation of their country). He had a very definite mounted superiority over Ömer (who was virtually without cavalry), and this superiority might have been used decisively in the marshes and forests of Mingrelia, now that the rains had broken. But Bagration did not now even propose to await the Turks behind the Tskhenis-dzqali, and on the night of 19–20 November ordered the large reserves of stores accumulated at Marani to be burnt, as impossible to evacuate. This last foolish action so infuriated Muravyev that he ordered the dismissal of his distinguished subordinate (who was, incidentally, the direct heir of the Mukhranian line of the Georgian kings). Bagration was replaced by the more balanced General Brunner, but it was not possible to replace the stores which had been collected at Marani under conditions of the greatest difficulty.[1]

[1] P. K. Uslar, the pioneer of Caucasian linguistic studies has left an account of events in western Georgia in *Kav. Sbor.* v, pp. 235–334, 'Guriyski otryad v 1855 g.', and xxiv, pp. 1 (*bis*)–104, 'Zapiski Generala Uslara o voyennykh deystviyakh Guriskago otryada v 1855 g.'

At the beginning of December, Bebutov, following the fall of Kars, had been sent to Kutaisi to prepare an offensive against the Turks in Mingrelia. But difficulties of climate and supply—the latter largely created by Bagration's premature destruction of depots—impelled him to advise that an advance was impracticable. However, he re-occupied Ozurgeti and disposed his troops in cantonments round Kutaisi and along the Tskhenis-dzqali.

Meanwhile, Ömer continued his retirement, covering it by rear-guards and still continuing to hold Zugdidi. Bebutov proposed to start a counter-offensive in March, but in February he was informed that the Turks had evacuated Zugdidi, Khopi and Kwaloni and were concentrating on the coast round Redutkale. Then it became known that the Turkish forces were being evacuated by sea to Batum. Ömer Paşa had long since returned to Istanbul.

Peace negotiations stopped Muravyev's preparations for a Caucasian campaign in 1856. He had proposed to liquidate Ömer Paşa's force in March and April and, in May, to march on Erzurum from Kars. He planned to have 30,000 men (including 8000 cavalry) concentrated at Kars, and anticipated that Erzurum (where the Turks were reported to have less than 15,000 men with bad equipment and low morale) would be in his hands by June.

To counter Muravyev's expedition the Turks must have transferred from the Crimea their relatively good troops amounting in numbers to about 30,000, but to conduct a successful campaign for the defence of Erzurum and Trebizond they would have required the support of at least 20,000 British troops (the attitude of the French government was such that it was not likely that French troops would have been made available).

Even with the intervention of strong Allied forces based on Trebizond (and advancing along the long track to Erzurum as their only line of communication), Muravyev would have been in a position to fight a defensive war of manoeuvre over the terrain between Erzurum and Kars. In the event of the continuance of the war through the summer of 1856, the Allied and Turkish commands would have had to face a major and difficult problem in the defence of Asia Minor. It may be said therefore that Muravyev's pertinacity, in the adverse situation before Kars, had an important influence on the Russian position at the Peace Conference.

With regard to the Mingrelian expedition, it has often been claimed that had the Allied governments agreed to Ömer Paşa's proposals earlier in the year, the expedition would have achieved the object of forcing Muravyev to raise the siege of Kars and would have constituted a threat to the whole Russian position in Transcaucasia. But the position in the Crimea (complicated by the repulse before Sevastopol in June) made it difficult for the Allies to provide the necessary forces before the autumn.

In the previous year, when strong Turkish forces were in existence at three points along the Caucasian frontier, an expeditionary force landed in Mingrelia might have achieved decisive results. But here, as has already been indicated, the Allies had to choose between concentrating their main effort in the Crimea or in the Caucasus, and reasons of high policy imposed the Crimean alternative.

BOOK III

THE CAUCASIAN CAMPAIGN OF 1877-8

CHAPTER VII

THE RUSSIAN AND TURKISH ARMIES IN 1877. FAILURE OF THE RUSSIAN 'PROTECTIVE OFFENSIVE'

MAPS 1, 4, 7

UNDER the terms of the Treaty of Paris which concluded the Crimean War (1856), the Black Sea was neutralized and the Russian and Ottoman governments agreed not to maintain naval bases nor any considerable naval forces in those waters. This condition, introduced on the initiative of Great Britain, clearly favoured the Turks and the protection of the Turkish Black Sea coasts: while Turkey retained naval forces available in the Sea of Marmara and the Mediterranean, the Russian Black Sea fleet ceased to exist and Sevastopol and Nikolayev were reduced to the status of commercial ports. The Treaty of Paris, indeed, substantially improved the security of Turkish territories in Asia (as well as in the Balkans). Without control of the Black Sea it was still possible for a Caucasian army to conquer Kars and even Erzurum; but the experience of Paskevich's campaigns (1828–9) had proved that lack of naval support for combined operations made it difficult for a Russian army to advance in the direction of Trebizond, while effective operations against Sivas and the interior of Asia Minor required the establishment of forward bases at Trebizond and Giresun (Kerasund). Similarly, without naval control of the Black Sea, combined operations against Batum were impracticable and in 1877–8, as in 1853–6, the Turks were able to maintain themselves in Batum in a position flanking Russian communications from Caucasia into the interior of Asia Minor.

Turkish Black Sea ports, maintained from the sea and covered on the landward side by the formidable range of the Pontic Alps, not only threatened the right flank of any Russian advance into Anatolia; a Russian advance in the general direction of the upper valleys of the Tigris and Euphrates through Diyarbekir towards Iraq was also impracticable so long as the Black Sea ports remained unconquered in the rear. British opinion, under the influence of Palmerston, had been nervous of Russian plans in the general direction of the Persian Gulf. This nervousness had been stimulated first by the predominance of Russian influence on the Black Sea and at Istanbul during the decade following the Treaty of Hunkâr Iskelesi (1833). As a check on Russian expansion in the Middle East, the naval clauses of the Treaty of Paris gave certain definite results, and British sacrifices in men and treasure during the Crimean War seemed to have had some justification. It was not long, however, before Russian policy reacted strongly to the restrictions imposed by the Treaty of Paris. Instead of the rather problematic danger of a Russian advance in the direction of Diyarbekir and Iraq, a more real threat was developed during the sixties in the Russian penetration of the independent khanates of central Asia and a general projection of Russian political and military influence towards Afghanistan.[1]

In 1871, Prince Gorchakov took advantage of the general disturbance of the European balance of power, following the fall of the Second French empire, to call a conference which met (in London) to agree to the abolition of the naval clauses of the Treaty of Paris. It then became clear that it was only a matter of time and opportunity before Russian expansion at the expense of the Ottoman empire would be resumed.

During the decade following the Treaty of Paris, the Russian government proceeded to liquidate the resistance of the Caucasian mountaineers—a process which had been delayed by the events of the Crimean War. As has been indicated, the existence of wide unconquered areas where militant tribes were free at any moment to attack Russian interior lines of communication had produced serious difficulties during 1853–6; about half the total Russian forces available in Caucasia had been detained on internal security duties, while Ömer Paşa's diversion in Mingrelia, although it had produced no important practical result, had greatly impressed the Russian general staff with the extent to which the Caucasian dominion was exposed to the threat of combined operations along the Black Sea coast.

In 1856, following the termination of the war, forces amounting to three armies were concentrated in Chechnia and Dağistan with the object of achieving the final reduction of the Murids. The popularity of Shamyl had declined as a consequence of his iron rule and of the

[1] For an interesting commentary on this process, see the article by Harold T. Cheshire, 'The Expansion of Imperial Russia to the Indian Ocean', in *Slavonic Review*, Aug. 1943. But cf. also A. Popov, 'Anglo-Russkoye sopernichestvo na putyakh Irana', in *N.V.* XII.

For a recent survey of Russo-British relations in central Asia during the nineteenth century, see Sir W. K. Fraser-Tytler, *Afghanistan: A study of political developments in central Asia*, Oxford, 1950; also review by W. E. D. Allen in *Asiatic Review*, July 1951.

losses and devastations of which the people of Chechnia and Dağistan had been the victims during the war which had now lasted for a quarter of a century. Nevertheless, the final campaign was not easy. General Evdokimov spent two years in the reduction of Chechnia, and it was only in the spring of 1859 that he occupied the stronghold of Veden which gives access from Chechnia to the valley of the Andi-köysu. Shamyl fled to his last stronghold, Gunib, above the gorge of the Kara-köysu. Only 500 Murids remained loyal to the old imam, but the new viceroy, Prince Baryatinski, concentrated 40,000 men with forty-eight guns to ensure the capture of a man who had proved himself to be the greatest guerrilla leader in the history of war.

With the exile of Shamyl to Kaluga the war against the Murids had at last been brought to an end. On the Black Sea coast, however, the Cherkesses still continued to maintain a precarious independence. Economic as well as military and political considerations imposed not only the reduction but the expatriation of a people who had attained probably the most 'civilized' tribal culture in Asia. In Chechnia and Dağistan the Russians were satisfied with the submission of the local population to Russian law, but on the Black Sea coast they intended to gain possession of the wide and fertile Cherkess lands to provide for a part of the great wave of Russian peasant migration which had been one of the results of the emancipation of the serfs in 1861. Every year Cossacks and peasant migrants from central Russia were penetrating farther and farther up the affluents of the Kuban, Laba, Belaya and Urup. These new Russian villages and Cossack *stanitsy* were subject to frequent raids by the Cherkesses who (not unnaturally) resented this settlement of territories which they regarded as their own tribal lands. A paradoxical situation was to be observed in the region of the lower Kuban about 1860, where, within thirty miles of the capital of the Kuban region, Ekaterinodar, at the source of the Afips the Cherkess Bzhedukhs were still living independent of Russian rule, while the neighbouring Natukhais had pretensions to the exclusive possession of all lands between the mouth of the Kuban and the new port of Novorossiysk. The Russian government decided to enforce the mass migration of the Cherkesses to other regions of the empire or, if they preferred it, to Turkey. General Evdokimov, who in Chechnia had proved himself a firm but moderate and efficient administrator, was entrusted with the execution of this drastic policy. He proceeded to form a number of mobile columns made up of Rifle battalions of the Black Sea line and Kuban plastuns (Cossack foot), supported by mounted Cossack *sotni*. During 1860, 4000 Cherkess families, without resistance, abandoned their lands along the left affluents of the Kuban and sailed to Turkey. Evdokimov, however, met a stubborn reaction from the Cherkess tribes farther to the south-east. Three of the more important of these—the Abadzekhs, the Shapsughs and the Ubykhs—formed an alliance and tried to create an organ of national government under the form of a great *Meclis* (or assembly) which sat in Sochi. The *Meclis* applied in vain for help from Turkey and Great Britain. In September 1861, the Emperor Alexander II, on a visit to Ekaterinodar, received a delegation of the Cherkess chieftains. The Abadzekhs were inclined to accept the lands offered them on the northern side of the Kuban[1] in exchange for their own, but the other tribes remained firm.

Military operations against the remaining Cherkess tribes were undertaken in the spring of 1862, but it was two years before a Russian column, marching from Maykop by the valley of the Pshish, penetrated to Tuapse, and achieved the conquest of the Shapsugh tribal lands. Another Russian column, moving up the valley of the Laba and across the watershed into the valley of the Mzymta, completed the pacification of the Ubykhs. The last shots were fired in the long history of Cherkess resistance in May 1864. More than 600,000 Cherkesses migrated to Turkey, where their descendants are to be met with in scattered settlements from the Balkans to Transjordan. The Kuban region was thickly settled by Russian immigrants; but along the Black Sea coast the numerous Russian, German, Greek and Bulgarian colonists who were endowed with the tribal lands of the Cherkesses proved unable to support the humid climate and the forest environment, and to-day the wilderness has conquered the wide acres of orchards and gardens once cultivated by prosperous Cherkess communities, while the *dachi* and proletarian institutions of the 'Red Riviera' occupy the sites of the former coastal villages.

The reign of Alexander II (1855–81), which began with the emancipation of the serfs (1861), was a period of rapid economic development and administrative reforms; these reforms included the complete reorganization of the Russian armed forces. As a result of the lessons of the Franco-Prussian War, conscription was introduced in 1874 and a great expansion in the numbers of the Russian army was undertaken. By 1877 the establishment comprised, in addition to

[1] Their descendants are to be found to-day in the Cherkess-Adyghe autonomous district along the middle Kuban in the region of Krasnodar (Ekaterinodar). Cf. *Istoriko-ekonomicheski obzor Adygeyskoy Avtonomnoy Oblasti* (Krasnodar, 1923) and *Sovetskaya Adygeya* (Krasnodar, 1925).

the three guards divisions and three grenadier divisions and several rifle brigades, forty-one divisions of infantry of the line. Four of the newly formed divisions (38th, 39th, 40th and 41st) comprised regiments with Caucasian names and they were stationed in Caucasia, where the old Caucasian army had a standing establishment: the Caucasian Grenadier Division, and the 19th, 20th and 21st Infantry Divisions of the line. (The 40th Division remained on the lower Volga, from where it could reinforce the Caucasian army at short notice.) The strength of the regiments comprising the Caucasian divisions was four battalions each, instead of the three battalions which was the normal strength in other Russian divisions (with the exception of the Imperial Guard). Four Caucasian rifle battalions were also created.

The Franco-Prussian War had destroyed the delicate balance of power established by the Treaty of Paris. At the same time the growth of the Pan-Slav movement in Russia was a phenomenon comparable to the contemporary nationalist movements in Germany and Italy. The outbreak of insurrections in Bosnia and Hercegovina in 1875, and the Serbo-Turkish War of 1876 stimulated Pan-Slavist sentiment in Russia to fever point, and the complicated diplomatic manoeuvres which followed failed to prevent the outbreak of war between Russia and Turkey in April 1877.

The experience of Paskevich's campaigns and the easy conquest of Anatolia by the Egyptian army of Mehmet Ali during the thirties had indicated that the Ottoman empire was probably most vulnerable on its Asiatic frontiers. On the other hand, the concentration of Russian sentiment on the Balkan Slavs imposed on the Russian general staff the necessity of treating the Balkans as a major theatre of war. Further, the weakness of Russian naval forces in the Black Sea influenced the Russian general staff to direct their main offensive as far as possible to the west.

Six years had passed since the London Conference of 1871, and the Russians had not been slow to re-establish their positions on the Black Sea. Fortifications and heavy batteries defended Kerch, Sevastopol, the mouth of the Dniepr and Odessa. These fortifications were reinforced by minefields. But the actual naval force of the Russians on the Black Sea was limited to four armed steamers. The Turks, therefore, with more than twenty armoured monitors and gun-boats, mostly built in British yards, and commanded by Admiral Hobart (ex-Captain R.N.) and other British and foreign officers, completely dominated the Black Sea. The possibilities of Turkish naval action on the Caucasian coast and at the mouths of the Danube and along the lower Danube were not only fully appreciated but rather overestimated by the Russians. As a result, Russian strategy in the Balkans avoided the lower Danube and the quadrilateral of Turkish fortresses which could be supplied and reinforced from the Black Sea, and was directed towards Trnovo and Sofia and the passes of the western Balkans. Equally Russian strategy in the Caucasus was influenced to overcaution by the fear of combined operations against the Caucasian coast under the protection of the powerful Turkish fleet.

Russian communications in Transcaucasia had been improved since the Crimean War by the construction of a railway from Poti to Tiflis, but Turkish command of the sea made communication between the Russian Black Sea ports and Poti impracticable, and the Caucasian army remained dependent on the Georgian Military Road from Vladikavkaz to Tiflis and on the sea route from Astrakhan to Baku.

Fear of Turkish naval diversions was combined, in spite of the recent forced migration of the Cherkesses, with an exaggerated nervousness of insurgent action among the mountain tribes.

The peace stations of the Caucasian army were as follows:

Station	Unit
Northern Caucasus	19th, 20th, 21st and 38th Divisions
Lower Volga	40th Division
Eastern Georgia	Caucasian Grenadier Division
Western Georgia	41st Division
Akhaltzikhe	39th Division

The famous Caucasian Dragoon Division and ten Cossack regiments were also stationed in Transcaucasia. During the course of 1876 the 19th Infantry Division was transferred from the northern Caucasus to the Erevan district and, at the beginning of 1877, it was decided to reinforce the troops in Transcaucasia with the 38th Division and the rifle battalions. Thus approximately half the Russian forces available in Caucasia were retained on internal security and coastal defence duties—an interesting illustration of the extent to which the mere threat of combined operations or insurgent action can immobilize substantial proportions of a defending army.

In spite of the astonishing weakness of Turkish forces in Armenia in the spring of 1877, concentration on the European theatre of war and the predominance of the Turkish fleet in the Black Sea inhibited the Caucasian command from any serious offensive action. An active advance was proposed if circumstances should prove favourable to it; this active advance was limited to action against Batum and Ardahan—where work on the new Turkish fortifications was known to be incomplete—and to the blockade of Kars.

The Caucasian command was singularly lacking in strong personalities like Paskevich, Muravyev and Bebutov who had made a success of previous campaigns. The viceroy was the Grand Duke Michael Nikolayevich, a brother of the tsar, extremely benevolent and well liked but quite lacking in military capacity. His principal adviser was General Prince Svyatopolk Mirski, an officer who lacked both energy and military talent. General Count Loris Melikov was in command in the field. A distinguished and enlightened soldier (who subsequently became the last chancellor of Alexander II), he was an Armenian by birth with a profound knowledge of conditions in Caucasia, but he lacked initiative and the power of decision. In Tiflis during the winter of 1876–7 an atmosphere of reserve and caution predominated. This atmosphere tended to an exaggeration of the strength of the 'new' Turkish army.

During the two decades which intervened between the Crimean War and the Near Eastern crisis of 1875–7, military reforms in Turkey had been proceeding at a fair *tempo*. Prussian influence, dating from the military mission of von Moltke, had been strengthened by the prestige of Prussian victories in western Europe, and German advisers and German theories were replacing the British advisers and Polish and Hungarian *émigré* officers of the period of the Crimean War.

The population of the Ottoman empire in 1876 was approximately 22,000,000. Of these only the Muslims (12,000,000 in Asia and 4,000,000 in Europe) were, in fact, liable for military service. Under the conscription laws of 1869–76 a man was liable for four years' service in the nizams (regular army); for a further two years he was considered to be on leave and he then passed to the redifs (reserve), where he was liable for service for a period of another nine years. The Asiatic provinces provided the bulk of the man-power, but the centre of gravity of the distribution of forces lay in Europe. The European territories were divided into army corps districts, I, II and III. Erzurum was the headquarters of army corps district IV and Damascus of army corps district V. The organization of army corps districts VI and VII in Iraq and Arabia was incomplete. A special coastal district had been organized to cover ports between the Russian frontier and Samsun.

The system of army corps districts was copied from the Prussians, but instead of divisional subdistricts as in Prussia, regimental subdistricts had been introduced in Turkey. Army corps district IV consisted of six regimental subdistricts and the coastal district of three regimental subdistricts. On peace establishment, army corps district IV (Erzurum) consisted of six regiments of three battalions each and six rifle battalions. Each regimental subdistrict was supposed to mobilize twelve battalions of redifs of category 1, and twenty-five battalions of redifs of category 2 (the latter, the *müstahfiz*, equivalent to *Heimwehr*). Thus the nominal strength of army corps district IV (Erzurum) was 24 nizam battalions and 72 redif battalions (category 1) and 150 redif battalions (category 2), i.e. 246 battalions. The coastal district had no nizam establishment but was expected to mobilize 100 battalions of redifs. The total potential thus represented about 350 battalions, or approximately 250,000 men. As reinforcements from army corps district V (Damascus) might be expected, the total strength of the Turkish army on the Caucasian front should have attained, in theory, some 300,000 men with 50,000 mounted irregulars. Such were the official figures which circulated in Erzurum and which were evidently well known in Tiflis where they produced a great impression. The Russian intelligence service, strangely enough, took quite seriously a purely theoretical calculation of the Turkish military potential. In fact, when war broke out, the total force available in Armenia, scattered in small detachments over a wide area, amounted to about 15,000 nizams, 50,000 redifs and some 25,000 irregulars, a total of 90,000, about 25 % of the supposed figure.

Unlike Muravyev with his acute sense of Oriental psychology, the Russian commanders of 1877 appear to have taken little account of the obvious and traditional defects of Turkish organization. Great stores of food and clothes had been concentrated in Erzurum and Trebizond but distribution was very defective; transport and animals were almost completely lacking; medical services—in a long view as important as supply—were, and continued to be, virtually non-existent.[1] Artillery was seriously below strength and guns, provided under war establishment, were scattered without transport between Sivas, Trebizond and Erzurum.

The officer commanding army corps district IV was Ismail Hakki Paşa, an influential chieftain of the Haydaranli Kurds, who was as incompetent as he was corrupt, but who was supported by powerful influences at court which protected him from removal even when Ahmet Muhtar Paşa arrived in April 1877 to take command of the forces in Armenia. Only thirty-eight years old, Ahmet Muhtar was typical of the new professional officer who had been produced by the reforms of the last two decades. The son of an old Turkish family of

[1] See C. B. Norman, *The Times* correspondent; and reports of H.M. consul at Erzurum.

Bursa, he had served with some distinction in the Yemen and in Bosnia. He was to prove an energetic and courageous commander with a strategic sense superior to that of his opponent, Loris Melikov, but he lacked tactical adroitness and officers capable of supplying this deficiency.

The Turkish army was well armed; the infantry was provided with modern Martini-Peabody rifles which were superior to the Russian berdans (*berdyanki*). The Turks had also purchased numbers of the new steel guns manufactured by Krupps, while the Russians were still armed with the older model brass guns. The greatest weakness of the Turkish army was the lack of competent trained officers, in which respect the Russian army was infinitely superior. The weakness of the Turkish staff was notorious, and the few capable officers were in most cases foreign soldiers of fortune, who were the victims of continual obstruction and jealousy. This inadequacy in the higher ranks of the Turkish army explains the inability to create large field units (divisions and brigades). The *tabor* (battalion) and the battery were the only effective field units; brigades were sometimes created but they were generally only temporary formations of six to eight battalions improvised for a particular action.

The partial mobilization of the redifs began in the autumn of 1876. The Russian Intelligence was so badly served that nobody was aware in Tiflis that the first two complete corps of redifs mobilized in army corps district IV (forty-eight battalions) had been sent from Erzurum to Europe. When Ahmet Muhtar visited Kars (18 April 1877) he found that the garrison consisted only of eighteen battalions, to which he added three which he had brought with him from Erzurum.

At Ardahan there were eight battalions to which Ahmet Muhtar was only able to add two. There were three battalions at Hasankale covering the main road along the Aras valley, and in the Eleşkirt valley eleven battalions were completing mobilization. Out of the theoretical army of 350,000 men, Ahmet Muhtar only had available as a mobile field force nine battalions, six squadrons of cavalry and one battery. Other forces (not under Muhtar's command) were available in the coastal area for the defence of Batum, where Hüseyin Paşa had about 12,000 men. Efforts were being made to raise the Kurds in the regions of Erzurum, Muş and Van, and Ismail Hakki boasted that he was preparing to invade Transcaucasia with 40,000 horsemen.

The Turks remained obsessed with the importance of Kars and Ardahan for the defence of the Armenian highlands; the fortifications of both strongholds had been improved by German engineers, and modern long-range positional artillery had been procured from Krupps. The Hungarian veteran of the campaigns of 1853–6, Kolman (Feyzi Paşa), had prepared excellent defensive lines covering Batum; and a *chaussée* connecting that port with Ardahan and the upper Kura valley had been begun but remained unfinished thirty to forty miles south-west of Batum. In fact, the function of both Kars and Ardahan was that of advance bases for the invasion of Transcaucasia, and their utility in a purely defensive strategy was only that of strong points which might delay the advance of the enemy and force detachment of part of his forces. The real defence of the Armenian highlands and Erzurum lay in the complex of mountains known in different sections as the Soğanli-dağ and the Deve-boyun, and the coming campaign was to prove that the Turks, with their weaker forces and limited capacity for organization, were mistaken in concentrating on the defence of Kars and in neglecting the natural mountain defences on which their main forces might have been based to greater advantage.

Three years before the outbreak of the war, the Russian general staff had drawn up plans for a rapid invasion of Asia Minor with the object of occupying Erzurum within six weeks from crossing the frontier and thus making impossible the organization of any large Turkish army within striking distance of the Transcaucasian frontier. But this bold conception of the defensive offensive was abandoned by the Grand Duke Michael in favour of a cautious policy which retained two divisions on internal security duties in north Caucasia (20th and 21st) and one on coastal defence (41st). At the last moment the threat of troubles in Chechnia and Dağistan decided him to detain there also the 38th division. Thus only 3½ divisions remained available for offensive operations. These forces were disposed in three principal strategical directions, and each group appeared too weak for any prompt or decisive action. In the centre at Aleksandropol were 1½ divisions—one of grenadiers and a half of the 19th. The other half of the 19th Division was on the extreme left at Erevan covering the Aras valley. Half the 38th Division was at Akhalkalaki opposite the Turkish force at Ardahan. The doubts of the Russian general head-quarters were expressed in the creation of a strong reserve round Tiflis—a half of the 19th and a half of the 38th. At the beginning of 1877 this reserve was suddenly moved in an unexpected direction, four battalions to Akhaltzikhe and twelve battalions to the Rion valley, where the 41st Division, reinforced by three rifle battalions, was already in station. Thus in the Rion valley and western Caucasia there were the equivalent of three divisions at the outbreak of war—numbers in excess of those of the main striking force at Aleksandropol. The Russian staff, which had already overestimated the danger of Muslim revolt in Dağistan, was now evidently preoccupied with the phantom of Ömer Paşa's invasion of Mingrelia and the possibility of Turkish combined operations against the Caucasian coast. The mere threat of these diversions had rather more than halved the forces available for offensive action against the fortresses of Asia Minor.

The decision to concentrate towards the coast and in the western regions, and the necessity, at the same time, of taking some action to satisfy the warlike enthusiasm of the troops themselves, dictated the strategy of the Russian staff which decided to concentrate on two limited objectives in proximity to the area of defensive concentration. The original conception of an advance on Erzurum and a smashing blow against the Turkish armies, before they had had time to concentrate and organize, was now abandoned in favour of reconnaissances in force against Ardahan and Batum, while on the centre and left—the most promising operational areas in the experience of earlier campaigns and the only areas in which a decision could be expected—action was limited to the observation of Kars and a demonstration against Bayazit. The fear of an internal front and the exaggeration of the danger to be expected from Turkish naval action had, in fact, imposed a reversal of Russian strategy and abandonment of the only logical plan of decisive action against Erzurum.

War was declared on 24 April 1877, and on the same day the Russian troops crossed the frontier. The Russian order of battle was as follows:

Group	Commander	Strength	Remarks
Aleksandropol	Gen. Heimann	24 battns infantry 2 dragoon regts 2 Terek Cossack regts 2 regts Dağistan horse 9 field batteries, 2½ horse batteries	Tasks: observation of Kars; reconnaissance in force against Ardahan Total: 18,000 infantry, 7000 cavalry, 92 guns
Akhaltzikhe Akhalkalaki	Gen. Devel	8 battns infantry 2 regts Kuban Cossacks 3½ batteries	Task: reconnaissance in force against Ardahan Total: 6500 infantry, 2500 cavalry, 28 guns
Erevan-Kulp	Gen. Tergukasov	9 battns infantry 1 regt dragoons 4 regts Cossacks 3 field batteries 1 mounted battery	Task: reconnaissance in force against Bayazit Total: 7000 infantry, 3500 cavalry, 32 guns
Kobuleti	Gen. Oklobzhio	16 battns infantry Some Cossacks 6 batteries	Task: reconnaissance in force against Batum Total: 11,000 infantry, 1000 cavalry, 48 guns

When Ahmet Muhtar received information that Russian forces had crossed the frontier he immediately ordered his modest field forces at Subatan and Haciveli to withdraw on Kars. On 28 April the Russians reached Poldervan and Kurudere. The same night Muhtar, leaving twenty-one battalions to hold the fortress of Kars, withdrew his field force (nine battalions, two *süvari* regiments and one battery) in the direction of the Soğanli-dağ. His decision to leave Kars to withstand a blockade and to use his limited field force to cover the concentration of Turkish reinforcements at Erzurum was the reverse of that taken by Fenwick Williams in 1855, but in both cases the decisions were, in the different circumstances, correct. By the evening of 30 April Muhtar's troops, whose retreat might have been endangered by the powerful Russian cavalry, were already in position in the Soğanli passes.

On 2 May the Aleksandropol force was concentrated at Zaim where the road from Aleksandropol to Kars meets that coming from Ardahan. Devel, advancing from Akhalkalaki in the direction of Ardahan, had information from Armenian citizens of that town that the garrison was by no means prepared for immediate hostilities.

Ardahan is situated on both banks of the Kura with the principal part of the town on the left (north-west) bank. The river which is here narrow but rapid and coursing between steep banks, runs through the town in a direction west-east, but on leaving the town turns north and is overlooked along its eastern bank by the heights of Guylaberti. Sloping from the western bank to the north of the town, a plain rises to an isolated and rather steep hill on which had been built the modern fort of Ramazan Tabia. Another fort, Kaya Tabia, covered the left bank of the river opposite the Guylaberti heights and overlooking the road to Akhaltzikhe. The Turks clearly considered this sector to be the most dangerous, although the best line of approach for an attacking force proved to be from the south-east. The Guylaberti heights above the right bank overlooked Kaya Tabia, while from the south the slopes of Alagöz mountain offered a series of commanding positions within reach of the defences along the southern sector. The heights on which stood the villages of Gürcik and Tain-kilise were only two miles from the Turkish fortifications. In this sector the fort of Emir-oğlu had been built to strengthen the line but it was unfinished. Of ninety guns in position only half were modern; the garrison was adequate enough: two battalions of nizams, nine battalions of redifs, one *süvari* brigade and three field batteries and 1500 irregulars. The commandant, Hasan Sabri Paşa, was extremely inefficient, and the only competent officer at his disposal was Colonel Mehmet Bey, a German who commanded the sector from the Guylaberti heights to Emir-oğlu Tabia.

On 11 May, General Heimann, with a grenadier brigade and a brigade of cavalry and five field batteries, joined forces with Devel, while Loris Melikov arrived to take command. Loris Melikov concentrated his artillery and his main infantry force on the heights of Gürcik and Tain-kilise on either side of the Kars road. From the north Devel, with only six battalions, was deputed to storm the Guylaberti heights.

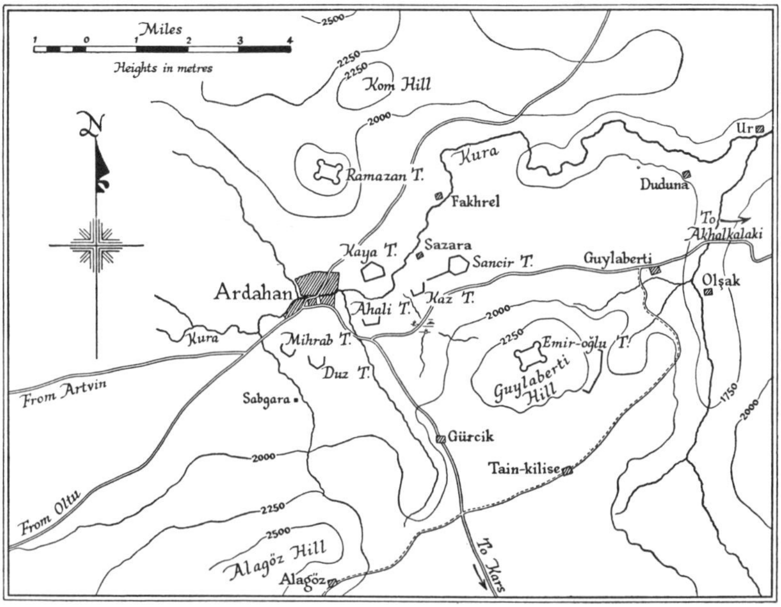

Map 7. The Ardahan operation, 1877

A general attack was fixed for dawn on 16 May, and Devel sent three battalions of the Elizavetpol Regiment against the Guylaberti heights before the guns had opened fire against the south-eastern sector. The Turkish fire control proved hopelessly inefficient, and the Elizavetpol men, with the advantage of some dead ground, stormed the trenches with only moderate losses. Their further advance, however, was held by the fire from Emir-oğlu Tabia. Meanwhile the Russian batteries on Tain-kilise kept up a bombardment which, although not very effective, demoralized the untrained Turkish troops who began to abandon Emir-oğlu and run in disorder towards the river. Loris Melikov then launched his infantry attack, but the

Elizavetpolskis with flags flying and band playing were already storming into the fort from the Guylaberti heights.

Loris Melikov now displayed those characteristics which were to mark his actions during the rest of the campaign. Cautious and systematic, he failed to exploit the initial success and adhered to his plan of an assault on the following day. Forces were regrouped and Devel passed to the left bank of the Kura to observe the fortress from the north; troops were concentrated on the Guylaberti heights for a main attack against the two smaller forts of Sancir and Ahali which covered the approaches of the town from the east. The Turkish forts were strangely silent but the Russian commander was impressed by the strength of the fortifications and showed himself still doubtful of rapid success. He even decided to defer the storm for three or four days to await the arrival of reinforcements from Akhaltzikhe and of siege guns from Akhalkalaki.

The Turks in the meantime were panic-stricken; and Hasan Sabri, the commandant, after having upbraided the German, Mehmet Bey, for the loss of the Guylaberti positions, on the night of 16–17 May withdrew from the fortress marching along the Batum road with two battalions of nizams, all the regular cavalry and such field pieces as he could move. The irregulars deserted and Mehmet was left to hold the fortress with eight battalions of redifs and the positional guns. The Russian patrols failed to observe the withdrawal of Sabri, but Armenians on the following morning carried the news to the Russian camp.

Loris Melikov still hesitated to order a premature storm and he only yielded to the insistence of General Heimann in time to order an attack on Sancir and Ahali forts about 6 o'clock of the evening. The grenadier battalions easily carried the two positions, and the Turks poured in disorder over the two bridges into the town. Meantime Loris Melikov's staff had neglected to warn Devel of the decision to attack, and the two forts on the left bank of the Kura, Ramazan and Kaya, were evacuated by the Turks without any attempt on the part of the Russians to capture them.

All the fortress guns and 1000 prisoners fell into the hands of the Russians; the Turks lost 2000 men in the two days' fighting, the Russians 500, mostly among the Elizavetpol battalions. Melikov was as shrewd as a politician and administrator as he was over-cautious as a commander; he sent all the redif prisoners to their homes with gifts of corn seed for the spring sowing.

The Ardahan operation was remarkable for the ineptitude of the command on either side. The Turkish regimental officers, led, it is true, by an incompetent coward, had shown themselves lacking in such elementary training as fire-control, while the Russian timing of tactical movements had proved very faulty.

The operation had important psychological effects, not altogether fortunate for either side. The Turks in spite of their new and expensive weapons were discouraged, and their sense of inferiority in the face of Russian troops was increased. On the other hand, the easy success at Ardahan had encouraged officers like Heimann of the old Caucasian school, who continued to ignore the significance of the increased fire-power of new weapons and adhered to their old faith in the bayonet.

For the Turks the defeat was ominous of the disastrous results which flowed from the interference of court influence in the appointment of general officers. Even Hasan Sabri, protected in Istanbul, escaped any serious punishment. Meanwhile men of like kidney in Erzurum were infected with his panic and were only checked by Ahmet Muhtar from ordering the evacuation of that city. European observers were impressed by the contempt in which Turkish troops and junior officers held their military leaders at this time.

The Turks were fortunate only in the personality of Ahmet Muhtar, who proved to be a leader of character and of strategic insight. The news of the fall of Ardahan reached him about the same time as information that the Erevan force was threatening the defences of the Erzurum region from the south-east. Tergukasov had continued his advance to the west: on 11 May he was at the Armenian monastery of Surp Ohanes[1] at the entrance into the Eleşkirt plain. Muhtar Paşa quite reasonably supposed that Tergukasov's advance and the occupation of Ardahan were both parts of a general Russian operational plan directed towards the capture of Erzurum. In view of the slow course of the organization of the Turkish field army, the situation seemed very serious to the Turkish commander. He was first concerned for the defence of the direct ways of access from Ardahan to Erzurum, and he sent three battalions with a battery to create a defensive position in the Gürcü-boğaz, the historic Georgian Gate, which gives access from the upper valley of the Kura by the valley of the Oltu-çay to the Erzurum plain and makes it possible to turn both the Soğanli and Deve-boyun positions. A 'brigade' of six to eight battalions and one battery was moved to Oltu with a view to delaying the expected Russian advance from Ardahan. Muhtar established his

[1] In 1945 this magnificent building still remained standing, although in a ruinous condition.

main force (only twelve battalions and two batteries) in the western part of the Soğanli on the Çakir-baba ridge over against Bardiz, from where connexion might be established with the detachment at Oltu. Two battalions were stationed at Köprüköy as liaison with the eight battalions and two batteries which had been sent to hold Tergukasov's advance from the Eleşkirt valley. Muhtar's total available forces amounted to 18,000 infantry and thirty-six guns. His cavalry was very weak, mostly comprised of Kurd and Cherkess irregulars.

Fortunately for Ahmet Muhtar, his Russian opponent had not seen the strategical situation so clearly. Following the capture of Ardahan, Loris Melikov returned to the neighbourhood of Kars and began to concentrate at Zaim the bulk of both Heimann's and Devel's forces. Colonel Komarov was left to hold Ardahan, and, although he was 'allowed' to advance to Oltu, he was not told to do so and to do so quickly. On 3 June Komarov occupied Penek with four battalions and two regiments of Cossacks, and on the 4th he reached Oltu ahead of the troops sent to hold it by Muhtar. On the 5th Muhtar, sensible of the importance of this highland village, sent orders to attack, and the Russians withdrew without resistance. Komarov had, in fact, on the previous night not only received orders from Melikov to return to Ardahan but also to detach four of his six battalions to the camp at Zaim. On 4 June the Russian commander-in-chief had at last taken a decision—and this decision was to be attended by fatal results during the following three months.

Loris Melikov had received reports of the course of events in western Transcaucasia which caused him some alarm. Oklobzhio had, in his advance on Batum, encountered unexpected difficulties of terrain—difficulties which seemed to have been ignored by the topographical section of the general staff at Tiflis. There was also news that Sukhum had been evacuated and that Turkish troops, supported by Abkhazian insurgents, were being landed at points along the Black Sea coast. The Kutaisi reserve on which Oklobzhio depended was sent towards the Ingur. Melikov did not command in western Transcaucasia and he was not in a position to assess the situation. He accordingly awaited the arrival of the Grand Duke Michael in Aleksandropol. When after some delay the viceroy arrived, Melikov learnt that in Tiflis the position was regarded as grave, and that the rising in Abkhazia had already had dangerous repercussions in Chechnia and Dağistan. The advisers of His Imperial Highness found it necessary to hold in northern Caucasia the reinforcements promised for Transcaucasia (one brigade of the 38th Division). It was considered that Oklobzhio's operation against Batum should be held up until the position had been re-established along the Caucasian coast and that Melikov should undertake no major operation which would remove the bulk of his forces from the neighbourhood of the Russian frontier. Melikov was ordered to limit himself to the attempt to reduce Kars by artillery. He had available only one full infantry division and one cavalry division to protect his batteries, and he realized well enough that Muhtar's forces behind the Soğanli were growing in strength from day to day, and that in a few weeks the Turks would be able to move to the relief of Kars with a relatively powerful field force. Melikov therefore sent to Tergukasov an order to continue his advance, for he felt that the threat to the Pasin valley would divert the attention of the Turks from the forces round Kars. It was an inconsiderate order coming somewhat strangely from so cautious a commander as Melikov, and it proved very dangerous for the modest force of Tergukasov.

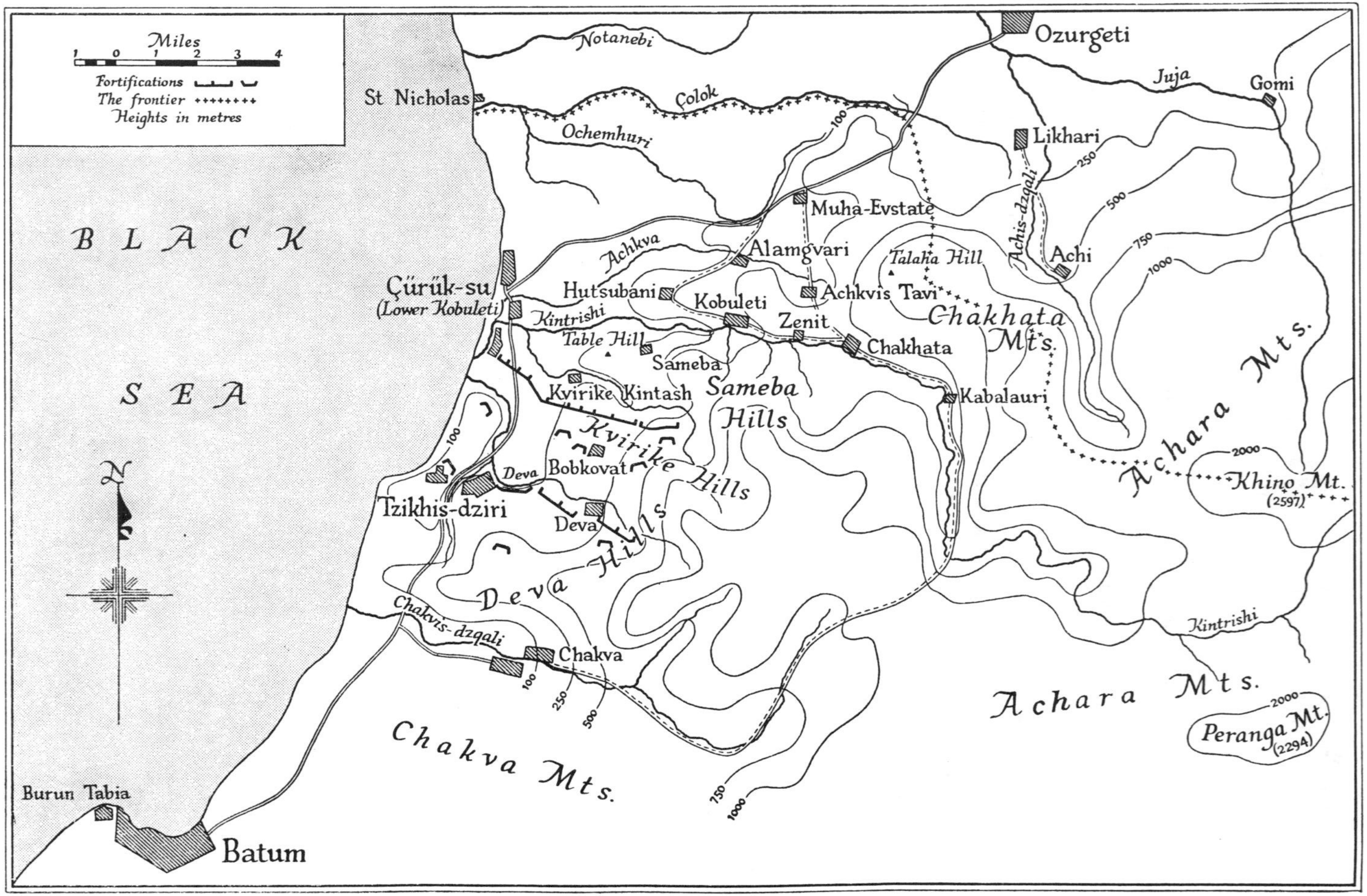

Map 8. The Batum operations, 1877–8

CHAPTER VIII

THE BLACK SEA COAST. OPERATION AGAINST BATUM AND TURKISH DIVERSION IN ABKHAZIA

MAPS 1, 4, 8, 9

THE port and fortress of Batum occupy a very strong natural position. Surrounded on the east and north-east by the precipitous forested mountains of Acaristan (G. *Achara*), Batum is approachable only from the north along the narrow foreshore of the Black Sea. This foreshore, nowhere more than fifteen miles in width, is intersected by numerous streams. The intervening ground is marshy and in many places thickly wooded. The natural difficulties of the approach were in 1877 increased by the warlike character of the inhabitants of the adjoining mountains. The Acars, who speak a Georgian dialect resembling that of the neighbouring Gurians and Mingrelians, had been converted to Islam during the latter part of the sixteenth-century and, like the Bosniaks in the Balkans, they remained fanatically loyal to the sultan-caliph. During Paskevich's campaigns the Acars had constituted a formidable irregular force which proved to be the main support of the Turks on the Black Sea coast and round Akhaltzikhe. It is curious that the Russian general staff, which had so overestimated the dangers of action by the Muslim tribes in Dağistan, had taken no account of the Acars.[1] Some attempt was made by Russian intelligence officers to negotiate with the elders of Kobuleti but without result; and in April 2000 Kobuletians and 1000 Acars, attracted by the free distribution of new Martini-Peabody rifles, joined the Turkish garrison at Batum. This reinforcement to Hüseyin Paşa was appreciable, since he had available for the defence of Batum only two battalions of nizams and four of redifs. During the first week of the war 6000 Laz irregulars joined him from Trebizond and his force reached 12,000 men. If three-quarters of these men were not disciplined troops they were excellent sharpshooters, and the peculiarities of the terrain round Batum favoured the development of irregular tactics.

[1] Before the outbreak of war D. Kazbeg and D. Bakradze had made detailed studies of Guria and Achara (see Bibliography, I (*e*) *Regional*). Possibly their reports were overlooked by the general staff of the Caucasian army.

A Russian advance against Batum was only possible along a narrow front between the sea and the mountainous massif of Chakhata. On the left of the Russian advance the massif itself was inaccessible, and on the right, along the seashore, the Russians came under the fire of Turkish warships.

The massif of Chakhata, about 1000 m. in height, gives rise to a fan of streams—the Çolok, the Ochemhuri and the Achkva—which are easily swollen by the frequent rains. Their narrow, deep ravines are overgrown with dense subtropical vegetation, and lianas and thorn plants are a serious obstacle to advancing troops. The heights between the rivers are easily transformed into defensive positions. The Muha-Evstate[1] position between the upper waters of the Ochemhuri and the Achkva had more than once been defended by the Turks in previous wars.

To the south of the Achkva, the Hutsubani heights offer an even stronger alternative position. To the south of Hutsubani the considerable stream of the Kintrishi runs through a deep ravine with steep banks. The Kintrishi is flanked to the east by the slopes of the high and rock-strewn Peranga massif which finds its twin peaks in Khino (2597 m.) and Peranga (2294 m.). The slopes of the Sameba hills (G., 'Trinity'), between the Kintrishi and its affluent, the Kintash, form natural defensive terraces.

Another offshoot of the Peranga ridge, the Kvirike heights, forms a second rocky terrace, and to the south-west of the Deva stream a third terrace runs almost to the seashore, falling in perpendicular cliffs to the beach. At this point stood the old frontier fort of Tzikhis-dziri (G., 'Bottom fort').

All the heights to the south of the Kintrishi, and particularly the cliff-like terrace above the Deva, had been transformed by the Turks, under the supervision of the old Hungarian *émigré*, Feyzi Paşa, into fortified lines of trenches covered by well-placed batteries, some of which, like those on the Tzikhis-dziri heights, had been armed with modern heavy Krupp guns.

The Batum area represented a really formidable defensive complex, certainly incomparably stronger than the Plevna defences improvised some months later by Osman Paşa in Bulgaria.

The weakness of the defence lay in the lack of a sufficiently strong garrison but Hüseyin Paşa was awaiting the arrival of from twelve to twenty-four redif battalions mobilized in the Black Sea coastal district, and he could rely on the support of the new Turkish ironclads.

[1] *Muha-Evstate*, 'Oak of St Eustace', patron saint of the Gurians.

Very sensibly he concentrated behind the line of the Kintrishi and disposed his irregulars to the north of that river with the object of fighting a delaying action. General Oklobzhio had some 12,000 troops and forty-eight guns for active operations, while a further 7000 with sixteen guns were held in reserve at Ozurgeti and Kutaisi. Oklobzhio, whose strategy was strictly dictated by the narrow terrain, ordered his force in two columns. On the right, General Shelemetev with the 163rd Regiment, the Imeretian militia and four batteries, was ordered to move along the coast road where the flat and relatively open terrain might allow him to deploy his guns. The stronger left column, under General Denibekov, composed of the 164th Regiment, the 1st and 2nd battalions of the Caucasian Rifles, one Kuban plastun battalion, the Gurian militia and two batteries, followed a very bad road which wound up from the frontier stream of the Çolok through the wooded hills of the upper Kintrishi. The Caucasian Rifles and the Kuban battalions had been specially trained for forest warfare, but the 163rd (Lenkoran) and 164th (Zakatali) Regiments were newly raised and belonged to the recently formed 41st Division. On 25 April the troops passed the frontier under a downpour of spring rain, and on 27 April, after some skirmishing with Acar bands, they crossed the Ochemhuri with the loss of 100 men and occupied the Muha-Evstate heights. Here the left column was forced to halt, with the object of preparing the road and constructing bridges and bringing up supplies from Ozurgeti. A decision had suddenly been taken in Tiflis to strengthen Oklobzhio, but no provision had been made for the supply of a larger force. The Russians were, therefore, immobilized on the Muha-Evstate heights for about a fortnight—a delay which did not help to raise their prestige among the Kobuletians and Acars who gathered in growing numbers to oppose a further advance beyond the Achkva.

The march was renewed on 11 May. The passage of the Achkva was not an easy task, since the upper course of this river runs through a narrow ravine with a very steep left bank, and the lower course was covered with dense forest reaching almost to the seashore. Under these conditions of terrain Oklobzhio could only deploy relatively small forces. Beyond the Achkva some 4000 irregular riflemen awaited the Russians; the Acars showed great tactical skill, and single riflemen were scattered in depth over a wide area. On the left Denibekov advanced with difficulty, and after heavy losses in men and horses he halted. To cover his right flank Oklobzhio moved up one battalion of the 75th (Sevastopol) Regiment. The forest had to be cleared step by step at the point of the bayonet. The passage of the river with its steep left bank proved even more difficult. But by mid-day the Acar irregulars began to leave the heights of Hutsubani and withdraw beyond the Kintrishi. It was not, however, until 20 May that the Russians had finally cleaned up the Hutsubani area. It was nearly four weeks since Oklobzhio had crossed the frontier, and his advance had been entirely held by the irregulars who had not received the support of any regular Turkish troops or guns. The Russians were only now, after a month, approaching the first line of the Turkish fortified positions, and Oklobzhio reported to Tiflis that the passage of the Kintrishi would be a difficult task and would require some time for preparation. At this moment the attention of Russian general headquarters became absorbed by events farther north along the Black Sea coast. The misgivings of those who had emphasized the danger of Turkish naval action now seemed to be justified.

As early as 26 April three Turkish monitors had demonstrated against the Russian frontier post of St Nicholas (at the mouth of the Çolok). The Turkish ships next appeared off Poti but they did not approach the port which was defended by twenty-four guns. The same three monitors participated in the fighting of 11 May for the Hutsubani heights. On that day it was reported that a squadron of six ironclads had been observed off Ochemchiri steaming northward. On 12 May the Turks landed troops at Gudauti to the east of Pitsunda Point, some distance to the northward of Sukhum. This was the beginning of the Turkish diversion along the Abkhazian coast which produced such a vivid impression in Tiflis and which induced the revision of Russian plans. In spite of the expatriation of the Cherkess population, the Russians had reason to fear serious action, and they anticipated that the Turks would not only land large numbers of those Cherkesses who had taken refuge in Turkey, but that they would also make an effort to stimulate and support insurrection in Abkhazia and the eastern Caucasus. Musa Kundukov had formed in Turkey twenty-eight squadrons of Cherkess horse; but these irregulars proved relatively useless and quite failed to maintain the reputation which they had won fighting against the Russians in their own country. It was decided, therefore, in Istanbul to limit the diversion to the Abkhazian coast (to the east of Pitsunda Point) and to direct the principal effort to Sukhum as had been done in the last year of the Crimean War. Although the Russians had not developed an official policy of expatriating the Abkhazians, several thousand Mussulman Abkhazians (*muhacirler*) had fled to Turkey at the time of the Cherkess emigration. Presently between 2000 and 3000 Abkhazians and Cherkesses had been collected at Trebizond and later embarked on Turkish transports in company with Turkish regular troops. The transports carried also 30,000 Snyder rifles for the arming of the population. Turkish regular forces, made available for the Abkhazian expedition, amounted to ten to twelve battalions and four batteries under the command of Fazil Paşa. The expedition sailed in two groups: an advance force of three monitors and three steam frigates, followed in a few days by the main force of four large transports protected by four ironclads. At Gudauti on 12 May the Turks landed only about

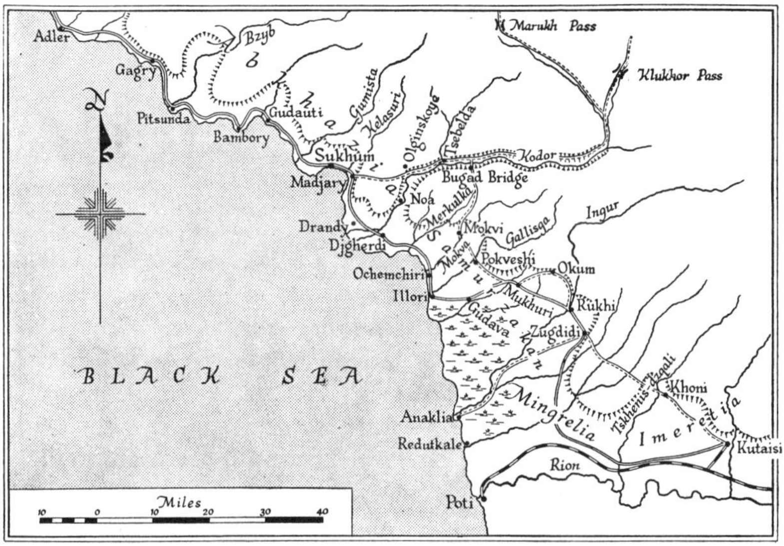

Map 9. The diversion in Abkhazia, 1877

1000 *muhacirler* who were quickly joined by Abkhazians of the district, including the local militia. A highly coloured account of this 'invasion' reached Sukhum where the Russian commandant, General Kravchenko, appears to have lost his head. His panic became complete when, on 14 May, six Turkish warships approached Sukhum and bombarded the port and town. Some fires broke out, and during the night the local roughs began to loot shops and private houses. There were no Russian troops in the town, since on the previous day Kravchenko had withdrawn to a defensive position on the hills behind Sukhum. On 15 May a Turkish squadron appeared off the coast to the south of Sukhum, and Kravchenko concluded that his line of retreat was threatened. In consequence, he ordered complete evacuation of the Sukhum region and immediate retreat into the interior.

No regular Turkish troops were landed at Sukhum and the whole effect was produced by the appearance of six Turkish warships and the landing of a few *muhacirler*. The main force for the planned diversion did not sail from Trebizond until 18 May—three days after the evacuation of Sukhum by Kravchenko.

In justice to General Kravchenko it must be said that despite all the administrative nervousness the Black Sea coast had remained, strangely enough, very lightly protected by troops. There was not a single coastal battery along all the stretch of coast between Novorossiysk and Poti. Small detachments amounting in all to about 4000 men (including 500 Abkhazian militia who were quite unreliable) were scattered between Pitsunda, Sukhum, Ochemchiri and the mountainous district of Tsebelda. Detachments at Novorossiysk, Sochi and Adler amounted to a further 4000 men under Colonel Shelkovnikov, who, fortunately for the Russians, had a more sturdy character than his superior.[1]

The retreat of Kravchenko soon assumed the character of a general panic, since his troops were joined by the administrative and hospital personnel of Sukhum and by thousands of refugees, including Russians, Greeks and Bulgarians, from the villages newly established along the coast. An enormous caravan moved slowly towards the Kodor River. However, Kravchenko thought it too dangerous to continue along the coast road as there were rumours that the Turks had landed at Ochemchiri. He turned inland a few miles south of Sukhum along a difficult track (now a modern road) through a defile leading to the middle course of the Kodor and to Tsebelda. On 16 May the Russian troops with the evacuated hospital personnel and some 2000 refugees were safe behind the strong natural position of Olginskoye, fifteen miles from the coast. No Turks had landed and the only actual enemy were the insurgent Abkhazians, who were particularly interested in the looting of the Sukhum shops and villas and the neighbouring Christian villages. They could not have numbered more than 3000.

Kravchenko now received a report (which subsequently proved false) that the Turks had captured Ochemchiri and were preparing to occupy Samurzakan (the region between the Kodor and the Ingur). Kravchenko saw himself cut off from Kutaisi. He decided to evacuate all Abkhazia and to cross the Kodor. He decided also not to use the bridge at Noa which seemed to him to be too near the coast, but sought to cross the river somewhere in the wild Tsebelda district. On 17 May

[1] Shelkovnikov (i.e. Ipekjiyan) was an Armenian (see *R.B.S.*).

PLATE III

A. THE TURKISH FLEET BOMBARDING SUKHUM, MAY 1877

B. THE FORTRESS OF ERZURUM IN 1877

he crossed the Kodor by the primitive bridge at Bugad, a construction only 3 ft. in width thrown from rock to rock across the 60 ft. ravine of the river. He was forced to abandon all his transport and stores; many horses were lost in swimming the river, but the gunners managed to save their four guns. On 29 May, Kravchenko's exhausted and famished column camped to the south of the Kodor where they were within reach of General Alhazov's detachment which had been moved up from Kutaisi.

When the news of the evacuation of Sukhum had reached Tiflis, Alhazov had been ordered to move by forced marches towards the Kodor with the reserve of troops at Kutaisi which amounted to about 3500 men and eight guns. Alhazov's march was not very swift, and it was only on 27 May that he reached Okum, a village some forty miles from the Kodor. Kravchenko's troops, demoralized by the behaviour of their commander, now marched in to meet Alhazov's men.

Meantime the Turks were extending their coastal operations northward. On 23 May their ironclads bombarded Adler and landed about 1500 *muhacirler*. Colonel Shelkovnikov withdrew the plastun battalion in Adler to Sochi and concentrated there two further battalions and two *sotni* of Cossacks. On 2 June the Turkish ironclads bombarded Sochi, but an attempt at landing failed. There was, in any case, little sense in effecting landings at Sochi or Tuapse since there was no local population available to support the invader; but their naval mobility enabled the Turks to threaten the Russians at a number of points and thus to impose a continual dispersion of forces.

In Tiflis the impression created by the bad news from the Black Sea coast was considerable. All the troops of the strategic reserve were held in the west, and both Loris Melikov and Tergukasov were deprived of the possibility of support at a critical moment on the main front. Reinforcements continued to be sent to General Oklobzhio, and for some reason which remains obscure it was considered that a success in front of Batum would re-establish the whole situation in the Rion valley and on the Black Sea coast. Oklobzhio, with his forces greatly increased, was ordered to act most energetically. He had spent a fortnight between 13 and 27 May in laboriously fortifying his position on the Hutsubani heights and in bringing forward the stores necessary for an attack on the Turkish lines beyond the Kintrishi. This time, however, had not been wasted by the defenders of Batum. Reinforcements continued to arrive from Trebizond.

A new commander, Ibrahim Derviş Paşa Lofçali,[1] who had served with some distinction under Muhtar in Montenegro, had now at his disposal about 20,000 infantry, of whom half were regulars, and thirty to thirty-five guns, some of which were of heavy calibre and long range. This was a serious force on the well-prepared Kintrishi and Deva positions.

The action undertaken by Oklobzhio on 28 May was well planned and proved successful. Its object was limited to securing his left flank from the irregulars who remained in considerable numbers around the upper course of the Kintrishi. A bold operation was planned: to cross the Kintrishi and to capture the Sameba heights which occupy the triangle between the Kintrishi and its left affluent, the Kintash, and thus to divide the irregulars in the district of upper Kobuleti from the Turkish forces on the lower Kintrishi. A surprise attack in the early morning of 28 May gave the Russians the Sameba heights at slight cost, but the action was not complete, since the Kobuleti men retreated farther into the mountains, while several hundred Acars came to their help and occupied a very strong natural position to the rear of the Russian left on the Talaha mountain, near the village of Zenit and the source of the Achkva river. On 2 and 3 June there was fierce fighting round Zenit and the neighbouring village of Achkvis-tavi,[2] but the Russians failed to scale the steep rocks of the Talaha position. The Russian left flank still remained exposed to the renewal of irregular attacks, while the situation on the Sameba heights was insecure. A major action was unavoidable and required a certain amount of preparation. The rapid decision demanded by Tiflis general headquarters was further delayed.

The absorption of all strategic reserves by the Rion and Black Sea region was in part a result of the Turkish diversion. At the same time Turkish action had increased nervousness at Tiflis in regard to Chechnia and Dağistan. As a result Tiflis was unwilling to move troops from the northern Caucasus to reinforce the fighting fronts.

In spite of the surrender of Shamyl in 1859 the Russians still, twenty years later, retained very strong forces in Chechnia and Dağistan; the 20th and 21st Divisions were garrisoned there, while in the adjoining Terek Cossack region the newly formed 38th Division was held in readiness to reinforce them. In addition to this considerable

[1] Military reputations live long in Turkey. In 1946 I was offered in the Istanbul bazaar a gold embroidered fur robe, priced high because it was said to have belonged to Derviş Paşa. (W.E.D.A.)

[2] *Achkvis-tavi*, G., 'Head (source) of the Achkva'.

infantry force, the majority of the Terek Cossack regiments were retained in the eastern mountains and about half the Kuban Cossack mounted regiments and foot battalions remained in the Kuban region. As against this serious freezing of forces by the threat of insurrectionary action, the Russians tried to weaken the potentially active element in the mountains by drafting the young men into regiments of irregular horse for service outside Dağistan. These troops, like the Muslim recruits from eastern Transcaucasia, proved reliable and excellent soldiers when mixed with Russian units on the Turkish front.

Nevertheless, the exaggerated reports of a Turkish invasion of the western Caucasus could not be without its repercussions in Dağistan. During the previous two years Turkish agents had been active in the mountains, but while Chechnia was always ready for revolt, there were in Dağistan important elements which were pro-Russian and an even larger part of the population which was indifferent. In Chechnia a certain Haci Ali Bey proclaimed himself imam, and the wild Chechens brought out the muskets which they had buried twenty years before. The old fanaticism flickered through the auls and there were a number of encounters between the tribesmen and Russian troops in which the former suffered greatly under the fire of the new Russian rifles; Russian losses were negligible. In a fortnight the movement was broken, but the Russian military governor of Vladikavkaz, General Svistunov, was not satisfied. He sent small columns into Ichkeria (the heart of Chechnia) with the demand that the inhabitants should deliver up Haci Ali, Sultan Murat and other rebels. These punitive columns provoked further resistance and the disorders spread to Dağistan. By 1 June the situation seemed to be serious, and the reports of Svistunov, following those from the Black Sea coast, caused increased anxiety in Tiflis and prevented any further dispatch of reinforcements to the main front.

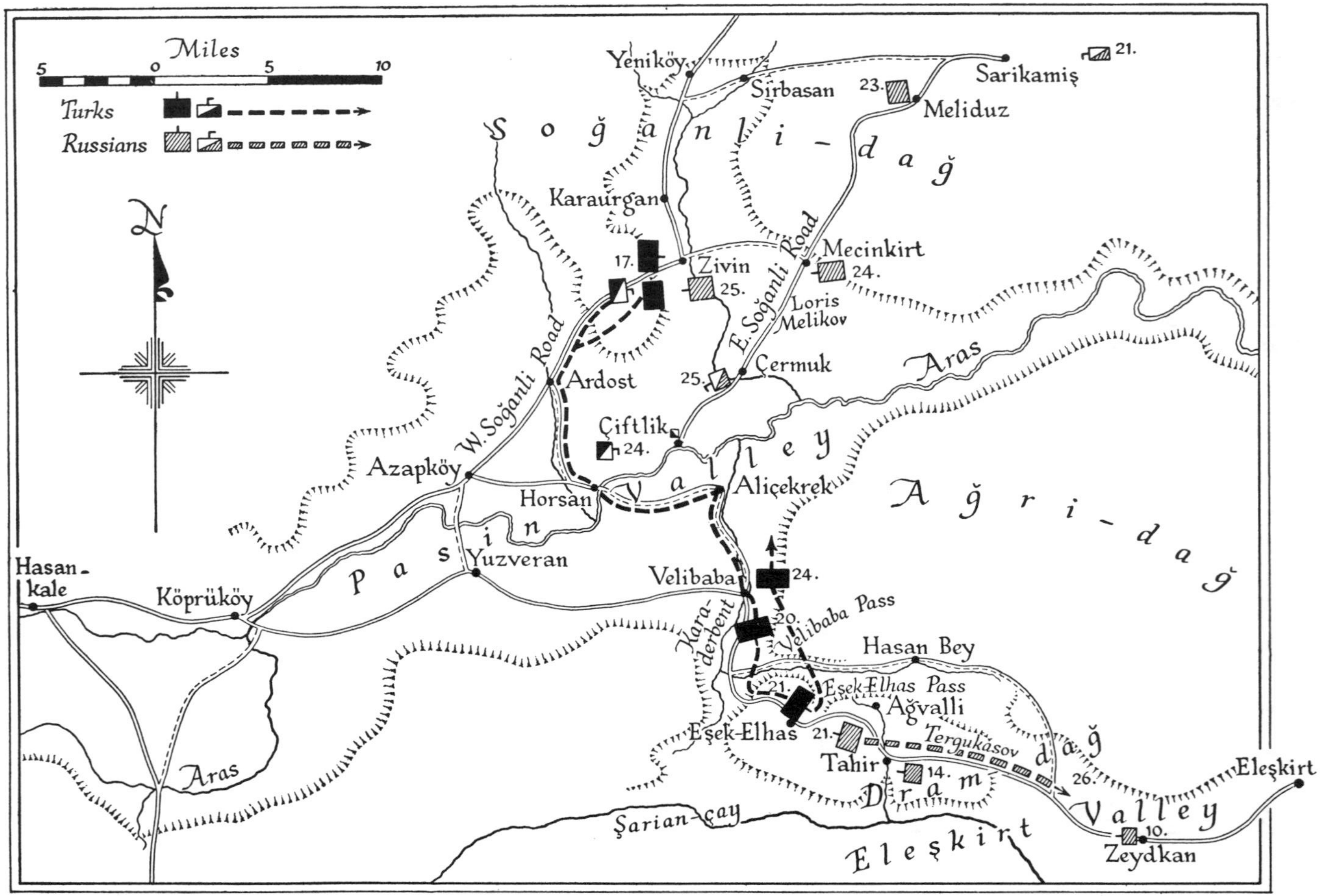

Map 10. The operations covering the Aras valley, spring 1877

CHAPTER IX

TERGUKASOV'S OPERATION IN THE ELEŞKIRT VALLEY. THE TWO ACTIONS AT TAHIR (JUNE 1877)

MAPS 1, 4, 10, 11; cf. also 16, 13

DURING the first fortnight of June, Loris Melikov had opened the siege of Kars. The fortifications prepared by Fenwick Williams in 1854–5 had been strengthened and modernized, and a new fort, Karadağ Tabia, had been built on the Karadağ heights on the right bank of the Kars-çay. On the left bank along the Çakmak heights—which in 1855 had been defended only by the trenches called the 'English lines'—two modern forts had been constructed, Williams Paşa (or Ingiliz) Tabia and Teesdale (also Muhlis or Bluhm Paşa) Tabia. These two forts on Karadağ and Çakmak were armed with heavy Krupp guns, and their field of fire covered the open country to the east of the fortress. Preliminary operations indicated that the Russians had no chance of gaining the easy success which they had had at Ardahan; they prepared, therefore, for systematic siege operations and brought up from Aleksandropol 120 heavy guns and several scores of 6 mm. mortars. Meantime, a successful cavalry action at Benliahmet dispersed a small Turkish mounted force which was trying to maintain contact between the Kars garrison and the Turkish field army in the Soğanli.

On the left wing of the Russian front, the Erevan force under General Tergukasov operating from Bayazit had, on 20 May, occupied Karakilise. On 28 May Loris Melikov had ordered Tergukasov to continue his advance as a diversion designed to prevent any attempt to raise the siege of Kars on the part of Muhtar's forces on the Soğanli. This diversion was in fact very risky, since the Russian forces in the Eleşkirt valley were small (5000 infantry, 3000 cavalry and thirty guns) and Ahmet Muhtar, who had just begun to pay serious attention to Tergukasov's movement, was not by 1 June so helpless as he had been on 10 May.

The organization of the Turkish field army was progressing slowly. The troops mobilized in the coastal district had been diverted for the operations along the Caucasian coast, but six battalions of redifs had arrived in Erzurum from Istanbul and eighteen more were expected from Damascus. After Komarov's withdrawal to Ardahan the Turkish detachment at Oltu had joined the main force on the Soğanli, which now consisted of twenty-one battalions, twenty-four guns and 1000 horse. The capable old Hungarian officer, Feyzi Paşa, who had organized the defences of Batum, was appointed chief of staff to Muhtar, and he was preparing a strong defensive position at Zivin in the region where in 1828 Paskevich had gained his notable victory.

Mehmet Paşa's force, covering the approaches from the Eleşkirt valley over the Dram-dağ into the Pasin valley, had been increased to twelve battalions with twelve guns and 1000 Cherkess horse. As a central reserve Ahmet Muhtar had some seven battalions of redifs between Erzurum and Köprüköy; the recruits for seventeen new battalions were waiting in Erzurum, but they had not yet been armed. There were also eight or nine new field batteries available but without the necessary animals. By the end of the first week in June Ahmet Muhtar disposed of some 20,000 men and thirty-six guns. It was a force not strong enough to threaten both Heimann (who was covering the siege of Kars from the west) and Tergukasov; so Muhtar wisely decided to turn against Tergukasov's weaker force.

Tergukasov was already concerned as to the renewal of his relatively unprotected communications. The Armenian part of the population of the Eleşkirt valley was friendly, but to the south the Kurds of the Ala-dağ, though not very loyal to the Turks, were quite unreliable. Tergukasov also feared action by the Turkish garrison of Van against his rear at Bayazit. He sent the Georgian Prince Amilakhvari to reconnoitre in that direction; this officer returned with a reassuring report which was not justified by immediate developments. Before 20 May, Ahmet Muhtar had ordered Faik Paşa commanding at Van to march towards Bayazit and join forces with the two Turkish battalions retreating from that place. By the end of May, Faik had collected in the Ala-dağ six battalions with six guns and several thousand Kurds ready to operate against Tergukasov's unprotected lines of communication.

Reassured by Amilakhvari and under orders from the commander-in-chief whom he assumed to be well informed on the general situation, Tergukasov continued his westerly advance across the Eleşkirt plain. On 10 June he reached Zeydkan, while the Turks withdrew before him towards the Kara-derbent[1] into the Dram-dağ. Ahmet Muhtar decided to stop Tergukasov and ordered Mehmet Paşa to take up a

[1] *Kara-derbent*, 'Black pass' (or gorge). Kara-derbent is marked *Kizil-* ('Red') *derbent* on modern maps.

defensive position covering the Kara-derbent and the Eşek-Elhas Pass.

The road leading from the Eleşkirt valley into the Pasin valley, after passing Zeydkan, runs roughly parallel to the small river Şarian,[1] an affluent of the upper Euphrates. From Zeydkan, across the three ridges of the Dram-dağ, the road forked, the upper road passing Tahir village while the lower ascended the narrowing ravine of the Şarian. The two roads united again at the village of Eşek-Elhas on the Dram-dağ watershed, and the road descended through the Kara-derbent, reaching the open country of the Pasin valley at Velibaba.

On receipt of Muhtar's order, Mehmet Paşa, who had withdrawn to Eşek-Elhas, turned back in the direction of Zeydkan, intending to take up a position on the most easterly of the three ridges of the Dram-dağ. He found, however, that the Russians were already in occupation of the first ridge, and on 14 June he took up a defensive position covering the village of Tahir on the most westerly of the three ridges. Here he concentrated ten battalions and two batteries with two battalions and one battery at Eşek-Elhas to secure his line of retreat. Ahmet Muhtar sent him two battalions and one battery with 1000 Cherkess horse as reinforcement. Only the horsemen arrived in time for the action.

The left and right flanks of the Turkish positions were protected by high and steep mountains, on each of which Mehmet Paşa posted one battalion of infantry and two guns. The centre was covered by hastily prepared trenches occupied by four battalions and six guns. Two battalions were held in reserve at Tahir village, while two other battalions with the cavalry and two guns were stationed to the rear of the right flank. Mehmet Paşa's dispositions were satisfactory, but he had made the mistake of failing to occupy the middle ridge of the Dram-dağ which lay between his own position and the eastern ridge already occupied by the Russians.

At dawn on 16 June the Russians established a battery on this middle ridge which began to shell the Turkish positions, while columns of Russian infantry descended the steep ravine to attack both Turkish flanks. As at Ardahan, Turkish fire control was bad and the Russians advanced with slight losses, particularly over the dead ground in front of the Turkish positions which protected them from the fire of the Turkish artillery. The principal Russian attack was directed against the Turkish right flank, but the Turks resisted stubbornly

[1] *Şarian* is an Armenian name. The river carries the Turkish name *Güzel* on modern maps. *Eşek-Elhas*, 'Elijah's ass', may have a biblical connotation.

until Tergukasov sent in his reserve (only one battalion of the Crimean Regiment) to turn the position from the south. The Russian cavalry in the meantime had penetrated the Şarian ravine but was held by the fire of the two Turkish guns in reserve on the right flank and was only able to proceed when reinforced by a mounted battery.

Mehmet Paşa moved up all his reserves to strengthen his right flank and led in his Cherkess cavalry in an effort to check the Russian advance. The Paşa himself was killed, and at the same time a counter-attack from the Turkish left flank failed. The retreat of the Turks began at a moment when Tergukasov only had two companies left in reserve. The Turks were saved from a complete rout by the action of the two guns stationed at the head of the Şarian ravine. At the same time Amilakhvari proved as casual in the field as he had in his reconnaissance, and the Russian cavalry which might now have dominated the battlefield only approached Eşek-Elhas after the mass of the retreating Turkish infantry had already by-passed this point. The two battalions and six guns left in reserve at Eşek-Elhas were strong enough to cover the further retreat of the Turks, and late in the evening two battalions and one field battery, despatched by Muhtar as reinforcements before the action, arrived to strengthen the rear-guard. The Turks left 1000 casualties and 1000 prisoners on the field and they lost ten guns; Russian losses were less than 300. Tergukasov's victory, although incomplete, was a definite success, since the Turkish group covering the Pasin valley was defeated and disorganized.

On 17 June, Tergukasov sent an account of the action to Loris Melikov. He reported at the same time his anxiety with regard to his rear, since he had received information of Faik Paşa's concentration in the Ala-dağ which seemed to threaten Bayazit. On the same day Loris Melikov, who had also received disturbing news from Bayazit, called a military council which met before the receipt of Tergukasov's dispatch. Loris Melikov, who had designed Tergukasov's advance as a diversion to cover the siege of Kars, now decided to dispatch a force under General Heimann to demonstrate against the Soğanli positions with a view to creating a diversion in favour of Tergukasov's threatened force. At the same time, General Devel was ordered to continue the siege operations against Kars. The idea of a serious bombardment preliminary to an attempt to storm was abandoned in favour of demonstrative artillery action. Thus Loris Melikov, in spite of his victory at Ardahan and of the superior force which he could still concentrate for a single stroke, had been reduced, as a result of the Turkish diversionary demonstrations against the Caucasian coast, to

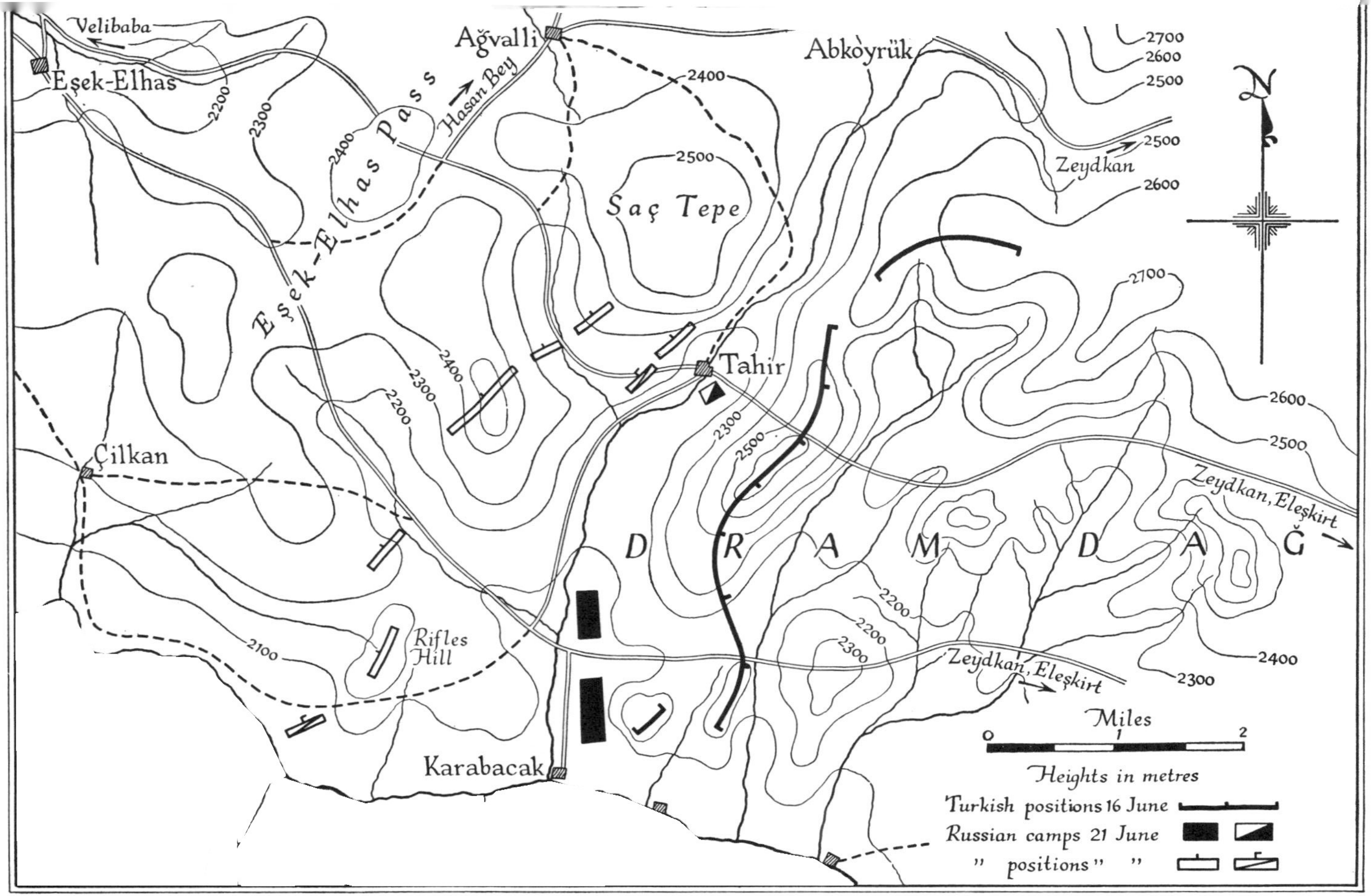

Map 11. The actions at Tahir, June 1877

adopt a strategy himself which amounted to nothing more than a series of rather ineffectual defensive diversions. The Russian commander had, in fact, lost the initiative and was allowing Ahmet Muhtar this same initiative with weak forces during the vital weeks which were necessary for the concentration and organization of the Turkish army at Erzurum.

Ahmet Muhtar, on the other hand, was a natural and sturdy optimist, and his strategic courage was really remarkable in view of the weakness and poor quality of the forces at his disposal. Not only was he not greatly impressed by the defeat of Mehmet Paşa, but he found in this defeat the stimulus for an energetic and bold action. Leaving half the Zivin force to cover the tracks over the Soğanli, he himself marched to Velibaba, where by 20 June, four days after the defeat at Tahir, he had concentrated approximately 10,000 infantry, four batteries and 3000 horse.

Meantime Tergukasov remained encamped on the battlefield of 16 June. He was unaware of Muhtar's concentration and had received no information of Heimann's proposed diversion against the Soğanli. At dawn on 21 June, a strong reconnaissance of two companies of infantry and four *sotni* of Cossacks, under Colonel Medvedovski, moved north from Tahir in the direction of Ağvalli village with the object of exploring the track—an alternative to that through Eşek-Elhas and Kara-derbent—by the pass of Hasan Bey. Beyond Tahir, Medvedovski observed masses of Turkish horse moving along the road from Eşek-Elhas to Tahir. He immediately sent warning to Tergukasov and took up a position covering the gorge and village of Tahir.

Before dawn on 21 June, Ahmet Muhtar had moved thirteen battalions, his cavalry and three batteries from Velibaba in the direction of Eşek-Elhas, leaving the defeated troops of Mehmet Paşa, which had been reformed into six battalions with eight guns, as a strategic reserve at Velibaba. Descending through Eşek-Elhas towards the east, the Turks reached a point where the two forks of the road from Zeydkan rejoin. From here half the cavalry, five battalions and one battery proceeded towards Tahir, while the main force of eight battalions and two batteries and the remaining horse, under Reis Ahmet Paşa, followed the road descending into the ravine of the Şarian. Muhtar Paşa supposed that Tergukasov's camp was at Tahir, and he planned to immobilize the Russians with his left column while his right column, advancing along the Şarian, turned their left in depth.

Tergukasov's main force, however, was camped not at Tahir but two and a half miles farther south in the Şarian valley. Warned by Medvedovski's message, about 10 a.m., Tergukasov disposed his men in a defensive position covering more than four miles—too much for six and a half battalions. But, fortunately for him, the Turkish attack was directed only against both extremities of his position, and the enemy was acting in two separate and unconnected groups.

The first Turkish attack was directed up the Tahir defile where five battalions and 1500 Cherkesses, under Musa Paşa Kundukov (a Cherkess, formerly an officer in the Russian army), was met by one and a half battalions of infantry and several Cossack *sotni*. The Turks captured the rocky Saç hill but were unable to advance farther owing to the fire from the Russian battery. By noon the situation here appeared to be stabilized. In the centre between the Tahir defile and the Eşek-Elhas road the ground was defended by one battery, one and a half battalions of infantry, two companies of rifles and a few dragoons and Cossacks. On the left flank, between the road and the Şarian stream, there were one battalion of infantry, two companies of rifles, three squadrons of dragoons and one mounted battery. Soon after midday Reis Ahmet's infantry strongly attacked 'Rifles Hill' occupied by two companies of rifles, while *süvari* and Cherkesses tried to advance down the Şarian stream. Fierce bayonet fighting continued round the hill, and the Cherkesses charged the Russian infantry but were beaten off by a counter-attack of the dragoons. Tergukasov threw in his last reserve of two companies and one dragoon squadron. The continued Turkish attacks, although strong, were badly timed and badly directed, and the Turks failed to capture any vital point of the defensive line. On the Russian right flank General Bronevski, noting that the enemy were getting tired, managed to collect the strength of four battalions for a counter-attack. The Turks were driven out of the Tahir defile and lost the Saç hill.

Muhtar Paşa had meanwhile sent from Velibaba three battalions and one battery across the Hasan Bey Pass in order to turn the Russian right flank from the north. But this movement took all day, and the advance-guard of the Velibaba column only appeared near Ağvalli village after the fighting in the Tahir valley was at an end. In the evening Medvedovski's Cossacks occupied Ağvalli and the Turks withdrew.

Ahmet Muhtar had, in fact, demonstrated on this day all his qualities and defects as a military leader: a considerable strategic ability, combined with an incapacity to direct a battle in detail and to change his plans on the field according to circumstances. His plan had been based on an erroneous supposition, and the main Russian force was not where he had expected to find it. As he was not himself on the field he left the tactical direction to Reis Ahmet, who failed to observe that the weakest point of the Russian position was in the centre, which might easily have been broken through. Instead of being present in person and ready to change his plan on the field, Ahmet Muhtar had relied on the manoeuvre through the Hasan Bey Pass which was theoretically attractive but in practice abortive.

Turkish losses on 21 June amounted to about 2000 killed and wounded, 20% of the strength engaged. Russian losses were 700, 10% of the troops engaged. Ahmet Muhtar had suffered an expensive repulse; but the Russian troops were exhausted. Their small strength was further diminished and they were running short of ammunition; and Tergukasov anticipated a renewal of the Turkish attack on the following day.

So if Muhtar Paşa had failed to attain a tactical victory, he had scored a strategic success. Muhtar knew that Bayazit was already threatened by Faik and that Tergukasov was in danger of being cut off from his base. He rightly judged a Russian retreat to be inevitable, and with his own limited forces he wished to avoid the cost of a renewed attack. Tergukasov was, therefore, surprised and pleased when, on the following day (22 June), the Turks proposed twenty-four hours' armistice to bury the dead and remove the wounded. The Russian commander still had no news from Loris Melikov, but that night Muhtar received information that Heimann was approaching the Soğanli positions.

Muhtar knew the strength of the Zivin camp and was confident that it could be defended, but he was apprehensive of a Russian movement between the two groups of the Turkish army; the Russians might by-pass the Zivin position and descend from the Soğanli by the tracks into the Aras valley at Horsan (or to the east through Çiftlik and Aliçekrek) and attack Muhtar's base at Velibaba. The Turkish commander, therefore, decided to reform his forces at Velibaba, and he sent most of his cavalry to Horsan to strengthen the three battalions and one battery stationed at Köprüköy—which force constituted a liaison between the groups at Velibaba and Zivin.

CHAPTER X

BATTLE OF ZIVIN-DAĞ. TERGUKASOV'S RETREAT AND THE RELIEF OF BAYAZIT

MAPS 4, 10, 12; cf. also 13

IN all the history of the Caucasian wars there is no other operation so badly conceived and executed as the advance of Loris Melikov to Zivin in June 1877.

The force available for the diversion under the immediate command of General Heimann consisted of sixteen battalions of Caucasian grenadiers, one battalion of sappers, one brigade of Caucasian dragoons, three Cossack regiments, four regiments of Muslim irregular horse, six field and two mounted batteries—a total of 18,000 infantry, 6000 cavalry and sixty-four guns. It was a considerable force, and the commander-in-chief personally accompanied it.

However, the aim of the movement was not clear beyond the general intention to help General Tergukasov. It was possible to effect this object in two ways: Loris Melikov might join forces with Tergukasov or he might attack the main Turkish army known to be at Zivin and thus compel the diversion of the Turkish troops concentrated against Tergukasov. If an attack against the Zivin position were intended, the most natural line of advance was the road which climbed the Eşekmeydan Pass and leading through Yeniköy and Karaurgan threatened the left (northern) flank of the Zivin position (which faced east). This, the old main road from Kars to Köprüköy, was the direction followed by Muravyev's column in the operation of 1829. Alternatively, if the object were to join forces with Tergukasov as soon as possible, the obvious route was by the easterly road ascending from Sarikamiş over the Meliduz col and descending to the Aras at Horsan or Çiftlik-Aliçekrek. If this route were used it would be necessary to cover the advance by a strong detachment at Mecinkirt ready to oppose a probable Turkish attack from Zivin. (There was a connexion between the two alternative roads by way of Sirbasan on the Soğanli plateau, a route along which Paskevich had himself led a column in 1829.)[1]

With his troops already advancing on the Soğanli plateau, Loris

[1] After the establishment of the new frontier in 1878, the route Sarikamiş-Sirbasan-Yeniköy-Karaurgan became part of the principal road between Kars and the Russian frontier at Karaurgan. The track of the (new) Erzurum-Kars railway follows this route.

Melikov, characteristically, had arrived at no decision. He at first intended to take the second (eastern) road through Mecinkirt to the Aras, but, when already on the plateau, he changed his plan and decided to attack the Zivin position.

On 23 June the Russians camped on the plateau. Cavalry patrols advanced beyond Mecinkirt without coming into contact with the Turks. During the night no news came from Tergukasov, and a Karakalpak courier sent to him had failed to return. Loris Melikov became nervous and gloomy; there were rumours that there were 40,000–50,000 Turks at Zivin. The commander-in-chief called a council of war, and all the Caucasian veterans, led by General Heimann, insisted on an attack on the Zivin position. Loris Melikov acquiesced rather reluctantly. Information received from spies and from Armenian villagers was more or less correct: some twenty battalions of Turks and three batteries on the Zivin heights; another group at Velibaba (Muhtar's force which lay south of the Aras between Melikov and Tergukasov); cavalry (Musa Paşa's Cherkesses) at Horsan.

The Zivin position had been laid out by Feyzi Paşa (Kolman) along the heights which form the western side of the Zivin-çay ravine. The main road from Karaurgan to Erzurum crossed these heights, and a track to Mecinkirt branching from the main road below Zivin village followed up the bed of the Zivin-çay. The position thus barred the main (western) road from Kars to Erzurum, and it lay on the flank of the eastern road passing through Mecinkirt to the Aras bridge-heads at Horsan and Çiftlik. Feyzi's position extended along the ravine to a point five miles south of Zivin village and to within seven miles of the eastern road. Thus, by the establishment of the camp at Zivin, Muhtar hoped to protect both the roads leading down to the Aras valley and to Erzurum. This ambitious aim, however, determined the most important defect of the Zivin position—its very considerable length. When Muhtar drew off one 'division' for the operation against Tergukasov, only twelve battalions remained on the Zivin heights. These had been reinforced by four battalions from the reserve, but sixteen battalions with eighteen guns and scarcely any cavalry was an inadequate force with which to hold a front six miles long.

Turkish soldiers had been working for more than two months on the Zivin fortifications; and the Anatolian redifs were born diggers, accustomed to terracing the mountain side and carrying irrigation canals across stony plains. By 25 June the entrenchments were finished except along the left flank of the position overlooking the

Karaurgan-Köprüköy road. This flank was practically open, and the exposure was the more dangerous since slopes here were not steep and were suitable to mounted action.

The Russians might have been attracted to the left flank of the Turkish position since, in 1829, the Turks had been outflanked by a Russian move to the west of the main road. The Russian staff reconnaissance, however, was now concentrated on the front of the Turkish position and on the possibility of outflanking the position from the south where the terrain was very steep and rough and intersected by many ravines.

Loris Melikov, or rather his enthusiastic subordinate, Heimann, was planning a complete destruction of the Turkish army. Attacked on the left, the Turks might have been able to retire along the road to Köprüköy (as had happened in 1829), but a successful attack on their right held promise of cutting their line of retreat. In the result the Russian operation developed into nothing more than a straight frontal attack by sixteen battalions against very strong positions held by an equal number of Turkish battalions. It is true that the strength of a Russian battalion was 25% in excess of that of a Turkish battalion, but as Loris Melikov retained four battalions in reserve, effective infantry strengths were, in fact, equal. The Russians were much stronger than their opponents in cavalry and artillery, but Loris Melikov's plan really excluded the effective use of cavalry, and the steep ground made it difficult for his guns to be brought up to support the infantry.

Early in the morning of 25 June Loris Melikov struck camp at Mecinkirt and moved in the direction of the Zivin-çay, leaving at Meliduz his train under the protection of a rather strong composite force. The right column, including all the infantry and five batteries with three regiments of cavalry, moved along the track to Zivin village, ten miles distant. The bulk of the cavalry under Chavchavadze took the track leading from Mecinkirt to Horsan. At 8 a.m. Melikov at last received news from Tergukasov (brought by the Karakalpak courier who was given a reward of 2000 rubles). Tergukasov, after reporting the actions of 16 and 21 June and his serious losses and shortage of ammunition, stated that he was retiring in the direction of Erevan and asked that superior Turkish forces should be prevented from concentrating against him. Always irresolute, the Russian commander-in-chief now stopped the movement of both columns at a moment when infantry was already approaching the ravine of the Zivin-çay. Troops remained immobile until 2 p.m. while Melikov discussed the situation with his generals. He questioned whether the attack on Zivin was necessary, and favoured a cavalry demonstration against Horsan. But the old Caucasian commanders, Heimann and Komarov, ambitious to complete the double manoeuvre of Paskevich in 1829, urged an attack on Zivin to be followed up by a move against the Turkish group supposed to be at Horsan. (In fact, only the Cherkess irregular cavalry dispatched by Muhtar from Velibaba were there.)

Meanwhile Feyzi Paşa, who in effect commanded at Zivin although his superior, the incompetent officer commanding army corps division IV, Ismail Hakki Paşa, was present on the field, had had time to dispose his forces to meet the Russian attack.

Loris Melikov set three batteries on the heights along the eastern side of the Zivin-çay, and left with the guns his general reserve (the four battalions of the Erevan Grenadiers). Only one regiment of Terek Cossacks was in observation along the Karaurgan road. The attacking force consisted of the Mingrelian and Tiflis Grenadier Regiments with three battalions of the Georgian Grenadiers and two batteries.

The Mingrelskis were ordered to cross the Zivin-çay and attack the right of the Turkish position which was separated from the centre by a long and deep intersecting ravine. Their left was covered by two regiments of Dağistan horse who had also to keep touch with Chavchavadze's flanking column. The Tiflisski and Gruzinski (Georgian) Grenadiers were directed against the Turkish centre.

As the valley was narrow where the Mingrelskis had to cross, the guns of their supporting battery were able to dominate the single line of Turkish trenches. Heimann ordered two battalions to the attack while the third moved towards an isolated hill to the south of the Turkish lines. In spite of the steep and difficult ground the Mingrelian Grenadiers reached the summit and after fierce bayonet fighting took all the Turkish line of trenches, the enemy taking flight into a ravine in their rear. This success was attained by 5 p.m. with comparatively small losses.

In the meantime, against the centre, the Gruzinskis crossed the Zivin-çay in the lead and marched along the track at the bottom of the ravine in the direction of Zivin village, two miles away. General Komarov commanded this column which was to attack the northern part of the central sector of the Turkish defences, while only one battalion of the Tiflis Grenadiers stormed the southern part. The marching column came under a concentrated gun and rifle fire. The Russians at the same time lacked artillery support because it was impossible to deploy in the valley the battery which was accompanying

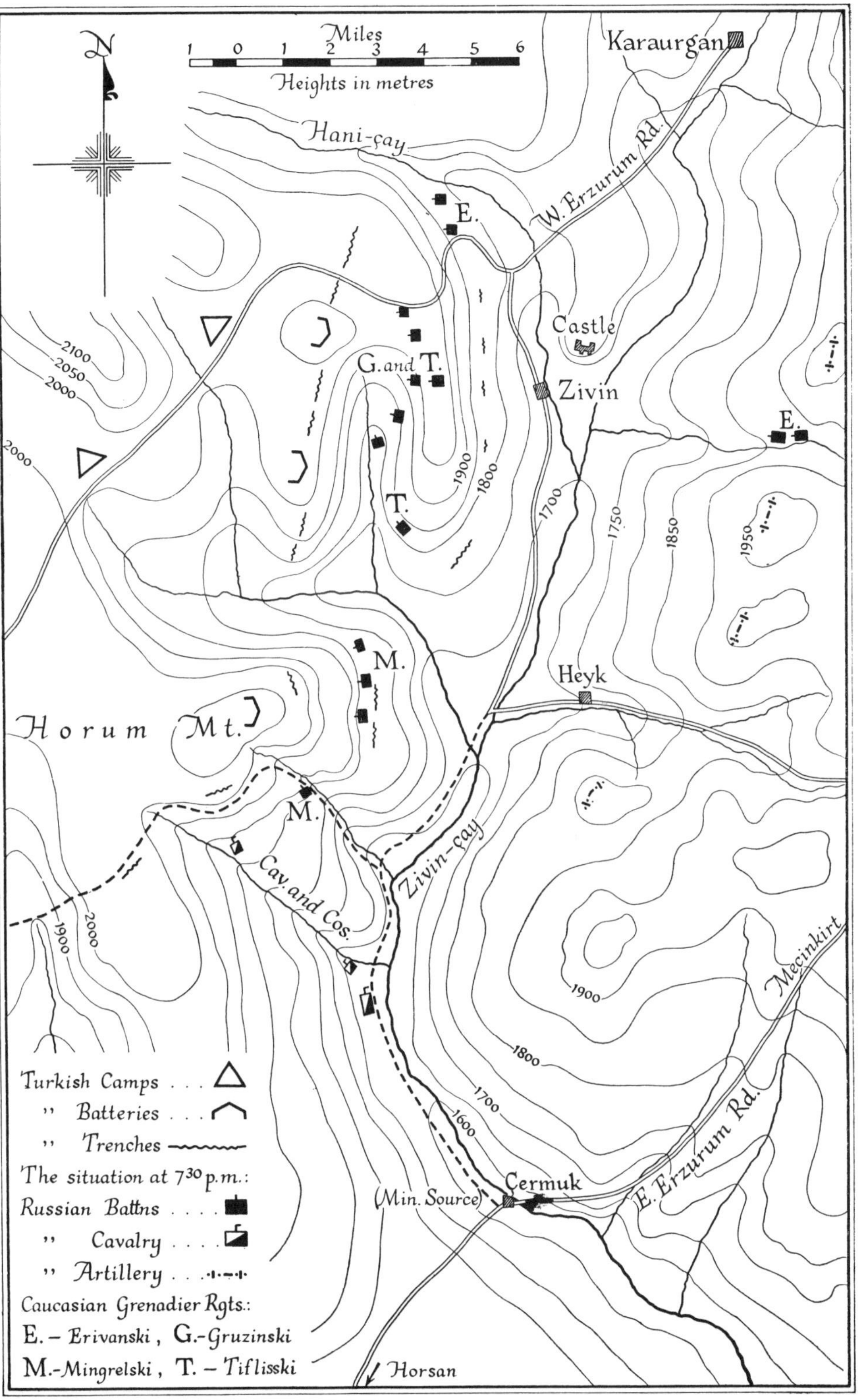

Map 12. The battle of Zivin-dağ, June 1877

the column, while the three batteries which had been left on the heights above Zivin were out of range of the Turkish positions (two and a half miles). The Georgian and Tiflis Grenadiers, however, attacked and carried the first of the three lines of trenches on the rocky terrace above Zivin. Their losses were heavy, particularly in officers, and Komarov was seriously wounded. They were further held by an intervening ravine under the fire of the second line of Turkish trenches. The single battalion of Tiflis Grenadiers attacking the southern part of the central sector had also been successful in taking the first line of Turkish trenches, but they also were held up by a second line on a higher terrace, and by the fire of guns on the heights above.

The situation of the attacking regiments exhausted by their long march and their heavy losses became momentarily more critical. Some companies tried to turn the triangular height forming the bastion of the Turkish centre on the north, but when they approached the Karaurgan road they came under the fire of two Turkish battalions stationed on the height above the ravine of the Hani-çay. The Russian commander sent two battalions of the Erevan Grenadiers to support the attack in this direction. He had in fact found the weak part of the Turkish defence. The two battalions advanced along the Karaurgan road but they were too late, and Feyzi Paşa was able to deploy four battalions of his small reserve in a counter-attack which pushed the Erivanskis back into the valley of the Zivin-çay.

It was now 6 p.m. but there was no news of the turning movement undertaken by Chavchavadze's cavalry. Awaiting the cavalry, the Mingrelski battalion which had taken position on the isolated hill to the extreme south of the Turkish position remained inactive. The other Mingrelski battalions stayed immobilized in the trenches taken by them at 5 p.m. As the evening drew on the failure of the Russian attack became clear. At 7.30 p.m. rifle fire was heard at the southern extremity of the front. Dismounted Severski Dragoons and Cossacks were approaching the isolated hill already occupied by the Mingrelskis. But a development of this attack was impossible without artillery support, and the difficulties of the ground proved so great that Chavchavadze found himself unable to bring forward his mounted batteries. As dusk came on, the cannonade and rifle fire began to die away all along the front. At 9 o'clock Melikov gave the order to cease fire, and with the darkness the Russians began to withdraw behind the Zivin-çay.

On the morning of 26 June the Russians stood to arms on the heights to the east of the Zivin-çay ready to meet a counter-attack from the Turks. But the Turks remained passive; before leaving the Zivin camp, Ahmet Muhtar, who knew the weakness of his men in manoeuvre, had forbidden them to leave their trenches under any circumstances. He preferred to be satisfied with a defensive victory. On the Russian side there were some voices in favour of a renewal of the attack. But Loris Melikov gave a categorical refusal. Russian losses were heavy, 1300–1500 killed and wounded. Compared with the losses at the unsuccessful storm of Kars in 1855 they were certainly not catastrophic; but the troops had suffered a serious shock to their morale. The grenadier regiments had lost confidence in their leadership; every soldier understood that a frontal attack by infantry against precipitous and well-defended positions without any effective artillery support was a desperate and stupid affair. Junior officers particularly were very sharp in their criticism of the commanding officer and staff;[1] and the grenadiers were gloomy and resentful.

On the following day (27 June) Loris Melikov drew off to his train at Meliduz where he remained in position during the 28th and 29th with a view to covering the retreat of Tergukasov. On the 29th, Melikov had news that the latter had successfully extricated himself from his difficult situation, and on the same day an order came from Tiflis authorizing withdrawal to the Kars positions.

In the meantime Ahmet Muhtar arrived at Zivin where he found that Turkish losses had not exceeded 500 killed and wounded. His first concern was to rid himself of the difficult and mischievous Ismail Hakki, whom he deputed to take command of the troops acting against Tergukasov. The optimism of Ahmet Muhtar now knew no limits, and he began to prepare with his usual energy for an advance to the relief of Kars. On 30 June it was reported that the Russians were retiring from the Soğanli. Muhtar, who considered that Tergukasov could scarcely escape the forces which were on his track, decided to follow Loris Melikov with the troops he had available at Zivin.

The situation of Tergukasov had become really grave by 25 June. Already by 10 June, Faik Paşa had concentrated in the Ala-dağ to the south-west of Bayazit six and a half battalions of infantry, two squadrons of *süvari* and twelve guns with 2000 mounted and 600 foot irregulars (mostly Kurds). Tergukasov's line of communication was protected by one Cossack regiment at Surp Ohannes and two weak regiments of irregular horse nearer to Bayazit. On 17 June Faik overwhelmed a rash reconnaissance from Bayazit, under Colonel

[1] See letters captured later by the Turks and quoted by C. B. Norman, *The Times* correspondent, in *Armenia and the campaign of 1877*, pp. 185–7.

Patsevich, at Teperiz. When hordes of Kurdish irregulars descended upon the town of Bayazit, Captain Stokvich, who had been left in command of the old Turkish citadel, could scarcely collect a thousand combatants for its defence.

On the 20th, after the Kurds had looted the town, Faik moved up two battalions and three guns to attempt to storm the citadel. The paşa himself, mistrusting his wild allies, remained with the rest of his regular troops at Teperiz. In the confusion of the attack the white flag was raised in the outer court of the citadel, but before Turkish officers could intervene in the mob of refugees, irregulars and Cossacks, the Kurds had cut down about 200 of the panic-stricken crowd. The incident attracted the attention later on of European correspondents, and his incapacity to control the Kurds broke the career of Faik Paşa.

The Turkish commander, after stormy conferences with the Kurdish chiefs, finally withdrew all his forces except for the crews of two guns—which continued to fire spasmodically into the citadel—and left the Kurds to maintain the blockade of Bayazit. The blockade—it can scarcely be called a siege—lasted for twenty days. The garrison was reduced to sore straits by lack of food and water, and epidemics reduced from 1000 to 400 the number of men ready to man the walls.

Meantime, on 26 June, Tergukasov began to retreat in the direction of Zeydkan. His force, after the actions of 16 and 21 June, was reduced to 5000 men and they carried with them 1000 wounded. Great crowds of Armenian refugees, following the retiring troops, added to their difficulties. From Velibaba, Reis Ahmet Paşa with sixteen battalions and three batteries was following the Russians with the object of bringing them to action; and on the flank of the retreat Faik Paşa appeared with a force of about 10,000 regulars and irregulars and ten guns. The Russians, already weakened and ill provided, were in fact threatened by forces nearly four times their own strength. The ineptitude of the Turkish commanders alone saved them from disaster. Reis Ahmet allowed the Russians a day's grace before he moved himself; he was weak in cavalry since he only had about 1000 Cherkesses and the superior Russian cavalry prevented adequate reconnaissance. All the same the Turkish commander showed little stomach for a fight, and on the night of 29–30 June he finally lost contact with Tergukasov. Ahmet Muhtar, so ill served by the political careerists who were usually forced on him as subordinate commanders, was furious.

On 1 July the incompetent Kurdish chieftain, Ismail Hakki, arrived in the Eleşkirt valley to take command. He brought with him four fresh battalions from Erzurum.

Tergukasov meantime, embarrassed by 3000 Armenian families from all over the Eleşkirt plain, passed Karakilise and on 2 July reached Surp Ohannes. Faik Paşa, with a force double the Russian strength, was in the immediate vicinity. On the evening of the 3rd, the Russians, protecting their long column of fugitives and covered by a devoted regiment of Kuban Cossacks, moved east of Balik-göl (T., 'Fish lake') towards the Russian frontier. They marched all night and by the morning had left the Gürünsaray Pass behind them. By the same evening Tergukasov had all his troops and refugees safe on Russian territory. His losses during the retreat were only one officer and thirty men.

Both Turkish paşas, Ismail and Faik, appeared quite satisfied with the 'disappearance' of their enemy. Ismail Hakki gave his troops a rest in the region of Surp Ohannes and Diyadin. Faik with his regular battalions went into camp a few miles nearer to Bayazit at Teperiz. He was expecting daily the capitulation of Bayazit, but to accelerate it he sent two battalions and two more guns to take part in the siege. The command under Bayazit was confided to Münib Paşa.

Tergukasov wasted no time in preparing for the relief of the place. By the end of the first week of July he had concentrated his troops at Iğdir and on the 8th he moved on Bayazit with six battalions of the Crimea and Stavropol Regiments, a rifle battalion, the Pereyaslav Dragoons and several *sotni* of Kuban and Terek Cossacks, a total of about 6500 men with twenty field and four mounted guns. To keep his movement secret he chose the most difficult and seldom-used passage across the Ağri-dağ, the Zor Pass (2544 m.),[1] and descended into the valley of the Balik-çay at Musun. By the evening of the 9th Russian cavalry patrols were within five miles of Bayazit.

In spite of all secrecy, Tergukasov's march was known to the Turks on the same day. Ismail Hakki sent orders to Faik to join Münib before Bayazit, but he himself preferred to stay at Diyadin. Faik moved only after dawn on 10 July.

Tergukasov's troops were already in action by 5 a.m. of the 10th. The main force moved directly on the town from the north-west, while cavalry with the four mounted guns covered the road to Diyadin. An attack by two Crimea battalions was directed against the 'new town', north of the citadel, while two of the Stavropol battalions

[1] *Zor*, T.: lit. 'difficult'.

marched against the 'old town', south of the citadel. The Russians early occupied the important heights to the east of the town, which commanded not only the upper part of the town but the citadel itself. The Turks appeared to have no standing force posted there. By 7 o'clock strong Turkish columns under Faik Paşa appeared from the south, from the direction of Teperiz, but two Russian battalions in reserve on the ridge behind Maryamana village checked their advance.

The crisis of the fight for the town was reached at 10 a.m. The Krimski men had captured the 'new town', but the Stavropolskis were held up by the stubborn resistance of Turks and Kurds in the stone houses of the 'old town'. Faik Paşa, meantime, with four battalions and several thousand Kurds, was trying to turn the Russian flank at Maryamana and he brought five guns into action against the four of the Russians. Cavalry patrols reported that new enemy forces (under Ismail Hakki) were moving in from the direction of Diyadin.

However, the issue in the town was finally decided by the two rifle companies which had occupied the heights to the east of the citadel. The 'old town' was now attacked from three sides and the resistance of the defenders finally broke. The Russian garrison of the citadel could open the gates to their deliverers.

Faik Paşa was evidently impressed by the turn of events in the town; he stopped his attack on Maryamana to await orders from Ismail Hakki. Amilakhvari's cavalry opposing the advance of Ismail Hakki from Diyadin was now reinforced by guns and infantry from the town and the perspicacious Kurd broke off the engagement.

At dusk, Tergukasov concentrated all his force, including the exhausted garrison of Bayazit, at the village of Zangezur, north-west of the town. Next morning he began his withdrawal on Iğdir without any attempt at interference on the part of Ismail Hakki. Tergukasov's losses during this really brilliant operation of the relief of Bayazit did not exceed 100 men.

Tergukasov's success, in the true tradition of Russian arms during the earlier Caucasian wars, was badly needed to raise the prestige of Russian commanders and the spirits of troops in the field.

Devel's siege operations before Kars had met with no success, and the powerful Turkish batteries covering the fortress had inflicted on the Russian batteries, which had been set up, rather more damage than they had themselves sustained. At the same time a further offensive by Oklobzhio against Batum had failed, and the situation north of the Rion and along the Black Sea coast was causing anxiety.

On 9 July, the Grand Duke Michael, following an exchange of

telegrams with imperial headquarters on the Danube, ordered the withdrawal of the Russian army before Kars to the site of the old camps of former wars in the region of Başgedikler and Kurudere. The Danube had recently been crossed with the loss of only 850 men, and imperial headquarters seemed to consider prospects in the Balkans as favourable and all offensive action on the Caucasian front as unnecessary and even dangerous. Orders were given to abandon the operations against Kars and Batum, to achieve the liquidation of the Turkish threat to the Caucasian coast, to ensure the protection of the Erevan region by reinforcing Tergukasov, and to take all steps for the full maintenance of internal security in Chechnia and Dağistan. The 40th Infantry Division, twelve battalions of which had already been ordered to the Danube, was allocated as reinforcements for the Caucasian army; it was expected to arrive by the middle of August.

The discovery on 10 July that the Russians were withdrawing from round Kars was a surprise to Muhtar Paşa. The force with which he had advanced from Zivin was certainly not stronger than Heimann's corps and definitely weaker than the combined forces of Heimann and Devel. By the 20th, Muhtar had available in the Kars region, in addition to the garrison of the fortress (about 8000 all arms), fifty-two battalions of infantry, 5000 cavalry—mostly Cherkesses of doubtful value—and forty-eight guns. Hasan Paşa, the new and efficient governor of Erzurum, was organizing another twenty battalions and three batteries in his rear which should soon be ready for further operations.

The strategic initiative had, in fact, now passed to the Turkish commander, but, in spite of a natural optimism, he realized the limitations of his situation. His troops, stubborn and courageous on the defensive, had not been adequately trained to manoeuvre in the field. His numbers were limited and he could rely on no substantial reinforcements, since even recruits from the coastal district were being dispatched to the Balkan theatre. On the other hand, as the year advanced, the grand duke would undoubtedly receive reinforcements. Ahmet Muhtar could not utilize the initiative, which had so unexpectedly passed to him, otherwise than by taking advantage of a period of grace to prepare for a renewal of the Russian offensive. Thus by 15 July, both armies on the Caucasian frontier were unable to undertake an offensive and both commanders chose to assume for an indeterminate period a definitely defensive attitude.

CHAPTER XI

THE DEFENCE OF BATUM AND THE DIVERSION IN ABKHAZIA. SECOND PHASE, JUNE–AUGUST 1877

MAPS 4, 8, 9

IN the Batum area General Oklobzhio had, during the fourth week of June, renewed his attempt to obtain a decision. Following the capture of the Sameba heights, the Russians had spent ten days in preparing a further forward movement. The natural difficulties of bringing forward artillery and supplies up to the Sameba position were very great. Moreover, in a territory only ten miles in depth, the troops had to construct eighty miles of roadway and thirty bridges and to cut down thousands of acres of timber. By the evening of 22 June the batteries on the Sameba heights were ready. Two more batteries were put in position on the Table Hill in the angle formed by the Kintrishi and its affluent, the Kintash. The rest of his guns Oklobzhio had placed on the Hutsubani heights.

Oklobzhio divided his force into five groups:

(1) Colonel Gurchin's column (1st and 4th Caucasian Rifles, Cossack plastuns and Gurian and Imeretian militia, with two battalions of the 76th Kuban Regiment in reserve) had orders to cross the Kintash ravine and to capture the Turkish positions on the Kvirike ridge.

(2) General Shelemetev (2nd Caucasian Rifles, two battalions of 163rd Lenkoran Regiment and Imeretian militia, with three battalions of 101st Aleksandropol Regiment in reserve on Table Hill) was to advance on Gurchin's right.

(3) A general reserve (151st Pyatigorsk Regiment and two battalions of 75th Sevastopol Regiment) was stationed on the Hutsubani heights.

(4) On the left rear, two battalions of the 163rd and one battalion of the 161st were held in observation.

(5) The 164th Regiment watched the lower Kintrishi.

This distribution of forces indicated that General Oklobzhio had not definitely decided what kind of action he intended to undertake against the Turkish positions. Officially, he proposed a reconnaissance in force; but if the attacking columns of Gurchin and Shelemetev proved successful, he intended to develop the reconnaissance into a major offensive.

Derviş Paşa had made his preparations to meet the Russian attack. He had four four-gun batteries supported by ten battalions of infantry, disposed along the Kvirike ridge. His right flank in the dense forest on both sides of the Kintrishi ravine was covered by 3000 irregulars. The Kvirike ridge represented the second line of the Kintrishi position, the third and principal line being along the Deva heights to the south of the stream of that name. Here two batteries of Krupp positional artillery had been installed supported by two batteries of field guns. The field of fire of the heavy pieces covered the Kintrishi valley up to Table Hill. There were other heavy and field batteries in the Tzikhis-dziri region. The 9 in. guns of a monitor could be brought to bear on the Hutsubani heights. The bulk of his infantry Derviş Paşa held in readiness in a fortified camp in the area between the Deva and Tzikhis-dziri.

Early on the morning of 23 June a strong Russian artillery bombardment silenced the Turkish guns on the Kvirike ridge. The supporting Russian infantry attack, however, progressed very slowly. One line of Turkish trenches was taken and during the afternoon a second, but the Turkish infantry stubbornly held on to the ridge. Gurchin hoped that his plastuns and militia, moving along the Kintash ravine, would be able to outflank the Kvirike position, but in the valley the advancing troops came under a hot fire from parties of irregular sharpshooters and the fighting assumed the character of numerous skirmishes between small dispersed groups. Moreover, the Turkish artillery on the Deva ridge prevented the Russian field batteries from advancing to support the infantry. By the evening Russian losses had reached 1000 killed and wounded. Troops were tired and a further advance could only be pressed by reinforcements from the reserve. Gurchin and Shelemetev both reported to Oklobzhio that the Deva positions were extremely strong and well defended. About 7 p.m. Oklobzhio decided to treat the fighting of 23 June as nothing more than a reconnaissance in force, and he ordered the troops to withdraw during the darkness to the positions on the Sameba heights and Table Hill. On the following day (24 June) the Turks counter-attacked. The dense undergrowth allowed them to approach the Russian positions so closely that the Russian batteries were for some time in danger of capture and were for hours firing grape. The Turks carried through several attacks with the greatest spirit and they were only held back at the point of the bayonet. In the afternoon Derviş Paşa drew off his men with losses approximating to those suffered by the Russians on the previous day.

Oklobzhio certainly could not consider the results of these two days' fighting as satisfactory. In his report to general headquarters at Tiflis, he described the strength of the Turkish positions on the Kvirike and Deva heights, and he explained the dangerous situation of the Russians on the Sameba line with their left flank exposed to action by partisans round Achkvis-tavi and their right under fire from Turkish warships. Oklobzhio's report was received in Tiflis at the same time as the news of the Zivin defeat, and he was ordered to retreat to his original position at Muha-Evstate. By 30 June, Oklobzhio had completed his withdrawal to the Muha-Evstate line where his troops had first concentrated two months before. During all this period, considerable forces had been diverted from the principal front; and 2000 killed and wounded, with an equal number sick from the enervating coastal fever, were the price of this miscalculated and ill-prepared offensive.

While Oklobzhio was threatening Batum along the coast, a movement against the town had been made from the direction of Ardahan on the initiative of the commandant of that place, Colonel Komarov. On 24 June, Komarov marched with two and a half battalions against Ardanuch which commands the ravine of the Çoruh and lies fifty miles from Ardahan on the Batum road. He completely defeated the local Turkish force of 3000 redifs and irregulars, but, when he asked permission to march on Artvin, he was ordered to return to Ardahan without delay. (Cf. Map 1.)

At the beginning of June, General Kravchenko's detachment, retiring from Sukhum, and General Alhazov's column, advancing from Kutaisi, had united at Okum. The combined force amounted to 6000 infantry, 1500 cavalry and twenty guns, and should have been adequate to liquidate the bands of insurgent Abkhazians (who did not amount to more than 3000 or 4000 operating on both sides of the Kodor). Alhazov, however, had no definite plan of action. He sent small detachments towards the Kodor, but when some *sotni* of Samurzakan and Imeretian militia were surprised by the Abkhazians he was so impressed that he decided to concentrate his forces behind the River Gallisga at Illori at the mouth of the river and at Pokveshi, twelve miles from the sea. Ochemchiri, ten miles to the north of Illori, as well as the northern part of Samurzakan, lying between the Gallisga and the Kodor, was left to the enemy. After this concentration, Alhazov received reports that the Turks were preparing to attack Illori and Pokveshi.

Three Turkish steam frigates bombarded Illori noisily but rather harmlessly, and the Abkhazians swarmed round Pokveshi where the

Imeretian yeomanry, armed only with flint muskets, were left to fight them off. The Abkhazians now insisted on more effective Turkish support, and between 15 and 17 June the Turks landed at Ochemchiri four battalions of infantry with one field battery. Transports, frigates and two monitors continued to lie off the town. The Turks began to send detachments along the coast towards Illori; and Alhazov, urged to action by orders from Tiflis, decided to attack Ochemchiri. He planned an operation in three columns for 17 June. Two battalions with half a battery were ordered to cross the Gallisga at Illori and engage the Turks, while two battalions of yeomanry and militia, under a loyal Abkhazian noble, Prince Sharvashidze, marched on Ochemchiri with the object of outflanking the Turks who were advancing on Illori. A third column, led by Alhazov, acted between the Illori force and Sharvashidze's local troops. Sharvashidze moved too early and approached Ochemchiri before the Illori column had crossed the Gallisga. He was met by gunfire from the warships and a hot fusillade from the Turkish infantry. He halted, awaiting orders, got none, and slowly retired. The Illori column crossed the river and also halted, awaiting the two other columns, which failed to appear. Alhazov, leading the third column from Pokveshi, was ambushed by the Abkhazians in the marshy undergrowth along the right bank of the Gallisga. The Russian losses in this confused and indecisive fighting amounted to 350 killed and wounded.

The Abkhazian insurgents, encouraged by an action which they regarded as a victory, immediately received a large accretion of strength from the surrounding countryside. They crossed the Gallisga ten miles above Pokveshi and threatened Alhazov's base at Okum. The Russian commander was saved from serious embarrassment by the approach of reinforcements which had already crossed the Ingur. Headquarters at Tiflis had decided to effect a definite liquidation of the situation along the coast. The arrival of reinforcements quickly raised the number of troops under Alhazov's command to 9000 infantry, 1500 cavalry and twenty-eight guns. He was ordered to take the offensive, while at the same time Colonel Shelkovnikov, who had three battalions and one Cossack regiment at Sochi, advanced along the coast, and General Babichev, with two regiments of Cossacks and two plastun battalions, was ordered to move from the Kuban by the Marukh Pass. Thus about 17,000 troops were now concentrated to liquidate a diversion which had been initiated by a few Turkish battalions and a few thousand Abkhazian insurgents who had really no military value.

Between 10 and 20 July small Russian detachments took the offensive against the Abkhazians in Samurzakan. There were successful actions at the villages of Tkvarcheli, Gubp and Eshketi. The insurgents dispersed and the local population quickly submitted. Alhazov then moved three battalions, one regiment of Cossacks and four guns to the River Merkulka where the Abkhazians were defeated on 22 July. Djgherdi, at the head of tracks leading to the lower Kodor, was then occupied. This movement appeared dangerous to the Turkish command, and on 1 August the Turks evacuated Ochemchiri by sea and landed a detachment which took station at Drandy, facing Djgherdi on the northern bank of the Kodor. By 10 August Alhazov held all the left bank of the Kodor, and the Turks on the right bank withdrew to the line of the Kelasuri river, only three miles south of Sukhum.

In the meantime Shelkovnikov had on 10 August begun to move along the coast from Sochi, while Babichev started his march over the Marukh Pass. In order to hide his movement from the Turkish warships in observation along the Coast Road, Shelkovnikov followed the difficult mountain tracks which run parallel with it. On 13 August he reached a point one and a half miles from Adler. He had to cross the River Mzymta and then the River Bzyb, and between these two streams a major difficulty awaited him in the Gagry defile. At this point the rocky and precipitous flanks of the coastal range descend to the beach, and there was no passage other than the coastal road which had been blasted out of the rock ledges. This road was under fire of a Turkish monitor which had been lying off Gagry since the Turks became aware of Shelkovnikov's movement. An abandoned Russian stone fort situated to the north of the defile was held by several hundred Cherkesses and Abkhazians. Shelkovnikov, however, secured the co-operation of Russian naval officers at Adler. On the night of 17–18 August in heavy rain his column advanced towards Gagry. The fort was taken by a surprise attack, and at the moment when the Turkish monitor was heavily bombarding the fort and the road, the fast Russian sloop *Konstantin* appeared on the horizon drawing off the monitor in pursuit. The diversion gave Shelkovnikov several hours in which to pass his column through the defile. On 20 August he reached the River Bzyb which he crossed ten miles inland. On 23 August he was already approaching Gudauti. Here a fortified position was occupied by one Turkish battalion and 1000 irregulars with four guns, supported by three Turkish warships off the coast; but it soon became evident that the Turks were only resisting long enough to cover their evacuation. They had, in fact, already decided to abandon Sukhum, and they were principally concerned to remove their regular troops and the Abkhazian insurgents who feared to remain in the country.

To gain time necessary for the evacuation of Sukhum, the Turks concentrated their forces on fortified positions covering the town in a semicircle. Their trenches were supported by six four-gun batteries, and three monitors and two frigates gave covering fire from the sea. On the night of 23–24 August, Russian naval officers, taking advantage of an eclipse of the moon and operating from the sloop *Konstantin* with converted launches, undertook a gallant torpedo attack on the Turkish ships. The attack damaged but failed to sink a monitor, but it was sufficient to shake further the confidence of the Turks. By 27 August, Shelkovnikov was marching on Sukhum. Babichev, meantime, had crossed the Marukh. On 24 August he was at Tsebelda, and on the 29th he was within four miles of Sukhum. Alhazov's column had crossed the Kelasuri and approached the Sukhum entrenchments from the south. Shelkovnikov insisted on an attack, but Alhazov preferred to allow the Turks to withdraw without serious action. On the night of 31 August–1 September the Russians found that the Turks had already gone.

During the same period the Russians had been successful in liquidating the simmering revolt in Chechnia and Dağistan. The Turkish operations on the Caucasian coast and the news of the Russian defeat at Zivin had led to a dangerous spread of discontent and fanaticism throughout the eastern mountains, but armed revolts in different areas were spasmodic and unorganized. Conditions in the mountains continued to make necessary the retention of the 20th and 21st Infantry Divisions and a number of Cossack regiments, and such was the nervousness of general headquarters at Tiflis that in August only two regiments of the 38th Division in the northern Caucasus were moved from the mountains to reinforce the army on the Turkish frontier.

CHAPTER XII

THE ARAS VALLEY AND THE PASSES OF THE AĞRI-DAĞ. BATTLE OF KIZIL-TEPE. AUGUST–SEPTEMBER 1877

MAPS 4, 13, 14

CIRCUMSTANCES imposed a lull in operations during the best months of the campaigning season on the main front along the Transcaucasian frontier. The Russians hesitated to undertake any offensive movement pending the liquidation of the Turkish threat to the Caucasian coast, and they were awaiting the release of troops from this area and from Dağistan and the arrival of reinforcements from Russia. On the other hand, the Turkish commander, while realizing that the initiative had passed to him, was under no illusions as to the capacity of his troops to defeat the main Russian army in an offensive action; and he knew that with the autumn the arrival of reinforcements would alter the balance of forces in favour of his opponent. Nothing was left to Ahmet Muhtar but to develop a defensive strategy, and this strategy had two aspects: the occupation and fortification of a strong defensive position from which to resist the anticipated offensive of Loris Melikov, and the development of a threat to the relatively weak Russian forces under Tergukasov in the Aras valley which might have the effect of diverting reinforcements from the main Russian army near Kars to the Erevan front.

Ahmet Muhtar chose for his defensive positions the Alaca-dağ massif which forms a part of the mountain complex bounded on the south by the Aras, on the north by the Kars-çay and on the east by the Arpa-çay. Alaca-dağ rises to more than 2440 m., sloping in a number of bluffs to the upland plain—which itself lies at a level of 1500–1800 m.—between Kars and Aleksandropol. The saddle of Alaca-dağ, about five miles across, forms the watershed of the Mavrik-çay (modern Kuru), flowing north-east to the Arpa-çay, and of the Mağarcik-çay, flowing west to the Kars-çay. The deep ravines of these two streams give protection to troops on the western flank of the main massif. The ravine of the Mavrik-çay is overlooked by the shoulder of Çift-tepe ('Twin hill') which itself forms part of the ridge called Naharci-dağ falling steeply south-east to the valley of the Arpa-çay at Alaca village two and a half miles from the river. From the southern flank of Alaca-dağ the Teymahlik ravine runs down to the Arpa-çay. The whole of the Alaca-dağ massif constitutes a very strong natural fortress overlooking the plain between Aleksandropol and Kars on the north, and the valley of the Arpa-çay ravine forms a *glacis* which is further strengthened by the situation of two isolated hills which represent natural bastions, Kizil-tepe ('Red hill') and Üç-tepe ('Three hills').

Çift-tepe, on the left flank of the Alaca-dağ position, is thirteen miles from the fortress of Kars (the nearest fort, Hafiz Tabia) with which it was connected in 1877 only by a bad track leading from Haciveli (south-west of Subatan on the Aleksandropol-Kars road) up the ravine of the Mavrik-çay through the Yayla[1] Nalband hamlet and round the southern flank of the isolated mountain of Sivri-tepe ('Steep hill'). Sivri-tepe, with its twin Avliar, became necessarily the advance left-flank defences of the Alaca-dağ position—together with Vezinköy village (half-way between Çift-tepe and Hafiz Tabia on the 2000 m. contour). But to ensure the defence of Vezinköy and the maintenance of communications with Kars it was necessary to hold also two other isolated heights to the north, known as Büyük ('Great') and Küçük ('Little') Yahni ('Stew').

From the above description it will be clear that the greatest defect of the Alaca-dağ position was its length; it was too long to be adequately held by the forces at the disposal of Ahmet Muhtar. The distance from Inah-tepe to Vezinköy is fifteen miles, but to defend this front the Turkish commander had only his army corps, forty battalions strong, reinforced by a weak composite division from the Kars garrison—a total of 30,000 infantry with sixty guns. His problem became even more difficult and virtually insoluble when he found it necessary to occupy and fortify as advance bastions the Great and Little Yahnis to the west of the Subatan-Vezinköy road and Kizil-tepe and Üç-tepe to the east of it. In the end his total front extended over an area of twenty-five to thirty miles. Moreover, there were other serious defects. There were no lateral communications along the front through Inah-tepe and Çift-tepe, while the deep ravine of the Mavrik-çay made difficult contact between troops on the Alaca-dağ position and units in the Avliar-Vezinköy area. In practice the ravine divided his forces into two badly connected sectors of defence. There were few natural springs on the heights of Alaca-dağ

[1] *Yayla*, T., a summer settlement of pastoralists; as opposed to *Kişlik*, a winter settlement.

and it was difficult to maintain supplies from Kars. At a height of over 2400 m., troops suffered from the heat of the high noon and the sharp cold of the nights. There was a shortage of greatcoats and blankets, and the bare slopes of Alaca-dağ provided little firewood or even scrub.

By the end of July Loris Melikov had determined, through cavalry reconnaissance, the main outline of the Turkish positions, and he knew that the Yahnis with Avliar-tepe and Sivri-tepe were not fortified and only weakly held by irregulars. The Russian commander was relieved of his anxiety that Muhtar might move to recover Ardahan, but he still remained uneasy as to an offensive directed from Bayazit against Erevan, and he thought that Muhtar might move with the object of placing himself between the main Russian army and Tergukasov's force at Iğdir. Melikov therefore stationed a strong cavalry force at Kuyucuk, where the Mavrik-çay joins the Arpa-çay, and on 30 July he pushed a column under Tsitovich down the Arpa-çay as far as the ruins of the medieval Armenian capital of Ani, north-east of the village of Alaca. When on 4 August Russian headquarters got the news that Ismail Hakki had crossed the Ağri-dağ and invaded Russian territory, Tsitovich's column was ordered to reinforce Tergukasov. Further reinforcements were passed on through Ani to Iğdir, while Melikov himself drew new troops from Ardahan and the Rion basin. Muhtar, by means of Ismail Hakki's diversion on the Aras, was thus achieving his object of keeping down the numbers available to Melikov for a main offensive.

By the end of July reinforcements sent from Erzurum had brought the number of troops concentrated in the Diyadin-Bayazit area up to thirty-six battalions (about 20,000 men), six batteries (of six guns each), 2000 *süvari* and Cherkesses and 4000 Kurds. To oppose an advance by Ismail Hakki and to cover the borderland between Mount Ararat and the point where the Arpa-çay joins the Aras, Tergukasov had a force which had been increased now to about 8000 infantry, 4000 cavalry and forty-eight guns.

Several mountain passes give access across the Ağri-dağ into the valley of the Aras and general headquarters at Tiflis expected Tergukasov to be able to cover all these. The events which follow are instructive, for they demonstrate that the defence of a mountain line with numerous passes is even more difficult than the defence of a river line with several passages across it. The efficient Tergukasov with his excellent troops was unable to prevent the incompetent Ismail with numerically superior but indifferent troops from crossing the mighty ridge of the Ağri-dağ.

PLATE IV

OLD BAYAZIT: THE RUINED CITADEL AND THE MOSQUE OF IŞAK PAŞA

Tergukasov's difficulties were increased by the particular character of the terrain which he had to defend. The borderland between the Ağri-dağ and the Aras, sixty miles in length from the Persian frontier to the neighbourhood of Kulp (Tuzluca), is only twenty miles wide; from eight to twelve miles of this depth are occupied by the steep slopes of Ağri-dağ; from twelve to eight miles form the plain through which the Aras runs. The plain itself has considerable altitude, more than 900 m. above sea-level; but the Ağri-dağ rises to heights of between 2100 and 3350 m. (the peak of Great Ararat reaches 5100 m., so that the northern slopes fall rather abruptly to the plain. A great number of small streams flow towards the Aras and in the plain they unite and form wide stretches of marshy ground. This marshy belt had been exploited by the Armenian peasantry for the cultivation of rice, and the ground was intersected by a network of irrigation canals. Since the Armenian immigration following the war of 1828–9, Iğdir had become the centre of the numerous prosperous villages of the rice region, and the Russian government found it important for political reasons to ensure the absolute protection of this peaceful population: both Loris Melikov and Tergukasov, as distinguished Armenians, were particularly concerned for the safety of their countrymen.

Iğdir became necessarily an advance base for Tergukasov, although this administrative centre, which was connected with Erevan, some forty miles to the north-east by a bad road, was very inconvenient and, surrounded as it was by marshes to the north and east, an unhealthy station for troops.

Several tracks led from the Aras across the Ağri-dağ to Bayazit and Diyadin. These were, from east to west:

(*a*) a mule track from Iğdir through the Harbahacar gorge (3355 m.) round the western flank of Ararat to Bayazit;

(*b*) the usual road from Iğdir by Argaci and Orkov to the Çengel Pass (2074 m.)—where the Russians had fought a successful action in 1854—descending to Arzap and Bayazit;

(*c*) a track from Iğdir by Halfali, Alikoçak, Zor village, through the Zor Pass (2500 m.) into the valley of the Balik-çay ('Fish river') at Musun: this track proved passable for wheels;

(*d*) across the eastern shoulder of Oyuklu-dağ ('Hollow mountain') (twelve miles west of Iğdir) over the Gürünsaray Pass (2208 m.) to Musun: this pass had been used by Tergukasov earlier in the year during the retreat from the Eleşkirt plain;

(*e*) a track from Göllükçü on the Iğdir-Kulp road to the Aslanli Pass (2257 m.) and Musun;

(*f*) a track from Göllükçü by the Abbas-göl Pass (2440 m.) to Lake Balik and Surp Ohannes.

Of all these tracks, that over the Çengel Pass was deemed to be the best and the shortest and Tergukasov directed his principal interest to it. He placed the bulk of his troops in echelon along the route Iğdir-Çengel: five battalions with two regiments of Cossacks and four batteries in position covering the pass, four battalions with six guns at Orkov, and one battalion with two guns at Iğdir. The Zor Pass was watched by six *sotni* of Cossacks with four squadrons of dragoons and four *sotni* in reserve at Halfali. At Kulp were three battalions, fourteen *sotni* but no guns.

Tergukasov supposed that Ismail's main force was concentrated at Bayazit, but he was wrong. Reasonably enough Ismail had chosen Diyadin as his principal base, and most of his men were in the well-watered upland valley of the Balik-çay near where numerous tracks cross at Musun. The Kurds of the district had informed Ismail that the Zor Pass was not too bad for the passage of troops and, accordingly, at dawn on 4 August 4000 Kurds, supported by six battalions of infantry and six guns, attacked the Cossacks stationed on the Zor and advanced to Alikoçak four miles to the north of the pass. Here the Turkish infantry and guns took up a position, but the Kurds advanced a further six miles and descended to the plain at the village of Halfali, only five miles from Iğdir. At this point they were counter-attacked by the Pereyaslav dragoons strengthened by a mounted battery, and they withdrew to Alikoçak where they remained. Meanwhile another band of Kurds, 1500 strong, had appeared in the Gürünsaray Pass, where they met no opposition; they penetrated across the Oyuklu-dağ, but later turned back to the pass. Ismail made no further move, and nothing had happened beyond the incursion of some Turkish advance elements a few miles over the Russian frontier. But a great impression had been created. The Armenian population took flight in panic across the Aras. Tergukasov regarded his main force in the Çengel as outflanked and fell back in haste on Iğdir; at the same time he sent to general headquarters for reinforcements. For a second time, the Turks, by an ingenious and inexpensive diversion, had caused serious anxiety to the Russians and imposed a redistribution of forces at the expense of the main army in the field.

Ismail did not think it opportune to move all his forces across the Ağri-dağ, but the detachment (strengthened by four battalions and one battery) remained on the track from the Zor Pass to Halfali. The Turks found a series of good defensive positions on the natural terraces between Zor and Alikoçak. Their patrols remained in view of Halfali and a small force stayed in the Gürünsaray Pass. Tergukasov's troops, concentrated in the region Iğdir-Halfali-Argaci, had to organize a defence in the most unfavourable conditions of terrain and climate—in the fever-ridden flats of the Aras in high summer. The Turks were meantime enjoying the excellent grazing and water of the Ağri-dağ highland. Never was the importance of the tactical use of local climate and terrain for the purpose of imposing maximum strain and wastage on the enemy better demonstrated.

Russian general headquarters apprehended that the Turks would advance from the Gürünsaray Pass towards Kulp and the valley of the Arpa-çay and try to cut in between the forces of Tergukasov in the Aras valley and of Loris Melikov at Kars, and Tergukasov received orders not only to protect Erevan but to secure Kulp and the approaches to the Arpa-çay. But Ismail Hakki had all the failings of a Kurdish chieftain—and a lazy one at that; he had shown himself incompetent as an organizer and irresolute as a leader in the field. When he had the possibility of further disrupting Melikov's autumn plans to an extent which might have altered the subsequent course of the campaign, he preferred not to move and remained passive with two-thirds of his army at Musun. He had the mountain chieftain's dislike of initiating operations in the fall of the year, and he must be credited too with an appreciation of the extreme difficulty of supplying his forces if he pushed the bulk of them across the Ağri-dağ and fought the Russians on the northern slopes of this wild and desert range.

Meantime Tergukasov, with a view to covering Kulp and the Arpa-çay and getting on the flank of Ismail's positions in the Zor and Gürünsaray Passes, gradually moved a part of the Iğdir troops to the west. When General Tsitovich arrived at Kulp, Tergukasov directed half his own force to Göllükçü, west of Iğdir and on the road to Kulp. From Göllükçü tracks forked south-east over the Aslanli Pass to Musun and south-west over the Abbas-göl to Surp Ohannes in the Eleşkirt valley. Only seven battalions, with two regiments of Cossacks and sixteen guns, remained between Halfali and Iğdir. The Turks, observing a reduction in the Iğdir force, made two attempts (18 and 24 August) to reoccupy Halfali, but they were repulsed.

The Russians now intended an offensive movement by Tergukasov against Ismail Paşa's forces, and while reinforcements were sent to the Kulp-Göllükçü area, Loris Melikov undertook a reconnaissance in force against the positions of Ahmet Muhtar on the Alaca-dağ. It might, in fact, be claimed that the Turkish demonstration in the

Ağri-dağ had been so effective as to deflect temporarily the direction of Russian offensive action. The Russian attack directed against the Turkish left-flank positions round the two Yahni heights failed to achieve any object and was beaten off with losses, on either side, of about 500 men. A Russian reconnaissance round the extreme right of the Turkish positions in the neighbourhood of Alaca village elicited, however, the important information that there were neither fortifications nor troops in this area, and that the emphasis of the Turkish defence was on the positions to the west of Inah-tepe. At the same time the Russians came to the conclusion that any attack on Inah-tepe would not be easy and that even its capture would not endanger the main Turkish positions on Alaca-dağ.

On 23 August further reinforcements under General Devel were sent to Tergukasov, who continued to show some reluctance to launch his troops over the difficult paths of the Abbas-göl and Aslanli passes against the main force of Ismail Hakki in the upland valley of the Balik-çay. Loris Melikov only retained thirty-two battalions in his camp at Başgedikler, and Ahmet Muhtar on Alaca-dağ soon observed the reduction of Russian strength on his left flank.[1] At the same time the failure of the Russian attack on 18 August had heartened the Turks, while the hardship imposed on his men by the rainstorms of the heights and the cold, which had begun to be uncomfortable since the beginning of August, decided the Turkish commander to move the bulk of his forces forward into positions at a lower altitude: along the slopes of Alaca-dağ (east of the Mavrik-çay ravine) and along the forward slopes of Avliar-tepe (west of the ravine). His new front was thus designed to run from Haciveli through Kerhane to Dünyalik. On the left this line would be secured by the two heights of Great and Little Yahni, and on the right Ahmet Muhtar chose as the objective of his local offensive the isolated height of Kizil-tepe, lying in a direct line about half-way between the Turkish position on Inah-tepe and the Russian camp at Başgedikler.

On 23 August Muhtar observed Devel's column moving out of the Başgedikler camp towards Kuyucuk, and on the next day it was reported moving along the left bank of the Arpa-çay towards Kizil-kilise opposite Ani, where it camped.

At 3 o'clock on the morning of 25 August, eight picked battalions under the German Mehmet Bey surprised the battalion of Imeretin-

[1] Norman, *The Times* correspondent, remarked that from the heights of Alaca-dağ it was possible to observe all movements of Russian troops in and around the camps in the plain below.

skis on Kizil-tepe; these, after losing half their strength, fled in disorder towards the Başgedikler camp. The alarm spread to the principal Russian camp round Kurudere and Poldervan, where the crowds of Armenian stall-holders began to pack their goods and flee towards Aleksandropol; and the panic infected the train personnel in the rear cantonments. Loris Melikov with his staff took station on the Karayal hill; he was awaiting reports from his left flank before issuing orders. Here, Russian units, without orders, were in some confusion; but Chavchavadze took the important step of moving up his cavalry and a horse battery to occupy the height of Üç-tepe between Kizil-tepe and the Arpa-çay. The Turks meantime, always quick to dig, had entrenched themselves on the three summits of Kizil-tepe and repulsed repeated attacks by the Vladikavkazskis and Imeretinskis under Colonel Romanov; the latter, however, finally gained a footing on the eastern summit of the hill.

On the left of the Turkish line, strong columns moved from Haciveli and Kerhane against Subatan. They were driven back by three Sevastopol battalions and two rifle battalions under Colonel Komarov, but another Turkish column moving along the right bank of the Mavrik-çay, by-passed Subatan village and placed Komarov's men in a very dangerous position. The Mingrelski Grenadiers with one battery, however, checked this new Turkish advance.

The Russian front had been stabilized, but Ahmet Muhtar continued to manoeuvre with some success against the impressionable Loris Melikov. A movement of Turkish infantry from the Yahni positions and the appearance of Turkish cavalry on the Parket road was sufficient to confirm the Russian commander's suspicion that the attack on Kizil-tepe had been only a feint and that the main Turkish attack was yet to be delivered against the Russian right flank. Hence Loris Melikov continued to immobilize, between Karayal and Kurudere, most of his artillery and cavalry, the 1st Grenadier Brigade and two regiments of the 40th Division, while thirty Turkish battalions, active on the Subatan-Arazoğlu front, were only opposed by fifteen Russian battalions (Komarov's men and the 2nd Grenadier Brigade). Without news from Devel's column (which he had recalled from its march down the Arpa-çay) Loris Melikov became more and more nervous and, having first ordered Komarov's troops to concentrate at Kulveran village in order to maintain liaison between his left and right flanks, he at noon decided to withdraw the whole of his left flank behind the Mavrik-çay on to the triangle, Başgedikler-Oğuzlu-Bayraktar. Melikov's order came at a moment when the Tiflis

Grenadiers and the Vladikavkazskis had almost recaptured Kizil-tepe after desperate fighting, and while Mehmet Bey was only able to hold on to the southern summit. Even the Nizhegorodski Dragoons were ordered to leave Üç-tepe and reinforce the cavalry on the Russian right (which had not yet been engaged), and only a few Cossacks were left to hold this vital point.

At this juncture, Devel's column suddenly and unexpectedly appeared in the neighbourhood of the hill. Devel, who seems to have been in some confusion as to whether he was to obey orders from Loris Melikov or from general headquarters at Tiflis, had camped at Kizilkilise on the night of 24 August; here an order from Tiflis reached him to proceed on the morrow to Kulp. Early in the morning of the 25th, the sound of firing to the north-east was heard, where Mehmet Bey was opening his attack on Kizil-tepe. Devel's troops stood to, but their commander did not know what to do. His officers insisted that it was necessary to cross the Arpa-çay immediately and proceed to the scene of battle. At 6 a.m. Devel got news from General Dukhovski that Kizil-tepe had been taken by the Turks. Loris Melikov's chief of staff added that Devel must return and act against the Turkish flank if he was not bound by any other order (i.e. from Tiflis). General Devel (like many other officers) evidently feared his superiors more than the enemy, and he considered himself bound by the Grand Duke Michael's order of the previous evening. But his officers were insistent, and he decided to march to the sound of the fighting. It was considered more expedient not to return to the Kuyucuk crossing, but to organize the passage of the Arpa-çay on the spot at Kizilkilise. The decision proved a mistake, and Devel was only able to cross the river, at 11 a.m., with six battalions of infantry, the Cossacks and four mountain guns. The rest of the column—twelve guns, the trains and two battalions of infantry to protect them—remained on the left bank of the Arpa-çay. Soon after midday Devel approached Üç-tepe. The Turks at Dünyalik and Arazoğlu did not move but awaited the attack of Devel's troops; Devel, with only four guns, showed no eagerness to attack.

During the afternoon there was a lull in the fighting round Kizil-tepe and Subatan; only the guns on either side continued firing while the Turks were busily digging in. On their left flank they drew in their troops to the Yahni positions, thus abandoning any threat which Melikov might have thought was intended to his right.

Devel had sent a runner to headquarters at Karayal, and by 4 p.m. he got a reply from General Dukhovski that guns and cavalry were on their way to reinforce him; the chief of staff added a disconsolate appreciation of the position and indicated that, in the event of the evacuation of the Poldervan-Kurudere camp, Devel was to retire on Aleksandropol along the left bank of the Arpa-çay. Observing that, in fact, the Russian infantry on his right was retiring beyond the Mavrik-çay, Devel, on the arrival of the Severski dragoons and one battery as reinforcements, still thought it better not to attack, and he withdrew to the Kuyucuk crossing over the Arpa-çay. The Severskis and the battery, however, were left at Üç-tepe, where they were later joined by two companies of Vladikavkazskis coming up from Bayraktar. Thus this important position was actually retained—but not on the order or initiative of Melikov's staff. At Kuyucuk, Devel was joined by some officers who tried to induce him to start an attack on the exhausted Turks at Kizil-tepe; but Devel lost patience and refused abruptly, since his troops were tired and hungry and had no train with them. With the bitter observation that on the morrow they would say that he was responsible for all that had happened, he gave the order to cross the Arpa-çay.

The Kizil-tepe battle of 25 August 1877 was undoubtedly a victory for Ahmet Muhtar and not, like the action of the 18th, an imaginary one. It had, in fact, been only a half-day battle, and during ten to twelve hours' fighting—at times, as at Kizil-tepe and round Subatan, fiercely contested—each side had lost about 1500 killed and wounded. On the Russian side, half this loss had fallen on the 158th (Imeretinski) Regiment, which had suffered 30% casualties and was badly shaken. But as a whole the morale of the Russians was little affected and they remained merely bewildered by the confused handling of the day's fighting by Melikov and his staff. On the following morning the Grand Duke Michael arrived in person at Kurudere, accompanied by General Obruchev, a close collaborator of the Russian Minister of War, General Milyutin, by whom he had been sent from Petersburg to Tiflis as an official adviser to His (rather ineffectual) Imperial Highness. Dukhovski was made the scapegoat of a generally inefficient staff and was particularly reprimanded for having communicated with Devel by runner instead of making use of the new field telegraph (first used by the British ten years earlier during the Magdala campaign).

While still awaiting the arrival of reinforcements released from the western Caucasus and newly mobilized from European Russia, the Russian front was established along a line of about ten miles. A strong cavalry force at Parket covered the main Aleksandropol-Kars road and the routes leading from Kars to Çildir (Zurzuna) and Ardahan. Another cavalry detachment covered the Russian left at the Kuyucuk crossing of the Arpa-çay where passed the main line of supply from Aleksandropol. The main body of the Russians (forty battalions of infantry and 160 guns) was concentrated on a front of six miles from the camp at Kurudere to Üç-tepe hill, which became the outlying locality of two new fortified camps at Oğuzlu and Bayraktar. The Başgedikler camp was thus abandoned as being too exposed to fire from the Turkish guns on Kizil-tepe. The general Russian position assumed a direction north-west to south-east covering Aleksandropol.

The newly advanced front of Ahmet Muhtar ran from Cale past Dünyalik and Kizil-tepe through Subatan village past the Yahni heights to Halefoğlu village and Taşni-tepe. This front was thirteen miles long and infantry was dispersed over an area of ten miles. Further, Muhtar had to keep small garrisons in the fortifications of Inah-tepe, Avliar and Vezinköy. He could not put into the front line more than thirty-five to forty battalions, and these battalions were 25% below the strength of the Russian battalions. He had fifty to sixty guns, of which eighteen were mountain pieces. This new front was good for the deployment of an offensive action, but he had not enough troops available for the effective defence of such an extended line. Muhtar tried to secure reinforcements, but both Derviş Paşa at Batum and Ismail Hakki in the Ağri-dağ refused him troops, and the vicious system whereby all three commanders depended on the intrigues of their respective supporters in Istanbul made it impossible for Muhtar to assert his superior authority. Seconded by Hasan Paşa, the new and efficient governor of Erzurum, Muhtar obtained, during September, twelve fresh battalions of redifs and three batteries; but this was all, and it was clear that, with the prior claims of the Balkan front to be satisfied, the recruiting resources of Asia Minor were approaching exhaustion. Muhtar took all but five infantry battalions from the Kars garrison and organized five more field batteries from the guns of the fortress. He hoped that Russian reinforcements might not be so strong as was reported, but he realized that there was no chance of successful resumption of offensive action against the Russians even before their reinforcements arrived. He set his troops to strengthen the field works on the Yahni heights, Subatan and Kizil-tepe, and, since he could secure no reinforcements from Ismail Hakki, he pressed him to resume the offensive in the Ağri-dağ.

Although large numbers of his Kurds were returning to their villages with the approach of autumn, Ismail Hakki could dispose of a strong

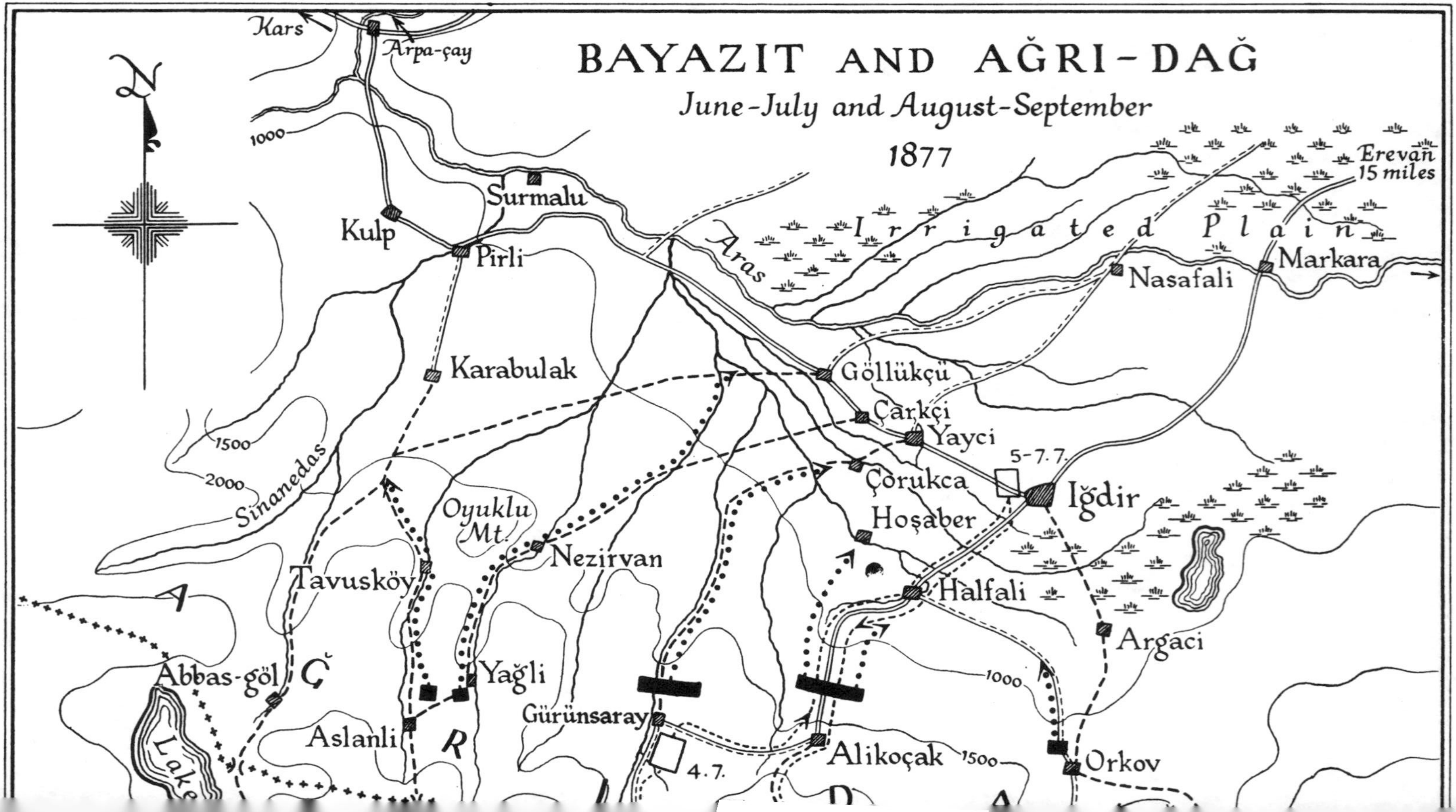
BAYAZIT AND AĞRI-DAĞ
June-July and August-September
1877
Kars
Arpa-çay
N
1000
Surmalu
Kulp
Pirli
Aras
Irrigated Plain
Erevan
15 miles
Markara
Nasafali
Karabulak
Göllükçü
Çarkçi
Yayci
5-7.7.
Iğdir
Çorukca
Hoşaber
Halfali
Argaci
1500
2000
Sinanedas
Oyuklu Mt.
Nezirvan
Tavusköy
Abbas-göl
Yağli
Gürünsaray
Aslanli
Alikoçak
Orkov
4.7.
Lake

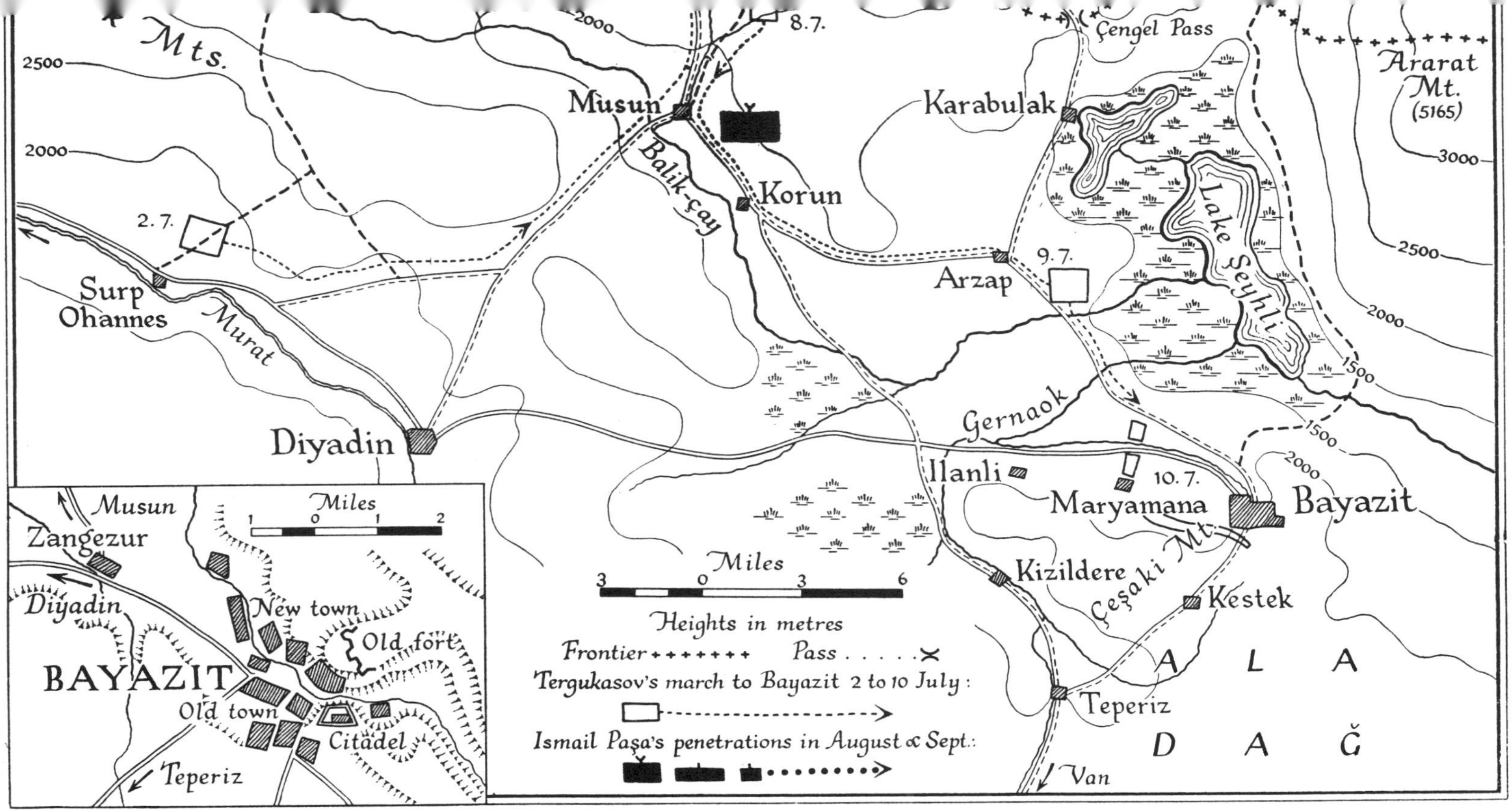

Map 13. The passes of the Ağri-dağ

force in the Ağri-dağ—divided in two 'divisions', each of twelve battalions with two field batteries, and a reserve of six to eight battalions with two batteries. About 4000 *süvari* (regular cavalry) and mounted Kurds and Cherkesses were available. Apart from Muhtar's appeal, the same conditions which had forced Muhtar to move down from the Alaca heights now forced Ismail to undertake a forward movement which would release his men from the increasing cold and wet of their positions in the Zor and Gürünsaray passes. The time was propitious, in fact, to descend into the more clement Aras plain. Without any great enthusiasm, therefore, the Kurdish chieftain decided, about 15 September, to begin an offensive against Tergukasov.

At this date Tergukasov, who had recently been reinforced by two regiments under Devel, maintained a strictly defensive position. His strength had been seriously affected by the fevers which overtook his men in the malarial Aras valley. His force was divided into two groups: the westerly forward in the Kulp-Surmalu area, and the easterly, encamped along a line five miles in length between the villages of Halfali and Çorukca, covering Iğdir. Between the two groups was a gap of about fifteen miles, and here a bad track found its way into the plain from the Gürünsaray Pass. Ismail planned to use this track in an attempt to get between the two groups and outflank the Russians at Çorukca while he made a frontal attack on Iğdir. His movement, however, was badly timed. On 19–20 September he made an attack on the Halfali position which was pressed hard. It was not until 27 September that another Turkish column descending from the Gürünsaray Pass made a strong attack towards Çorukca. General Devel, who commanded (Tergukasov was in the Kulp-Göllükçü area), counterattacked and the Turks retreated into the mountains, fighting a rather desultory rearguard action. Ismail henceforth remained quite passive, concentrating about three-quarters of his troops in the sheltered Balik-çay valley round the track-junction at Musun. His adversary, Tergukasov, during the third period of the campaign which was now approaching, was successful in convincing grand ducal headquarters that the maintenance of a defensive attitude was all that could be reasonably asked of the sorely tried Erevan force. Great events were impending on the main front round Alaca-dağ, and even Ismail Hakki agreed at last to reinforce Muhtar with six battalions and two batteries.

CHAPTER XIII

BATTLES FOR KARS: YAHNI AND ALACA-DAĞ, OCTOBER 1877

MAPS 13, 14; cf. also 4, 10

BY the end of September Loris Melikov had received the considerable reinforcements promised from Russia: 1st Grenadier Division, one Astrakhan and two Orenburg Cossack regiments, one battery of 21st Artillery Brigade and two Don Cossack batteries; further, five battalions, one battery and one Cossack regiment had been drawn from Ardahan, and one dragoon and one Cossack regiment with one battery from Tergukasov. The total of Loris Melikov's army now reached fifty-six battalions, twelve squadrons of dragoons, eighty-four *sotni* and thirty batteries, making 42,000 infantry, 12,000 cavalry and 240 guns.

With reinforcements drawn from Erzurum and the Kars garrison, Ahmet Muhtar could put into the field sixty-four battalions (34,000 infantry), 2000 *süvari* and 6000 irregular horse with ninety-six guns. He was therefore already substantially weaker than the Russians.

Approximately half his strength Muhtar held as a mobile force in the Kerhane-Dünyalik area. The rest was distributed on the fortified heights from Little Yahni to Kizil-tepe and Inah-tepe. A small reserve lay at Avliar (two battalions with two guns) and Vezinköy (five battalions with twelve guns). The bulk of his rather weak and unreliable cavalry lay on the right at Cale.

The distribution of Muhtar's forces was more or less known to the Russians, and it was clear that the weakest point in the Turkish line lay on the left centre between Little Yahni and Subatan. Both the heights of Avliar and Great Yahni were lightly held, and between them ran the main track to Vezinköy. Vezinköy was clearly the key to the Turkish front; with Vezinköy taken, Muhtar would be cut off from Kars whence he drew the whole of his supplies.

So definite appeared the situation that many officers (including Heimann) insisted that the best plan was to ignore the Turkish force on Little Yahni (eight battalions with four guns) and march with a strong column straight on Avliar and Vezinköy. The grand duke, with Obruchev, and Melikov preferred more systematic action; they saw an operation in two consecutive stages—the capture of the Yahni hills to be followed by an attack on Avliar and Vezinköy. During such action, a strong force had to cover the Russian left flank against counter-attacks from the Turks on Alaca-dağ.

Another and more imaginative element had been brought into strategic discussions by the arrival at headquarters of General (formerly Colonel) Shelkovnikov, the hero of the recent advance along the Black Sea coast which had liquidated the Turkish diversion in Abkhazia, who was now applauded as the Skobelev of the Caucasian army. This officer became the chief supporter of the idea of a manoeuvre designed to outflank in depth all the Turkish positions on Alaca-dağ by an advance from the east. The Russians, since 10–12 September, had been reconnoitring the valley of the Arpa-çay along all its length from Kuyucuk to Kulp. A convenient crossing for troops with trains had been found at Post Kambinski, twelve miles south of the Ani ruins. Officers' patrols had undertaken reconnaissances along the right side of the Aras to the south of the Alaca massif. They reached the important track junction at the village of Diğor; from there tracks ran to Kağizman and Kulp on the Aras, to Pazarcik and farther to Vezinköy west of Alaca-dağ, and to Cale east of the massif. Russian patrols even approached Pazarcik (eight miles south-east of Vezinköy and only six miles south of Avliar) without interference. Turkish patrols were not met with anywhere in this region, and Muhtar Paşa appeared to have ignored the possibility of any Russian move round the rear of Alaca-dağ.

Finally, on 23 September, Shelkovnikov received permission to take five picked battalions with two companies of sappers, two Cossack regiments, and twelve guns to act in the rear of the Turkish flank and when possible to develop his action.

Muhtar was made aware of a Russian movement along the Arpa-çay, and on 25–26 September he sent a detachment under Ahmet Paşa (six battalions, three squadrons of *süvari*, six guns, several hundred Kurds) which took up a position in the gorge of the Diğor-çay, mid-way between the village of Diğor and Post Kambinski. At the same time he ordered the fortifications on Sivri-tepe to be strengthened and earthworks to be thrown up on both sides of the Pazarcik saddle.

Since 15 September, Ahmet Muhtar had not been so confident as before. He knew that considerable reinforcements were reaching Melikov; at the same time his own troops were suffering from lack of warm clothing and supplies, and the inadequately equipped hospitals at Kars were crowded with sick and wounded. The Kurds and Cherkesses, with the mountaineers' instinctive dislike of winter operations, were deserting in increasing numbers, and the taste for desertion was spreading to the redifs.

Many of Muhtar's staff were critical of his decision to await attack by a reinforced Russian army on his extended front; and on the eve of the Russian attack, the Turkish commander, after a conference with Ömer Paşa (another German) and Hüseyin Paşa, both newly arrived from Istanbul, telegraphed for permission to withdraw to the Soğanli, making the reservation, however, that his troops were ready to die on the spot. He received no reply; but the proposal had anyhow been made too late, for a few hours later the Russians attacked.

The plan of battle of 2 October was created by Obruchev, confirmed by the grand duke, and executed by Loris Melikov. The battle, intended as a movement to capture Avliar, was limited by events to an attack on the two Yahni hills. The attacking force was divided into three strong columns. The right wing constituted the spearhead of attack, and it received protection from the supporting left wing. It was natural that the right wing should have been the stronger (thirty battalions compared with ten allocated to the left); but it was unfortunate that the third column, representing a general reserve, should have been concentrated eight to ten miles from the attacking units. The right wing was divided into five separate columns, and the orders given to their respective commanders do not seem to have been clear.

The defects of the original dispositions were increased by the bad organization of the movement of troops. The terrain was waterless, but no arrangements had been made to supply water to the troops. The columns attacking the Yahni hills had to make a night march of thirteen or fourteen miles from the camps at Poldervan and Kurudere and they were tired before the fighting began. Maintenance of direction in the dark, as might have been anticipated, proved difficult; and timing, as usual in this campaign, was defective.

The confusion of the Russian night march and the faulty deployment of the columns of Komarov and Grabbe, as morning broke, gave the excellent German commander on Little Yahni hill time to prepare to receive the attack. He retained four battalions in the fortifications on the hill, and distributed another four battalions between the hill and the Kars road to cover his rather vulnerable western flank. Cavalry, with a mounted battery, were along the Kars road, and Hüseyin Hami Paşa, warned at 5 a.m., was moving up seven battalions and two batteries from Vezinköy.

Russian cavalry appeared two miles north-west of Halefoğlu and engaged the Turkish horse and, as these were supported by infantry,

Grabbe and Komarov detached four of their own battalions west to sustain the cavalry action.

At 8 a.m., the colonel of the Pyatigorski Regiment, having had no orders, attacked Little Yahni from the east and was driven off with heavy losses. Grabbe then sent the Pernovski Regiment against the hill from the north, but in spite of the support of two battalions of the Tiflis Grenadiers the attack could not be developed. By 10 a.m. Mehmet Bey, reinforced by Hüseyin Hami's battalions from Vezinköy, was able to counter-attack the Russian right. *Süvari* and Kurdish horse even turned the flank of the Russian infantry, but they were driven back by the Severski Dragoons and Cossacks who pursued the Turkish cavalry to within range of the guns of the Kars forts.

Komarov, meanwhile, had resisted the Turkish counter-attack with his usual *élan*, and it gradually spent its force. Soon after midday came the last chance for a successful storm of Little Yahni. Mehmet Bey's battalions on the hill were suffering badly from the concentrated fire of twenty-four Russian guns, while the troops repulsed in the counter-attack against Komarov had retired exhausted to the Kaleköy ravine (one and a half miles south-west of the hill). By 2 p.m. officers in the groups round Komarov were shouting 'Cavalry! Bring up the cavalry!' Colonel Prince Eristov (a Georgian) with two *sotni* of Kuban Cossacks galloped against the western slopes of Little Yahni, while with even greater dash two *sotni* of Chechens and one *sotnya* of Akhaltzikhe Tartars rode over the trenches on the lower ground sabring the Turkish infantry. This was the moment for the storm, and two of the Pernovski battalions rushed the trenches to the left of the irregular horse. But the effort was made with inadequate force, and fresh cavalry arrived only in time to cover the withdrawal of the exhausted Pernovskis. The regiments of Komarov and Grabbe had been on the move for twenty hours and in battle since 5 a.m. But no units from the reserve were available to drive home the assault.

The reserve columns, during the night, had moved south along the ravine which leads from Yayla Mesko (by Lake Mesk) towards Great Yahni hill. On the hill two miles north-east of Great Yahni and three miles east of Little Yahni, their commander, General Solovyev, stopped at dawn and was for several hours forgotten. General Solovyev had no orders and no initiative to move elsewhere. Later in the day some of Solovyev's units were used rather ineffectively to support General Heimann's operations round Great Yahni, and at one time the Georgian Grenadiers were actually behind Little Yahni without their commander being aware that their support was so badly needed nor Komarov knowing that they were available.

On the extreme left of the Russian right wing, the column of General Shelemetev had reached the base of Great Yahni hill at dawn. This high and cone-shaped eminence was lightly held by only one battalion of nizams, and by 7.30 a.m., after desperate fighting, the position was in the hands of the Russians (only 150 of the defenders survived to surrender).

General Heimann arrived at this juncture to take command. With the self-confidence which usually characterized this commander, he ordered Shelemetev's infantry and horse to advance towards Avliar and Vezinköy. He observed that there was no activity on Avliar, and rightly concluded that its fortifications were not manned by infantry nor provided with guns. He could see also that Turkish infantry were being hurried westward from Kerhane and Bulanik. Muhtar was, in fact, preoccupied with the dangerous situation on his left and had ordered the occupation of Avliar with several battalions and four guns. Heimann thought it possible to stop this regrouping of Turkish forces by a demonstration in the direction of Subatan and Haciveli. He therefore ordered General von Schack to attack with five battalions in that direction. Meantime, Shelemetev reported that his men who had established themselves on Great Yahni had been on their feet since the previous evening and were beginning to suffer from thirst under the increasing heat of the morning sun. Moreover, they had few cartridges left, their reserve ammunition having, in error, been sent with von Schack's column.

Heimann was, in fact, right in his supposition that at 8 a.m. there were no troops in Vezinköy. At that moment there was a wide gap in the Turkish positions, about six miles wide between Avliar hill and the Kaleköy ravine. Two *sotni* of Terek Cossacks actually penetrated as far as Vezinköy and rested quietly in the empty Turkish lines, until 3 p.m.

To cover the left of an attack which Heimann found himself unable to pursue, von Schack attacked as on the parade ground. On his left the Mingrelskis were held up by intensive fire from the Turkish trenches round Subatan, but on his right the 157th (Kutaisi) Regiment took Haciveli at the point of the bayonet and, in the heat of their success, crossed the Mavrik-çay ravine and began to advance up the slopes towards Kerhane. But here they attracted the fire of strong defending artillery and they were counter-attacked from three sides. They were thrown back with heavy losses and only half the regiment recrossed the Mavrik-çay. Simultaneously with the Turkish counter-attack towards Haciveli, Turkish cavalry and infantry were seen moving from Bulanik towards Great Yahni hill. The newly raised Nesvizhski Regiment at the foot of Great Yahni, without cartridges and exhausted by their first experience of fire, began to withdraw, but the Tver Dragoons and a Cossack battery rode up in time to stop the Turkish advance, while the 2nd Rostovs, advancing up a ravine between Haciveli and Great Yahni, took the Turks in flank. This sudden attack threw back the Turkish infantry and ended the Turkish attempt to retake Great Yahni. After 10 a.m. Turkish counter action in the region of Subatan, Haciveli and Great Yahni slowed down, for, to the surprise of both Turks and Russians, the troops of Shelkovnikov had suddenly appeared on the principal ridge of Alaca-dağ.

Shelkovnikov, informed that the general attack was fixed for 2 October, crossed the Arpa-çay on the night of 30 September/1 October. His somewhat daring plan was to make an attack against the Turkish position at Kizilkule in the ravine of the Diğor-çay; to feign a retreat to Post Kambinski, and then, during the night, to recross the river and make a surprise march up the eastern flank of Alaca-dağ.

On the morning of 1 October Shelkovnikov engaged Ahmet Paşa at Kizilkule. His troops grumbled when their commander ordered the 'Cease fire' and began to retreat after the loss of only about 100 men. By 3 p.m. Shelkovnikov had retired across the Arpa-çay, and a Turkish cavalry patrol observed his passage to the eastern bank. In the evening, reinforcements from Tergukasov (two battalions of the Bakinski Regiment) came into the camp. Only a few hours' rest could be given to the troops; Shelkovnikov spent the time visiting units and explaining his plan. After 9 p.m. Shelkovnikov began to pass his troops back across the river under conditions of utmost caution. The column marched in the darkness along the right bank of the Arpa-çay, then turned west to the hamlet of Alam, four miles from Alaca village (the terrain had earlier been thoroughly explored by daylight). The movement was carried out successfully and no Turks appeared, although the Turkish camp at Kizilkule was only four miles to the south-west of Alam. The most difficult part of the march lay ahead: the climb along the steep ridge of Naharci-tepe which forms the south-east shoulder of Alaca-dağ. This proved practically impossible in the darkness, so Shelkovnikov took the risk of awaiting the dawn.

He had the good luck not to be discovered. By 8 a.m. the Russians reached the summit of Naharci-tepe and moved along the shoulder towards the main ridge of Alaca-dağ. Between 9 and 10 o'clock, to the complete bewilderment of the Turks, Russian mountain guns opened fire from the summit of Alaca-dağ in the direction of Kerhane. Shelkovnikov saw the fighting going on between Haciveli and Subatan, but he could not ascertain what was taking place round the Yahni hills. He decided not to descend into the battle, but ordered his troops to occupy the empty summer fortifications of Muhtar Paşa on Alaca-dağ. In the event of Russian success against Avliar he proposed to defend himself here.

The Russian troops were greatly encouraged by the appearance of Shelkovnikov on Alaca-dağ, but only General Lazarev (another Armenian commander in the Russian service), who was in observation along the Mavrik-çay and opposite Kizil-tepe with orders to remain strictly on the defensive, thought to send him reinforcements—one battalion and one regiment of Dağistan irregular horse from Üç-tepe, who failed to get beyond Ani.

The crisis of the battle of Yahni, between 10 a.m. and midday, came too early to be rightly understood by either the Russian or Turkish headquarters staffs. Both sides proved unable to profit by the advantages which were open to them: empty spaces between groups (Turkish side) and between columns (Russian side). Only Heimann had grasped the possibility of exploiting the gap which led to Vezinköy, and he had deflected his available reserve to von Schack's attack against Subatan.

In the same way Muhtar, who was hurrying troops across his front from east to west to meet the threat to Vezinköy, became distracted by the appearance of Shelkovnikov on the summit of Alaca-dağ. He had to stop the movement of reinforcements to his left—which might, in fact, have seriously threatened the exhausted Russian forces round the Yahnis—and sent first six and finally twelve battalions to climb Alaca-dağ.

Soon after midday Shelkovnikov was engaged by Turkish infantry advancing from the three directions of Çift-tepe, Kerhane and Inah-tepe. Two hours later the fire of a dozen Turkish field pieces was making his position difficult, but he resisted successfully until 4 p.m., when Cossacks, who were patrolling in his rear, reported that the Turkish brigade at Diğor was moving to cut his line of retreat. Shelkovnikov realized that his last chance of withdrawal had come. While Cossacks held the Turkish earthworks as a rearguard, his infantry began to move back along the Naharci ridge and then down the steep sides of the mountain. With the pursuing Turks firing from the upper heights while the Russians with their wounded crowded through a narrow gorge under Naharci, the situation of Shelkovnikov became critical, and he expected to be caught at any moment by Ahmet Paşa's Diğor brigade moving up from the south. Daylight was fading as he reached Alaca village and in the darkness the Turks lost contact with the Russian column. Shelkovnikov, who had brought away all his four guns, bivouacked for the night in the Arpa valley, east of Alaca village, and in the morning crossed the river at Post Kambinski. His losses in this desperate expedition were 800 killed and wounded (about 15% of his strength), but his movement had succeeded in attracting about one-quarter of Muhtar's infantry during the most critical hours of the fighting below Alaca. As a strategic diversion Shelkovnikov's seizure of the Alaca ridge is full of interest; as a feat of military endurance, both the night march and the withdrawal down the mountain demonstrate the possibilities of a small and independent command under a capable and resolute leader.

Shelkovnikov's diversion certainly contributed to an early standstill in the fighting of 2 October. By 5 p.m. Loris Melikov, seeing that there was no further hope of capturing Little Yahni, withdrew Komarov's and Grabbe's columns. They retired to a hill called Kabak-tepe ('Pumpkin Hill'), about two miles east of Yayla Mesko and five miles north-east of Little Yahni, where the exhausted troops passed the night in digging themselves in—a practice which the Russians were at last beginning to copy from the Turks. The first necessity of the men on Kabak-tepe was to secure water from the not very distant Kars-çay—and in this they were more fortunate than the units left at Great Yahni.

As Melikov intended to resume the offensive on the following morning, Heimann's troops remained in position, at the end of the fighting, on a line from Great Yahni, along a ravine north of the Mavrik-çay, and then along the north side of the Mavrik-çay, past Haciveli and through Kulveran to Bayraktar (at which two latter points the line was held by Lazarev's left-wing troops). Melikov had at last been convinced by Heimann that the capture of Little Yahni was not necessary, and he proposed to adopt on the morrow Heimann's plan for a direct attack on Avliar and Vezinköy.

However, this plan was changed by the Grand Duke Michael. His Imperial Highness was shocked by the losses which already amounted to 5000, while all reports testified to the extreme fatigue of the troops and their sufferings owing to the waterless character of the terrain over which they were expected to operate. The renewal of the offensive was deferred to 4 October.

Meantime Ahmet Muhtar, with a view to strengthening the position on his left, where the loss of Great Yahni hill made him uneasy, ordered a strong attack from Subatan in the direction of Kulveran on the afternoon of the 3rd. By 4 p.m. the Turkish infantry had taken Kulveran, but they were driven back by a strong counter-attack which routed the nizams and carried the Ekaterinoslav Grenadiers into the village of Subatan. At the same time, the Elizavetpol Grenadiers penetrated between Subatan and Kizil-tepe and approached the Turkish position at Kerhane.

Heimann's troops on Great Yahni did not participate in the fighting of the 3rd, but their position was becoming desperate from lack of water; hot food reached them only in the evening after a wait of forty-eight hours. The grand duke, considering that the men required a really serious rest before they could undertake further action, cancelled the orders for the 4th and decided on a general withdrawal to the Kurudere and Oğuzlu camps. The Russians accordingly evacuated all the positions occupied with the exception of Kabak-tepe.

The victory of Yahni was acclaimed in Istanbul, and Ahmet Muhtar, already a marshal (*müşir*), was accorded the rare style of *gazi* ('victor of the faith'). He had repulsed an army better equipped and disciplined and stronger in all arms, particularly in artillery and cavalry. But the credit for the tactical ability and resolute command in action rested really with the German Mehmet (now Paşa), and Ahmet Muhtar's merit had been of a negative character; he had avoided all mistakes and he had benefited from the miscalculation, the bad organization and the faulty leadership of the Russian commanders.

The new gazi realized well enough that he was victorious, but 'without a morrow'. His own losses were serious (with the unsuccessful attack on the 3rd, they were not far short of the 5000 of the Russians), and they had fallen most heavily on the picked battalions of Mehmet Paşa. After regrouping his units, Ahmet Muhtar was left with forty to forty-two weak battalions, and 5000 mediocre horse to hold the fifteen miles of front running from Cale to Kizil-tepe and through Subatan to the Yahni hills. He could hope for no reinforcements beyond the six battalions on their way from Ismail Paşa. Muhtar was pressed by many of his advisers to withdraw now to the Soğanli, as he had himself proposed before the battle of Yahni. But the gazi was now reluctant to retreat on the morrow of his 'great victory', and he felt himself strong enough to smash a renewed

Russian attack and then to spend the winter quietly in the face of an enemy doomed to inaction. He decided, however, to withdraw to his original fortified positions on Alaca-dağ.

On the Russian side, the lessons of the recent battle had been appreciated. It was now proposed to resume the offensive by an attack in the direction of Avliar and Vezinköy, accompanied by demonstrations on both flanks. At the same time a deep turning movement via Diğor-Pazarcik was planned, to be started several days before the frontal attack and to be carried out by a considerable force. The command of the flanking force was given to Lazarev with Shelkovnikov as his chief of staff, and preparations for the new offensive began already on 5 October.

Lazarev was to start his turning movement on 10 October, but the withdrawal of the Turks from their advanced positions hastened the course of events. On the night of 8–9 October the Turks removed their heavy guns from Kizil-tepe and evacuated their positions near Subatan and Haciveli. In the morning the Cossacks found also Great Yahni hill empty. It was clear to Russian headquarters that Muhtar was withdrawing to his higher fortified positions on Alaca-dağ, but they supposed that he intended only a temporary halt preliminary to withdrawal either into Kars or westward towards his old spring positions in the Soğanli. In either case the Turks would escape encirclement. Lazarev was ordered to march immediately (9 October) and to complete his turning movement to the rear of Alaca-dağ by the night of 12–13 October. The attack on Avliar and Vezinköy was fixed for the 14th.

The force allotted to Lazarev amounted to seventeen battalions (10,000–11,000 men), 2500 sabres and seventy-two guns. His movement was naturally slower than that of Shelkovnikov's smaller and more composite force during the first October battle. A special signal unit accompanied the force for the purpose of laying a field telegraph, and Lazarev thereby remained in constant touch with general headquarters.

Lazarev crossed the Arpa-çay at Kuyucuk and recrossed it, in the tracks of Shelkovnikov, at Post Kambinski. The march of the long Russian columns, only six to eight miles distant from the Cherkess outposts at Alam on the extreme right flank of the Alaca-dağ position, was not observed by the Turks.

On 11 October, Russian cavalry passed through the Kizilkule defile and in the evening reached Diğor. Here they were joined by General Tsitovich who had marched from Kulp along the track by way of

Nahçivan village with the 154th (Derbent) Regiment, two *sotni* and a mounted battery. Parallel to him on the west was marching the brigade (six battalions) sent by Ismail to reinforce Muhtar (which appears to have taken the main road via Kağizman to Kars, forking at Ardos to Hacihalil).

Late in the evening of the 12th, Lazarev's infantry began to concentrate, without their trains, at Diğor. The worst part of the track lay ahead of them: from Diğor it ascends one of the southerly ridges of Alaca-dağ and then from Akriyak village gradually descends to the upland plain of Pazarcik, which is about four miles wide and surrounded by mountains. The cavalry of Loris Melikov[1] the younger had reached Akriyak on the 12th; on the 13th it moved towards Pazarcik, from where a squadron of Nizhegorodski dragoons and a *sotnya* of Dağistan horse were sent westward towards Hacihalil to track the whereabouts of Ismail's battalions from Kağizman.

Loris Melikov's cavalry discovered a Turkish force in position on the Borluk heights which overlook the Pazarcik plain from the west. Lazarev moved up Tsitovich's composite brigade to reinforce the cavalry, with orders to act against the Turks (Ahmet Paşa's brigade) next morning.

Meantime the advance cavalry reconnaissance under Major Witte had had an adventurous ride. On entering Hacihalil they found that the Turks had camped there the night before and had moved towards Mağarcik, evidently making for the Vezinköy road. Witte decided to follow the enemy as far as the deep Mağarcik ravine and then turn east towards Pazarcik. The cavalry reached Mağarcik village where no Turks were seen. They followed the ravine within two miles of its end, but when they climbed east ascending the track to Pazarcik they found themselves suddenly in the middle of the marching Turkish column. They cut their way out with forty men and seventy-five horses killed and, after wandering through wild mountain country, joined Russian troops on the 15th.

The Turkish brigade, which Witte had run into, had in fact received an order to march not on Vezinköy but on Pazarcik. On the night of 12–13 October, Muhtar had had information that the Russians were approaching Pazarcik. When, on the evening of the 13th, General Lazarev made a personal reconnaissance of the Turkish position above Pazarcik village, he found that the enemy had guns and infantry on the heights two miles to the north and north-east of the village, but that they had neglected to occupy the Satir ('Cleaver')

[1] A son of the general, died later in camp before Erzurum.

Mountain two miles south-west of it. As this south-west direction was of the greatest importance to the Russians, Lazarev ordered Tsitovich to occupy the Satir position at dawn on the 14th. But late in the night of the 13th the Turkish brigade, which had had the fray with Witte, was approaching by the ravine leading to Pazarcik from the north-west, and at dawn on the 14th Turkish infantry began to climb Satir. Tsitovich decided to attack Satir without further delay. Round Pazarcik village he deployed guns and cavalry to engage the Turks on the heights to the north-east of the village. At the same time, while Ahmet Paşa moved three battalions against the village to help the Turks on Satir, Tsitovich led his own Derbent Regiment up its slopes. He had only four battalions and cavalry against twelve Turkish battalions, but he took advantage of the separation of the Turks into two groups to press home his attack on Satir. The Turks had had no time to dig in and they broke before the bayonet, making up the slopes of Borluk-dağ, two miles to the west of Satir.

Shelkovnikov now came up with the 4th Caucasian Rifles, two battalions of the Sevastopol Regiment and four mountain guns. He decided immediately that the Turks must be dislodged from the steep and lofty Borluk-dağ. In the view of Shelkovnikov, this mountain was the key to all the Alaca-dağ front. It is one and a half miles to the south of the ruins of Borluk castle (which commanded the long ravine leading to Mağarcik), three miles south of Sivri-tepe (the strong point of the higher Vezinköy position), and only six miles from Mağarcik, where through a difficult and steep ravine passed the track from Vezinköy to the Soğanli—the main rear road of Muhtar's army. The attack of the Derbent regiment was carried out brilliantly and after the spectacular manner of bygone wars—with pipes blowing and standards fluttering in the wind. Borluk-dağ, where the Turks had had little time to dig the trenches which always gave them confidence to withstand the Russian charge, was taken by storm, and nine Turkish battalions with a battery drew off towards the north to take refuge behind the fortifications of Sivri-tepe. There they were joined later in the day by reinforcements sent by Muhtar.

Even on the 13th the Turkish commander had obstinately refused to believe that the Alaca-dağ positions had been dangerously turned by a strong Russian force. He preferred to treat the movement as a demonstration undertaken by a light force and was convinced that Ahmet's brigade with the reinforcements from Kağizman were sufficient to guarantee his rear. His view was strengthened by the fact that the Cherkess general, Kundukov, commanding six battalions and

2000 Cherkesses on his extreme right flank, had reported nothing of the Russian move, although he had patrols at Alaca village and even at Alam, only nine miles from Diğor. However, by the morning of the 14th Muhtar's illusions were dissipated; he sent further reinforcements of six battalions and a battery from Yayla Nalband to Pazarcik; but tracks were bad and the Turks moved slowly. When the officer commanding heard that the Russians had taken the Satir and Borluk mountains, he proceeded to Sivri-tepe where, by the night of the 14th, fifteen battalions with two field batteries had accumulated.

On this same night the situation had become clear enough to Ahmet Muhtar. He called a council of war. None present had any doubt that the Russians would open a general attack on the following morning from both north and south against the triangle of Sivri-tepe, Vezinköy and Avliar. The situation of the main body of the Turkish army on Alaca-dağ had become critical. Muhtar's chief of staff, Kâzim Paşa, considered the situation desperate and advised negotiations for a capitulation. But the gazi was determined, and felt obliged, to fight. Some twelve hours of darkness remained, and this time might allow for the partial evacuation of troops and munitions to Kars. Muhtar hoped that naturally strong and well-fortified positions held by an obstinate rear-guard would allow him to continue the movement of troops from east to west during the following day and that he might be able to escape with a part of his troops to the Soğanli while the rear-guard took refuge behind the fortifications of Kars. Such a plan was practicable only in theory; the line along the Turkish front from east to west had always been deficient in tracks, and the only serviceable road lay by Yayla Nalband and Sivri-tepe along the northern flank of Borluk-dağ. Borluk was already in the hands of the Russians, and an attack on Sivri-tepe might be anticipated for the following morning.

Ahmet Muhtar now had twenty-five battalions (more than 10,000 troops) facing south against Lazarev. There remained to hold the main Russian attack from the north thirty-five battalions (including four at Vezinköy), 2500 cavalry and twenty-four guns. Thus about 15,000 infantry had to hold a front of fifteen miles. Russian forces, not less than 15,000 strong, supported by powerful artillery, might be expected to attack on the narrow front of three miles between Çift-tepe and Avliar.

At 7 a.m. on the 15th, the forty-eight guns of Heimann and the sixteen guns of von Schack opened a terrific bombardment of the Avliar fortifications. The Turkish guns were soon silenced, but the infantry continued to sustain the bombardment in two lines of trenches which ringed the height. At 10 o'clock an attempt by Heimann to turn the position from the north-west was thrown back by Turkish battalions in reserve behind the mountain, and it was not until 1 p.m. that the Georgian and Pyatigorsk grenadier battalions finally stormed Avliar; the two Turkish battalions on the height fought to the last man, while the reserve battalions retired on Sivri-tepe. The fight had been decided by the Russian weight in guns. Heimann lost 700 men, mostly among the Erevan Grenadiers who had made the attack from the north-west. By 2 p.m. the first act of the battle was over: the Russians had taken Avliar.

About the same hour a no less decisive success had been attained by Lazarev, who had a general instruction not to attack before the development of Heimann's attack.

The principal fortifications on Sivri-tepe faced north, and Lazarev ordered Shelkovnikov to attack them from the rear with two Baku and two Sevastopol battalions supported by twenty guns. At the same time Tsitovich had to attack the southern end of the hill with his Derbent Regiment covered by the fire of twelve guns. General Gurchin with the 4th Caucasian Rifles and the other two Sevastopol battalions protected Shelkovnikov's right flank and rear from any Turkish counter-attack from the direction of the heights north-east of Pazarcik or from Çift-tepe. Two regiments of the 40th Infantry Division with twenty guns were in reserve near Pazarcik village, while the other two regiments of the division protected the communications with Diğor and the trains left at that village. Lazarev thus used only eleven battalions in action and held twelve inactive; but he preferred to rely on the skill in mountain fighting of the Caucasian battalions (which were armed with Berdan rifles, while the 40th Infantry Division still had only old weapons of the immediate post-Crimean period).

Some time after Heimann's attack began, the Turkish brigade with the battery which had stood on the Borluk-dağ and then retired on Sivri-tepe moved to Vezinköy village, leaving on Sivri-tepe the garrison facing north and nine field battalions with a battery facing south. Ahmet Muhtar himself remained on Çift-tepe until 10 a.m. with the intention of directing his troops at Sivri-tepe and Çift-tepe. But before 11 o'clock, when the fall of Avliar seemed imminent, he rode with his staff and cavalry (which had been at Yayla Nalband) to Vezinköy village. These movements of infantry and artillery followed by numerous horse towards Vezinköy village were observed by Lazarev from Borluk heights and he took them as a signal to attack.

By 11 a.m. orders to attack reached Shelkovnikov and Tsitovich. The Derbent battalions met with very feeble resistance; from the southern side of Sivri-tepe the now demoralized askers simply fled towards Kars leaving their six guns behind them. On the northern side the garrison tried to hold the fortifications against Shelkovnikov, but they were attacked from their open rear by the Nizhegorodski Dragoons and the wild Dağistan horse who galloped along the steep slopes of the mountain and rushed the trenches. The garrison withdrew in disorder along the Vezinköy road pursued by the Russian cavalry. By 1 p.m. Sivri-tepe had been taken—actually before the fall of Avliar.

Immediately after the taking of Avliar, Russian troops from the direction of Great Yahni marched on Vezinköy; but this village with its fortifications had been occupied, since 3 p.m., by the 154th (Derbent) Regiment—which in the last two days had had an extraordinary series of successes with a loss of only 300 men. The Turks were now thoroughly broken and the infantry at Vezinköy fled towards Kars, even before the Russians approached. But they did not escape the cavalry, and whole battalions were cut to pieces by the dragoons, Cossacks and irregulars.

Nearer to Kars, Muhtar Paşa with his staff tried to stop the fugitives, while Hüseyin Hami Paşa sent out some well-disciplined fortress battalions to Hafiz Paşa Tabia to stop the disorderly mob from penetrating the town and creating a general panic. Many units bypassed the fortress and scattered in groups or singly all over the surrounding countryside. The retreat had become catastrophic; and when Heimann's grenadiers met Lazarev's troops at Vezinköy the rest of Muhtar's army found itself surrounded by an iron ring in the Alaca positions.

General Kuzminski, commanding on the Russian left wing, had been ordered not to press his advance and to let the Turks on the Alaca heights remain where they were as long as possible; however, the Cherkess general, Kundukov, after 10 a.m. had no illusions as to the result of the battle; he evacuated Inah-tepe and decided to escape with his troops as best he could. His cavalry from the Cale region went south, while part of the infantry descended Alaca-dağ to Alam, from where they marched across the Diğor road in the direction of Kağizman. Thus four to six battalions and 1500 horsemen escaped destruction.

By midday Kuzminski's troops were on Inah-tepe, and cavalry coming up from the Arpa-çay through Kozluca reached the Naharci ridge. Before going to Vezinköy, Muhtar had ordered Haci Reşit Paşa to concentrate all the Alaca forces towards Çift-tepe, covering the withdrawal by a rear-guard under Hamdi Bey in the trenches over Kizilkilise (at the head of the Mavrik-çay ravine). Şevket Paşa's six battalions had time to join the brigades of Mustafa Cavit and Ömer Paşa on Çift-tepe. But Kuzminski's infantry, marching from east to west along the Alaca ridge, pushed back the remnants of Kundukov's battalions which had not had time to escape south, while General Roop moved the Ekaterinoslav Grenadiers up towards the Kizilkilise heights. Between 3 and 4 p.m. Hamdi Bey's rear-guard was surrounded and laid down their arms. On Çift-tepe, at Yayla Nalband and north of Pazarcik, the remnants of the broken army were crowded into a square of territory of which each side was rather less than three miles.

As evening fell the Russian columns awaited an order for the final attack on Çift-tepe. Before 7 p.m. Haci Reşit, as senior paşa, called the other generals to a conference. Kâzim and Ömer insisted on capitulation; the men were clearly desperate and in several cases had refused to obey orders. 'Let us preserve our men for the Turkish state' was the opinion of the German, Ömer Paşa. An officer with a white flag was sent to General Gurchin at Pazarcik, and 'Cease fire' was ordered. Meanwhile Ömer and Kâzim, as officers speaking European languages, rode in the darkness to seek the Russian commander-in-chief. They met a cavalry patrol and were accompanied to General Chavchavadze and then to General Roop. Later Loris Melikov arrived, and at 11 p.m. a capitulation was signed by Haci Reşit Paşa. Seven paşas, twenty-six battalions, 252 officers, one squadron of cavalry, in all a total of about 8000 men, laid down their arms. Only twenty-two guns were taken. With prisoners captured during the fighting the total reached 12,000.

In the meantime, on the Russian right flank, there was only slight cavalry activity. During the evening Mehmet Paşa was able to withdraw from Little Yahni his eight battalions with all their guns.

The Russian victory was complete. Turkish losses had been very great, particularly during the disorderly retreat from Sivri-tepe and Vezinköy. In addition to the prisoners, Muhtar had lost 6000 men in killed and wounded, so that his total loss of 18,000 amounted to three-fifths of his infantry. As for his cavalry, the majority, composed of irregulars, had deserted. Some 10,000 infantry and twenty-three to twenty-five field guns had reached Kars, while Kundukov, with 2000 infantry, four mountain guns and 1500 horse, had made his way to Kağizman.[1]

The Alaca-dağ battle was a great success for the Russians; it concluded the third period of the campaign and opened the fourth—which proved to be the last. Russian losses were comparatively small, less than 1500 killed and wounded. The battle was won by the manoeuvre of Lazarev who, together with Shelkovnikov, had proved a very efficient commander. But the battle was won rapidly and with relatively small losses as a result of the great Russian predominance in artillery. Massed gunfire made possible Heimann's capture of the really strong and well-fortified position at Avliar; and the Russian strength in artillery excluded the possibility of any successful Turkish counter-attack. In the Alaca battle, as at Ardahan, the Russians knew how to clinch the smashing artillery bombardment with a vigorous bayonet attack. Their neglect of proper artillery preparation against the obstinate Turkish infantry had been the cause of their defeats at Zivin-dağ, at Kizil-tepe and the Yahni hills.

During the night of 15–16 October, Ahmet Muhtar Paşa was informed, by stragglers who had escaped, of the capitulation of Haci Reşit. He called together his surviving senior officers, including Mehmet Paşa, the heroic German defender of Ardahan and Little Yahni, Hüseyin Hami Paşa, commandant of Kars, and Hüseyin Bey, the chief engineer of the citadel. In difficult circumstances, as always, Muhtar proved efficient and resolute. The situation of the troops at Kars was as follows; some twenty unbeaten battalions of the fortress garrison and Little Yahni and about fifteen battalions of the rout from Vezinköy—a total of 15,000 infantry. Muhtar decided to lead eight of the best battalions, with the *süvari* brigade and two batteries, towards the Soğanli-dağ where he expected to be joined by Ismail Hakki's troops from Ağri-dağ at Köprüköy. Hüseyin Hami was left to hold Kars with the remaining 10,000–11,000 infantry, who might be joined by 2000–3000 more stragglers as they came in from the surrounding countryside. Again, as at the beginning of the campaign, Ahmet Muhtar was taking the road from Kars to the Soğanli with little more than a brigade of men. He was already on the march at dawn on 16 October.

The more cumbrous mechanism of the victorious Russian command had not functioned so rapidly. The more impatient officers wanted to take advantage of the panic which was said to reign in Kars

[1] For a biographical note on Musa Paşa Kundukov, see Bibliographical and Supplementary Notes to Book III, pp. 546–7.

and to storm the fortress without delay, but the Grand Duke Michael and General Loris Melikov were of more cautious temperament and preferred systematic action. Something, however, had to be done immediately, since on the morning of 16 October cavalry patrols reported the movement of a Turkish column towards the Soğanli. A mobile column (six battalions, two regiments of Cossacks, two mounted batteries) under General Heimann was organized for the pursuit of the enemy. On the evening of the 16th Heimann moved towards Mağarcik; on the 17th he was on the Erzurum road, but here he had to await his supply train which had taken the wrong direction. Heimann remained immobile throughout the 18th, by which time Russian general headquarters realized that there was no chance of intercepting the Turks before they reached the Soğanli positions. New arrangements followed on the evening of the 18th. In all these circumstances Tergukasov had been completely forgotten by Russian general headquarters, and only on the evening of the 17th did he get news of the victory of Alaca-dağ.

Since 16 October, Tergukasov had information of the retrograde movement of Ismail's troops from the Ağri-dağ passes. With no information of events at Alaca-dağ, Tergukasov assumed that his opponent was merely regrouping his forces in anticipation of approaching winter conditions. However, during the night of 16–17 October, Ismail had completely evacuated his advance positions on the slopes of the Ağri-dağ; he maintained rear-guards only in the passes of Çengel, Zor and Gürünsaray and concentrated the bulk of his troops at Musun, ready to march westward.

On 17 October, Tergukasov observed the absence of Turkish patrols along the northern slopes of the Ağri-dağ, but he began to move only after 4 p.m., when he received a dispatch from the grand duke with an account of the Alaca victory and an order to advance and pursue the enemy most energetically. At dawn on the 18th, Tergukasov moved his advance-guards up towards the Ağri-dağ passes, when the Turks, left to hold them, withdrew towards positions covering the valley of the Balik-çay. The main force of Ismail Paşa was already on the march towards Surp Ohannes in the Eleşkirt valley, the point reached by the Turks on the 20th, while Tergukasov, delayed by the difficult tracks of the Ağri-dağ, had only been able to arrive at Musun. Ismail had the advantage of two days' march, and this usually lazy commander now proved capable of extremely rapid movement. He even succeeded in increasing his initial advantage, covering some thirty miles a day. On 24 October he reached Zeydkan, on the 25th Tahir, and on the 26th he passed the Kara-derbent and entered the Pasin valley only a week after he had left Musun.

Tergukasov was at Zeydkan only on the 27th, and from here he telegraphed to the grand duke that bad roads, incessant rain and lack of transport were preventing him from continuing effective pursuit of the enemy.

On the same day Ismail marched into Yuzveran where he was on the rear of the Soğanli position and in contact with Muhtar.

CHAPTER XIV

THE BATTLE OF THE CAMEL'S NECK (4 NOVEMBER 1877). ATTEMPT TO STORM ERZURUM (8–9 NOVEMBER)

MAPS 4, 10, 15

WHILE Ismail Hakki Paşa was executing his fortunate and rapid retreat before Tergukasov, the main Russian forces were delayed, partly by the necessity of formulating a new operational plan and partly through the difficulties of collecting adequate transport trains.

The Grand Duke Michael might be well content; after a series of reverses the Caucasian army had achieved a decisive victory at a moment when the Russian armies in the Balkans were held up before Plevna and were engaged in repelling the attacks of Mehmet Ali Paşa on the Shipka Pass.

An important consequence of the Alaca-dağ victory was the re-establishment of internal order in northern Caucasia; as soon as the news of the Turkish defeat reached Chechnia and Dağistan, the sporadic attempts at revolt which had held important bodies of Russian troops on security duties all through the summer and autumn came to an end. After the victory Loris Melikov found it good policy to release all the Muslim irregulars (with the exception of the 3rd Dağistan Horse), who returned to the mountains to recite their own contribution to the victory of Russian arms over the soldiers of the sultan.

As for an operational plan, it was decided to conclude the campaign by the conquest of the fortresses of Kars and Erzurum and, if possible, Batum. Since any operation against Kars required the participation of siege artillery, its preparation demanded time; meanwhile it was considered urgent to organize an expedition against Erzurum. Loris Melikov planned to cut off the forces of Muhtar and Ismail from their line of retreat on Erzurum, and for this purpose a special Soğanli force was prepared which, it was intended, should unite with Tergukasov in the Pasin valley. The task of this force, which was put under the command of General Heimann, was to clear the country between Kars and the Aras of such Turkish forces as remained

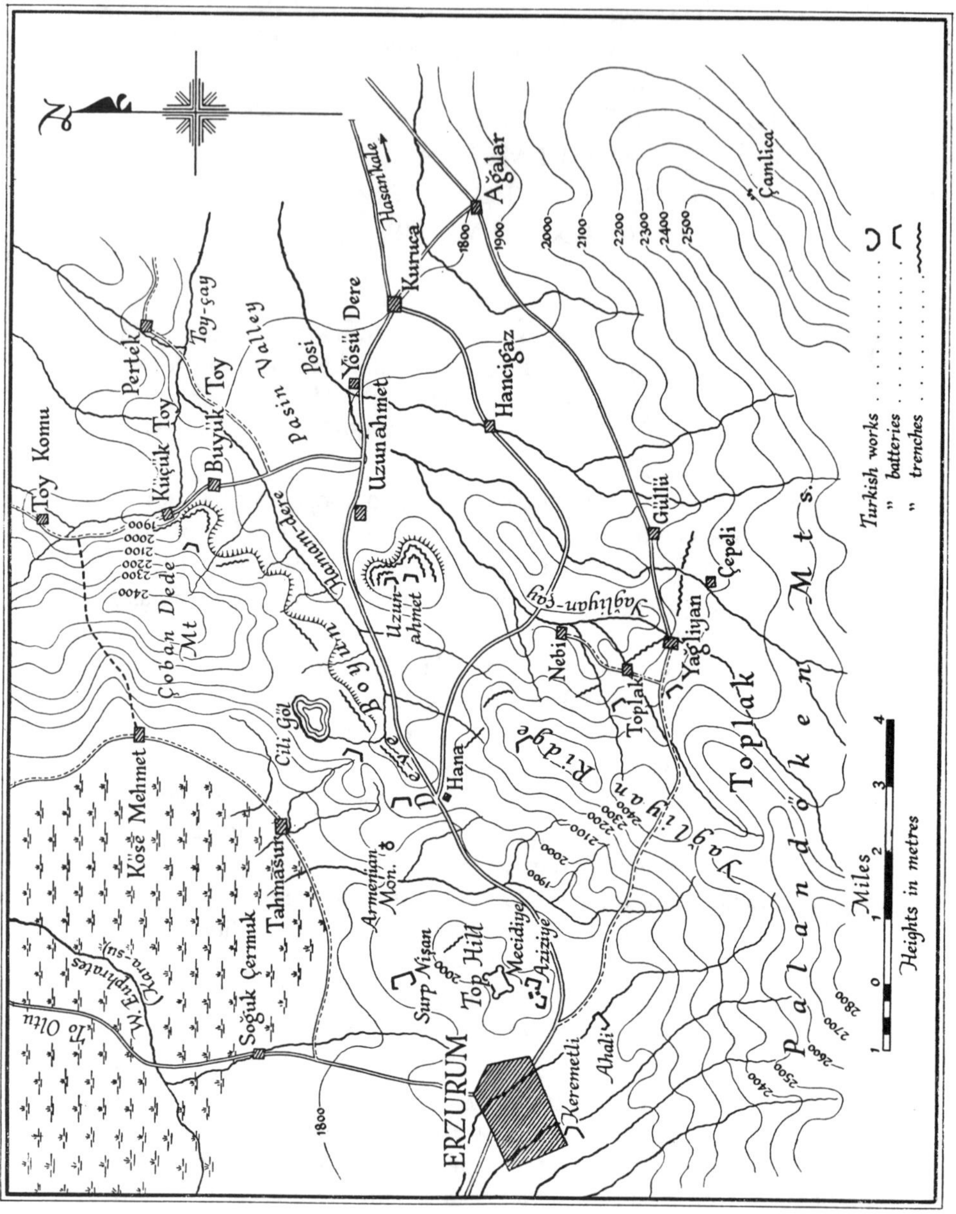

Map 15. The operations before Erzurum, winter 1877

(principally the detachment of Musa Paşa Kundukov round Kağizman) and to establish early liaison with Tergukasov. Heimann moved General Tsitovich (one and a half regiments, some Cossacks and artillery) to Kağizman, and with the rest of his force advanced on Sarikamiş. The ever-cautious Loris Melikov maintained at Kotanli (along the transverse which links the Kars-Erzurum and Kars-Kağizman roads), a reserve group under General Solovyev.

Supply difficulties prevented Heimann from moving before 22 October. On the 24th he reached Meliduz, where he got information that the Turks were occupying their old position on Zivin-dağ.

Although friendly Armenians stated that these Turkish forces were not strong, Heimann (who had with him only twelve battalions) was in doubt whether to turn against the Zivin position or to follow the road to Horsan and the Pasin valley. To avoid responsibility, this normally bold commander asked instructions from general headquarters. There were two days' delay (during which Solovyev's reserve was ordered to join him), and when Heimann finally reached the Aras valley at Horsan on 27 October, he learned that Muhtar's men from Zivin and Ismail's from Yuzveran were already marching up the valley towards Köprüköy. Heimann's fighting spirit revived. The Turkish forces, after their junction, were estimated at some thirty to forty battalions with forty guns, and when Heimann, at dawn on the 28th moved on Köprüköy, he had with him only one grenadier brigade, four field batteries and some dragoons and Cossacks. He was already too late; the Turks were in veritable flight towards Erzurum, and their commanders actually arrived there on the night of the 28–29th. Russian cavalry caught the Turkish rear-guard at Hasankale (Pasinler) and took 300 prisoners, but on the 29th the bulk of the Turkish forces reached the safety of the fortified position at Deve-boyun.

Heimann halted several days at Hasankale awaiting the arrival of other columns—Solovyev, Tergukasov and Tsitovich. Thus a certain lack of urgency (combined with extremely bad weather) had brought to nothing the plan of cutting off the two retreating Turkish groups.

Once more difficulties brought out the best qualities of Ahmet Muhtar's character. He was fortunate in having at Erzurum the old Hungarian, Feyzi Paşa, who had played no small part in the organization of the defences of Kars in 1855, and whose new fortifications at Batum had so successfully repulsed Oklobzhio. Since the beginning of October, Feyzi had been working on the defences of the Deve-boyun position and preparing fortifications round the city of Erzurum. Feyzi was well backed by the Erzurum commandant, Hasan Paşa, who had succeeded in forming there several new battalions of redifs and one extra field battery. Thanks to this competent support, Muhtar was now able to concentrate on the Deve-boyun positions forty battalions of infantry (of a weak average strength of 400 effectives), twelve squadrons of *süvari*, fifteen squadrons of Kurdish and Cherkess irregulars and eight batteries—a total of 16,000 foot, 2500 horse with forty-eight guns. At least half the infantry, however, was exhausted and demoralized by the long retreat and the night panic at Hasankale when Heimann's Cossacks had come up on their rear.

The forces, gathered by Heimann for the attack (excluding Tsitovich's detachment and several battalions left by Tergukasov on line of communication), did not exceed thirty-four battalions, thirty squadrons and twelve batteries, a total of about 20,000 foot and 3000 horse. But Heimann could dispose of nearly 100 guns.

Deve-boyun[1] is the name of a col across a ridge over which passes the main road from the Pasin valley to Erzurum (which city lies only four miles from the top of the pass). The ridge itself forms the watershed between the Aras basin (Pasin valley) and the western Euphrates basin (Erzurum plain). The altitude of the Deve-boyun ridge is rather over 2440 m., but it must not be forgotten that Erzurum itself lies at a height of 1882 m. and the altitude of Hasankale is 1694 m. The height of the saddle over which the road passes is some hundreds of feet lower than 2440 m.

To the north of the Deve-boyun ridge is the massif of Kargapazar which rises to 3050 m. and runs in a general direction north-east by east. Kargapazar is divided from the northern part of the Deve-boyun ridge, which here goes by the name of Çoban-dede[2] ('Holy Shepherd') by the steep and narrow gorge of Toy where a village of the same name is situated.

The southern extremity of the Deve-boyun ridge merges into the higher ridge of Palandöken (about 3350 m.), running in a general direction east-north-east. In the angle between Deve-boyun and Palandöken is the source of the stream named alternatively Yağliyan or Nebi (after two hamlets on the westerly slopes leading down to it). This Yağliyan stream follows a rather shallow bed running along below the Deve-boyun ridge until it turns eastward to the Aras.

The main road from Hasankale to Erzurum crosses the Yağliyan

[1] *Deve-boyun*, literally 'Camel's neck', but used topographically to denote a depression in a mountain chain; A. V. Moran, *Türkçe-Ingilizce Sözlük*.

[2] Not Çoban-*dere* ('valley') as on some modern maps. 'Valley' in this context does not make sense.

stream at the point where it turns eastward to the Aras; and four miles from here the road reaches the summit of the pass. In 1877 another and rather bad track approached the pass from the east running along the northern side of the Yağliyan-çay and between it and the Toy-çay. This track passed round the north of an isolated table-topped hill called Uzunahmet ('Long Ahmet') and to the west of the hill reached the pass by following up the Hamam-dere ('Bath valley') ravine. Another side-track forked from the main road in a southerly direction going up to the villages of Yağliyan and Çepeli on the slopes of Palandöken-dağ; then, after crossing the Toplak ravine and the southern end of the Deve-boyun ridge, it joined the main road at the gates of Erzurum.

The Turks had chosen and carefully fortified an excellent position along the Deve-boyun ridge running over a length of about five miles, partly to the north and partly to the south of the pass itself. Uzunahmet hill and the slopes of Palandöken near Yağliyan village gave the defenders the opportunity of creating advance defended localities which had an enfilading position in relation to the main road. On their right the Turks occupied entrenchments and batteries along the heights of Toplak, Yağliyan and Çepeli. On their left they had skilfully fortified the very convenient table-topped Uzunahmet hill, the flanks of which have a steep fall to the tracks passing round and behind it.

The Turkish troops on Deve-boyun were disposed in three groups: a centre group under Ahmet Muhtar occupied trenches on either side of the pass and were supported by three or four field batteries and a battery of fortress guns which had been dragged out from Erzurum; Ismail Hakki (with the aid of the Hungarian Feyzi) commanded on the right in the Toplak-Yağliyan-Çepeli sector; on the left the German, Mehmet Paşa, commanded the troops on Uzunahmet hill as well as a few battalions and guns which were in position on Çoban-dede and at Toy village.

After several days spent in reconnoitring the Turkish positions, Heimann decided to make his main attack against the Uzunahmet hill, at the same time developing a demonstration in force against the Turkish right flank along the lower slopes of Palandöken. He proposed, after the capture of Uzunahmet, probably on the following day, to force the central Deve-boyun entrenchments by directing an attack up the track which follows along the Hamam-dere ravine. The possibility of a turning movement up the Toy-çay was also discussed with the object of reaching a track which ran behind the Çoban-dede mountain. The task of the assault troops did not seem at all easy, since the flanks of Deve-boyun have a steep fall even at the points where tracks and paths lead up them; in many places it was impossible for even a company to deploy. Uzunahmet has almost vertical sides everywhere except on the north-western corner, where the Turks had made a track for the passage of guns. This joined the track which ran up the Hamam-dere and was under the fire of guns on Çoban-dede.

At 7 a.m. on 4 November, troops moved out from the camp at Kuruca and the artillery took up positions and opened fire. The Turks replied vigorously from Uzunahmet, from the batteries on Deve-boyun and, on the right, from the Toplak heights. Against the Turkish right a column under von Schack advanced towards Güllü village which the Tiflis Grenadiers occupied about 10 a.m., while on their right the Mingrelskis moved up towards Yağliyan and Çepeli. This demonstration was carried out with such effect that Ahmet Muhtar took it for the main Russian attack and he directed his reserves to Toplak, where by midday twelve battalions and eighteen guns had been concentrated. In the ravine of the Yağliyan-çay Turkish horse and infantry began a series of counter-attacks and von Schack was hard pressed with his ammunition running out, but the arrival of two battalions of Georgian Grenadiers enabled him to hold. Soon, however, a decision came in unexpected form and at an unexpected point of the battlefield.

Observing the course of the action on his left flank, General Heimann decided, by midday, that the demonstration in force had attained the expected result, for Turkish guns and infantry could be seen moving towards the Toplak heights. Meanwhile the Russian guns had been heavily bombarding the Uzunahmet position while, on the Russian right, Tergukasov had taken Toy village, dispersing enemy infantry in the ravine of the Toy-çay, and silenced the few Turkish guns active on the slopes of Çoban-dede. The right flank of Colonel Amirajibi's brigade, detailed for the assault of Uzunahmet, was thus covered, and at 2 p.m. the Elizavetpolskis and Bakinskis began the ascent of the slopes from the north-west end, where the Turks had built a track. These Caucasian regiments, carefully trained in climbing and taking cover, reached the table-top and attacked with the bayonet the trenches which were stubbornly defended by Mehmet Paşa's infantry, the only Turkish troops which had so far never been beaten. At the same time, while the result remained in doubt, General Bronevski sent skirmishers of the Krymskis and Stavropolskis to climb the almost vertical ridges of the eastern scarp of the hill.

As the Uzunahmet fighting appeared to be going favourably, Heimann now sent forward Avinov's brigade to support the hard-pressed von Schack on the left. They deployed along the east bank of the Yağliyan-çay, with the Georgian Grenadiers on the left supporting von Schack and the Erivanskis on the right near the main road. As the Erivanskis were under fire from the Turkish heavy guns on Deve-boyun, the colonel placed his men behind the protection of a rocky spur which juts out into the plain along the right bank of the Yağliyan-çay almost as far as the main road. (It was at this point that, before 1914, the Turks built Kaburga, the most easterly of the Erzurum forts.) In this sheltered spot the Erivanskis remained under cover from gunfire and were invisible to the Turks on Deve-boyun; this chance circumstance had a sensational consequence decisive for the course of the battle.

General Amilakhvari's cavalry brigade had followed Avinov's infantry along the main road. When the Gruzinskis and Erivanskis left the road and took up positions along the Yağliyan-çay, Amilakhvari, who had no definite order, continued to advance. Someone had reported that Uzunahmet had been captured and he was preparing to catch the fugitive garrison as it poured along the main road. Amilakhvari appears to have been moved by that combination of obtuseness and daring which sometimes characterizes Georgian officers; without any particular object or orders he began to deploy his cavalry on the lower slopes of Deve-boyun and was soon within range of a hot rifle fire from the Turkish trenches.

Ahmet Muhtar who had no report of the progress of the fighting on Uzunahmet was on the summit of the Deve-boyun with his staff and a group of foreign officers (including members of General Sir Arnold Kemball's military mission). He suddenly ordered the infantry to leave their trenches along the rocky terraces on the slopes of the mountain and to charge the Russian cavalry. With shouts of 'Allah, Allah' eight Turkish battalions rolled down the slopes of Deve-boyun, while two mounted batteries limbered up and followed them along the road. Amilakhvari began to retire—his retreat gradually gaining speed under heavy rifle fire. In enthusiastic pursuit, the Turkish infantry, who had now lost all regular formation, advanced in disorderly groups through Nebi village. Then the Erivanski Grenadiers suddenly appeared on the crest of the spur beyond the Yağliyan-çay where they had been waiting out of range of the Turkish guns on Deve-boyun, and counter-attacked in dense formation across the stream towards the village. This move appears to have been made on the initiative of the Erivanski colonel, who saw Amilakhvari's cavalry pressed by the Turkish infantry. The whole issue was certainly not a part of Heimann's operational plan.

The Erivanskis were joined in their attack by a part of the Gruzinskis, while Bronevski—who was the general nearest to the spot—moved the Stavropol battalions from behind Uzunahmet hill against the left flank of the advancing Turks. The rout of the Turks along the left bank of the Yağliyan-çay was immediate and complete. Amilakhvari's Pereyaslav Dragoons and Cossacks turned and began to sabre the fleeing nizams. The two Turkish batteries, galloping along the road, were captured before they could unlimber. The rush carried the Russian infantry along the road and up both slopes of Deve-boyun, and panic spread to the Turkish gunners who abandoned their pieces covering the pass. By 5 p.m. the Deve-boyun position was in the hands of the Russians and Muhtar's forces were cut into two parts.

The situation of the defenders of Uzunahmet had meanwhile become desperate. The cavalry of the Khan of Nahçivan had got round Mehmet Paşa's left and penetrated into the Hamam-dere ravine. The troops on Uzunahmet ceased resistance and tried to escape in small groups towards Çoban-dede. Only on the Turkish right was some order still maintained, and Feyzi Paşa succeeded in withdrawing his infantry and fourteen guns into Erzurum.

Turkish losses were appalling: forty-two guns including eight heavy howitzers; 3000 killed and wounded and 1000 prisoners; 4000 deserters scattered over the countryside.

General Heimann had gained his extraordinary victory in an extraordinary way and with a loss of only 1200 men. The legend of an 'ambush' prepared for the Turkish infantry by Heimann had no foundation in fact, although welcome to the reputation of that somewhat spectacular officer. The operational plan establishes that there was no planned ambush, and the result of the battle arose from the reckless—and in itself quite foolish—advance of Amilakhvari and from the prompt initiative of the Erivanski colonel (whose name has not even been preserved in accounts of the battle). Once more Muhtar Paşa had demonstrated his helplessness on the field of battle, his lack of ability to grasp the essential developments of the fighting and his tactical incapacity. His sanguine character on this occasion served him badly; as late as 4 p.m. he told foreign officers and war correspondents that the situation on all sectors was favourable and quite stable.

The rout of Muhtar's army had been so complete and confusion in Erzurum was so bad that there is little doubt that if Russian troops had followed through with the streams of fugitive Turkish soldiery they could have penetrated to Erzurum without great difficulty. Heimann has been reproached with his failure to take advantage of the situation, but in his favour it must be said that the Russians only came into possession of the pass between 5 and 6 p.m., and then only some elements of the Erivanskis and the Stavropolskis, with dragoons and Cossacks, had reached this point. It would not have been possible to organize them for a further advance before 8 o'clock; and the November darkness falls early in the Anatolian highlands. Two-thirds of the Russian forces, exhausted by the hard day and the difficult tracks, were still miles from the pass. In such circumstances it was probably impracticable to improvise the capture of Erzurum even on the heels of the fleeing Turks. As it was, Heimann worked all night to concentrate his troops on the pass, and in the morning he sent an officer with a white flag to demand the surrender of the fortress. The gazi refused.

It was not so easy for Muhtar on 5 November to refuse a capitulation. He now had not more than 6000–8000 reliable infantry at his disposal, to which might be added 2000–4000 irregulars and armed inhabitants of doubtful value. Only fifteen to twenty field guns remained. The fortress itself in 1877 had little in common with the powerful defensive complex which had been organized before the outbreak of war in 1914.

In 1877 the modernization of Erzurum had been in process for several years, but, since the beginning of the war, the attention of such excellent officers as Feyzi and Hasan Paşas had been concentrated on the fortifications of the covering position of Deve-boyun. The old town wall, more than four miles round, was high and wide and was reinforced by numerous bastions, and it had a moat, in places twenty-five feet in depth, which could be a serious obstacle to an infantry storm; but it required a numerous garrison to man it adequately. New forts had been built on the principal approaches to the town: Mecidiye, to the north of the Kars road; Surp Nişan, covering the north-east angle of the city walls between the Kars and Oltu roads; Keremetli, at the south-west angle of the walls. As the eastern approaches were the most dangerous, two advance works had been built on both sides of the Kars road: Aziziye to the north and Ahali to the south of the road. Muhtar proceeded to garrison all these works by the best of his remaining infantry who were provided with a few field guns to supplement the positional pieces already established there.

On 7 November Heimann held a conference at which were considered possible alternative lines of action. All knew that to attempt a siege was impracticable, since the siege park of the Caucasian army was needed at Kars. A blockade was also impracticable for a force of 25,000 men. The country was vast and there was the possibility of an enemy move from Trebizond (since, in spite of reinforcements sent to the Balkans, Derviş Paşa still had a considerable force at Batum). There was the further factor that communications were already long and, during winter, might become unsafe. Only two decisions remained: to storm Erzurum or to keep the garrison under observation from the Deve-boyun ridge, awaiting the fall of Kars and the reinforcements and train necessary for an effective siege. Heimann proposed not a real storm but a surprise attack (there were not even storming ladders available); and the best chance for such a surprise would be at night. As the November nights were long it was considered that the night of 8–9 November would be adequate for the troop movements necessary for the attack and for the surprise itself. Heimann's staff produced a long and detailed disposition where all moves were precisely timed, and recent experiences of muddled timing and mistaken direction were forgotten. The plan explained that a daylight attack would be accompanied by serious losses and that only a night operation could give a chance of success—'not one shot and straight in with the bayonet'. The maximum precautions were expected of the officers commanding the columns.

At 6 o'clock on the evening of 8 November General Avinov was ordered to move with the Gruzinski and Erivanski Grenadier Regiments and two companies of sappers by difficult mountain paths along the slopes of Palandöken to a point within one mile of Keremetli Tabia; arrived at his rendezvous he had to await the sound of attack from the Kars road. The conception was daring enough, but Avinov's men were expected to cover ten miles of unknown mountains in the pitch darkness of a November night. At 2 o'clock on the morning of the 9th, Colonel Krusenstiern, with the 3rd Caucasian Rifles and one Crimean battalion, had to turn the Ahali position and attack it, without a rifle shot, from the rear where it was unprotected. A rocket from Ahali was to signal its capture. Amirajibi, with the 156th (Elizavetpol) and the 153rd (Baku) Regiments had to infiltrate his men by small groups into the ravine in front of the Aziziye fort and storm it at precisely 2 a.m. Von Schack, with the Tiflis and Mingrelian Grenadiers, was to concentrate in the Deve-boyun Pass, ready at 2 a.m. to march decisively on the Kars gate. The reserve, under General

Tergukasov, was made up of two Crimean, three Stavropol and three Taman battalions and the artillery. The cavalry was to move only after 5 a.m., and then in the direction of the Oltu and Trebizond roads to cut off the beaten enemy.

The attack against Mecidiye Tabia, the strongest and most important of the forts, was designed as a secondary stage of the action after the capture of the Ahali and Aziziye works and the entry of Avinov's troops through a natural breach in the walls near Keremetli Tabia.

The difficulties of timing and movement control in the dark became apparent when the Elizavetpolskis (of Amirajibi's column), who had only four miles to march, mistook their direction and left the Bakinskis to storm Aziziye alone. This work was brilliantly captured; the Turks were completely surprised. The moment was favourable for a surprise attack on Mecidiye, only half a mile distant; but the Bakinskis had no orders and the Brigadier, Amirajibi, had disappeared in the darkness with the Elizavetpolskis. Time passed and dawn approached. Mehmet Paşa, in Mecidiye, counter-attacked, and at the same time a great mob of armed inhabitants made a sortie from the Kars gate.[1]

Krusenstiern, meantime, after having marched twice round the Ahali redoubt—as it subsequently turned out—and having failed to find it, had begun to march back to his starting point. Von Schack, awaiting Krusenstiern's rocket, failed to move against the Kars gate at the moment of the sortie of the townsmen. The Bakinskis in the Aziziye redoubt were left alone to sustain gunfire from the Mecidiye and the attacks of Mehmet Paşa's infantry and the mob from the gate. With the loss of 600 men they managed to extricate themselves, bringing with them some 600 Turkish prisoners; 130 Russian prisoners were taken by the enemy. After the recapture of Aziziye, Mehmet Paşa tried to advance, but with dawn Amirajibi had found direction and the Elizavetpolskis drove back the triumphant Turks. By 9 a.m. the complete failure of the surprise night attack was quite clear.

Avinov was marching all night. Intentionally or not, his native guide led him by a long detour and his exhausted column appeared before Keremetli Tabia in broad daylight—to be met by Turkish gunfire. Avinov withdrew and marched back almost all the day without having achieved any result. Russian losses amounted to more than 800 officers and men, three-quarters of which were suffered by the

[1] This sortie is still the subject of popular legend in Erzurum. In 1947, on the seventieth anniversary of the episode, recollections of survivors were collated by Bay Doğan Nadi and published in the Istanbul newspaper, *Cumhuriyet*.

gallant Bakinskis. The Turks lost the three battalions in Aziziye, a total of 1600 men.

The failure of the night attack and all the errors in movement control and timing infuriated the temperamental General Heimann. At Russian headquarters several days passed in heated discussions. Heimann insisted on a second attempt (to be fixed for 14 or 15 November), but he met with the determined resistance of Tergukasov supported by Devel (who had recently arrived). As both these commanders were of the same grade and authority as Heimann, the latter found himself unable to enforce his view.

Tergukasov and Devel had based their objections on the fatigue and bad health of the troops and the lack of proper supplies of food and equipment which now began to be felt. Winter was approaching—severe enough on the open upland of Deve-boyun, 2440 m. above sea-level. Although part of the troops might go into billets and cantonments in the villages of the Pasin valley, it was necessary to continue to man the ridge. The cavalry, moreover, were suffering from lack of forage and regiments were able to horse only 30% of their strength; this weakness in cavalry was decisive against any attempt even to enforce a blockade of Erzurum. As a result the communications of Erzurum with Erzincan and Trebizond remained open, and Muhtar Paşa was bringing in munitions and supplies. An infantry brigade (six battalions) arrived, having been sent by Derviş Paşa from Batum via Trebizond. The spirit of the defenders of Erzurum was high during the first weeks after the success at the Aziziye redoubt; soon, however, epidemics were to break their spirit in a way which no fighting could.

CHAPTER XV

THE STORM OF KARS. WINTER OPERATIONS IN THE ÇORUH VALLEY, NOVEMBER 1877 TO JANUARY 1878

MAPS 4, 14, 16, 8

IN a study of the military topography of the Transcaucasian frontier, an account of the storm of Kars, seventy-four years after the event, may be considered of merely academic interest. At the same time the affair has its interest as a feat of arms; an American historian of the war of 1877–8 expressed the view that the battle of Alaca-dağ and the storm of Kars were among the most brilliant feats of arms in Russian military annals. In World War II storm tactics again became of actual interest, and the tactics adopted by the Russians at the storm of Kars, worked out as the result of previous failures at Erzurum and elsewhere, merit study.

Since the Crimean War a serious attempt had been made to modernize the defences of Kars—under the supervision of German engineers. The forts on the right bank of the Kars-çay had been strengthened and new forts had been built on the Karadağ heights. On the left bank, where in 1855 only one real fort had been built—Veli Paşa Tabia—new forts had appeared on the heights of Çakmak and Şorah (an angle of the defences the importance of which had been demonstrated during Muravyev's attempt at storm).

For the Russians the capture of the fortress could be the only logical conclusion to the campaign of 1877; at the same time success would release large forces for operations against Erzurum and Batum, and the taking of the last place was, for political reasons, considered essential. The strength of the fortress, defended by more than 300 guns—of which about half were modern positional pieces—had been adequately appreciated by the Russian commanders. General de Courcy, French military attaché with Muhtar Paşa, passing through the Russian lines on his way to Europe, had declared that the fortress was impregnable and that an attempted Russian assault would result in enormous losses ('ils ne réussiront prendre un seul fort et périront tous').

One factor alone favoured the Russians: the garrison left there by Muhtar Paşa was inadequate to man the elaborate defences. The

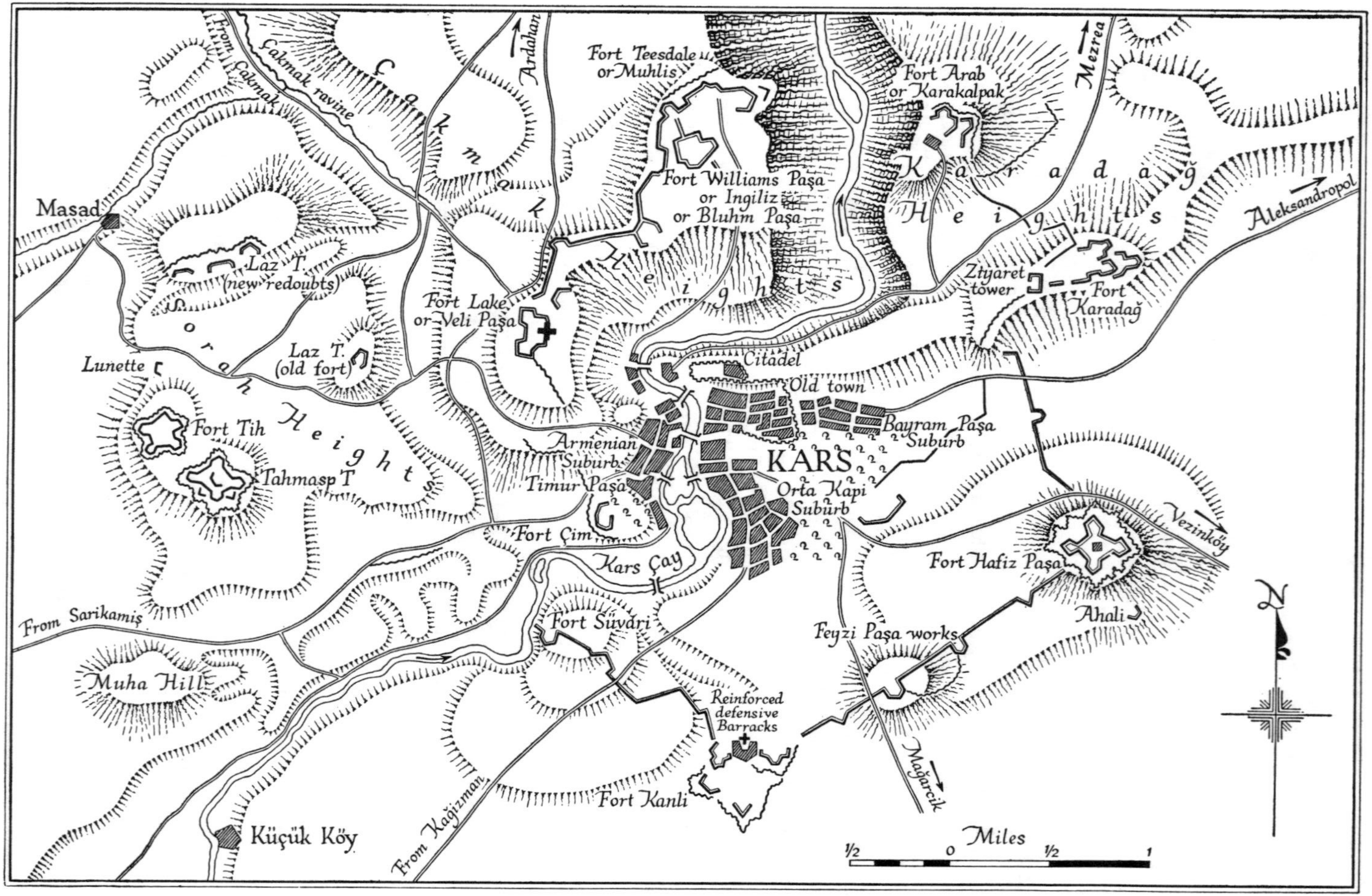

Map 16. The storm of Kars, 1877

Russians rightly calculated the strength of the garrison at about 15,000 men, and it was known that their spirit was not good. At the same time a blockade could hardly be effective since, in contrast to 1855, Kars was known to be well stocked with food and munitions.

A blockade had, in fact, been established a few days after the Alacadağ victory. A blockade corps under command of General Lazarev with a strength of about two and a half infantry divisions, 6000 cavalry and 138 guns was stationed on the north in the Mezrea-Meliköy area, on the south-east between Vezinköy and Mağarcik, and on the west and south-west between Arvartan and the Erzurum road. Cavalry detachments maintained contact between the three groups.

After discussions at Russian general headquarters it was decided, first, to test the spirit of the garrison and its commanding officers; secondly, to establish siege batteries with the object of maintaining fire against the forts and of impressing the civil population who were said to favour capitulation; and, thirdly, to storm a well-chosen sector of the Kars defences. It was estimated that these preliminary operations would require about a month.

Negotiations with the Turkish commandant, Hüseyin Hami Paşa, having proved futile, the bombardment of the forts began a few days after the news of Deve-boyun had been communicated to the Turkish lines. Since one of the aims of the bombardment was to discourage the civil population, the fire was principally directed against the three forts in the plain to the south and south-east of Kars; for from this direction the southern suburbs of the town could be brought under fire, while from the other directions the town was protected by the fortifications on the heights of Karadağ and Şorah. By 3 November, forty-eight siege guns were in action echeloned along a front of about four miles against the forts of Süvari, Kanli and Hafiz Paşa. The bombardment gave some results, although the Turkish forts, supported by the guns of Tahmasp Tabia on the Şorah heights, replied often with effect. At the same time the Turks made sorties with some spirit, and managed to build during darkness a small redoubt one and a half miles to the east of Hafiz Tabia where they established a field battery which did considerable damage to Russian working parties.

This action, however, had an unexpected result. On 8 November, the Russians undertook an operation for the destruction of the redoubt and a force under General Alhazov, operating from Karaurgan, forced the Turkish battery to withdraw. At the same time two battalions of the Kutaisi Grenadiers, under Colonel Fadeyev, had advanced from the east from the direction of Vezinköy. Fadeyev's original task had been to cut off the retreat of the battery into Hafiz Tabia. During the course of their movement the Russian infantry descended into a kind of depression, about a mile from the fort, where they were hidden from the view of its garrison. Fadeyev knew that the guns had already been withdrawn from the redoubt as a result of Alhazov's action, and he was aware at the same time that he must be under observation from the Turkish forts on the Karadağ height. But evening was drawing on and he decided to wait for darkness.

When darkness came on, Fadeyev cautiously approached the fort from the northern side. Hafiz Tabia, like other forts built before the Crimean War, had a rather shallow ditch, and Fadeyev determined to scale the nine-foot earth wall and take the place by surprise. The surprise, however, was not complete, because the Russians had been observed, as night fell, from the Karadağ height, and the Turks occasionally threw a shell in their direction.

Leaving two companies to protect his rear, Fadeyev led his other six companies right under the walls of the fort and then, in the sudden light of a shell thrown over from Karadağ, led his men to the attack, shouting 'Come on, boys, here is glory for the Kutaisi Grenadiers'. The Russians were across the ditch and over the wall before the sleepy garrison began to fire. The gunners were killed as they ran to serve their pieces, and the one infantry battalion in garrison—a unit badly shaken at Alaca-dağ—fled panic-stricken from the fort. One of the strongest forts of Kars had been captured in a few minutes by a handful of men.

Fadeyev was under no illusions as to the chances of holding the fort; he ordered the guns to be spiked, and before dawn the Kutaisi men left Hafiz Tabia bringing with them ten Turkish officers and seventy men. Their total casualties had been forty-two killed and wounded.

This episode raised great enthusiasm among the Russian troops before Kars, and the show of slack Turkish morale determined the grand duke to attempt the storm of the fortress without further delay. The Kutaisi men, he told his officers, had shown them the way into Kars.

On 9 November General Lazarev was ordered to prepare the storm. With reinforcements from Ardahan, he had available, at the beginning of November, thirty-eight battalions, fifty-four squadrons and *sotni* and fifteen mounted and field batteries—total 26,000 infantry, 6000 cavalry and 120 field pieces. In 1855, Muravyev had had 25,000 all arms against a garrison, weakened by starvation rations, of 20,000.

Now the Kars garrison did not exceed 15,000 fit men, but they were well fed. The Turks had far more powerful fortress artillery than in Muravyev's time, but the form of storm proposed by Lazarev precluded the use of positional artillery.

Despite all previous examples and despite the recent failure against Erzurum, Lazarev firmly believed in the efficacy of a night storm, and his view was supported by the courageous affair of the two Kutaisi battalions.

Lazarev had given careful attention to the tactics to be employed, and he introduced several new and important corrections into current practice: to avoid, as the experience of Erzurum suggested, a very dark night; and to choose a night in which moonlight was available at least for the opening moves; to avoid long night marches for the attacking troops and to concentrate troops gradually at rendezvous three-quarters of a mile from the points of attack; to attack simultaneously at several points of the perimeter in order to keep the enemy guessing as to the main objectives of the attack; to send out parties of night skirmishers during several nights before the storm in order to get the enemy accustomed to skirmishing in the neighbourhood of the forts and to the appearance of Russian patrols at all hours; and to increase the strength of the bombardment during the last three or four days before the storm but not particularly so on the last day.

These measures were well worked out, and their results favoured the action on the night of 17–18 November. The organization of skirmishing patrols proved particularly useful as the men engaged became familiar with the approaches to the forts. During the actual storm there were only a few cases of wrong direction taken by attacking columns or reserves.

The three forts on the right bank of the Kars-çay in the plain, Hafiz, Kanli and Süvari, with Çim Tabia covering the ravine of the river on the left bank, were chosen as the main objectives of the attack. This principal attack was to be carried out by 15,000 infantry out of a total of 26,000 engaged.

The method intended was to capture the four forts which gave access to the town from the east and south-east and then to take the town itself. On the following day an attack from the rear was to be developed against the forts on the left side of the Kars-çay and against the powerful fortifications on Karadağ. In a supplementary instruction from the grand duke it was indicated that the columns demonstrating against other points might, in the event of easy success, act decisively, but only if it were possible to do so with the strength available on the spot. A special note on Karadağ pointed out that a surprise attack on this position would be extremely difficult and should only be undertaken in the event of complete confusion in the enemy ranks. The attacking force was divided into seven columns of uneven strength.

At 8 o'clock on the clear and frosty night of 17 November under an almost full moon, the attacking columns started their march from points chosen one and a half miles approximately behind the semi-circle of siege batteries. They marched in complete silence with skirmishing patrols, already familiar with the ground, at their head. They were not discovered until they were less than a mile from the enemy positions.

The Kanli ('Bloody') fort consisted of two redoubts covered by a strong lunette behind which was a stone-built barracks backing on to a small ravine which led down to the Kars-çay between Süvari Tabia and the suburbs of Kars. Count Grabbe moved two battalions of Pernovski Grenadiers against the front of Kanli, while he sent the 1st Caucasian Rifles against the trenches connecting Kanli with Süvari. Grabbe, at the head of his men, was struck down and the command fell to Colonel Belinski who was leading his Sevastopol battalion against the western redoubt of Kanli. Some of the Pernovskis managed to climb the parapet, but the majority took cover in the ditch. The Sevastopolskis were more successful and captured the western redoubt. But the Turks held on with determination in the easterly redoubt and in the great lunette and barracks. The struggle had desperate fluctuations. The Turks recaptured the western redoubt but were driven out. Belinski was killed.

The Russians were expecting support on their right from Colonel Vozhdakin's column which had been sent against the trenches connecting Kanli with Hafiz Paşa Tabia, where a field-work known as Feyzi Paşa had been thrown up as an enfilading cover between the two forts. Vozhdakin's column was the only one which on that night mistook its direction, and it was still marching while the struggle was raging in Kanli. Some time after 9 p.m. fire from Feyzi Paşa enabled Colonel Vozhdakin to orientate himself. He was wounded and Colonel Karasev took over. Feyzi outwork was easily taken by the Imeretinskis, and the Russian infantry, having here broken the Turkish line, entered the camps behind and penetrated to the first houses of the suburbs. Meantime, Colonel Karasev, collecting three battalions, marched along the line of captured trenches towards Kanli Tabia.

Between 10 and 11 p.m. a difficult struggle was in progress for the possession of the eastern redoubt of Kanli which was captured and recaptured several times. At 11 o'clock Turkish reserves began to arrive and the crisis was reached in the battle for the fort. There were the remnants of eight Russian battalions inside and around the fort; some groups were firing down from the parapet against the Turks resisting from the lunette, but many men had taken cover in the ditch. Different units were mixed up and no common command was functioning. Turkish reinforcements strongly counter-attacked both inside and round the fort. At this moment the sapper colonel, Bulmering, sent by Loris Melikov to take over command, arrived and directed an attack, reinforced by dismounted cavalry, against the big lunette from both the western and eastern ends. Cossacks and Dağistan irregulars, who climbed up and fired in through the windows of the lunette and the barracks, encouraged the exhausted infantry. Other Cossacks, with Dağistan and Tush (Georgian mountain) irregulars, made a brilliant attack on the Turkish infantry outside the fort. By 1 a.m. all the works, with ten officers and 200 prisoners, were in the hands of the Russians, but the Turks continued to resist in the solidly built stone barracks. However, the fort was completely surrounded and isolated by the attackers.

Meanwhile more important and, in fact, decisive results had been attained farther to the north. General Alhazov's column opened an attack on Hafiz Paşa Tabia at 8.30 p.m., and after 11 p.m. the fort finally fell to the Russians following a desperate contest. Colonel Fadeyev had been sent with 400 picked volunteers and two battalions of the Kutaisi Regiment to repeat his feat of the 8th and turn the fort from the north. But his attack failed to develop and he seemed to have disappeared in the darkness. In fact, in executing the manoeuvre ordered, he had come under fire from the line of trenches which connected Hafiz Tabia with the Karadağ forts. This line was defended by a small battery, and Fadeyev had naturally turned to silence it before attacking Hafiz. The Turks in the battery and trenches broke and fled up the hill towards Karadağ hotly pursued by the Russians. Fadeyev made the sudden decision to scale the Karadağ slopes—leaving in his rear the Bayram Paşa suburb of Kars. The Russians attained and overran a terrace, where nine field pieces were in position, and reached a large but empty camp where Fadeyev halted his men for a short rest. Here the assault troops came under a hot rifle fire—harmless in the moonlight—directed on them from the two forts of Karadağ Tabia and Arab Tabia and from a twenty-one-foot stone-built tower called Ziyaret ('Pilgrim') covering the gorge below

Karadağ Tabia. This tower was armed with one long-range gun and two field guns, but its garrison was only a few score of men; furthermore, the garrison of Karadağ Tabia itself, which had been manning the trenches connecting it with Hafiz Tabia, had already been weakened and somewhat demoralized by the rout up the mountain. Fadeyev sent Lieutenant Arkhangelski with one company to capture Ziyaret tower and himself led an attack on Karadağ. Having blown in the gate with dynamite, Arkhangelski's men captured the tower and from it directed a fire against Karadağ which quickly produced confusion among its defenders. Many of them fled over the parapet towards Arab Tabia and Colonel Fadeyev entered Karadağ Tabia from the gorge below it. By 11 p.m. this powerful fort—one of the principal defences of Kars—had fallen to the assault of 700 or 800 braves. Fadeyev, however, fully understood the difficulties of his situation. He sent an officer to General Alhazov to ask for reinforcements, and at the same time moved up the rest of his own men from the trenches connecting with Hafiz. Alhazov sent what remained in reserve of the 157th Kutaisi Regiment with a message to keep on Karadağ and try to turn the Turkish guns against Arab Tabia and the town. He promised further reinforcements.

By midnight the Turkish defensive line had been broken everywhere between Kanli and Karadağ and Russian troops had penetrated into the town, where fierce fighting began in the streets of the southern suburbs. All the Russian reserves were now in action, and Lazarev, who had come forward himself into the middle of the street fighting, seeing that Kanli and forts on Karadağ were continuing resistance, ordered a concentration in the Turkish camp between the suburbs and Feyzi Paşa redoubt.

The situation of Fadeyev on Karadağ became for a certain time dramatic. The Turkish guns in Arab Tabia and from Ingiliz Tabia (formerly Williams Paşa Tabia across the river) were concentrated against him while the Turkish infantry in Arab Tabia, supported by cavalry, made no less than six counter-attacks over the ground between Arab and Karadağ forts. Fadeyev's men were exhausted and they would have been overwhelmed but for the timely intervention of General Rydzevski's column which had originally been ordered merely to make a demonstration against the Karadağ forts. Rydzevski's column had been firing spasmodically on Arab Tabia from the north when at 2 a.m. an order came from Lazarev to attack. Dawn was near when the 160th (Abkhazian) Regiment moved to the attack. This very strong fort was really weakly defended, and its garrison tired and discouraged by the failure of the counter-attacks on Karadağ Tabia. The sound of firing in the town below and the general confusion finally broke the spirit of the men who began to make their way down towards the Kars-çay. With the first dawn at 5 o'clock Russian columns could be seen flowing down the hill from both the forts on Karadağ mountain.

In Kanli Tabia, Davut Paşa with 300 men (and 500 more lying dead around them) had continued to resist until 4 o'clock of the morning behind the walls of the stone barracks. Finally Bulmering, having with great difficulty raised a gun on to the parapet, persuaded the Turks that further resistance would only result in their being blown to pieces. The remnants of the garrison laid down their arms amid the general acclamation of their conquerors.

The fate of Kars was really decided along the eastern sector of the defences. Along the southern sector the Russian assault proved far less successful. The column of Loris Melikov the younger captured Çim Tabia with slight loss, but they were driven back from Süvari with heavy casualties. A simultaneous attack made by Komarov's column on Tahmasp Tabia was also repulsed. The failure of the southern attack had no great importance, as it was a demonstration designed against the weaker forts of Süvari and Çim in order to facilitate the principal assault on Kanli and Hafiz by threatening the rear of these strong points.

The demonstration directed against the fortifications on the western and north-western side of Kars had more importance. Here Colonel Chereminisov had led four battalions against the Çakmak heights. He deployed his infantry and some field pieces against the forts of Ingiliz (Williams Paşa Tabia) and Muhlis (Teesdale) and then, dissatisfied with his passive role, led five companies of the Nesvizhski Grenadiers against Laz-tepe Tabia, which was rather an outwork than a fort in spite of its name. Against definite orders given him before the action, Chereminisov tried to improvise an assault on Laz-tepe (which was defended by two battalions with sixteen guns). His men climbed the ice-covered parapet with the greatest intrepidity, but they were beaten back. They remained, however, in the neighbourhood of the fort and kept up a steady rifle fire; and this demonstration had a certain importance, since it attracted the attention of the Turkish commander, Hüseyin Hami Paşa.

By daybreak of 18 November the Russians were masters of all the right bank of the Kars-çay with the town and citadel. Some looting began, but General Lazarev checked it with a firm hand. During the storm Hüseyin Hami had lost not only the initiative but all possibility of controlling the battle. His greatest handicap was the lack of telegraphic communication even between the principal forts surrounding Kars. The prolonged resistance of Kanli had also had a rather harmful effect; for, together with the successful repulse of the Russian demonstrations against Tahmasp, Süvari and Laz-tepe, it had promoted the illusion that the Russian assault had failed. Until late in the night, Hüseyin Hami thought that the Russians had penetrated his lines only at Hafiz Tabia. About 2 o'clock in the morning he became aware of the real situation, and from his headquarters at Veli Paşa Tabia he tried to organize the defence of the left bank area, but the realities of the situation made continued resistance virtually impossible. A great mass of infantrymen who had escaped from the right bank were crowding over the ground between Tahmasp and Laz-tepe. Gradually a general movement began in the direction of the Erzurum road where only a weak Russian force was stationed in observation. Individual officers organized groups of infantry into some sort of order, and a retreat was made towards the villages of Bozanli, Arvartan and Samova. At the same time the Turkish cavalry made towards the west through Çakmak village and were later joined by Hüseyin Hami and some of his staff.

The rapidity of the Turkish retreat took the Russians by surprise, and the Rostov Grenadiers at Bozanli were so impressed by the weight of troops approaching them that they withdrew before them. Russian infantry marching on Arvartan and Samova were too late, and the pursuit was left to Prince Shcherbatov's dragoons and Cossacks. Overtaken by the horse the Turkish infantry hastened to lay down their arms, and only one detachment of Turkish cavalry succeeded in breaking through in the direction of Oltu. The Cossacks rode after them for fifteen miles and sabred about 100 of them. Thirty or forty escaped, thanks to the quality of their horses, and among them was Hüseyin Hami Paşa.

The Turks left 2500 dead on the battlefield; and 17,000 prisoners, with five paşas and 800 officers, were counted in Kars; this number included all the administrative and hospital personnel of the fortress, and in the hospitals the Russians found another 4000 sick and wounded. Captured guns numbered 303, of which 200 were serviceable fortress guns. Russian casualties scarcely exceeded 2000, but these included a high proportion of officers; and the storm of Kars was notable for the number of field officers who lost their lives. The Russian army in 1877, as in 1855, remained an officers' army.

The fall of Kars completed the campaign of 1877. On the Balkan front, Plevna had not yet fallen; and in decisive successes the Caucasian army had outdistanced the armies of the Grand Duke Nicholas on the Danube. Only two objectives remained: Erzurum and Batum. Reinforcements were accordingly despatched without delay to the forces in Guria, and Komarov returned with a strengthened group to Ardahan, since a combined action against Batum from Ozurgeti and Ardahan was now planned. The 40th Infantry Division was sent to join Heimann before Erzurum.

All through November Muhtar had been collecting reinforcements at Erzurum which arrived often by single battalions from the interior of Anatolia. By the beginning of December he could muster 17,000–18,000 infantry and several thousand Laz and armed Muslim inhabitants of the city. When, in mid-December, epidemics broke out, Muhtar, anxious not to lose all which he had managed to recover, and wishing after the news of the fall of Plevna to conserve some strength for the further defence of Anatolia, withdrew his best troops (6000–8000 infantry with twenty guns) to Bayburt. It was a sensible decision. Old Ismail, supported by Hasan Paşa, was left to hold Erzurum with a garrison of 15,000 regulars and irregulars.

By 10 December the Russians had concentrated three divisions in front of Erzurum: the Caucasian Grenadier Division, the 39th and 40th with the 150th (Taman) Regiment, the 3rd Caucasian Rifles and strong cavalry. Accord between Heimann and the other generals was bad and Loris Melikov arrived to take personal command. During the bitter cold of December on the Armenian highlands, the Russians slowly extended their lines round Erzurum. On 5 January 1878, the occupation of Ilica village on the Trebizond road completed the blockade of the city which, with its garrison of 15,000 and its 50,000 inhabitants, was now cut off from the outer world. Food was known to be getting scarce, and it was calculated that Erzurum must capitulate in not less than two months' time. Consequently there was no object in undertaking a costly storm. The blockading army was, however, in almost as parlous a condition as the beleaguered city. The frost became increasingly severe; blizzards swept the roads and made transport almost impossible; some units remained without bread and even biscuits for ten days at a time. Horses were in a desperate state. Finally, the spotted typhus which had been ravaging the city spread to the besieging army. In both the Russian and Turkish ranks some units could not muster 100 fit men. More than 3000 Russian soldiers and officers died of the epidemic; among them Heimann, Shelkovnikov and Loris Melikov the younger. These terrible conditions came to an end only with the signature of the armistice on 31 January 1878. In accordance with the conditions announced, Ismail Hakki Paşa left Erzurum with the regular garrison and such armed inhabitants as wished to remain with the army; and the Russians occupied the city on 8 February.

Operations against Batum had begun at the end of November. The Russians planned an advance from Ardahan which, when it had gained a certain impetus, was to be co-ordinated with a renewal of the offensive along the coast from Ozurgeti. The operational plan was theoretically logical, and the Russian forces available were far stronger than the Turks; but the difficult climatic conditions of the Pontic Alps, through which runs the valley of the lower Çoruh and the way from Ardahan to Batum, had been ignored. The ambitious and enterprising Komarov, commanding at Ardahan, seems himself to have been to blame, for he convinced Russian general headquarters that it was practicable to make the passage of the high pass over the Yalnizçam-dağ in winter.

Komarov left Ardahan on 14 December with thirteen battalions (the 151st, 152nd and the 1st Caucasian Rifles), one regiment of Cossacks, one regiment of irregulars and twelve mountain guns. A Turkish detachment of two battalions of nizams, and 2000 Laz irregulars with three guns occupied a strong natural position at Hane village covering the road to Artvin. This position was taken after a short fight and on 16 December Komarov entered Ardanuch, a small and ancient town which stands on a quadrilateral table of rock overlooking the ravine of the Çoruh. Here, for three weeks, powerful blizzards made further advance impossible; the road leading along the right side of the deep ravine of the Çoruh, which in places falls a sheer 1000 ft. to the river, was impassable.

Political reasons, however, compelled Russian general headquarters to insist on the development of the operation against Batum. In the first days of January 1878, the Russian armies in the Balkans were entering Thrace, and it was obvious that an armistice could not be delayed long; and St Petersburg was concerned to have Batum 'in the bag' at the coming peace discussions. At the same time it was well known that the forces of Derviş Paşa at Batum had been greatly reduced by drafts to Europe.

On the Batum front there had been a long lull which had continued through the autumn and early winter. Since July, Derviş Paşa had concentrated the bulk of his forces behind the Kintrishi, maintaining only an advance force covering several batteries on the Hutsubani heights. The opposing Russian detachment on the Muha-Evstate position had definite orders to remain on the defensive, and military action had been limited to some artillery duels and skirmishing. Several lively frays took place, between Gurian militia and Kuban plastuns on the one side and Acar and Kobuletian partisans on the other, in the region of the source of the Achkva (Achkvis-tavi) and the passes leading into Achara.

In October Derviş had been obliged to send eight battalions and one battery to Erzurum, and in November he was ordered to despatch six battalions to Istanbul. He decided to shorten his front and on the night of 27–28 November evacuated the Hutsubani heights. He withdrew to the Kvirike-Deva positions covered by the lower Kintrishi and its affluent the Kintash. He retained an advance force, mostly of irregulars, on the Sameba heights.

After the fall of Kars, General Denibekov (who had succeeded General Oklobzhio in command of the Guria force) received orders to prepare an attack, when Komarov's advance down the valley of the Çoruh should have been sufficiently developed. But Komarov was only able to move from Ardanuch on 13 January. The Turks had taken up a very strong natural position at Dolis-hane—an easy six hours' pony trot from Ardanuch. On 21 January Komarov dislodged the Turks from the Dolis-hane position by a sharp bayonet attack, with a loss of one officer and sixty men, and the road to Artvin lay open. However, more than fifty miles lay between this pleasant town, above the main gorge of the Çoruh, and Batum. Along all this route the defence of the Laz and Acar irregulars was likely to be even more stubborn and formidable than it had been between Ardanuch and Dolis-hane. These circumstances increased the impatience of the grand duke's staff, since it was obvious that an armistice could not be delayed for more than another fortnight. It was therefore decided to make a frontal attack on the Batum positions which, during the summer, had repulsed Oklobzhio. Komarov was selected to take command. Russian general headquarters held the view that Turkish morale at Batum would be completely broken by the news of the disasters in the Balkans, but they underestimated the determination of the local elements who had conducted all along at Batum what was in effect the guerrilla defence of a defended locality and who were of that primitive psychology which is only impressed by the course of events in their immediate environment.

Komarov arrived at Kobuleti on 28 January; on the same day he made a reconnaissance of the Turkish positions (in a region completely new to him). Next day he discussed the situation with the generals, Denibekov and Shelemetev, who had been fighting on the Batum front since April 1877. The attack was fixed for 30 January. The forces at the disposal of Komarov were considerable: 41st Infantry Division, 76th (Kuban) Regiment, 2nd Caucasian Rifles, several battalions of Kuban plastuns, Gurian, Imeretian and Georgian militia, several Cossack regiments and strong artillery. If a part of this force had to be left in observation along the coast and in the Achkvis-tavi area, Komarov had at least 15,000 infantry available for the attack against the 12,000 infantry and irregulars which Derviş Paşa could put into line. At the same time the Turkish fortified positions were intact; batteries were stationed along the ridge between Deva and Tzikhis-dziri, and four ironclads gave supporting fire off the mouth of the Kintrishi.

Komarov planned to attack along a rather wider front than Oklobzhio had tried. On the left, irregulars, supported by one infantry battalion, were to cross the upper Kintrishi and capture the Sameba and Table hills, where guns were to be mounted without delay. Then, supported by these guns, the 2nd Caucasian Rifles and the 163rd Infantry Regiment were to cross the Kintrishi and force the Kvirike-Deva position. On the right General Shelemetev was to advance along the coast and, with strong artillery support, to cross the river. The whole plan was based on the supposition that Turkish resistance would be definitely weaker than it had been at the time of Oklobzhio's attack.

Unfortunately for the Russians, the first stage of the fighting seemed to justify this illusion. The Sameba and Table positions, where there was only a weak force of irregulars, were captured easily. As the passage of the guns across the Kintrishi—a more formidable task in January than it had proved in June—required some time, the intrepid Caucasian riflemen in the centre, followed by companies of the 163rd, began to cross the river without artillery support. The Turkish guns on the Deva heights remained silent, and the Russians began to suspect that the enemy had evacuated his fortified camp. But these notions were soon dispersed when from several rows of trenches along the Deva and Kvirike heights rifle and gun fire poured down upon the attacking troops. At the same time along the coast the guns at Tzikhis-dziri and the Turkish ironclads opened a heavy long-range fire on the area where General Shelemetev was trying to establish his batteries.

Shelemetev was killed and there were some losses among gunners and horses; but the situation of the infantry in the centre, where about two battalions had succeeded in crossing the Kintrishi, rapidly became tragic. At many points they found themselves unable to recross the river which was treacherous—deep and rapid in unexpected places. The 2nd Caucasians suffered heavy losses: almost all their officers were killed or wounded. The remnants who escaped only managed to do so by making their way up the left bank of the river towards Table hill. Komarov, seeing what a bad turn the fighting had taken, withdrew his men to their positions of the morning. Losses were relatively heavy: more than 1200 killed and wounded—little less than those of the big Alaca-dağ battle. The onslaught proved to have been quite useless, since on the following day, 31 January, an armistice was signed. The Turks remained in possession of Batum during the armistice period, and this town with its potentially valuable port passed to the Russians only after the Berlin Congress.[1]

In the last period of the Caucasian campaign of 1877, from October to November, the Russians had won a decisive victory. This victory might have been reached earlier (possibly in the months June to August), if a certain confusion in strategic direction had been avoided. The Russians further allowed themselves to be effectively influenced by the Turkish diversions against the Caucasian coast and the Aras valley and by an exaggerated fear of an internal front; they had let the strategic initiative pass to the Turks, but the latter had not been strong enough to make use of it. Turkish political strategy, in the conception of the diversions, had been clever and had constituted an appropriate reply to the Russian exploitation of political and military diversions (Serbia and Montenegro) in the Balkans; in Ahmet Muhtar the Turks had a strategist superior to his antagonist; but the Turkish army (in spite of its modernization since the Crimean War) was an instrument definitely inferior in quality to the Caucasian army opposed to it, and ultimately all Turkish strategy was forced on to the defensive.

The defects of the Russian army were the defects of the high command, increased by some failures of subordinate commanders. It was

[1] The successful defence of Batum proved to have strengthened the position of the Turks at the Berlin Congress and its fate was one of the principal subjects of dispute between the British and Russian delegates: Medlicott, *Congress of Berlin*, pp. 101–25. Müşir Ibrahim Derviş Paşa Lofçali (1817–96), more fortunate in battle than the gazis, Osman and Ahmet Muhtar, never received that highest title of the Turkish soldier. The long defence of Batum was a remarkable feat, yet it is not even mentioned in the biographical note on Derviş Paşa in the official *Türk Meşhurlari* (Famous Turks). But see a short article in *I.A.*

a top-heavy structure: at the head the grand duke with his (irresponsible) advisers, Svyatopolk Mirski and (later) Obruchev; under the grand duke, the commander-in-chief, Loris Melikov, wholly responsible but at the same time wholly dependent upon the grand duke (who fortunately was a benevolent and reasonable, if not brilliant, man). Staff work was bad everywhere and always—except later on at Alaca-dağ and Kars, under the firm and efficient Lazarev.

As to personalities, the campaign of 1877 affords some interesting psychological material having perhaps an eternal meaning for the study of military characters in any epoch. The defects of Loris Melikov have been demonstrated more than once. He became the subject of many criticisms in the Russian military literature of the time, but at least it was allowed of him that he never sacrificed his men's lives for the sake of prestige. In this he may well be contrasted with Heimann, who often sought to dominate his superior by his dynamic and insistent character. This figure of Heimann was rather extraordinary: of Jewish origin he made a rapid career in the Russian army by his flamboyant personal courage and his remarkable good luck. There was something of the gambler in him, as well as (to a lesser extent) in Komarov. Quite different was Tergukasov, extremely serious, firm, yet beloved by his troops. Lazarev, another general of Armenian origin, was no less firm (even something of a disciplinarian), efficient and with a spark of real talent in him.[1] He worked well with the young Shelkovnikov who was the only one among the senior officers who combined the qualities of a good staff officer with those of a leader of men.

Quite remarkable in this campaign was the long series of intrepid colonels, among whom the most impressive were Fadeyev of the Kutaisi Grenadiers, Amirajibi of the Elizavetpolskis, and (Brigadier) Tsitovich with his Baku and Derbent Regiments. Battalion and company commanders, often quite young officers, were very good. The campaigns of 1877–8 were the first trials for the new Russian army based on the general conscription of 1874. The results proved

[1] As in 1828–9 and 1853–6, the Caucasian army retained the strongly Caucasian flavour in its command: besides the Armenians (the two Melikovs, Tergukasov, Lazarev, Avinov, Shelkovnikov), there were the Georgians, Chavchavadze, Oklobzhio, Sharvashidze, Amirajibi, Amilakhvari and Eristov, and the Tartars, Denibekov and Nakhichevanski. The territorial names of Caucasian regiments have little significance; e.g. the Bakinski, Derbentski, etc. were largely Russian. At the same time young Georgian officers, particularly, preferred such regiments as the Nizhegorodski and Severski Dragoons. The Caucasian regiments with their long periods of service in Caucasia, and their mixture of Russian and Caucasian nationalities had, at the same time, a Caucasian and a truly imperial character.

satisfactory on the Caucasian, as well as on the Balkan, front. Newly created divisions, like the 38th and 40th, returned from the battlefields with the same fame as the old Caucasian grenadier regiments or the regiments of the veteran 19th Division. Only the Moscow Grenadier Division proved rather less satisfactory. The Caucasian Rifle and Dragoon Regiments justified their reputation as crack troops and good service was done by the Cossacks and Dağistan irregular horse.

On the Turkish side the soldier proved excellent and—what was encouraging for the future—the redifs were as good as the nizams. Well armed, Turkish units seemed still to lack adequate instruction. Lack of good officers persisted—a result of the peculiarly decadent social influence of the court and of Istanbul. In comparison with previous campaigns, the Turks had a greater number of officers approaching a European standard: Ahmet Muhtar had been one of the first instructors at the newly founded Military College in Istanbul, and officers like Hüseyin Paşa (commander of engineers at Kars) had passed through Woolwich. (Numerous naval officers also had done courses in England.) But the best officers under Muhtar remained men of foreign origin: the Hungarian Feyzi and the German Mehmet (who from Ardahan to the Aziziye redoubt had never been personally defeated). There were plenty of decent men in the army like Derviş at Batum, Hasan at Erzurum and Hüseyin Hami at Kars; but the majority were still representatives of the worst Turkish traditions: Hasan Sabri at Ardahan, Faik at Bayazit, and old evil Ismail Hakki.

As for Ahmet Muhtar, the Anatolian army was fortunate in having as commander-in-chief one of the best Turkish commanders of the time, not so good on the battlefield as Osman Nuri of Plevna fame, but certainly much superior as a strategist to Mehmet Ali or Süleyman. Undoubtedly Muhtar was an intelligent leader with a capacity for quick decision and rapid action. He had a cool and sanguine temperament and did not lose his head in moments of disaster. His principal mistake, the delay in carrying out the decision to retreat from Alaca-dağ, may be attributed to an unfortunate sentiment of *noblesse oblige* which obliged the new gazi to fight a battle of prestige.

BOOK IV

THE CAUCASIAN CAMPAIGNS OF WORLD WAR I, 1914–18

CHAPTER XVI

THE TRANSCAUCASIAN FRONTIER, 1878–1914. THE GERMAN FACTOR IN THE MIDDLE EAST

MAPS 1, 4, 17, 19

AT the Congress of Paris which closed the Crimean War, the western maritime powers (Britain and France—with Sardinia) had imposed a solution of the Eastern Question which had a definitely maritime emphasis; the neutralization of the Black Sea and the maintenance of the *status quo* round the shores of the eastern Mediterranean. The Germanic powers (Austria and Prussia), on the eve of an internal struggle within Germanic Europe, assumed a comparatively passive role.

Twenty-two years later the next attempt to order the affairs of the Near and Middle East—significantly enough—was negotiated in Berlin. Prussia had emerged as the master of central Europe, and the Treaty of Berlin which ostensibly reconciled the conflicting interests of Great Britain and Russia in the Near and Middle East, with Bismarck in the sympathetic role of honest broker, in fact marked the emergence of imperial Germany as a world power with carefully veiled plans for domination over all the area of the dissolving Ottoman empire.

The Treaty of Berlin was signed on 13 July 1878. Lord Beaconsfield returned to London bringing with him peace with honour. A few days earlier, on 8 July, H.M. Government had announced the occupation of Cyprus and the convention with Turkey. The Prime Minister met with severe criticism from Mr Gladstone, Lord Derby and others; and, as usual, the admirers of the Beaconsfield policy allowed their enthusiasm to overcome their discretion. Lord Sandon, for instance, declared that, during his wanderings in the valley of the Nile and over the plains of Syria, the cry had always been: 'When are you coming?' 'And now', he added, 'England has come at last.' Such words might help to explain to their lordships the significance of the occupation of Cyprus, but in Istanbul this and other hints and indiscretions left no illusions as to the value of the territorial guarantees which the representatives of the Ottoman empire had so recently secured from the Powers assembled at Berlin.[1] Indeed, very few years were to pass

[1] The progressive deterioration of Anglo-Turkish relations after the Berlin Congress is described in detail by Medlicott, *Congress of Berlin and After*, chap. VIII.

before it became clear that the convention of defensive alliance concluded between Great Britain and Turkey on 4 June 1878 had, in fact, terminated the era of British protection over the Ottoman empire, an era which had endured for the greater part of half a century. Clever enough as politicians, the Turks understood the implication of the development of the European situation since the rise of Prussia during the sixties and seventies and, even before they left Berlin, their representatives had already been courting another protector.[1]

During the eighties, British policy (especially during Gladstone's periods of office) was drifting more and more away from Turkey. Meantime after the occupation of Bosnia and Hercegovina by Austria and the conclusion, in 1879, of the Dual Alliance between Austria-Hungary and Germany, the new political situation in continental Europe came to be understood in Russia as a menace far more real and dangerous than the traditional antagonism of Great Britain in the East. The newly acquired independence of the Balkan States did not prevent them from falling one after the other under the influence of the Austro-German alliance. Likewise the Russians had no illusions as to the future political orientation of Turkey. When, after 1886, the prospects of a mighty coalition against Russia took form, the eventual role of Turkey in such a coalition became quite clear. If war broke out along the western frontiers of Russia, the Turks might be expected to concentrate their whole force against the Transcaucasian border.

The Treaty of Berlin had confirmed substantial changes in the military geography of the Russo-Turkish frontier. The Russians had acquired Batum, and, despite all the declarations in London and St Petersburg made during the period of the Berlin Congress, providing for the neutralization of Batum as a free port, the Russians, after a few years' delay, began to establish there a naval station protected by fortifications. 'This was a period of secret understandings and there is reason to believe that liberty to fortify Batum was privately conceded, although this proceeding has since been a popular ground for accusations of bad faith against Russia.'[2]

To the south of Batum, the new frontier had been traced along a line running parallel to, but ten to twelve miles south-east of, the lower Çoruh. Turning west to the Çoruh below Artvin, the frontier then ran south to the valley of the Oltu-çay and crossed the upper course of this river in the neighbourhood of Nariman. Oltu itself became a

[1] One of them, Mehmet Ali Paşa, was a German renegade who was treated with scant courtesy by Bismarck.

[2] Colonel Sir G. Sydenham Clark, *Russia's Sea Power: Past and Present* (London, 1898).

Russian town. The difficult mountain country of the Pontic Alps, through which the Çoruh cuts a deep ravine to the sea, was deemed to give sufficient protection in depth to the port of Batum. The railway was soon extended from Poti to Batum and two important military roads connected Batum, first, with the Borjom defile via the Acharis-dzqali and Akhaltzikhe and, secondly, via Artvin-Ardanuch-Ardahan with Kars.[1] From Akhaltzikhe, another military road was constructed up the Kura valley through Ardahan to Merdenik, whence it forked to Kars and Oltu.

These roads connected the western and central sectors of the borderland. From Oltu ran one of the principal tracks from the Russian frontier to Erzurum: Oltu-Nariman-Id-Gürcü-boğaz—very important since it outflanked the Deve-boyun positions from the north-east. In this direction, from Akhaltzikhe-Akhalkalaki to Erzurum via Oltu, the Russians had, for a future campaign, gained two-thirds of the distance.

In the central sector the Russians acquired Kars, and the new frontier was traced over the Soğanli-dağ between Zivin and Karaurgan; as the crow flies Erzurum was now only fifty-two miles from the Russian frontier, and here again two-thirds of the distance had been gained. While the Zivin positions of Muhtar Paşa remained in the hands of the Turks, Karaurgan (two miles from Zivin) became the Russian frontier station, and the frontier crossed the Soğanli plateau between the villages of Upper and Lower Mecinkirt. From Upper Mecinkirt the Russians built a road descending to the Aras valley west of Karakurt. Kars had thus become a Russian fortress and advance base, making extremely difficult any invasion of Transcaucasia from across the Turkish frontier. At the same time the new frontier deprived the Soğanli-dağ of its defensive unity, and the Turks were under the necessity of organizing a new defensive position nearer to Erzurum—in the Köprüköy area on both sides of the Pasin valley.

Under the terms of the Treaty of St Stefano, the Russians had received the Eleşkirt valley and the town of Bayazit. After crossing the Aras the new frontier had been designed to cut across the Ağri-dağ and the Eleşkirt valley and then to follow the Şarian-dağ to the eastern Euphrates (Murat-çay), reaching the Hamur defile (between Şarian-dağ and Ala-dağ) where this river turns sharply to the south. Beyond the Euphrates the frontier followed the watershed of the Ala-dağ to the Persian frontier.

At the Berlin Congress the British representatives raised objections

[1] Already planned and partly built by the Turks prior to 1877.

to the acquisition by Russia of the Eleşkirt valley and Bayazit on the ground that through them passed the principal caravan route from Trebizond (Trabzon) and Erzurum to Tabriz. A strategic issue was also involved, since the Murat-çay defile (the key to Turkish Kurdistan) gives access via Muş to Diyarbekir and the upper valley of the Tigris. The Russians yielded rather easily to the British objection, and the new frontier, under the Treaty of Berlin, crossed the Aras to Kokse-dağ, the western shoulder of the Ağri-dağ, and then followed the watershed of that ridge to the line of the old frontier in the region of Lake Balik. (Cf. Map 22.)

Thus along the eastern sector of the frontier, Russian acquisitions were inconsiderable. They increased their hold on the Aras valley from a point half-way between Horsan and Karakurt to Kulp (Tuzluca), and including the town of Kağizman. This place (first occupied by Muravyev in 1855 and then by Tsitovich after the battle of Alaca-dağ) became an important Russian advance base for the eastern sector of the frontier. The importance of Kağizman (connected by road via Kulp with Erevan) lay in its proximity to the Ahtalar Pass (which became, with the development of a road, the best of all the passes over the Ağri-dağ), whence a track led to Karakilise, distant twenty-five miles, the most populated centre in the Eleşkirt valley and itself only a few miles distant from the defile of the Murat-çay.

The advance of the frontier on the eastern sector implied that the way to Erzurum was no longer from Erevan through Iğdir, but from Kağizman (connected by road with Kars and by another road along the Arpa-çay valley with Aleksandropol); thus the distance from the Russian frontier to Erzurum on this sector was shortened by one-third, and, with the Ahtalar and Çat passes in their hands, the Russians could at any time invade the Eleşkirt valley with a view to developing an operation in the direction of Muş and Van. (Cf. Map 22.)

During the three and a half decades between the campaigns of 1877–8 and 1914–17, the Russians improved the communications of the eastern and central sectors of the frontier. During the nineties a railway was built from Aleksandropol to Kars, and later to Sarikamiş. Another railway was constructed along the Arpa-çay valley past the slopes of Alagöz and through the Erevan region down the Aras valley to Nahçivan and Julfa on the Persian frontier. From Alagöz station a good road was built to Kağizman.[1]

[1] The Transcaucasian railway (Batum-Tiflis-Baku), connecting via Derbent-Petrovsk and the north Caucasian lines with Rostov and the main Russian railway system, had been completed during the 1880's.

From the eighties onwards, however, the anxious attention of the Russian general staff was directed not to the Caucasus but to the western frontier of the empire. New dispositions for the Russian armed forces were worked out in the early eighties; almost the whole of the Russian peace establishment was concentrated in the western provinces. Only two army corps were retained in Caucasia, each made up of two infantry divisions: the Caucasian Grenadiers, and the 20th, 21st and 39th. The Caucasian troops of the old 19th Division were stationed on the Austrian frontier; the 38th, 40th and 41st Divisions, despite the Caucasian names of their units, also left the Caucasus for European Russia. At the beginning of the twentieth century two new Caucasian divisions were created, 51st and 52nd, and went to form part of the new Third Caucasian Army Corps.

After 1910–12, the Second (Caucasian Grenadier Division and 51st) and the Third (21st and 52nd Divisions) Caucasian Army Corps were allocated to the western frontier.

The decision of the Emperor Alexander III, in 1882, to recreate the Russian Black Sea fleet, had really effected far more important changes in the balance of power in the Ponto-Caucasian area than any modifications of the Transcaucasian land frontier. Naval docks were constructed at Sevastopol and Nikolayev and, in 1886, two battleships were launched, *Chesme* and *Catherine II*, followed, next year, by *Sinope*. By 1888 the Russians had in the Black Sea a powerful battle squadron: three battleships of 10,000 tons each with an armament of six 12-inch guns and a speed of fourteen to fifteen knots. These were supplemented by three military transports of 5000 tons each, a score of big commercial steamers which could be used for the same purpose, one 3000-ton cruiser, six gunboats and sixteen torpedo boats (speed 15–16 knots). It was a formidable naval force, certainly far stronger than the obsolete Turkish ironclads of the 1877–8 period, and it gave the Russians complete control in the Black Sea.

The new situation created by the development of Russian naval power in the Black Sea seems to have changed radically the operational plan of an eventual war against Turkey. At this time (about 1888), a curious pamphlet was circulating among the 'activist' groups in the Russian fighting services, envisaging a surprise attack on Turkey. The basis of this plan was a sudden assault on the Bosporus by the Russian Black Sea fleet whose action was to be combined with the landing of a picked expeditionary force on both shores. The pamphlet, widely circulated, was printed and published anonymously in Russia, but its source of inspiration was probably German and *provocateur* in character.

The last Anglo-Russian crisis—the Merv and Penjeh incidents (giving rise to the phase of russophobia known as 'Mervousness' in England)—assumed an acute form for a short period in 1885; but it passed quickly and, somewhat strangely, contributed to a rather deep and durable relaxation of the Anglo-Russian tension, as if both sides had become tired of the long period of irritations and counter-irritations. The Russians were becoming increasingly preoccupied with the developing Austro-German threat. After 1886–7 the structure of Russian influence in the Balkans, at once political and cultural, sagged badly. Ferdinand of Coburg appeared in Sofia—the former objective of so much Pan-Slav enthusiasm. On both sides of the western frontiers of Russia, military preparations were in progress and 1888 seemed the year destined for a military crisis. Bismarck went so far as to declare that the whole of the Eastern Question was not worth the bones of a single Pomeranian grenadier. And the chancellor was at pains to convey to his Russian friends that he had a perfect understanding of their impatience with regard to Constantinople and the Straits. At the same time the German leader was busily occupying himself with the idea of a Mediterranean convention: it was proposed that Great Britain, Austria-Hungary and Italy, with the assent of Turkey and Rumania, should guarantee the *status quo* in the Mediterranean, which implied clearly the existing regime on the Straits. The intention was that any attempt by Russia to take possession of the Straits should constitute the *casus belli* for a mighty coalition in which Great Britain, Austria-Hungary and Italy, with Turkey and Rumania, should function in a first *tempo*, with Germany certainly, and possibly Sweden, Bulgaria, Serbia (and even China) following in second *tempo*.

It was a fine play of Bismarckian Machiavellism, but the Russia of 1887–8 was not propitious ground for it—despite the anonymous propaganda among the activist groups. Alexander III was not a man of adventurous initiative; he wished to 'freeze' his empire a little, and frozen empires are unlikely to move. Then by a curious turn of events the accession to the German imperial crown of the warlike and ambitious William II saved the Russians from the consummation of Bismarck's scheme (to which, with variations, Adolf Hitler was to return fifty years later). If William II were inclined to war, it was in his own way and not after the way secretly prepared by the intolerable chancellor. The young emperor was a man of the new industrial era and of *Weltpolitik*. A certain time is necessary for the establishment of the broad outline of a new policy. And in the interval the Franco-Russian alliance was concluded and the Russian army began to be armed with a new rifle (Lebel), not inferior to the Mauser of the Prussians and the Mannlicher of the Austrians.

The Russians continued to strengthen their naval position on the Black Sea. In the period 1890–3 three more large battleships (*Twelve Apostles*, *St George* and *Three Saints*) were added to the existing strength. To meet these the Turks had only the three ironclads built in 1874 (*Turgut Reis*, *Hayrettin Barbarossa* and *Mesudiye*), which were quite helpless against the new Russian strength. In the difficult financial circumstances which followed the war of 1877–8 the Turks had neither the opportunity nor the initiative to attempt to compete with Russian building. The situation of their land forces was equally parlous. The internal situation of the empire absorbed all the attention of Sultan Abdul Hamit II, and the army was used merely as an auxiliary police force. New German elements were arriving in Istanbul in increasing numbers, but, under Bismarck, they had not gone beyond the stage of preliminary study of conditions—a prelude to considerable financial and economic help which soon took shape in the manifold activities of the Deutsche Bank.

This activity began to assume a military significance after the formation, in 1888, of the Anatolian railway company, which had the backing of the Deutsche Bank. In 1892 the railway connexion between Istanbul and Ankara was completed, and in 1896 another line was built from the Eşkisehir station on the Ankara line to Konya. Konya represented the starting point of the Bagdad-Bahn which soon began to take form. The concession had been acquired by the Deutsche Bank in 1902; in 1904 the line had already been finished over the 150 miles between Konya and the northern slopes of the Taurus. The progress of this railway policy was one of the principal factors determining a change in British policy towards Turkey, with its sequel, the Anglo-Russian Agreement of 1907.

During the nineties, after the arrival of a military mission under General (later Field-Marshal) von der Goltz, the Germans began to participate openly in the reorganization of the Turkish army. The first results of their work were in the successful campaign of the Turkish army in Thessaly in 1897.[1] The easy victory then obtained

[1] During the eighties the forts of the Dardanelles and the Bosporus were modernized under the supervision of German engineers, and in the nineties Germans elaborated and put into action plans for the modernization of the Erzurum forts.

drew the more serious attention of Berlin to the military possibilities of the Turks. But Abdul Hamit, who feared the emergence of army officers' 'cells' as an active factor in Turkish politics, was unwilling to go more than a certain distance in measures for reorganization; and this reluctance probably sealed his fate with the Young Turk groups in the army, who were in intimate contact with the officers of the German mission in Turkey.

After the abdication of the Sultan in 1909, the new Young Turk triumvirate (Enver, Talaat, Cemal) were free to reorganize the army and to accept all the proposals of their German friends. Many drastic changes were introduced in the higher grades of the army; many protégés of the Sultan's clique were dismissed, doubtless to good effect; but new protégés of the Committee of Union and Progress, often men of the worst type, took their places. During 1910–11 some limited results were achieved in the reorganization of the troops in the capital and in Adrianople; but, as a whole, the Turkish army remained badly organized, particularly in regard to transport and supply, while the rearmament of the artillery had only begun.

The Austrian annexation of Bosnia in 1908 brought to an end the Austro-Russian political armistice which had tacitly existed since 1897. After the defeat in Manchuria had checked their plans in the Far East, and following agreement with Britain on Middle Eastern matters, the Russians began to renew their interest in the Near East, particularly in the possibilities of checking the Germanic *Drang nach Osten* by a support of the southern Slavs. The Balkan League of 1911, favoured by Russia and actually negotiated by the Englishman, Bourchier, *The Times* correspondent in Sofia, was an effective reply to the audacious railway politics of the Germans in Anatolia, to the masterful attitude assumed by the German ambassador in Istanbul, and to the open supervision of the Turkish army by von der Goltz's mission; it was at the same time a suggestive example of Anglo-Russian policy acting in unison.

It is a curious fact that at this time the German government appears to have been badly informed on the subject of the respective strengths of the Turkish and Balkan armies, and Berlin and Vienna were more or less completely surprised by the result of the First Balkan War. In his *Diplomatic Reminiscences*,[1] M. Neklyudov, then Russian minister in Sofia, recalls a conversation in 1912 with M. de Giers, his colleague in Istanbul, when the latter told him that 'the Ottoman army was quite different from what it had been in Abdul Hamit's time:

[1] Published by John Murray, London, 1920.

admirably equipped and perfectly trained under the command of German generals and senior officers and Young Turk officers, it constituted a real power which might cause disagreeable surprises to Serbian and Bulgarian optimists'.

But slow mobilization of the Turkish forces and delay in their concentration, combined with an ill-judged forward movement taken against the advice of von der Goltz, proved fatal to Turkish arms. The campaign was decided before Turkish reserves and complements had had time to arrive.

The so-called Second Balkan War helped the Turks to recover, if not materially, at least morally. In 1913 the atmosphere was propitious for a more comprehensive reorganization under German control. In the autumn of 1913 a special German military mission was despatched to Turkey, at the head of which was General Liman von Sanders who was destined for the post of officer commanding I Army Corps, stationed in the capital. He was accompanied by forty-two officers, including Colonel Bronsart von Schellendorf, recommended as chief of staff over all the Turkish armed forces. The Entente powers protested at a step which openly converted the Turkish army into an auxiliary of the German empire, and a diplomatic crisis threatened when it became clear that the Young Turks were not prepared to yield. A host of German engineers, economists and financial agents followed in the wake of the military mission, and their role in the reorganization of the production and communications of the Ottoman empire was perhaps more significant and threatening than that of the soldiers.

The result of the Balkan wars was that the centre of gravity of the Turkish State had been removed from Europe to Asia; this was well emphasized in the emergence of the Pan-Turanian theory as a counter to the Pan-Slavism which had been exploited with such effect against the Sultan in his Slavonic provinces. The Young Turks now promulgated vague but vast pretensions to all the Russian territories in Asia inhabited by Turkish-speaking peoples, and, where appropriate, supplemented Pan-Turanian with Pan-Islamic incitements. Turkish propaganda in the Caucasus, dating from the days of Sheikh Mansur in the last quarter of the eighteenth century, was revived to attract into the modern Turkish state the new generation of the Tartar *bourgeoisie* of Baku and the still warlike and fanatical inhabitants of Dağistan.

Among the great European powers, only Austria-Hungary remained really interested in the Balkans. The interests of Britain and Germany alike were focused on Asiatic Turkey; Russia, from sentiment and tradition, retained an interest in the Slav or Orthodox states in the Balkans, but her most vital political and economic interests were centred also in Asia Minor.

During the Russo-Japanese War the relations of Britain and Russia had passed through a period of strain, and at that time (1904) Colonel Younghusband's expedition to Tibet had caused some irritation in St Petersburg. Further the interplay of British and Russian political interest in Persia had given rise to recurrent friction. There was, in fact, a purely practical and economic basis to the Russian interest in Persia, since, with the great industrial expansion of Russia in the nineties, trade with Persia grew naturally and rapidly. During the nineties the Russian railway from Aleksandropol to Julfa had been continued to Tabriz; and the Russians had built modern roads connecting Teheran with the port of Enzeli (Pahlavi) on the Caspian, with Kazvin and with Kum. There was also a constant interplay of psychological and cultural influences, due to the community of sympathies of Tartars, Persians and Armenians north and south of the Aras. The revolutionary troubles of 1905, so lively and widespread in Caucasia and in particular at Baku, were reflected in Persia in the revolutionary events of 1907–9.

This period saw the development of active German interest in Persia, running parallel with much Pan-Turanist activity in both Russian and Persian Azerbaijan.

The Anglo-Russian Agreement of 1907 therefore became the expression of several new trends: it was first a sequel to the admission of Britain into the Franco-Russian alignment following the establishment of the Entente Cordiale; it was the definite expression of the *détente* in Anglo-Russian relations throughout the Middle East; and it was a defensive reaction to the growing German control over Turkish life and to the German penetration into Persia. In the Anglo-Russian Agreement, Persia was divided into zones in which each power assumed responsibilities for the maintenance of order. At Teheran, the Russians had organized, on behalf of the shah, a brigade of Persian Cossacks as early as the middle eighties; in 1906 this brigade was strengthened by Russian elements. In 1910, disorders in Azerbaijan and troubles among the Kurds led to the establishment of a Russian force in that area: eight battalions of infantry, and a division of Cossacks, with twenty-four guns, were stationed in detachments at Tabriz, Urmia, Khoy and Dilman, with posts at Maku, Maragha and Suj-bulak.

All these factors—the Anglo-Russian Agreement, the distribution of Russian troops throughout Azerbaijan, and the existence of a railway connecting Tabriz with Caucasia—gave a new importance to the Tabriz-Urmia region and created new strategic problems in the event of war. The position offered brilliant possibilities for an offensive against Turkey; from Khoy (forty miles from railhead at Julfa) two tracks led to Van, one via the Kotur Pass and Erçek, the other via Dilman and Başkale. From Van a road south of the lake led to the upper Tigris valley and Diyarbekir.

Strangely, however, Russian plans in the event of war with Turkey were of a defensive character with only local and limited possibilities of offensive action. At the base of these conceptions was the view of the French general staff that 'all roads passed through Berlin', and Russian strategy in Asia Minor (as in the Carpathians) was aborted by the Grand Duke Nicholas's infatuation with the French official view. As for co-operation with British armed forces, nothing appears to have been planned or even discussed between 1907 and 1914.

In 1828–9 Paskevich had foreseen the possibility of decisive action through Asia Minor; his theories had been proved practical five years later by the Egyptian commander, Ibrahim Paşa. In 1877 the Russian general staff had, for political reasons, been compelled to focus their interest on the Balkans. But in 1914 the Russian neglect of the chances of an active offensive through Asia Minor remains the more difficult to understand in view of the fact that Alexandretta constituted a practicable military objective and a political objective more easily attainable than either access to the Dardanelles or a port on the Persian Gulf. The distances between Alexandretta and the principal Caucasian centres are more impressive than any detailed *exposé* of possibilities: Alexandretta to Tiflis as the crow flies is 600 miles; Alexandretta to Sarikamiş is 450 miles, about the same distance as from Erevan to Novorossiysk. Harput, which the Russian Caucasian army was within reach of attaining in 1916, is approximately half-way between Sarikamiş and Alexandretta and within easy distance of the principal route down the upper Euphrates.[1]

[1] 'In the spring of 1915 the assumption by Russia of power in Constantinople seemed inevitable in a few weeks. The Russian Government asked the Synod to give their verbal opinion on the question of administrating the Constantinopolitan Church. I had been called to come from Kharkov and take part in a private conference. During half an hour or more I tried to argue that Constantinople must be given up to the Greeks and the Byzantine empire must be re-established. As for Russia—*she must take the Holy Land and a wide strip of territory joining Jaffa with the southern Caucasus*. On this subject I published afterwards an article in the periodical, *Pastyr i Pastva* ("The Pastor and the Flock"), no. 1. It was reprinted in Volume VI of my works (Kiev, 1918).' These words were written by

Russian military preparations in the Caucasian area were relatively modest, for the general staff was rightly preoccupied with the formidable menace to the Russian western frontier. At the same time considerable attention was given to strengthening the naval forces in the Black Sea—forces which had a potential importance in both the European and Asiatic theatres. The Russians anticipated the possibility of a strong Austro-German naval force entering the Black Sea and operating from Turkish bases. The defences of Sevastopol and (to a lesser extent) of Batum were strengthened. At the beginning of 1914 the Russians had in the Black Sea two new battleships (*St Evstaphe* and *St John Chrysostom*) built in 1908, and armed with four 12-inch guns and having a speed of sixteen knots. There were also four older battleships, including *Three Saints* (1893), two modern light cruisers, twenty-six relatively new destroyers, twenty-four torpedo boats, eight submarines, and several scores of auxiliary ships. Under construction at Nikolayev were two 27,000-ton dreadnoughts (twenty-one knots and armed with twelve 12-inch guns each), *Empress Maria* and *Catherine II*;[1] the first was completed in 1915, the second in 1916. It was a really powerful fleet, and its existence had an important influence on the development of Turco-German war plans.

The Turks had no possibility of competing with the existing Russian Black Sea fleet, although certain orders had been placed in Britain. The alternative was the dispatch to the Black Sea of an Austro-German naval force, but doubt as to the attitude of Italy inhibited the Austrians from any adventurous naval policy.

On land, however, Liman von Sanders's mission worked with the utmost efficiency and, with all authority in their hands, achieved far more serious results in eight months than their predecessors had in double the number of years. It was too late to 're-educate' the senior personnel of the Turkish army (which had been one of the aims of von der Goltz). About 100 German officers simply took over the responsible posts in the Turkish army, principally as chiefs of staff of armies, corps and divisions. Large numbers of Germans and Austrians worked to organize the rear of the armies; commissariat and technical services were put in order and considerably improved, and a beginning was made with sanitary services. The first attention of the Germans, all the time, was directed to rail and road communi-

Metropolitan Antony (of Kiev and Galicia), who was in 1915 one of the most influential members of the Holy Synod. His point of view on Constantinople was supported by others among the Church dignatories and high officials. [Note by the late P. P. Muratoff.]

[1] A successor to the older *Catherine II* launched in 1886.

cations and to the working of mobilization. The main railways were finally put in the hands of German and Austrian officials. In eastern Anatolia the railway had not yet been extended east of Ankara (owing to Russian diplomatic opposition during the preceding decade), but the lines Ankara-Sivas and Sivas-Samsun were under construction. In

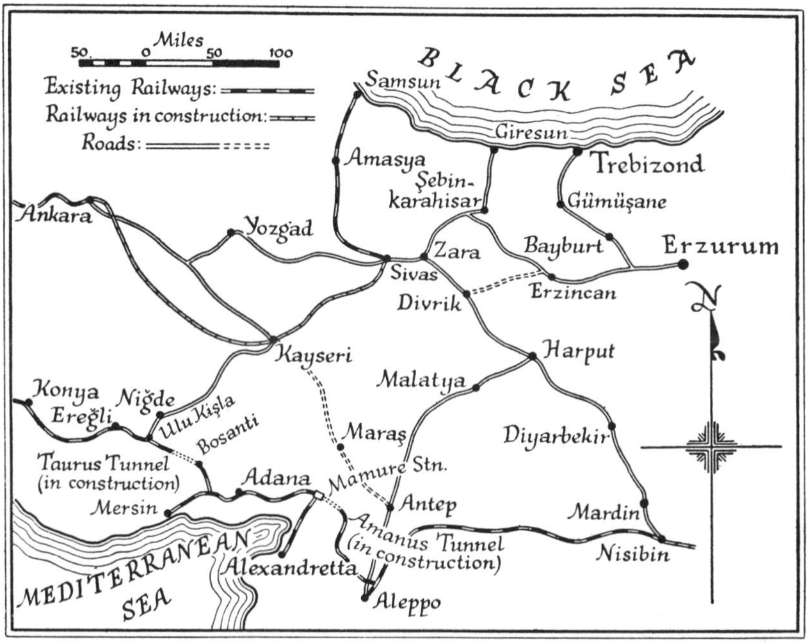

Map 17. The Anatolian railways in 1914

the region of Erzurum a series of strategic roads had been completed: Erzurum-Köprüköy, Erzurum-Bayburt-Trebizond (improved), Gümüşane-Erzincan, Şebinkarahisar, forking to Erzincan and Sivas, with branches to Giresun and Ordu, Ankara-Kayseri-Sivas, Kayseri-Adana. Kayseri, later to be the nodal point of the Turkish railway system, was already the most important cross-road of Asia Minor. Through Kayseri and Sivas troops might be moved from the Mediterranean coasts to Armenia or from Armenia to Syria and Iraq.

CHAPTER XVII

TURKISH PRELUDE TO WORLD WAR I

MAPS 1, 17, 19

THE progressive intimacy of Turco-German relations was consummated on 2 August 1914 by a secret alliance, to which two days later, the date of the outbreak of World War I, Austria-Hungary adhered. Nevertheless, the official attitude of the Young Turk government remained one of neutrality, and the mobilization which was proclaimed on the 2nd scarcely accelerated the military preparations which had been going forward throughout the earlier troubled months of 1914.

There were varied nuances in the interpretation of the alliance by different groups. To men like Baron Wangenheim, the German ambassador in Istanbul and the principal sponsor of the treaty, it was the reward of many years of concentrated and arduous work. But the German general staff remained sceptical of the prospects of Turkish collaboration and they were even reluctant to accept this *Sorgenkind.* They foresaw the danger of an Anglo-Russian *coup de main* which might open up direct communication between the western and eastern fronts. They appreciated that, neutral or belligerent, the first function of Turkey was to guard and to keep firmly closed the Straits.

Wangenheim himself was not satisfied with the secret alliance and he had pressed for an immediate declaration of war by the Turks on the Entente powers. But the Turks neatly refused on the grounds that they required at least thirty-six days to complete first-line mobilization. They were nervous about the defences of the Straits; and they feared even more the possibility of Russian landings on the coasts of Asia Minor or Bulgaria, since the modern Russian Black Sea fleet quite dominated the inland sea. The week which preceded the arrival of *Goeben* and *Breslau* in the Golden Horn was an anxious period for Baron Wangenheim.

In the meantime the more imaginative elements in both the German and Turkish ruling circles were weaving fantastic plans round the alliance. It particularly enflamed the fancy of the Emperor William II. He had for some time been inclined to exaggerate the possibilities of serious interior trouble in Russia—a mirage which had delighted and was to delight the souls of historical personalities so different as Benjamin Disraeli and Adolf Hitler. The emperor thought it possible to land parties of agitators and propagandists, well provided with arms, funds and leaflets, in order to provoke revolutionary or nationalist movements; in this connexion both Ukrainians and Tartars were regarded with favour. This adventurous plan presupposed the presence in the Black Sea of a strong naval force which could protect the line of supply of potential movements from the Russian fleet. Thus, even before the arrival of *Goeben* and *Breslau* in the Straits, William II had suggested the dispatch of strong Austro-Hungarian naval forces to the Black Sea. Enver and Cemal, for different reasons, urged the necessity of sending at least one Austrian dreadnought to the Bosporus. Both these proposals met with the refusal of the Austro-Hungarian naval commander-in-chief, Admiral Haus, who was supported by the Emperor Francis Joseph.

But some variation of the idea was still favoured by many influential persons in the high command of Austria and Germany. On 20 August Field-Marshal Conrad von Hoetzendorff himself raised the possibility of landing 50,000 Turkish troops in Bessarabia with a view to creating a diversion against the Russian rear in Galicia. The absence of adequate sea transport was an obvious obstacle, and in making his report the Austrian military attaché in Istanbul, General Pomiankowski, estimated that the total of mobilized troops in the region of the capital did not exceed the figure of 50,000—a minimum requirement for defence. But both Vienna and Berlin persisted, and a few weeks later the proposal was made to land 50,000 Turks, under command of the germanophil, Count Szepticki, in the Kuban region with the object of promoting an insurrection among the Kuban Cossacks (who were supposed to be under the influence of the Ukrainian separatists). Probably no one recalled the dreams of David Urquhart and the peregrinations of Sefer Bey and Behçet Paşa; but the absurdity of the project was clear enough to the Turkish leaders.

All these projects, however, are worth mentioning, since they illustrate the romantic conception of the possibilities of Turkish military collaboration as opposed to the more prosaic view which eventually formed the basis of Turco-German war plans; and this romantic conception had a certain influence. The prosaic view went no further than preparation for a strictly defensive policy on the part of the Turkish armed forces; it was represented by General Liman von Sanders and Ahmet Izzet Paşa, who emphasized the essential and important function of Turkey in blocking communications between Russia and her western allies.

In Istanbul, Wangenheim, who was on very bad terms with Liman von Sanders, remained the prophet of the romantic school, and he received frequent support from the Emperor William. Among the Turks the romanticists included all three members of the powerful triumvirate of Enver, Cemal and Talaat; Enver particularly played the part of the warlike political dreamer who set no limits to his fancy. There seemed to be complete mutual understanding between the dreamer Enver and the exuberant baron. But in reality there was a very essential difference between the vast ambitions of the Young Turks and the interpretation of these ambitions as intended by the Germans.

The triumvirate displayed before the admiring eyes of Baron Wangenheim the picture of a conquering Pan-Turanian movement expanding beyond the Caucasus to the Volga and Caspian and of old Abdul Hamit's resurrected Pan-Islamic movement conquering Egypt and Persia and threatening India. However, they did not disclose *le fond de leurs pensées*: that all these splendid achievements could only be attained with the full support of Germany, political, military and financial.

The conceptions of the romantic triumvirate were, however, exactly the opposite of what the professed admirers of this 'Three continents' plan among the Germans expected from their allies. Turkish 'dynamism' was merely considered by the Germans to be a fortunate instrument for the furtherance of German interests. It was estimated that Turkey should be able to put a million men into the field. Directed by a few hundred German officers and technicians, the Turks could certainly have a serious diversionary value in relation to the strategy of Britain and Russia. Even if Turkish hopes were exaggerated, the Turks could cause the maximum of embarrassment to the Entente with the minimum of expense to the Germans.

But suddenly the course of events during the first two months of the war converted these theoretical appreciations of the prospects of Turkish dynamism into a vital and immediate issue for the central powers. By the middle of September the military situation had become definitely and unexpectedly unfavourable; the Austrians had been ingloriously routed by the Serbs and utterly beaten by the Russians in Galicia, while with the check to the German armies at the Marne all hope of a rapid and victorious decision in the west had been lost. On the eastern front Russian reinforcements continued to arrive from the Caucasus, Siberia and Turkistan. The collaboration of Turkey therefore became an immediate necessity for the German general staff, who had previously regarded with lukewarm caution the enthusiasms of the interpreters of *Weltpolitik* in terms of big Turkish schemes. Telegrams from Berlin and Vienna poured into Istanbul urging Wangenheim and Liman von Sanders to accelerate the *tempo* of Turkish preparations, to fix a definite plan of action, and to ensure that the Turks made open cause with their German and Austrian allies.

But Turkish passive resistance to immediate intervention grew with the measure of Austro-German insistence—and for the same reasons; the moderate elements opposed to war had been encouraged by the issue of the battles in France and Galicia. Considerable nervousness reigned in Istanbul during the second half of September 1914. In the atmosphere of mutual suspicion and of contradictory reports and rumours, the outlines of military and even political plans became uncertain. In the meantime the first phase of Turkish mobilization slowly reached its end. The actual concentrations of troops expressed the confusion created by the existence of contrary tendencies; for some were suggested by the prosaic necessities of defence, while others were made in preparation for the vast 'dynamic' plans of the triumvirate.

The organization of the Turkish army on the Prussian model, already in embryonic development in the days of Ahmet Muhtar, had been effectively established during the decades preceding World War I by the military mission of von der Goltz. In 1877 the army corps had constituted little more than an administrative area, but by 1914, the growing cadres of trained officers with their German advisers had converted the Turkish army corps into field units.[1]

During the first months of the war the distribution of the army corps scarcely corresponded to the emergencies of the moment; out of thirteen army corps, five were located in an inner circle round the capital while three faced the Russian frontier; only five (and these were up to a strength of two and not three divisions each) were scattered over those more distant parts of the empire which were most exposed to attack by the maritime powers.

By the middle of September, thirty-six divisions were on a war footing and more or less up to strength.[2] The Turks were ready to

[1] The army corps comprised three divisional districts each of three regimental sub-districts. The territorial areas of the different army corps were as follows: I, Istanbul; II, Edirne (Adrianople); III, Rodosto; IV, Izmir (Smyrna); V, Ankara; VI, Aleppo; VII, Yemen; VIII, Damascus; IX, Erzurum; X, Erzincan; XI, Van; XII, Mosul; XIII, Bagdad.

[2] The average strength of a Turkish division was: three infantry regiments (each of three battalions and one machine gun company); one artillery regiment (of two or three sections, each composed of two or three 4-gun batteries)—total 9000–11,000 rifles, twenty-four to thirty-six field guns (part of them generally mountain guns), sixteen machine guns.

put into the field about 350,000 infantry with 1000 field guns. The mounted arm was limited to two regular cavalry divisions and some score of horsed reconnaissance squadrons attached to the infantry divisions. There were also under arms the Kurdish irregular cavalry of the Hamidiye formations organized in three or four divisions and distributed throughout the eastern army corps districts. Well-trained gendarmerie battalions and fifteen battalions of Frontier Guards formed of picked men were available to cover the extensive eastern frontiers. As in former wars real strength fell far short of theoretical estimates, but the army created by the German military missions and the Young Turks and recreated after the disasters of the Balkan wars was no negligible factor in the critical months of the autumn of 1914.

Minister of war only since February 1914, Enver Paşa had had no time to adapt the existing system of mobilization to his own 'dynamic' dreams, and the grouping of the bulk of the divisions in three armies was clearly based on established defensive concepts. The First Army, under Marshal Liman Paşa, was concentrated in European Turkey for the defence of Istanbul and the Straits; it comprised I, II and III Army Corps, to which it was intended to add VI—transported from Aleppo to St Stefano. At the same time the newly formed 19th and 20th Divisions were added to the strength of the First Army. The Second Army, composed of IV and V Army Corps, under the command of General Cemal Paşa, was based on the Asiatic shores of the Sea of Marmara. General Hasan Izzet Paşa was at the head of the Third Army which was made up of IX, X and XI Army Corps aligned towards the Russian frontier. Outside these three armies there remained four army corps which were entrusted with local defensive action: VIII at Damascus, XII at Mosul, XIII at Bagdad and VII in the Yemen. The same local defensive role was assigned to the 21st and 22nd Divisions in the Hejaz. The contrariety of view and design between the defensive and dynamic groups (represented in both the German and Turkish commands) continued to confuse all counsels and, despite long and impassioned discussions, the Turks appear to have entered the war without any coherent war plan.

About the middle of September when the exponents of political neutrality were being overborne by the triumvirate, the military protagonists of a defensive strategy lost ground. The idea of a Pan-Islamic 'dynamism' was adopted in principle; a Fourth Army was now created for Cemal Paşa, the partner in Enver's dreams. This Fourth Army was made up of VIII (Damascus) Army Corps, with the 8th and 10th Divisions added from the First Army in Europe and the

35th and 36th Divisions from XII (Mosul) Corps. The prospects of a successful defence of Iraq were further risked by the transfer of elements of XIII (Bagdad) Corps to reinforce Enver's 'Pan-Turanian' army on the Caucasian border. At the same time it was decided to retain X (Erzincan) Corps between Sivas and Samsun—probably as a gesture to satisfy the repeated German and Austrian demands for collaboration in the form of a diversionary demonstration against the Bessarabian or Crimean coasts.

At the beginning of October, the situation in Istanbul continued to cause concern and discontent in Berlin and Vienna: the war was in its third month; Turkish mobilization appeared to be complete; the promises of three-continent action were reiterated; yet hostile initiative against the Entente powers was delayed. With the nervousness produced by numerous disappointments in other theatres, Austro-German suspicions of Turkish double dealing grew from day to day. Wangenheim feared dismissal, while the German personnel of Liman von Sanders's mission and of Admiral Souchon's squadron threatened Enver that they would resign in a body and return to fight for their fatherland on active fronts. The Turks continued to claim that the 'Pan-Islamic' and 'Pan-Turanian' armies were not yet *Operationenbereit*. And the grand vizier seemed to be more and more dangerously entrenched behind his complaints that arms, equipment, clothing and money were lacking.

By 20 October, the representatives of Germany in Istanbul had definite instructions to force a *coup de main* in two stages. The first produced a secret meeting between Wangenheim and the members of the triumvirate, when a German loan of thirty million pounds in gold was proposed (with an immediate part payment of this sum) and a formal Austro-German undertaking was given to shatter the resistance of the Serbs, to overbear the Bulgarians, and to open direct land communication between Turkey and the central powers. The second stage saw the sortie of the Turkish fleet, early on 28 October, under sealed orders into the Black Sea. On the night of 29–30 October, Russian ports were bombarded and some small Russian war vessels sunk. The ability of a few German soldiers and diplomats, playing to the amateurish ambitions of the Young Turkish leaders, had succeeded in thrusting the empire into a death-struggle which the sagacity of traditional Ottoman statesmanship had always managed to avert: a simultaneous war with Russia and the western maritime powers.

CHAPTER XVIII

WAR ON THE CAUCASIAN FRONTIER, NOVEMBER 1914

MAPS 1, 17, 19

THE voluminous publications of the topographical section of the Caucasian military district are a monument to the study which, for decades before 1914, Russian staff officers had given to the problem of future operations in Asia Minor and the Middle East.

Yet when war with the central powers broke out in August 1914, the Caucasian frontier was almost ignored in the plans of the Russian general staff. There was a considerable doubt as to whether the Turks would in the event abandon their neutrality. At the same time it was believed that the war would be of short duration and that it would be decided by the conflict of great masses on the principal fronts. The commander-in-chief of the Russian armies, the Grand Duke Nicholas, particularly was under the influence of French military thought and he concurred in the popular view of the Russian staff that the road to Constantinople lay through Berlin.[1]

As it was, the realities of the situation on Russia's western front compelled the general staff to draw on both Caucasia and Turkistan. II and III Caucasian Army Corps with the Caucasian Dragoon Division and many Cossack units were in the west where they were

[1] The possibility of war with Turkey had been considered under three eventualities:

(*a*) *War with Turkey alone.* In this event the three army corps stationed in Caucasia were to be reinforced by four more army corps and a quick and decisive offensive was expected.

(*b*) *War with the central powers and Turkey.* In this event the Caucasian army was to send one army corps to the western front, as well as all the 'second call' regiments from the Kuban and Terek Cossacks; strategy was to be based on an active defence with possibly some local offensive. Reinforcements and replacements were to be provided by the formation of local reserve divisions.

(*c*) *War with the central powers; Turkey neutral.* Two Caucasian army corps for the western front, to be followed by the Caucasian Dragoon Division and the bulk of the Cossack regiments. When the first of two or three reserve divisions had been formed, a third Caucasian army corps was to be sent also to the western front.

In August 1914, the situation corresponded to hypothesis (*c*), but few doubted that (*b*) would become actual. But with Britain as Russia's ally it became possible to find a solution intermediate between (*b*) and (*c*) by utilizing troops in Turkistan which would have been immobilized had Britain been neutral or potentially hostile.

joined by I Turkistan Army Corps. II Turkistan Corps came to the Caucasus to reinforce I Caucasian Army Corps; and the political situation in Asiatic Russia subsequently made it possible to reinforce the Turkistan infantry units with the Siberian, the Transcaspian and, later on, the Transbaikal, Cossacks. On the Caucasian frontier a strategy of active defence was proposed in the event of the Turks deciding to abandon their neutrality.[1]

In October 1914, the strength of the Caucasian army was modest enough. I Caucasian Army Corps comprised the 39th and 20th Infantry Divisions, the 2nd Caucasian Rifle Brigade, the 1st Kuban Plastun Brigade, and the 1st Caucasian Cossack Division with their artillery. II Turkistan Army Corps was made up of the 4th and 5th Turkistan Rifle Brigades and the Siberian and Transcaspian Cossack Brigades with artillery. Units outside these two Corps formations were the 66th Infantry Division (with artillery), the 2nd and 3rd Kuban Plastun Brigades, the 2nd and 4th Caucasian Cossack Divisions and two odd Cossack regiments. There were also four sapper battalions and five battalions of Frontier Guards—the latter picked men.[2] The total strength of the Caucasian army was 100 battalions, 117 *sotni* and 256 guns. Since the regiments of the 39th and 20th Infantry Divisions were maintained at full strength in peace time, their battalions, after mobilization, became 25–30% above normal strength. On the other hand, the Kuban and Turkistan battalions were rarely above 800 men. At an average of 1000 men per battalion, the Caucasian army amounted to about 100,000 infantry, 15,000 cavalry and 256 guns. At the same time no less than 150,000 local reservists had joined the colours by 1 September, and these were in

[1] Compulsory military service was not enforced on the Muslim subjects of the tsar in the central Asian territories which had been conquered during the half-century preceding 1914. The Turkistan divisions were dependent on drafts from the Russian settlers in Turkistan and Semirechye. Nevertheless many Muslims served as volunteers. The 'Savage Division' (*Dikaya Diviziya*), famous on the Polish front, was recruited from Chechnia and Dağistan and from the Turkoman tribes of Transcaspia.

[2] Each infantry division consisted of four regiments of four battalions each, these being eight machine guns strong. Rifle brigades also consisted of four regiments (but only two or three battalions strong each). Plastun brigades were of six battalions. Cossack regiments had six *sotni* (squadrons) each. Each infantry division had two artillery sections, one of field and one of mountain guns; in each section three 8-gun batteries. The rifle brigades had one section each of three 8-gun batteries. (The 5th Caucasian Rifles had two field and one mountain battery; the 5th Turkistan Brigade had only two batteries.) The plastun brigades had no artillery. The Cossack divisions had two 6-gun horsed batteries; and the separate two regiments had one.

The I Caucasian Army Corps artillery included a section of twelve 4·8 in. howitzers.

process of formation into additional units besides providing complements and filling the rear services. The 3rd Caucasian Rifle Brigade was ready by the end of November; and local volunteer units were also being recruited: four Armenian *druzhiny*[1] took the field on the outbreak of hostilities and others followed later (and fought well). The Georgians contributed two similar bodies of volunteers.

The viceroy, Count Vorontsov-Dashkov, was the nominal head of the Caucasian army. He was a sensible enough man but, like his predecessor in the Crimean War, he lacked military experience and was in poor health. His deputy, who was practically commander-in-chief, was General Myshlayevski. He was a former professor of military history in the High Military Academy and the period of Peter the Great was his speciality. But neither his experience nor his character could justify his appointment. Fortunately, the Caucasian army had as chief of staff, General Yudenich, a man with great experience of the Caucasus and a tireless and efficient worker. The viceroy was a good enough judge of men to maintain Yudenich and to support him during several service crises; and Yudenich had succeeded in training a small team of staff officers who followed in his vigorous tradition.

In October 1914, the strength of the Caucasian army was laid out on the basis of an operational plan which envisaged an active defence against a local superiority of force. Five separate groups covered the five principal alleys of communication across the frontier.[2] The 39th

[1] *Druzhina* means literally 'fellowship'; practically it may be translated 'commando', a force about 1000 strong.

[2] (*a*) *Erzurum-Kars*: 39th Infantry Division; one brigade of 20th Infantry Division; 1st Kuban Plastun Brigade (five battalions); 1st Caucasian Cossack Division: twenty-nine battalions, thirty *sotni*, ninety-six guns (including the howitzer section).

(*b*) *Erzurum-Oltu*: one brigade of 20th Infantry Division; 1 Cossack regiment: eight battalions, six *sotni*, twenty-four guns. (Both groups under General Bergmann, in command of I Caucasian Army Corps.)

(*c*) *Erevan-Bayazit*: one brigade of 66th Division; 2nd Kuban Plastun Brigade; 2nd Caucasian Cossack Division; Transcaspian Cossack Brigade: fourteen battalions, thirty-six *sotni*, fifty-two guns. (All under command of General Oganovski.)

(*d*) *Batum region*: 264th Infantry Regiment (of 66th Division); one battalion of the 1st Kuban Plastun Brigade: five battalions with eight guns—to which were added one and a half battalions of Frontier Guards. The six battalions of the 3rd Kuban Plastun Brigade were echeloned along the coast of the Black Sea.

(*e*) *Detachment in Persian Azerbaijan*: 2nd Caucasian Rifle Brigade; 4th Caucasian Cossack Division: eight battalions, twenty-four *sotni*, twenty-four guns; General Chernozubov.

The garrison of Kars was composed of the 263rd Infantry Regiment (66th Division): four battalions.

General Reserve: II Turkistan Army Corps; Siberian Cossack Brigade: twenty-one battalions, twelve *sotni*, forty-six guns.

Infantry Division was concentrated round Sarikamiş in the region of the frontier, with a brigade of the 20th Infantry Division and the howitzer section at Kars. Supporting General Istomin's detachment at Oltu, the 1st Caucasian Cossack Division (General Baratov) watched the frontier as far as the middle Çoruh. On the left the infantry of the Erevan group were at Iğdir, while the cavalry and Kuban Plastun Brigade lay along the frontier. In Persian Azerbaijan, Russian detachments held Julfa, Tabriz, Dilman, Khoy and Urmia. The general reserve was stationed round Tiflis.

It had been decided on the outbreak of war to occupy the Eleşkirt valley, really as a defensive measure to protect the border districts along the Aras from the incursion of the Kurdish Hamidiye divisions. No other offensive move was intended by the Russians. It was known that the Turkish IX and XI Corps only had completed their concentration round Erzurum, and this force was considered insufficient for any offensive movement of the Turks against Kars. It was anticipated that the assembly of forces sufficient for a Turkish offensive would require at least four weeks from the beginning of October, and November was thought to be too late in the year for large-scale operations in the highlands of Armenia. The viceroy's staff in Tiflis, in fact, adopted a rather complacent view; no effort was made to work out a combined strategy with the powerful Black Sea fleet, nor was any beginning made to effect co-operation with the British.

The British, in the meantime, were far from inactive, and in the plans conceived for countering the entry of the Turks into the war the British were alone among the Entente powers in showing an immediate appreciation of the wide strategic perspectives which had opened up. As early as 15 October, Anglo-Indian troops occupied the Bahrein Islands, and on 22 November they took Basra and defeated the Turkish 38th Division—the only force which Enver had left to defend Iraq, since the 37th Division was already on its way to reinforce the Pan-Turanian army on the Caucasian frontier.

In Erzurum Hasan Izzet Paşa, who was *persona non grata* with the triumvirate, had not been informed of Admiral Souchon's sortie into the Black Sea and the war which followed in three days took him rather by surprise. Of the three army corps composing his Third Army, X, as has been indicated, was concentrated in the region between Sivas and the Black Sea. Hasan Izzet Paşa retained the balance of his forces in the Erzurum region, with the exception of one division of IX Corps at Ispir on the Çoruh and one division of XI Corps at Tutak on the Murat-su, where it awaited replacement by the

37th Division from Iraq. About 8000 Kurds of the Hamidiye formations were in the Eleşkirt valley. Hasan Izzet therefore had at his own disposal round Erzurum four infantry divisions and one division of regular cavalry.[1]

The Turkish commander rightly regarded these numbers as quite inadequate for offensive purposes. At the same time he was aware that the Russian strength was relatively modest and he did not anticipate any serious aggression. In the event of a Russian advance he was ready to meet the enemy on the immediate approaches to the fortified area of Erzurum—approximately on the line from Dumlu-dağ through Kargapazar and Hasankale by the Aras river to Dram-dağ.

On the Russian side, after the declaration of war, the formidable strength of the general reserve was still retained in the neighbourhood of Tiflis. All group commanders received instructions to act according to the orders given them. The absence of more detailed guidance from the high command could scarcely be criticized, since the theory of a limited advance had for long been firmly established. This obscurity of direction, however, proved dangerous in the case of General Bergmann, an officer who liked to imitate in appearance and manner the type of the old Caucasian hero-leaders. But if he followed the audacious tradition of General Heimann, he suffered the same misfortune as Loris Melikov in the abortive character of the offensive with which he opened the war. Bergmann had in fact none of the qualities which are necessary in a commander; he had had no experience of field operations, and was merely blindly obstinate when he thought to show strength of character.

At dawn on 2 November Bergmann moved his troops across the frontier. The 155th (Kubinski) Regiment, marching from Karaurgan, occupied the famous Zivin position after a few skirmishes with Frontier Guards. Simultaneously the 156th (Elizavetpolski) Regiment advanced from Mecinkirt on Horsan, while the 153rd (Bakinski) and 154th (Derbentski) Regiments moved up from Sarikamiş to the frontier. On the right flank, General Istomin was ordered to advance from Oltu on Id; and on the left, Baratov's Cossack division was sent to the right bank of the Aras, with the support of the 5th Kuban Plastun Brigade moving from Kağizman to Karakilise in the Eleşkirt valley. To the south-east troops from Iğdir crossed the Ağri-dağ by the Çengel Pass and occupied Bayazit and Diyadin. On 4 November,

[1] The total of the Turkish forces on the Caucasian frontier was 70,000–80,000 men with 160 guns; these numbers included fifteen battalions of the excellent Frontier Guards, holding posts along the border from Hopa to the Persian frontier.

Bergmann reached the line Ardos-Horsan (seventeen miles over the frontier), while Baratov was at Yuzveran and Istomin at Id. Everywhere only light Turkish patrols were encountered, and on the 5th Bergmann ordered his troops to halt and fortify the Sanamer-Horsan line. So far so good; Bergmann had accomplished the mission expected of him. But suddenly, on the evening of the 5th, he ordered a further advance on Köprüköy, since there was no enemy in the region except some advance patrols.

Bergmann's intuition might have proved fortunate, if the cautious Izzet had not, on the previous day, received a telegram from Enver suggesting that it might be useful to raise the morale of the troops by defeating Russian columns which had no liaison. It was a clear enough indication that the vice-generalissimo required a counter-offensive, and on the same day when Bergmann decided to occupy Köprüköy, Hasan Izzet ordered a general advance. Two divisions of XI Corps and the cavalry division advanced along both banks of the Aras towards Köprüköy, while two divisions of IX Corps moved up to occupy the so-called Köprüköy positions along the great Çakirbaba[1] ridge which flanks the north-west side of the Pasin valley, between Hasankale and the Soğanli region.

On the evening of 6 November contact between the two advancing armies was established along the banks of the Aras. There was some hot fighting during the night and, early on the 7th, Russian infantry and Cossacks occupied the village of Köprüköy and the historic bridge over the Aras from which the place takes its name. Bergmann reported his success to the viceroy, but this unexpected development worried rather than gratified the staff in Tiflis. Yudenich immediately telegraphed that he should ask the authorization of the commander-in-chief for any further advance. Yudenich was right: the same day was destined to disappoint the victorious General Bergmann.

On the morning of the 7th, in order to consolidate the position at Köprüköy, Bergmann sent forward six battalions to take the Badicivan heights which had, in fact, been already occupied by two divisions of the Turkish XI Corps. The Russian attack was beaten off by strong artillery and rifle fire. On the following day the experience was repeated. During the 9th and 10th Bergmann's force found itself in a difficult position in front of commanding heights occupied by superior numbers. Against twenty to twenty-two Russian battalions, Hasan Izzet had concentrated the full force of four divisions (thirty-

[1] *Çakir*, T., 'bluish grey' and *baba*, 'old man', 'father'; the sense of 'grey old man' may, perhaps, be rendered 'greybeard'.

six battalions); a fifth, summoned from Ispir, was beginning to gather on Bergmann's right flank. Istomin, who had only part of his brigade at Id, was unable to help. With no reserves at hand Bergmann ordered Przevalski's Plastun Brigade to join him by forced marches from the Eleşkirt valley.

On 11 November Hasan Izzet attacked. Two divisions of XI Corps marched along both banks of the Aras against Köprüköy, their right flank covered by the cavalry division and Kurdish irregular horse. At the same time the two divisions of IX Corps attacked to the north-east of Köprüköy and manoeuvred to turn the Russian right flank in the region of the Çilligül[1] mountain. Both Russian flanks were thus in danger. The Baku Regiment lost Köprüköy village and suffered heavily. Along the southern bank of the Aras the 33rd Division of XI Corps pressed back the Cossacks who had the support of only two battalions of the 79th Regiment. On the Russian right the 80th Regiment lost their trenches at Maslahat and retreated on Sanamer. By the evening of the 12th, Bergmann's army was on the line occupied on the 4th, Sanamer-Ardos-Horsan. But throughout the 13th and 14th the situation continued dangerous, particularly when the Turkish 33rd Division occupied Yuzveran, south of the Aras, and moved on Tayhoça; while against the Russian right the Turkish 29th Division by-passed Sanamer to the north and took Eğrek and Hoşap. The Turks were now threatening to cut both of Bergmann's lines of retreat—by the roads through Zivin and Mecinkirt; on the 14th, Bergmann's right flank was already in action with Turkish advanced forces in the area of Horum-dağ and Zivin.

The situation improved with the arrival of reinforcements. On the 15th, the 4th Turkistan Rifle Brigade, which had been railed to Sarikamiş from Tiflis, attacked the Turks at Ağviran and round Zivin, and on the following day Hoşap was recaptured. On the 17th, Izzet decided to draw off the 29th Division; the forces at his disposal were limited and the arrival of Russian reinforcements made him cautious. In the meantime, the 13th Turkistan Regiment had strengthened the Russian centre, and on the Russian left to the south of the Aras the arrival of Przevalski's plastuns had completely changed the situation. On the 15th, the Kuban infantry were at Aliçekrek and marching on Tayhoça. On the following day the Turkish 33rd Division was thrown back on Yuzveran. On the night of 16–17 November, Przevalski, crossed the Aras with almost all his mounted *sotni*, and at dawn attacked part of the Turkish XI Corps on the march along the

[1] T., 'Speckled rose.'

northern bank of the river. He checked the Turks and, later in the day, again forced the icy and fast-flowing river to organize the recapture next morning of Yuzveran. The Russian retreat had come to an end and after two more days the fighting died away.

Bergmann had suffered a setback for which he had been himself wholly responsible. He had lost more than 7000 killed and wounded (the Bakinski Regiment suffered 40% casualties). Izzet, despite rather higher losses, had succeeded in raising the morale of his troops as that of the Russians had fallen.

To strengthen his position in front of Erzurum, Hasan Izzet moved the last division of XI Corps from Tutak on the Murat-su (where it was in process of replacement by the 37th Division from Iraq) and impatiently awaited the arrival of X Corps from Anatolia. As winter broke over the Armenian highlands, Izzet Paşa looked forward to a period of relative inactivity, but, unfortunately for the Turks, the vice-generalissimo, Enver Paşa, was of a less patient nature.

On the extreme left the development of Russian operations had been more successful than in the centre. Having established contact with Baratov's Cossacks through Tahir in the Kara-derbent gorge, General Oganovski sent detachments of the 2nd Caucasian Cossack Division to occupy the Kiliç-gedik ('Sword Gap') Pass and cross the Şarian-dağ. Turkish forces in the Murat-su valley were weak, since the units of XI Corps stationed there were already moving towards Erzurum, while the 37th Division had not yet arrived from Iraq. Two temporary divisions (36 *bis* and 37 *bis*—only fourteen weak battalions, with a few guns) were hastily formed out of the crowds of reservists from the vilayets of Diyarbekir and Mosul and they were gradually concentrated to cover the Muş and Hinis region. Between 27 and 30 November, some elements from these two *bis* divisions approached the Kiliç-gedik Pass and Hamur, but they were easily dispersed by the Russians. (Cf. Map 22.)

Farther to the south-east, General Chernozubov's force in Persian Azerbaijan had moved towards the Turkish frontier and occupied the important pass of Kotur on the road Khoy-Van and Başkale on the road Dilman-Van. The turbulent local population of this area remained quiet, and even the Shahsevens of Karadağ maintained the allegiance which they had tendered in 1912. Along the whole of the left flank of the Russian front in the direction of northern Iraq the position seemed to be stable enough.

On the other extremity of the Caucasian front, the situation was far less satisfactory. The garrison of Batum was clearly too small and the Russian commander, General Elshin, proved absolutely helpless. It was a dangerous patch on the Caucasian frontier, and the Laz and Acars, who had proved formidable in the three Russo-Turkish wars of the preceding century, again began to show their qualities as guerrillas. The Turks organized a force of between 5000 and 6000 irregulars without any difficulty, and on 15 November they cut to pieces a column (one and a half battalions with two guns) sent to protect the copper-mines near Borchka.[1] Following the abandonment of Borchka, Artvin and Ardanuch, Elshin was replaced by the more competent Lyakhov, and reinforcements were hurried to Batum and Ardahan. The weakness shown in the Batum region and the evacuation of the lower Çoruh valley doubtless contributed to encourage the Turkish vice-generalissimo to undertake his audacious winter offensive. (Cf. Map 26.)[2]

Yudenich and his staff were humiliated by the unsuccessful début of the Caucasian army. The character of the viceroy's deputy, General Myshlayevski promised nothing good, and General Bergmann had proved even worse than had been expected. When Yudenich, about 20 November, visited the Pasin valley his impressions were bad: he found Bergmann in a state of depression, all services disorganized and troops rather gloomy and wondering why the high command was so far away from them in Tiflis. Following Yudenich's report to the viceroy, Bergmann's chief of staff was dismissed; but Vorontsov-Dashkov was unwilling to sacrifice a figure so spectacular as General Bergmann—personally known to the emperor.

Bergmann meantime had recovered from his despondency, and on 17 November (following the departure of Yudenich) he telegraphed asking authority to renew the offensive. The viceroy replied that their army was small and they ought to economize their troops and resources. They could not count on any reinforcements from Russia. In their situation they did not need to claim victories; they needed serious efforts to prevent the enemy from obtaining any success until their victory on the western front gave them the possibility to start decisive action in the Caucasus.

[1] On the Çoruh between Batum and Artvin.

[2] The Turkish action in Acaristan received some naval support. On 7 November (the night of the first Russian raid on Zonguldak) *Breslau* bombarded the port and railway yards at Poti for three-quarters of an hour. On 10 December, while the 3rd Division of the Turkish First Army was concentrated at Hopa, *Goeben* threw fifteen 280 mm. shells against the shore batteries at Batum. The Russians treated these attacks as demonstrative in character, but the officer commanding at Batum warned H.Q. that the guns of Batum would be inadequate to fight off a serious bombardment by *Goeben* (Novikov, *Operatsii flota protiv berega na Chernom more*, pp. 109 ff.).

CHAPTER XIX

SARIKAMIŞ: ENVER'S CAUCASIAN OFFENSIVE, DECEMBER 1914

MAPS 18, 19

SOME years before the war a personality unusual for a German consul had appeared in Tiflis. Traveller, sportsman and man of fashion, Count von der Schulenburg, an officer of the reserve of the Prussian army, took a particular interest in the ethnology and natural history of Caucasia. Well received in Tiflis society, the count soon enjoyed a certain popularity and he developed excellent relations with members of the civil and military administration, who gave him every facility to indulge his special taste for the chase. His expeditions frequently took him to the wild and mountainous frontier district round Oltu (where the Caucasian buck was reputed to abound). The German sportsman became a familiar figure to the native guides of this region; and it was credibly held in Tiflis, when war came, that he had made a special study of the terrain in preparation for the offensive which was undertaken in the late autumn of 1914.

Such was the legend, and Schulenburg himself was becoming a legendary figure who was in temperament quite sympathetic to the adventurous Enver.[1] In November the Pan-Islamic offensive had been entrusted to Cemal, but Enver had kept the development of Pan-Turanian action for himself. News from the Caucasian border stimulated his hopes; the Russian forces were obviously weak and the troops, so far, had shown little of their traditional fighting qualities. But Hasan Izzet was scarcely the man capable of obtaining a decisive result by bold and ingenious action. Enver had heard from the Turkish military attaché present with Hindenburg at Tannenberg that the Russians were particularly sensitive to outflanking manoeuvres. He was aware that the bulk of their Caucasian army was occupying

[1] Count von der Schulenburg became the patron of the 'Georgian legion' formed of Georgian political refugees and students in Berlin with some Lazes added. Its strength did not exceed 700 men (Pomiankowski, pp. 224–5). Active in the Caucasus in 1918, Schulenburg was German ambassador in Moscow in the years preceding World War II. After the failure of the Stauffenburg conspiracy against Hitler, Schulenburg, with many other distinguished Germans, was arrested by the Gestapo. He was hanged in Moabit prison in June 1944 in very disagreeable circumstances. Schulenburg was a fine soldierly man with a pleasant dry humour. He deserved a less indecent death. (Note by W.E.D.A.)

an advanced position to the south-west of the Soğanli plateau, and that their nearest base, forty or fifty miles away at Sarikamiş, was connected with the front by only one good road which ran parallel to the Çakir-baba ridge, winding along its lower slopes. The ridge of Çakir-baba itself was virtually a no-man's-land, and it was apparent that the Russians regarded it as a sort of wall against which their right flank was leaning and which at the same time protected their line of communications from any attack from the flank. They were relying on the winter when the mountains were deep in snow. In Enver's opinion these conditions gave the Turks the chance of effecting a complete surprise.

Beyond the Çakir-baba the Russian right flank at Oltu was protected by Istomin's single brigade. Oltu was an important road junction; if it were taken by surprise attack, an excellent road built by the Russians could be used by the Turks for an advance on Merdenik. The distance from Oltu to Merdenik is forty-five miles; from Merdenik, by a good road, it is only twenty-six miles to Ardahan and twenty-eight miles to Kars. Established in Oltu the Turks would be nearer to Kars than the Russians in the Pasin valley—seventy miles against ninety. Moreover, from Oltu a path ran through Bardiz to Sarikamiş. The imagination of Enver Paşa conjured up a seductive picture; after the surprise capture of Oltu, a strong Turkish group would be established between the Russians and their rear bases at Ardahan and Kars, and the road would lie open for an invasion of Georgia and an advance on Tiflis by the Kura and Borchalu defiles; the annihilation of the Russian army of the Caucasus would deliver all Transcaucasia to the Turks and the Turanian elements would rise in general insurrection. The phantoms of the Ingur and Abkhazian campaigns of the two preceding wars might assume substantial forms.

Such were the outlines of the grandiose scheme which the vice-generalissimo exposed, sometime in November, to Liman von Sanders. The latter was bewildered by the magnitude of Enver's plan and perhaps did not take it all quite seriously. However, Enver soon secured the moral support of several distinguished German officers, some of whom appear to have favoured his plan from purely personal motives (for instance, Bronsart von Schellendorf whom the friendship of Enver made chief of the Turkish general staff). Many German officers (including the forceful Major Feldmann) were consulted. The prevalent view was that the operation was not impossible but dangerous, and must be studied on the spot. The *mot d'ordre*, circulated perhaps by von Wangenheim himself, was: friendly reserve and complete responsibility on the Turkish general staff and especially on Enver himself. The general opinion was that, whatever the result might be, no harm could come to German interests—even in the event of the complete defeat of the Pan-Turanian army. Turkish success, on the other hand, might embarrass the Russians and attract forces away from the main Russo-German front.

There was no group in Istanbul powerful enough to oppose the wishes of the vice-generalissimo, but Enver anticipated opposition from the commanders on the spot. He decided to take over the command of the Third Army himself and to replace old and obstinate officers by younger and more imaginative men.

By the end of November, orders had been given to transport the 3rd Division of I Army Corps from the Bosporus and to land it secretly at the small port of Hopa which gave access to the lower valley of the Çoruh. The task of this division was to advance on Ardahan along the Artvin-Ardanuch road.[1] The 37th Division, which had now arrived in the valley of the Murat-su from Iraq, was to operate on the right flank of the Turkish Third Army and the 36th (which had been deflected from Syria) was to demonstrate along the Persian frontier. On 6 December, Enver, with a group of Turkish and German officers, arrived by steamer at Trebizond and from there proceeded to Erzurum. As he had anticipated, Enver failed to secure the agreement of Hasan Izzet Paşa—who resigned. The commanders of IX and XI Corps were dismissed. The officer commanding IX Corps, Ahmet Fevzi Paşa,[2] had served for many years in the Erzurum army corps district, and he knew the region over which operations were intended better than any man in the Turkish army. He gave it as his opinion that the operation might be possible only after the most careful preparation. Proper winter kit for the soldiers and the organization of advance bases were the prerequisites of success. The outflanking movement, traced on a map, seemed most promising; but the success of this movement must depend on the rapid advance of the outflanking force, and Ahmet Fevzi expressed the view that no troops could undertake this march in winter without special training and equipment. He considered further that if two corps were necessary to undertake the outflanking movement, the one remaining corps would not be sufficient to engage the Russians frontally on the Pasin positions. He proposed the

[1] Where Komarov had encountered such formidable natural difficulties in January 1878.

[2] Later Marshal Fevzi Çakmak, chief of the Turkish general staff, retired 1944, died 1950. Galip Paşa, commanding XI Corps, was replaced by Abdul Kerim Paşa.

transport of one more corps from Istanbul to drive home the frontal attack.

Before Enver left Istanbul it had been decided in principle that the outflanking manoeuvre should be carried out by IX and X Corps, while XI Corps, reinforced by eight or nine battalions of Frontier Guards and gendarmes and by the field artillery of IX Corps, made a frontal attack on the Pasin positions. The cavalry division was to demonstrate along the right bank of the Aras, where the 37th Division was expected to arrive from the direction of the Murat-su as a timely reinforcement.

IX Corps had already been somewhat weakened by losses, sickness and desertions, but the effectives of XI Corps were particularly strong—three divisions totalling about 35,000 men. The Third Army, including the division landed at Hopa and the expected 37th Division, totalled between 120,000 and 125,000 men—a considerable force compared with that at the disposal of Bergmann. Bergmann had replaced his losses and the 2nd Kuban Plastun Brigade had arrived from the Eleşkirt valley, but he could not dispose of more than 50,000 infantry and 4000 cavalry on his centre, while there were 8000 infantry and a few *sotni* at Oltu and only 2000 infantry at Ardahan; a total of 64,000 troops available with a small general reserve at Kars (14,000).[1] The Turks were thus about 50% stronger than the Russians in men; gunpower was about equal.

At the beginning of December, the 17th, 28th and 29th Divisions of IX Corps were concentrated on the Turkish left flank to the rear of the advance positions in the Köse-Portanos area. Through Köse-Portanos passes a track from Hasankale which, not far from the village of Köse, descends at Eğrek into the valley of the Oltu-çay. This track follows the Oltu-çay to Id and farther to Oltu. At Köse a mountain path forks right, gradually ascending, from the village of Lavsor, the slopes of Çakir-baba. This path follows the crest of the ridge which is in fact a narrow plateau about six to eight miles in width. The path is difficult enough but it has to cross no deep gorges and, as it follows the crest of the ridge, avoids all but the heads of ravines. It was suitable for infantry in column of march and for pack animals; hence the Turkish name *top yol* ('cannon way'), since such a track was more suitable to the transport of mountain artillery than a so-called wheel-track which might descend into and ascend out of steep gorges. In the winter there is less snow along this type of crest

[1] The newly created 3rd Caucasian Rifle Brigade (eight battalions), the 263rd (Gunib) Regiment and, still at Tiflis, the Siberian Cossack Brigade.

than on the lower slopes, since it is swept by the ferocity of recurrent gales; but the frost is naturally severe.

Twenty-five miles out from Köse the *top yol* descends to the village of Bardiz on the Russian frontier, lying in the valley of an affluent of the Oltu-çay. Above Bardiz the Çakir-baba ridge joins the Soğanli plateau. Two paths lead from Bardiz village to the main Erzurum-Kars road: the westerly crosses the Çakir-baba by the Hana Pass to Yeniköy; the easterly leads over the Soğanli to Sarikamiş by way of the Pass of Bardiz. From Bardiz to Sarikamiş the distance is about fifteen miles.

The 'Schulenburg legend' suggests that it was he who proposed to Enver the plan of an outflanking march by an unexpected, and even 'secret', mountain path; but the *top yol* was by no means secret and, although the Russians did not consider it practical for the movement of large bodies of troops, it was kept under observation by patrols of the Frontier Guard stationed at Bardiz.[1]

The Turkish 29th Infantry Division, followed by the 17th, had to begin their march on 22 December. They took with them only mountain artillery and baggage was cut to the minimum; the askers carried very limited rations, mostly the flat lightly baked wheaten bread of the country. There are no woods on the crest of Çakir-baba and fuel was not to be had. The askers had been supplied with no warm clothing and only a proportion carried greatcoats; they were shod in the peasants' footwear of the plains. Their vice-generalissimo delivered himself of the rather un-Napoleonic *mot*: 'Our supply base is in front of us.'[2]

It was intended that the 29th and 17th Divisions should reach Çatak at the northern edge of the Çakir-baba ridge by the night of 23 December. Two days before this force started, the third division of IX Corps (the 28th) had to move by the track through Id with the object of reaching Çatak on the *top yol* one day later than the 29th and 17th Divisions. The 28th Division had to make a particularly difficult march, but Enver planned, by dividing his forces, not only to relieve the traffic proceeding along the *top yol* but to thrust in between Istomin's brigade at Oltu and the main Russian group based on Sarikamiş.

[1] The *top yol* was described in *Karskaya Oblast*, published by the Topographical Section of the Caucasian General Staff about 1908.

[2] Gen. Pomiankowski, *Zusammenbruch des ottomanischen Reiches*, p. 103, states that the commander of IX Corps, Ihsan Paşa, had actually issued orders that greatcoats and knapsacks were to be left behind in order to facilitate movement through the snow.

Following behind the 28th Division, the 31st Division of X Corps from Eğrek had to move through Id along the valley of the Oltu-çay against the town of Oltu. This division had at the same time to synchronize its movement with that of the 30th and 32nd Divisions of X Corps which, several days earlier, had been secretly moved from the Erzurum region by a bad track to Tortum. On the 22nd these two divisions were to march along the ravine of the Sivri-çay in the direction of Oltu. To complete the planned encirclement of Istomin's force a special detachment under Ali Fethi Bey (one regiment of the 31st Division with two mountain guns) was sent still farther north up the Tortum-çay valley whence it had to cross the Ak-dağ and descend into the valley of the Oltu-çay at Hasköy, six miles to the west of Oltu. All these movements were timed with the object of effecting the capture of Oltu and the annihilation of Istomin's brigade on 23 December. Enver hoped to attack Sarikamiş with IX Corps on the 25th. According to his calculations, reinforcements from X Corps marching from Oltu should be available on the following day. At the same time it was left open to X Corps, after the capture of Oltu, to march on Merdenik, and afterwards on Ardahan. A separate movement against Ardahan was designed for a force under the German, Major Stange (two regiments of the 32nd Division with eight guns and 2000 Laz irregulars), which had to march along the Batum-Ardahan *chaussée* from Borchka through Artvin and Ardanuch and across the pass of the Yalnizçam-dağ.

All these moves were set out in an order of the day for 22 December, signed by Enver and dated 19 December.[1] In this order, its real authors, Colonel von Schellendorf and Major Feldmann, had badly miscalculated the timing of this extraordinary winter march.

In the Russian lines during the first half of December the lull in the fighting gave a false impression of stability. Bergmann's army occupied defensive positions along a line from the Kocut spur of Çakir-baba, through Maslahat to the Aras, then south to Yuzveran and the village of Ardi at the foot of the Dram-dağ.[2]

After the visit of Yudenich to the front there had been many improvements: troops had been provided with winter clothing, sanitary conditions were satisfactory and the commissariat was functioning relatively well under difficult conditions.

[1] Text given in full by Larcher, *Guerre turque*, pp. 383–4, citing Col. Şerif Köprülü.

[2] The Turkistan units were on the right; the 39th Division in the centre as far as the Aras; then the plastun brigades and two battalions of the 79th Regiment; the 1st Caucasian Cossack Division held the left flank; the Armenian volunteer companies were at Kara-derbent; and the 80th Regiment in reserve at Horsan.

The first big fall of snow came on 19 and 20 December, and with it the first alarming reports from Oltu. On the 19th General Istomin reported that the Russian post at Id had observed the appearance of a small force of Turkish infantry in the valley of the Oltu-çay at Eğrek. He added that a local informer had stated that a whole brigade was following and that the Turks intended an attack on Id. Bergmann treated the news with indifference, but on the same day he received a telegram from Tiflis inquiring his plans for helping Istomin in the event of an attack on him. The attention of Yudenich had already focused on a move which, at that season, seemed unusual. On the same day Yudenich ordered one regiment (two battalions) of the 3rd Caucasian Rifle Brigade (newly formed in Kars) to proceed to the remote district of Olor in the angle of the Oltu-çay and the Çoruh, where armed bands of Acar insurgents were showing increasing truculence. Further alarming telegrams came in to Bergmann's headquarters at Mecinkirt (with copies to Yudenich in Tiflis). On 21 December Istomin reported that after a fight a Russian battalion had recaptured Eğrek; but an Armenian agent had sent him a warning that on the following day the whole of the Turkish 28th Division was to attack Id. On the same day movements of Turkish columns were quite unexpectedly observed by Istomin's posts near Kotik in the valley of the Sivri-çay. (These were advance elements of X Corps.) Later on the same day, Istomin reported that the enemy remained passive on the Oltu-çay and that their principal line of advance seemed to be down the Sivri-çay. Later again he reported that the Turks had been observed still farther to the north in the vicinity of the frontier post of Cucurus (Ali Fethi's column). The last circumstance gave confirmation—if any were needed—that the Turks were making an outflanking movement against Oltu, and Istomin stated that he proposed to leave only a rear-guard at Id and to concentrate all his available forces for the defence of Oltu. He sent one battalion with two guns to reinforce the Cucurus post.

Istomin's dispatches of 21 December witness that the nature of the Turkish movement against Oltu had been accurately appreciated by the commander of the force at Oltu. His patrols had worked well, and the Turkish advance-guards (particularly the patrols of Ali Fethi) had somewhat prematurely exposed themselves. In Tiflis the imminent attack threatening Oltu was quite well understood by the evening of 21 December. A different view was taken by Bergmann who was Istomin's immediate superior. In answer to Istomin's request for a demonstration in the area Köse-Portanos (a few miles to the south of Eğrek), Bergmann refused on the grounds that the snow was too deep. He belittled the information brought in by Istomin's Armenian agents; he knew that the Turkish XI Corps was in front of his main body in the Pasin valley and he did not realize that X Corps had joined the Third Army. The Turkish demonstrations which began along the right bank of the Aras on 20 and 21 December seemed to him more significant than any local move in the Oltu area.

On 22 December Russian general headquarters informed Caucasian army headquarters that reports of foreign agents revealed the arrival of Enver Paşa at Erzurum. Combined with Istomin's dispatches this information seemed to confirm the beginning of some important Turkish offensive operation. Yudenich had a conversation with the viceroy, and both found cause for dissatisfaction with Bergmann's reaction to the situation. It was decided that General Myshlayevski, as the viceroy's deputy, and General Yudenich, as chief-of-staff, should leave for field headquarters at Mecinkirt (23 December).

It was high time. Istomin telegraphed in the late afternoon of the 22nd that his brigade was being attacked in the region of Id by one division, while two further divisions had appeared in the valley of the Sivri-çay. Istomin had decided to cover his retreat from Id by a small rear-guard and, during the night, to concentrate the bulk of his force on a position covering the town of Oltu. Early on the morning of the 22nd he added the information that the force operating down the valley of the Sivri-çay had been identified as two divisions of X Corps.

On the evening of the 23rd, General Myshlayevski (who was not very pleased to be leaving Tiflis) and General Yudenich took train to meet General Bergmann at his headquarters and, if necessary, to take charge of operations. Yudenich brought with him the assistant quarter-master-general, General Bolhovitinov, and the head of the operational section of the general staff of the Caucasian army, Colonel Maslovski —who was to become later the historian of the campaigns of 1914–17.

In the meantime the Turks, during the course of 22 December, had carried through the movements for that day outlined in Enver's order. The 31st Infantry Division appeared in the valley of the Oltu-çay, while the 28th were tramping along the paths which would bring them on to the tail of the other two divisions of IX Corps (29th and 17th) already marching along the *top yol*. Round Id, in front of the 31st Division, there was only a weak rear-guard of six companies with two guns under Colonel Kutetadze. Keri's *druzhina* of Armenian volunteers had been sent by Istomin to the hamlet of Virintap at the south-western end of the *top yol* in order to make contact with Bergmann's patrols; instead, the Armenians observed and reported two Turkish battalions of IX Corps moving along the crest of Çakir-baba. This was the first authentic news that the Turks were using the *top yol* route.

In the afternoon of the same day Kutetadze halted near the village of Nariman to rest his men. He neglected to put out pickets, and as a result he was suddenly attacked by advanced elements of the 31st Division. He lost his head and surrendered with 750 men. Two guns were lost, but some 500 or 600 of his men refused to lay down their arms and the early twilight allowed them to make good their escape. The Armenian *druzhina* was not involved in this small disaster.

The Turkish 31st Division, already very tired, bivouacked for the night at Nariman. Meanwhile, the 30th, followed by the 32nd, was struggling along the bad tracks down the valley of the Sivri-çay. Their advance was slow and they covered only ten miles during the day. Istomin, therefore, had plenty of time to occupy, at dawn on the 23rd, a previously prepared and fortified position covering Oltu. Two battalions of the 9th Rifle Regiment (3rd Caucasian Rifle Brigade) afforded a timely reinforcement. The Turkish 29th Division was already on his flank at Çatak on the *top yol*, where they passed a miserable night in the open without fires in a severe frost.

During the day Bergmann had received the reports of Istomin, including the information about the advance along the *top yol* brought in by Keri's *druzhina*. He also had reports from the Bardiz frontier post of the appearance of parties of the enemy at Norsin not far from Bardiz. To help Istomin he ordered some units on his right flank to attack in the direction of Köse-Portanos; but he sent in only a few companies and the move was no more than an obvious diversion. He reacted to the news from Bardiz by sending one Turkistan battalion to Yeniköy to protect the main road.

Throughout the day of 23 December, Istomin was fighting at Oltu. From 8 a.m. he was attacked on his right by the 30th Division and Ali Fethi's detachment. A fortunate accident due to the early fog saved him from a simultaneous attack on his left; units of the Turkish 31st Division mistook columns of their own 32nd for the enemy and heavy rifle fire and some artillery fire broke out, resulting in more than a thousand casualties between the two Turkish divisions. Istomin had only eight battalions against three divisions, but this circumstance of faulty recognition helped him to maintain his position for over three hours. He used his artillery to cover his retreat and four guns were left to the enemy. By midday Istomin had evacuated Oltu; he even managed to remove a part of the supplies stored there. He retired in good order to the Avçali salt mines, twelve miles from Oltu on the Ardahan road. The Turks did not follow; the tired and hungry askers were busy looting the small modern town which had grown up during the last forty years.

During the same day the few companies sent by Bergmann for the diversion against Köse-Portanos failed to make good an attack owing to the deep snow. The demonstration which the Turks made along the Aras front was also not pressed. Movement was more active on the icy wind-swept heights than in the snow-clogged valleys. The 29th and 17th Divisions were continuing their painful march along the *top yol*; the 29th was near Bardiz, while the 17th with the greatest difficulty, reached Çatak. The 28th was at the hamlet of Pitgir—between Nariman and Çatak—rather lost on the wild snowy flanks of Çakir-baba. The vice-generalissimo himself spent the night of 23rd–24th at Id.

On the next day (24th) a heavy fall of snow increased the troubles of the marching Turkish columns. The 17th Division suffered particularly since it was trapped in a blizzard on the *top yol* not far beyond Çatak; about 40% of the men became stragglers. The 28th found itself virtually unable to struggle up the terrible slopes of Çakir-baba. The 29th reached Bardiz, from where the Russian Frontier Guard detachment withdrew towards the Bardiz Pass. Enver, riding with Colonel Bronsart along the *top yol*, saw for himself the appalling conditions of the march. He arrived at Bardiz village in the evening. Bronsart advised him to concentrate there all three divisions of IX Corps, and only after that to continue the advance on Sarikamiş. But during the evening the vice-generalissimo was told by a captured Russian that at Sarikamiş there were no troops at all except a few rear units without artillery. Enver decided to march on Sarikamiş without delay with the 29th Division, while the 17th was to follow later in the day. He was in high spirits, and on the morning of the 25th issued an order of the day in which the victory at Oltu was proclaimed. While the 29th and 17th were to attack Sarikamiş, the 28th which was expected to arrive during the day in Bardiz, was to march on the Hana Pass in the direction of Karaurgan to cover the right flank of the advance from Russian elements which had been reported at Yeniköy (the one Turkistan battalion sent there by Bergmann). X Corps had to move up from Oltu to the east of Bardiz. Enver concluded his order: 'If the Russians try to retreat they are doomed; if they fight we will fight them with our backs to Kars.'

It transpired that the commander of X Corps, Hafiz Hakki Paşa, on his own initiative, had sent the 30th and 31st Divisions in pursuit of Istomin along the Ardahan road. Only the 32nd remained to move along the path from Oltu to the Eşek-meydan Pass; and the conditions of road and weather were so bad that in the evening this division (with several thousand men frost-bitten) had only reached the hamlet of Pertas, ten miles from Oltu and fifteen from Bardiz. On the front of XI Corps along the Aras, demonstrations were continued without any effect.

Early on the morning of the 24th, General Bergmann, impressed at last by the reports from the Frontier Guard at Bardiz and the patrols round Yeniköy, sent the whole of the 18th Turkistan Rifle Regiment (three battalions) with a battery under Colonel Dovgirt to Yeniköy. Later in the day he met at Mecinkirt the visitors from Tiflis. A lively discussion followed. Bergmann obstinately insisted that the events at Oltu had only a local significance, and he proposed a general attack in the direction of Köprüköy. All the officers of the general staff present pressed Myshlayevski to take over the command, but it was a responsibility which the professorial general was anxious to avoid. His bureaucratic training helped him to find a formula, to which, however, he was unwilling to put his signature. As it was urgent, he said, he would simply dictate it and they would write it down; and the deputy commander-in-chief produced his order of the day no. 1, aimed not so much at defeating the Turks as at satisfying the different contending personalities and the different views expressed. To placate old Bergmann the general offensive was not dismissed. But to quieten Yudenich and others who saw a growing menace to Sarikamiş, the Turkistan Corps on the right flank was to be strengthened by one and a half regiments of the 20th Infantry Division. Colonel Dovgirt had to defend Bardiz and the passes towards Sarikamiş. The 3rd Kuban Plastun Brigade was moved from the right bank of the Aras to strengthen the general reserve. The 3rd Caucasian Rifle Brigade was to march from Kars to Merdenik and reinforce Istomin.

This order of the day only remained valid for a few hours. Myshlayevski left Mecinkirt for Sarikamiş, and there the officer commanding Bardiz post reported to him personally that Bardiz village was already occupied by a strong Turkish force. Further reports of Dovgirt's patrols confirmed the uninterrupted movement of Turkish columns along the Çakir-baba ridge towards Bardiz. Deeply impressed, Myshlayevski, at 11.30 a.m., telephoned to Bergmann cancelling the proposed general attack against Köprüköy and ordering preparation for a withdrawal through Mecinkirt and Zivin. On the morning of the 25th, Myshlayevski preferred to leave Sarikamiş for Mecinkirt. On his way his car was fired on by an enemy patrol along the main road. During the preceding night all Russian commanding officers had been in a state of tense anxiety; but on that very night Enver, at Bardiz, had drawn up the orders which proved to be the first step in the destruction of his army.

Yudenich had been at Sarikamiş, with Myshlayevski on 24 December; and Myshlayevski had asked him to take command of II Turkistan Corps (the commander of which was absent). Yudenich had left Sarikamiş at 11 a.m. and gone by car to Sanamer. He realized that the situation at Sarikamiş was critical. As he passed through Karaurgan he stopped and ordered one battalion of the 18th Turkistan Rifles to proceed immediately to Sarikamiş. He also ordered the 17th Turkistan Rifles to join the 18th under Colonel Dovgirt near Yeniköy. The 80th Kabardinski Regiment (20th Infantry Brigade, but attached to II Turkistan Corps) was also ordered up to Sarikamiş. Yudenich himself prepared to fight a defensive battle on the Sanamer positions with only the 4th Turkistan Brigade and two battalions of the 79th Regiment[1] in reserve. He did not think that the Turkish force advancing on Sarikamiş could be stronger than one division; but he apprehended a combined attack against his front, right flank and rear by the divisions of the Turkish XI and IX Corps. As for X Corps, Istomin had reported that the enemy had appeared in the vicinity of the salt mines, and that he himself had taken up a strong position near Penek. It seemed to Yudenich that the objective of X Corps must be Ardahan. Yudenich felt that a defensive battle by four Russian brigades against the six Turkish divisions of XI and IX Corps presented a difficult but a possible task. He guessed perhaps that a part of the Turkish forces were already exhausted by the arduous march which they had undertaken. He feared only the orders which Myshlayevski might issue under the influence of his own pessimism.

After the shots at his car on the Mecinkirt road, Myshlayevski reached Bergmann's headquarters almost in a state of panic. He was convinced that Sarikamiş would be lost and that the bulk of the Caucasian army would be cut off from their line of retreat to Kars. All the attention of Myshlayevski was now directed to a side-road, a difficult track virtually impassable for wheeled traffic, from Mecinkirt to Karakurt on the Aras. From Karakurt there was a *chaussée* to Sarikamiş, which forked two miles south of that town; one branch

[1] Belonging to 20th Infantry Division.

went through Alisofu parallel to the main Kars road which it joined at Elkeçmez. If it proved too late to use this route, Myshlayevski thought it possible as a desperate measure to retreat by the bad track along the Aras from Karakurt through Başköy and Kağizman and then by the *chaussée* from Kağizman to Kars.

All these considerations were exposed to General Bergmann, who proceeded to order a general retreat to start on the night of 25–26 December. A special group only was detached to march along the main road and help in the defence of Sarikamiş.[1] At dawn on the 26th all the baggage trains and munition columns set out on their retreat. Three regiments of the 39th Division and one plastun brigade remained in the forward positions throughout the 26th, but in the evening they, in their turn, began to retreat. Myshlayevski suggested that the 1st Caucasian Cossack Division and the 2nd Plastun Brigade should retire from their positions to the south of the Aras through Basköy on Karakurt, and from there reach the Sarikamiş-Kars road somewhere between those two places. As commander of II Turkistan Corps, Yudenich was informed of the general retreat. He was told that his own line of retreat must lie along the main road and that, if necessary, he must fight his way through at Sarikamiş. Myshlayevski and Bergmann had in fact already accepted defeat without a fight as a result of the Turkish manoeuvre. Their only solution was to save the army or even a part of the army.

Sarikamiş, in the meantime, was preparing to fight, without even the shadow of a regular armed force. There were in this small town and military station only two *sotni* of Frontier Guards with two machine guns, two *druzhiny* of militia (*opolcheniye*), about 1000 railwaymen and some rear service details. Except for the Frontier Guards, this 'garrison' was armed with old Berdan rifles of the 1877 period. By extreme good luck two field guns were stranded on the station (on their way from the front back to Kars for an artillery section in process of formation). The mild old commandant manned some strategic points round the town with the militia; and a passing artillery officer, Captain Mushelov, put the two guns in position in the main square. Early in the morning of the 25th a detachment of the Frontier Guards and the thousand railwaymen marched out and posted themselves in the thick brushwood covering the approaches from the Bardiz Pass, five miles out from Sarikamiş.

During the same morning the 29th Turkish Division had begun their march from Bardiz on Sarikamiş. Something, however, was

[1] The 155th (Kubinski) Regiment and the 1st Zaporogian Cossack Regiment.

wrong with the Turkish maps; the staff of IX Corps expected to find Sarikamiş ten miles from Bardiz village, where in fact they came only to the Bardiz Pass. There was already considerable confusion, as units of the 29th Division were getting mixed up with elements of the 17th, who had begun to arrive from Çatak. Snow was falling heavily and the men of the 29th moved very slowly forward. By midday the Turks had reached the high Molokans' plain on Soğanli, where in summer cattle graze the rich upland grass among the brushwood. Here the advancing infantry ran into the Russians, and a lively fire began. The Turks deployed to attack, unaware of the insignificant numbers of their opponents. Only at 4 p.m. did the Russians begin to withdraw towards Sarikamiş. Dark was coming on, and the commander of the 29th Division thought it more prudent to break off the action and bivouac on the spot. With the shelter of Sarikamiş so near the Turks huddled for the night in twenty degrees of frost. They were fortunate only in that they had to hand plenty of brushwood for their fires.

CHAPTER XX

THE DEFENCE OF SARIKAMIŞ, 26–31 DECEMBER 1914

MAPS 18, 19

IN 1914 Sarikamiş was a small modern town with barracks for two regiments and a station which was the railhead for the line coming from Kars and Tiflis. Through the town ran the relatively new road (built after 1878) on the track of the old Erzurum road which Paskevich had followed in 1829; crossing the Meliduz saddle it forked north-west to Yeniköy and Karaurgan and south-west to Mecinkirt. North-east of Sarikamiş, an old track left the main Kars road and passing over the Eşek-meydan saddle and through the Hana Pass joined 'the new road' through Yeniköy; this was the route which Muravyev's column had taken in 1829.

Immediately to the north of Sarikamiş, the high ridge of Turnagel commands both the valley of the Sarikamiş-çay and the pass leading to Bardiz. Two tracks cross the flank of Turnagel and, four miles from the railway, reach the Bardiz Pass where they gradually descend with the general lie of the land to Yayla Bardiz[1] and then into the deep valley in which Bardiz village lies. These tracks pass the poor Cherkess hamlet of Upper Sarikamiş at the entry to the Bardiz ravine. To the west of the Bardiz ravine another ridge, Yağmurlu ('Rainy'), forms a counter-fort to Turnagel. Both Turnagel and Yağmurlu-dağ are thickly wooded with firs and brushwood. In some places the slopes of Turnagel throw up vertical rocks, the most notable of which is the 'Raven's Nest' above Upper Sarikamiş and the 'Eagle's Nest' overhanging Sarikamiş railway station.

During the night of 25–26 December, while the retreat was beginning on the main front, Sarikamiş was making urgent preparations for defence. Colonel Bukretov, chief of staff of the 2nd Plastun Brigade, who happened to be passing through the town, was appointed by telephone from Mecinkirt commander of all forces available on the spot. By the same train eight machine guns had arrived from Kars, and 100 young officers fresh from the cadet school in Tiflis. Bukretov also discovered on the station seven platoons of Turkistan riflemen, one platoon from each of the Turkistan regiments who were being

[1] The summer camp and pasture of the Bardiz people.

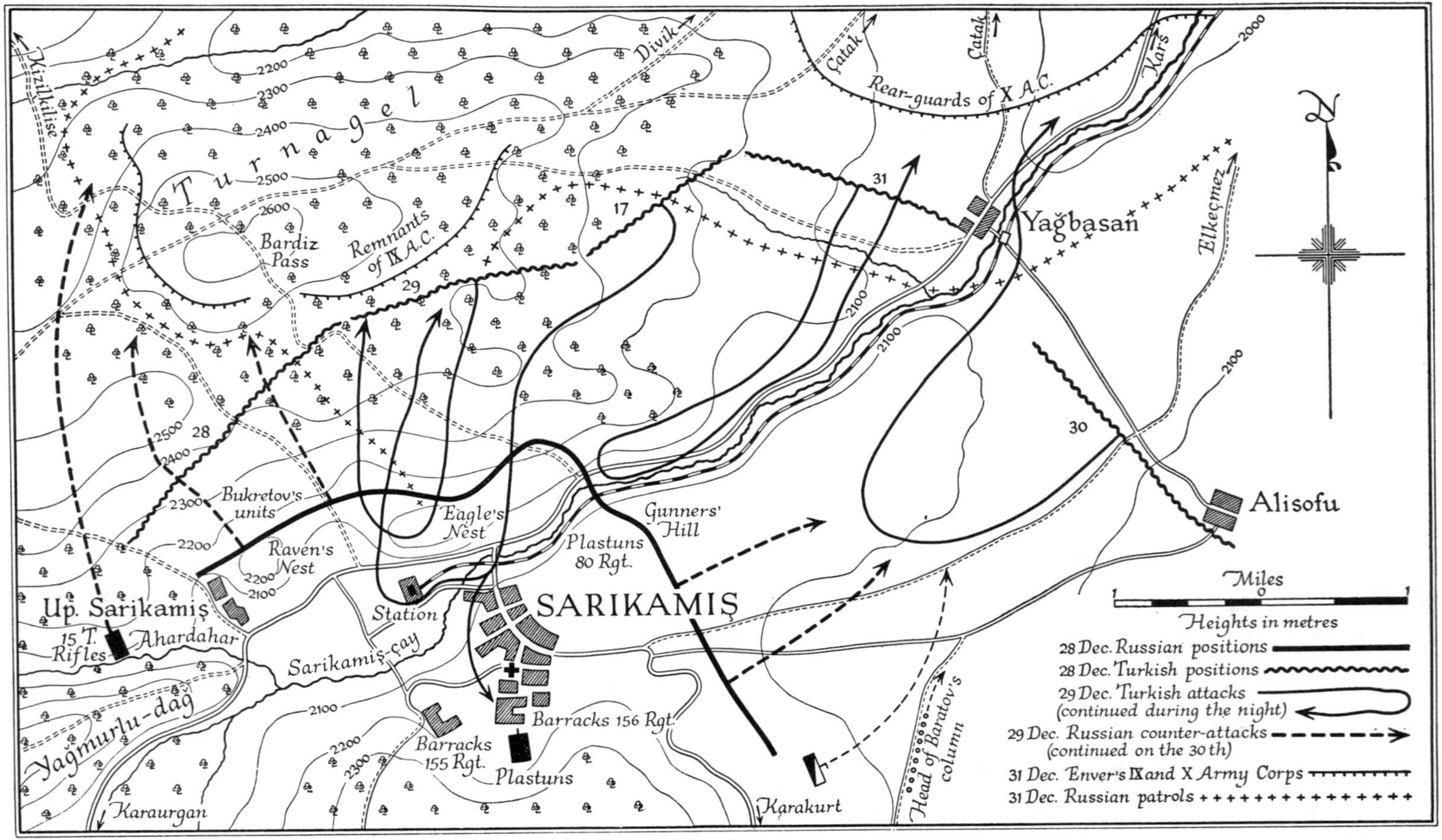

Map 18. Sarikamiş, 28–31 December 1914

sent back to the rear as training cadres for the 20th Regiment of the 5th Turkistan Brigade. Bukretov formed on the spot two companies of this embryo regiment. During the night the Turkistan battalion sent by Yudenich from Karaurgan came in; so in the morning Bukretov had available about 1500 seasoned troops, 1000 railwaymen, and 1000 militia with some hundreds of volunteers from the rear services.

Meantime Enver's unfortunate 29th Division was making a bad start. Despite the fires which the askers tried to keep alight from the brushwood on Turnagel, hundreds of the poor fellows lay frozen to death in the morning light and hundreds more had deserted to seek refuge in the few miserable hamlets scattered over the neighbouring mountain sides. The 29th which had started the march on 22 December 8000 strong could now put scarcely 4000 rifles into the firing line. But they had managed to bring with them eight mountain guns.

Enver, still exuberant, was quite confident that the Russians had no artillery in Sarikamiş. He therefore began the action of the 26th by deploying four of his own guns in an exposed position on the outskirts of the woods overlooking the town. Mushelov's two guns in the square near the church replied with such effect that three out of the four Turkish guns were soon silenced. That the Russians had artillery was an unpleasant surprise to the Turks; it was decided to await the arrival of the 17th Division. But only one regiment had arrived by midday, and an attack undertaken by elements of both divisions was burnt up by strong Russian machine-gun fire. In the afternoon the Turks occupied the Raven's Nest and Upper Sarikamiş, but they were unable to penetrate into the valley where the town lay. Ihsan Paşa, commanding IX Corps, began to lose heart and preferred to await the arrival of his third division, the 28th. The men of this division began to arrive towards the evening, but one regiment had lost so many stragglers that it could only muster 300 men. The three divisions of IX Corps—their strength reduced perhaps to not more than 7000 men—passed the night of the 25th in the deep snow of the lower slopes of Turnagel and Yağmurlu. Meanwhile the first serious reinforcements were arriving at Sarikamiş: the 1st Zaporogian Cossacks with a horsed battery came in in the evening; and during the night the four battalions of the 80th Kabardinskis.

With a little objectivity Enver Paşa might have understood, even on the evening of the 26th, that his bold plan had little chance of success. IX Corps stood before Sarikamiş; but, out of a strength of 25,000 rifles with powerful mountain artillery which had set out a week ago, only 10,000 men with twenty guns now lay between Bardiz and Sarikamiş; and these men were exhausted, hungry and demoralized by snow, frost and desertion. But Enver still counted on the arrival of X Corps and on an energetic offensive by XI Corps which should prevent the Russians from reinforcing the Sarikamiş position.

The temperament of Hafiz Hakki Paşa, commanding X Corps,[1] resembling in some respects that of Enver himself, explains the 'disappearance' of this corps during the critical days of 25 and 26 December. On 23 December, Hafiz Hakki had captured Oltu. His orders, for the following day, were to send a detachment in pursuit of Istomin and to march with the rest of his corps to Bardiz. But, informed that Istomin had taken up a position at the Avçali salt mines, not far out from Oltu, Hafiz Hakki marched against him along the Ardahan road. By 11 a.m. on the 24th Istomin had abandoned his position and retired on Penek, where he took up a position in a defile so naturally strong that he hoped to defend it with success. However, during the night of the 24–25th, with the appearance of Turkish patrols in the neighbourhood of his camp and with the arrival of militia who had only narrowly escaped being surrounded by Acar irregulars in the Panaskirt Pass, panic spread among the weary troops of Istomin's brigade. Early in the morning of the 25th, in a rather demoralized state, these troops, reduced in numbers to about 5000, began to move away in the direction of Merdenik. Here Istomin got a report that a strong Turkish force was attacking General Henike's detachment (three plastun battalions) at Ardahan. He considered his own troops for the moment incapable of fighting and moved from Merdenik along the Kars road where he halted (eight miles from Merdenik) at Demirkapu. From here Istomin reported his misadventures to general headquarters.

Hafiz Hakki occupied Penek on the 25th, and from Kosor he moved the 32nd Division via Kop to Bardiz. With the 30th and 31st Divisions he advanced on Merdenik which had already been evacuated by Istomin. Here an officer reached him with a categorical order to march immediately with his two divisions to the region of the Eşekmeydan Pass from where he was to unite with IX Corps in its attack on Sarikamiş. Hafiz Hakki was anxious to proceed to Ardahan, but the order was clear enough. He left one regiment to hold the Agundir Pass above Merdenik, and with the rest of his troops undertook the desperately dangerous march across the high Allahuekber massif[2]

[1] Originally as deputy for Colonel Bronsart von Schellendorf.

[2] A southerly arm of the Arsiani or Yalnizçam-dağ.

(at an altitude of nearly 3,000 metres) towards the villages of Beyköy and Başköy. The chief of staff of X Corps, the German, Major Lange, not only believed the somewhat fanciful distances indicated on the Turkish maps, giving Beyköy as fifteen instead of twenty-five miles away, but planned the march of two divisions over these wild mountains to be completed in five hours. The result was that the unfortunate men of the 30th and 31st Divisions, who, until now, had been in excellent spirits and condition, lost one-third of their effectives[1] during nineteen hours of terrible marching before they reached the two miserable hamlets which were their map-objectives. On the evening of the 26th, Hafiz Hakki arrived with his staff at Bardiz to meet Enver; the 32nd Division of X Corps had come in a few hours before, having made the much easier march through Kop. This division was now held round Bardiz to protect the Turkish rear from the movements of Colonel Dovgirt's detachment whose patrols had been observed in the vicinity of the Hana Pass. Enver therefore gained only one regiment of the 17th Division (now replaced at Bardiz by the 32nd) to reinforce his diminished strength at Sarikamiş.

The Russians at Sarikamiş were considerably stronger on the 27th than they had been on the previous day. The 155th Kubinskis and the 15th Turkistanskis were expected; Colonel Barkovski of the 80th Kabardinskis had taken over the command. Throughout the day the Turks fought courageously, led by their senior officers with revolvers in their hands. Upper Sarikamiş twice changed hands, and the Turks established themselves on the Eagle's Nest, a position which gave them a dropping fire on to the railway yards. Units of the Turkish 28th Division got as far as the main road, but in the evening they were driven back by the newly arrived Kubinskis. However, during the same day Turkish mounted patrols reached the Kars railway near Novo-Selim station and blew the rails. The incident made a considerable impression on the Russians. The attacks of the 27th on Sarikamiş had been driven off; but the number and state of the Turkish troops were unknown at Russian field headquarters, and during the night of 27–28 December, the danger to Sarikamiş seemed to be greater than it really was.

General Myshlayevski had been further alarmed on the 26th by the news that on the previous day General Henike's small detachment had been defeated at Ardahan and that that important point was in the

[1] Hafiz Hakki lost 7000 out of 20,000 men. Field guns could only move with the help of eight to ten bullocks per gun.

hands of a strong Turkish force. The anxiety of the Russian staff was again sharpened on the evening of the 26th by a curious chance which illustrates the extent to which accidents may affect strategy. The chief of staff of the Turkish 28th Division was captured by a Cossack patrol on the slopes of Yağmurlu. He was in possession of a copy of one of Enver's orders which indicated clearly enough the whole scope of the Sarikamiş operation; and Myshlayevski, being a man who had been trained to understand war on paper, was duly impressed. He now knew that the whole of IX Corps was attacking Sarikamiş and that X Corps was available to cut off that town and the Caucasian field army from Kars. The report of the demolition raid against the railway confirmed the Russian commander's worst fears. He realized, further, that a Turkish force of unknown strength was in Ardahan and that the fresh XI Corps was available for a frontal attack on the Russian positions along both sides of the Aras. The situation, to Myshlayevski, appeared to be desperate.

The deputy commander-in-chief pressed all measures to accelerate the retreat. At the same time he showed some positive quality which was in contrast to his subsequent conduct. He sent Przevalski's Plastun Brigade from the right bank of the Aras to reinforce, in two days' forced marches, the troops at Sarikamiş. He intended the complete evacuation of the right bank of the Aras and the concentration of whatever striking force could be assembled in an area between Sarikamiş and Kars, where they would be in a position to oppose a further Turkish advance from Sarikamiş and to threaten the Turks in Ardahan.[1]

During the night of 26–27 December, Myshlayevski conferred at Mecinkirt with Bergmann and Yudenich. The latter, commanding the Turkistan Corps, only had available in the front line his 13th, 14th and 16th Regiments, and two battalions of the 79th Regiment.[2] He protested, however, against an immediate and general retreat—particularly since the main road was already crowded with the baggage trains of Bergmann, which had been on the move since the previous night. Yudenich pointed out that such a retreat, in the face of XI

[1] The composition of this reserve group was to be: 2nd Plastun Brigade and 1st Caucasian Cossack Division, moving up from Karakurt; 3rd Caucasian Rifle Brigade and 263rd Regiment, based on Kars. The Siberian Cossack Brigade was, at the same time, railed from Tiflis to Mikhailovo station, with orders to march from there through Akhalkalaki in the direction of Ardahan.

[2] The 15th Regiment was on the march to Sarikamiş; the 80th Kabardinskis were already there; and the 17th and 18th Regiments were with Dovgirt near the Hana Pass.

Turkish Corps, might lead to disaster. He explained also that in the event of retreat the advance of Dovgirt through the Hana Pass must be abandoned; to this advance, however, he attached the greatest importance, because he had grasped the significant fact that the only line of Turkish communication with their rear, difficult and inadequate as it was, passed through Bardiz. Yudenich tried to convince Myshlayevski and Bergmann that the Turkish army was fighting under the worst possible conditions, and that in a few days it might not be able to fight at all. After heated discussion Myshlayevski agreed to defer the general retreat until the situation at Sarikamiş became clear. Yudenich returned to his Turkistan troops who, on the following day (27th), were more heavily attacked by XI Corps than they had been since the opening of the Turkish offensive on the 22nd.

The same day Myshlayevski decided that his own presence nearer the viceroy was necessary in order to organize the defence of Transcaucasia.[1] He went by car to Kars (via Karakurt) and took train to Erevan, having summoned Oganovski, the commander of the troops in the Eleşkirt valley, to meet him at Alagöz station. Myshlayevski ordered Oganovski to retire immediately behind the Russian frontier. The astonished and angry general (who had experienced no pressure at all against his own front) simply ignored the order. However, General Chernozubov, commanding in Persian Azerbaijan, who received the same order, had not the same independence of character; he began a general withdrawal, although he had only a few bands of irregulars in front of him. On arriving in Tiflis, Myshlayevski was so strongly convinced that the Turks had already won that he called the town council of the Caucasian capital together and warned them of the gravity of the situation. The alarm spread from the councillors to the street, and crowds swarmed round the railway station in panic anxiety to flee the coming Turkish massacres.[2]

In the meantime the situation at Sarikamiş had become definitely better for the Russians. The 28th passed quietly. Hafiz Hakki had convinced Enver that the troops of the 30th and 31st Divisions needed a rest, and that at least thirty-six hours were necessary to collect all the stragglers left along the route. But if this day gave some coherence to the troops of X Corps, one more day and one more night in the frozen woods of Turnagel diminished the strength of the remnants of

[1] The Siberian Cossack Brigade, sent from Tiflis towards Ardahan, was the last of the strategic reserve at the disposal of the general headquarters of the Caucasian army.

[2] In March 1915, Myshlayevski was dismissed the service by order of the Grand Duke Nicholas.

IX Corps.[1] On the morning of the 29th there were scarcely more than 6000 fit men out of all its three divisions.

On 29 December was fought the decisive battle at Sarikamiş. During the 28th the Turkish X Corps had been concentrated for attack. The 31st Division was on the slopes of Turnagel with its left flank along the railway at Yağbasan (about four miles to the north-east of Sarikamiş). During the night of the 28–29th, the 30th Division crossed the railway and occupied Alisofu on the southern side of the Sarikamiş valley and on the road from Karakurt to Kars. Except for the roundabout track to the south of the Aras through Başköy and Kağizman, the Russian army was actually isolated from Kars. In theory at least Enver's dream had been realized.

However, the two divisions of X Corps, after their difficult march from Merdenik, could put rather less than 12,000 men in the line; and their only connexion with IX Corps was along the difficult tracks in the region of Divik village winding through the woods and snow-drifts of Turnagel. IX Corps, at the same time, could muster not more than 6000 men and less than twenty guns.

General Przevalski was now commanding at Sarikamiş. He could dispose of twenty infantry battalions and one Cossack regiment; his own five battalions of Kubanian Plastuns were fresh troops; in all, he had between 13,000 and 14,000 men with thirty-four guns and several score of machine guns.

Przevalski planned to maintain an active defence. Colonel Bukretov, with six battalions[2] and two guns, was ordered to cross the Ahardahar ravine which divides the two massifs of Yağmurlu and Turnagel and attack the Turkish right flank; this manoeuvre involved the capture of the rocky Raven's Nest promontory at the south-west end of Turnagel, where the stone-built *saklyas* of Upper Sarikamiş were stubbornly defended by the Turks. Meanwhile seven battalions with strong artillery and machine-gun support held the line along the Sarikamiş-çay. Cossacks, supported by two Kabardinski battalions, observed the eastern approaches to Sarikamiş; and their patrols soon discovered the presence of Turkish infantry in force between Yağbasan and Alisofu. Believing that the most dangerous Turkish attack might come from the east, Przevalski kept all his Plastun Brigade in reserve in the eastern part of Sarikamiş town near the barracks of the 156th Regiment.

[1] On the 28th, the Russians had cleared the last elements of the 28th Division from the Yağmurlu ridge and removed all interference with communications between Sarikamiş and Yeniköy.

[2] Including the whole of the 15th Turkistan Regiment.

On the morning of the 29th, the Turkish 30th Division attacked with great *élan* from the direction of Alisofu. Przevalski counter-attacked with three plastun battalions and the Turks were thrown back. Outflanked by the 1st Zaporogian Cossacks, they abandoned Alisofu and retired beyond the railway line. The defeat of the 30th Division put a stop to the movement of the 31st Division which had attempted to advance along the railway line. During the afternoon, the 31st resumed their original positions on the slopes of Turnagel with their front to the railway and their left at Yağbasan. Meanwhile, from Turnagel, the remnants of the 17th and 29th Divisions, as on previous days, had proved unable to descend towards the railway; and one of the plastun battalions sent forward from the reserve had driven them from the Eagle's Nest. On the Russian left, Bukretov had had difficulty in crossing the Ahardahar ravine, where his men were held up by deep snowdrifts and suffered heavily under accurate Turkish machine-gun fire from the south-west slope of Turnagel. Here, as in other early battles of World War I, the machine gun was proving its efficiency as a defensive weapon. On his right flank, Bukretov failed to storm the stone saklyas of Upper Sarikamiş; but on his left some very tired companies of the 15th Turkistanskis managed to cross the ravine and by dusk were in the vicinity of the Bardiz Pass.

Enver, disappointed with the result of the day's fighting, decided to try a surprise night attack. At 10 p.m. units of the Turkish 17th Division suddenly attacked and recaptured the Eagle's Nest. They then rushed the bridge over the Sarikamiş-çay and penetrated into the town. Przevalski called out the last of his reserve—one battalion of the Plastun Brigade. There was a savage bayonet battle in the darkness; the Turks left several hundred dead lying in the frozen streets and 400 prisoners in the hands of the Russians. Three or four hundred determined askers took possession of the barracks of the 156th Regiment, but at dawn they were battered into surrender by close-range artillery fire. This episode marked the not inglorious end of the 17th Infantry Division of IX Corps as a fighting force.

In the morning the other two divisions of the same corps (29th and 28th) could not muster more than 3000 men. Even Enver could not insist on continuing the attack on the 30th. The offensive initiative had passed to the Russians who had just received six howitzers from Mecinkirt. Bombarded by 4·8 in. shells the Turks were driven out of the stone huts of Upper Sarikamiş which they had defended with such obstinacy. Bukretov, in the meantime, was within a thousand yards of the entry to the Bardiz Pass, but his exhausted troops proved unable to advance farther. The reproaches of Przevalski provoked him into resigning his command; but his successor, leading the attack on the following day (31st), had no better fortune, and a few hundred Turkish askers with machine guns at the head of the pass withered a determined assault. On the morning of 1 January 1915 exhaustion on both sides silenced the battle over the frost-bound mountains of the field of Sarikamiş.

However, 29 December had been a day of crisis not only on the Sarikamiş front. On the 28th Bergmann had again suddenly reverted to his plan for a general retreat; his troops began to retire, and Yudenich, on the right, was presented with the accomplished fact. Some reason for a partial retreat existed, since all troops had been withdrawn from the positions to the south of the Aras, and Bergmann had to look to the security of his left flank. Yudenich, however, continued to protest against the idea of a general retreat, since he saw that the main road, so vital for operational purposes, would become hopelessly blocked. On the night of 28–9 December he was able to persuade Bergmann to halt his troops, after their night march at Çermuk, eight miles south-west of Mecinkirt. Yudenich himself made a withdrawal to keep in line with Bergmann's positions and took up a new line along a spur of Çakir-baba known as Horum-dağ; his right flank covered the village of Hoşap.

Yudenich was maintaining his principal argument against too precipitate withdrawal: the necessity of awaiting the development of Dovgirt's operation against Bardiz. With true strategic instinct, Yudenich had appreciated clearly enough the importance of Bardiz to the whole position of the Turkish army. By the evening of the 28th, Dovgirt was in occupation of the Hana Pass and marching on Bardiz; and in the end Bergmann reluctantly agreed to defer further retreat until the evening of the 30th–31st.

During the 29th and 30th Yudenich's troops maintained their new positions against the attacks of XI Corps; and on the 30th occurred the success so impatiently awaited by Yudenich: Dovgirt's detachment (17th and part of the 18th Turkistanskis) took with the bayonet the height of Ahurezar and the hamlet of Yayla Bardiz. Both these positions overlooked Bardiz village from which they were three miles distant; and Russian artillery could now hold under fire the advance base of Enver's army in front of Sarikamiş.

With the continuing good news from Sarikamiş (after the fighting of the 29th), Yudenich held the view that the capture of Yayla Bardiz

PLATE V

THE SLOPES OF SARIKAMIŞ

had completely changed the situation and that the position of the main Turkish army at Sarikamiş was becoming desperate. But it proved difficult to convince Bergmann, who was ready to renew the general retreat on the night of the 30th–31st. Yudenich sent to field headquarters at Mecinkirt one of his officers, Colonel Dratsenko, with the mission to clarify the situation to Bergmann and, in the event of Bergmann's insisting on the continuance of the general retreat, to state that from that moment Yudenich, as chief of staff of the Caucasian army, took over the command of all the forces then operating under Bergmann. After a dramatic interview, Bergmann agreed to halt the 39th Division—which was already on the move—at the frontier. He proposed to leave all the rear-guard (the 39th and 4th Turkistan Divisions) at the disposal of Yudenich, and to proceed himself next day (31st) to Sarikamiş to direct personally the operations for a break-through and the re-establishment of communications with Kars.[1]

Yudenich was content enough with this arrangement. He withdrew his troops from Horum-dağ and took up positions running roughly along the frontier: his right on the Güllü-dağ spur of Çakir-baba and his centre along the heights between Yeniköy and Karaurgan; his left, covering Mecinkirt and the Soğanli plateau, stretched beyond the Aras to Başköy, eight miles to the south of Karakurt.[2]

Yudenich decided to stand firm along this front until such time as the issue should have been decided at Sarikamiş; his right at Güllü-dağ was only separated by six miles from Dovgirt's group at Yayla Bardiz, and it was the same distance from the *top yol*.

Yudenich appreciated clearly the possibility of surrounding the Turkish army in the triangle between Bardiz, Sarikamiş and the Eşek-meydan Pass. He knew that Baratov's Cossack Division and the 2nd Plastun Brigade were moving from Karakurt on Alisofu; and on the 31st he wired Vorontsov that the situation of the Turks was disastrous, they should and could be annihilated completely; and he asked for the removal of the 3rd Caucasian Rifle Brigade from Kars on Divik to operate against the Turks' flank and rear.

As early as 29 December the commandant of Kars, on his own initiative, had sent three battalions of the 263rd Gunibski Regiment

[1] Bergmann took with him to Sarikamiş the 154th (Derbentski) Regiment, leaving to Yudenich the 14th and 16th Turkistanskis and 2 battalions of the 79th Regiment. Yudenich thus had left 19 battalions with which to fight a defensive action against the 32 relatively fresh battalions of the Turkish XI Corps under Abdul Kerim Paşa.

[2] This flank was masked by two battalions of the 79th Infantry Regiment and a Cossack regiment (Kara-derbent detachment).

with a battery and a *sotnya* of Cossacks in the direction of Sarikamiş. Colonel Voronov, commanding this detachment, reached Novo-Selim on the 30th and captured a few prisoners from the Turkish 30th Division. On the following day he was at Elkeçmez, eight miles from Sarikamiş, and here he was joined by the 1st Zaporogians sent by Przevalski. On the night of 1–2 January 1915 the 3rd Caucasian Rifle Brigade was due to be transported from Kars to Novo-Selim station in response to Yudenich's telegram; and, on the 2nd, Baratov's Division was expected at Alisofu. Thus by 2–3 January a considerable force of Russians should have been available to manoeuvre against the Turkish left flank. But, fortunately for Enver, Bergmann, who had been installed at Sarikamiş since the 31st, wished to direct this offensive operation and did his best to introduce as much confusion into it as he could.

During these critical days of 1–2 January, Enver Paşa had lost that overbearing self-confidence which had proved so fatal to the Turkish Third Army. On the 31st, the chief of staff of IX Corps had reported that the remnants of this corps in the Turnagel woods were about 2500 strong with only 1000 men and fourteen machine guns in the line and fourteen guns fit for action. The corps with its entire staff was cut off from Bardiz. It was suggested that the vice-generalissimo should leave before it was too late to join XI Corps, and that he should press for such energetic action from Abdul Kerim as might yet save the survivors of IX Corps. Guns without gunners and regimental colours were to be removed, if possible, to the rear. Enver Paşa understood at last that he had failed. He accepted the advice of his officers and the same day left the Turnagel woods. He passed the next day with the troops of X Corps. The situation of this corps was also most difficult. Near Bardiz, the 32nd Division was isolated fighting against Dovgirt's group advancing from the Hana Pass. The 31st and 30th Divisions were concentrated in the Çatak-Divik region along the slopes of the Soğanli, where the track passes to the Eşek-meydan Pass. Hafiz Hakki had observed the appearance of Russian detachments from the direction of Kars, and on 2 January ordered a withdrawal from the position held. The same day Enver Paşa rode by a circuitous route round Bardiz; near Kizilkilise he narrowly escaped capture by a Russian patrol. On the 3rd he passed through Abdul Kerim's headquarters on his way to Erzurum.

Meanwhile various movements had taken place in the extreme north of the battle area. On 25 December, Ardahan had been captured almost without a fight by Major Stange's force advancing from the direction of Ardanuch.[1] Colonel Henike's three battalions of plastuns were taken by surprise and fled in disorder along the Akhaltzikhe road. But Stange did not pursue and remained strangely inactive after this easy success. On the same day (25th) Istomin, retiring from Oltu, had heard the cannonade from the direction of Ardahan but had considered his troops too exhausted and demoralized to intervene. On the night of 26th–27th, Istomin had been joined by Gabayev's 3rd Caucasian Rifle Brigade (from Kars) at Demirkapu. On the 28th, Gabayev marched against Merdenik and retook it; on the following day, after a lively action he drove the regiment of the Turkish 30th Division, which had been left there as garrison, back on to the Agundir Pass. But on the same day he got orders to return immediately to Kars and, leaving Istomin at Demirkapu, he reached the Russian base on 1 January, to be transported the same night to Novo-Selim.

Kalitin's Siberian Cossack Brigade had reached Akhaltzikhe from Tiflis on 27 December. On the 30th he was at Zurzuna, twenty-eight miles from Ardahan, where he was able to rally Henike's fugitive battalions. Two days later he was ready to advance on Ardahan;[2] Istomin was ordered to support him from the south.

[1] Two regiments of the Turkish 3rd Division with 2000–3000 Laz and Acar irregulars.

[2] Kalitin's force now amounted to three battalions, eighteen *sotni* and eight field and six horsed guns.

CHAPTER XXI

THE DEFEAT OF ENVER PAŞA

MAPS 18, 19

ON 1 January 1915 the complete failure of the Turkish offensive was clear to both the Russian and Turkish commanding staffs. But in spite of cold, weariness and hunger the troops on both sides were still ready to fight; and the battle was to continue for another fortnight.

Abdul Kerim Paşa, the commander of XI Corps, who, perhaps believing the exaggerated and optimistic reports of events at Sarikamiş, had not pressed the frontal attack by his corps, undertook on the last day of 1914 a fierce offensive. He had at his disposal more than thirty battalions and about 100 guns.[1] To oppose him Yudenich had only nineteen battalions and forty-eight guns.[2]

Abdul Kerim concentrated the attack of two divisions against the positions of the Turkistan Brigade between Güllü-dağ and Karaurgan. The Turks had no success on the 31st, but they renewed their attacks with increased insistence on 1 January. The situation became particularly critical at Karaurgan, where the position held by the 16th Turkistanskis changed hands several times. On 2 January, Abdul Kerim, informed that the Russian line seemed weak at Altin-bulak between the left flank of the Turkistan troops and the right of the 153rd Regiment, rushed a complete division in this direction; the hamlet was captured and the Turks took Height 808 at the head of a ravine giving access to the rear of Mecinkirt. Yudenich could only counter-attack with single companies of Turkistan rifles and the 153rd (Bakinskis); but the Turks were driven back and the Bakinskis captured a four-gun battery with 300 prisoners. On the 3rd the vice-generalissimo appeared at Abdul Kerim's headquarters and the attacks were renewed. On the 4th, one Turkish division made a decisive attack against the extreme right of Yudenich's positions; Güllü-dağ was lost by the Russians but the success was without a sequel. Abdul Kerim was suffering heavy losses, and after the 6th his attacks lost their *élan*. On the 7th, Enver left for Erzurum en

[1] The three infantry divisions and field artillery park of XI Corps with a few of the well-trained Frontier Guard battalions.

[2] Two regiments of the 39th Division; the three regiments of the 4th Turkistan Rifle Brigade; and 2 battalions of the 79th Regiment.

route for Istanbul, abandoning to the Armenian winter the shattered remnants of his 'Pan-Turanian' army. On the 10th, the Russians retook Güllü-dağ with six guns and 1000 prisoners.

During all these twelve days of tense fighting Yudenich never lost confidence in his admirable troops. Although the Turkistan rifles on his right flank were sadly reduced in numbers he still found it possible to send one battalion to reinforce Dovgirt who was in difficulties at Yayla Bardiz.[1] Unable to resist the repeated attacks of the Turkish 32nd Division he had been driven south, but he still held on to Çilhoroz mountain, distant only three miles from the extreme right of Yudenich at Güllü-dağ.

Przevalski had summoned the 17th Turkistanskis from Yayla Bardiz because he was planning for 2 January a movement aiming at the complete encirclement of the remains of the Turkish army in front of Sarikamiş. Although he was aware that only a shadow remained of the three divisions of IX Corps, he still supposed the Turks in the Turnagel woods to be 8000 and possibly 10,000 strong. In reality there were less than 3000 men left, but with a few mountain guns and a number of machine guns they were maintaining a stubborn resistance, skilfully adapting the defence to the difficult snow-covered ground. And Przevalski's own troops were almost completely exhausted by many days of fighting and many sleepless nights. In a particularly bad state was the detachment at the Bardiz Pass (originally under command of Colonel Bukretov): out of six battalions only 800 men remained fit for the line. Sarikamiş itself was overcrowded with 12,000 Russian sick and wounded and 3000 Turkish prisoners. The arrival of the 154th (Derbentski) Regiment, brought by Bergmann in person, carried the active fighting strength available to 7000 men.[2]

The complete encirclement of the Turnagel woods began on 2 January. That day the Bardiz Pass was captured with several hundred prisoners and six guns. On the 3rd, exceptionally dense fog made military operations almost impossible. On the 4th, the remnants of IX Corps were attacked simultaneously from the valley of the Sarikamiş-çay and from the rear. The brave askers still continued to fight stubbornly enough in spite of their desperate situation. With sixteen machine guns in the line they resisted from snow trenches on the edge of the woods. But they were taken from the rear, and the

[1] Przevalski had taken the 17th Turkistanskis from Dovgirt, who was left with only two weak battalions of the 18th Turkistankis—1200 men with two guns.

[2] Including the Plastun Brigade and the 80th Infantry Regiment.

Russians at last captured all that remained of the IX Turkish Army Corps: 200 officers (including Ihsan Paşa and the three divisional staffs) and 1000 men, with 20 machine guns and 30 mountain guns (most of which were out of action). That day and the next, Przevalski was mopping out the woods where several hundred more prisoners were captured. IX Army Corps of the Third Army had ceased to exist.

The liquidation of X Corps proved more difficult, much longer and was not so complete. The fault lay with Bergmann, who since 31 December had been trying to direct the operations of the shock group which had been concentrated between Kars and Sarikamiş.

It must be said, in justice to the unfortunate deputy commander-in chief, General Myshlayevski, that the conception of concentrating a shock group in the area between Kars and Sarikamiş to be used in an outflanking movement against the Turkish X Corps belonged to him. It had been his idea to move Baratov's Cossack Division and Gulyga's Plastun Brigade from Bergmann's extreme left, south of the Aras, to the extreme right in the area on both sides of the Kars-Sarikamiş railway.[1]

Myshlayevski had also been anxious to add to Baratov's and Gulyga's units, Gabayev's Rifle Brigade, which he had recalled to Kars after its march to Merdenik.[2] When Gabayev reached Kars from Merdenik, Myshlayevski had called him to the telephone and, in discussing with him the offensive against the Turkish X Corps, had outlined a very wide outflanking movement of considerable originality. Already on 1 January Myshlayevski rightly understood that Hafiz Hakki would try to retreat by the track through Divik, Verişan and the Eşek-meydan Pass. To cut this route he advised Gabayev to march from Novo-Selim station not on Divik but round the north of the massif of Emirhane-dağ through Başköy on Verişan. In this way the Turks would be cut off both from the Eşek-meydan Pass and from any possibility of retreat over the paths through Başköy and Beyköy into the Oltu region.

When, however, Gabayev arrived with his four rifle battalions at Novo-Selim,[3] he was informed that Bergmann had placed him under Baratov's command and that he had to participate in carrying out the operations north of Sarikamiş—planned by Bergmann himself. The

[1] Using for that purpose the road Karakurt-Alisofu.

[2] Myshlayevski proved that, if he were too impressionable in the face of events, he was competent enough in the quiet of an office and had an insight into strategic problems.

[3] The 9th Regiment of the brigade was with Istomin at Demirkapu and the 12th had been left in Kars.

operations were timed to start in a few hours and it was too late for Gabayev to appeal to Myshlayevski.

At dawn on 2 January the infantry of Gabayev and Voronov[1] were concentrated at Novo-Selim station. The Cossack Division was at Alisofu but the Plastun Brigade of Gulyga was still on the march. Baratov decided to postpone the attack until the following day.

Following the instructions of Bergmann, who was convinced that the left flank of the Turks was at Çatak and Divik, Baratov directed his principal attack (Gulyga's Plastun Brigade) against Divik, while the 80th Kabardinskis from Sarikamiş were ordered to attract the attention of the Turkish centre, believed to be at Yağbasan. As a deeper outflanking movement, Gabayev's infantry, supported by Voronov, were sent towards Upper and Lower Salut and Laloğlu, while Cossack units under Fesenko were dispatched to outflank the Turks marching on Çiplakli.

The plan failed completely. First, 3 January was a day of heavy fog which made troop movements impossible. What was more important was that Hafiz Hakki had well understood the danger of his situation when Enver left for Erzurum. On 1 and 2 January he had observed the movements of the Russian troops and on the night of 1st–2nd the men of the 31st and 30th Turkish Divisions were already beginning their retreat. The Turkish withdrawal continued all through the 2nd and the night of the 2nd–3rd. By the afternoon of the 3rd, the bulk of X Corps was already in the region of Verişan. Strong rear-guards still held Çiplakli and Upper and Lower Salut, but only two companies with machine guns were left in Çatak and Divik.

In the fog of the 3rd, Gulyga's patrols failed to discover that these two points were only occupied by a handful of men; and both villages remained in Turkish hands until nightfall. Gabayev, on the same day, by-passed Laloğlu, but his right flank was strongly counter-attacked in the fog. Realizing that the operation was badly planned, Gabayev decided to halt at Laloğlu for the night; Voronov bivouacked three miles behind him. On the morning of the 4th Fesenko's Cossacks penetrated as far as Çiplakli and even Verişan. But here they were suddenly counter-attacked by Turkish infantry and thrown back beyond Çiplakli. Both Gabayev and Fesenko came to the conclusion that instead of outflanking the Turkish left they were in touch with the main body of X Corps. The favourable moment had been missed owing to Bergmann's misinterpretation of the situation and Baratov's casual advance.

[1] Voronov had three battalions of the 263rd (Gunibski) Regiment.

On the 4th, Gabayev attacked in the direction of Upper Salut and was again counter-attacked on his right flank. By the end of the day he was in possession of both the Salut settlements, while, on the left, the action of Gulyga's brigade had been limited to the occupation of Divik. Fesenko again tried to march on Verişan and Başköy and was again driven back. A strong Turkish rear-guard recaptured Çiplakli and threatened Laloğlu. That evening, anxious for the right flank of the Russian line, Gabayev decided to act independently of Bergmann. He marched to Başköy where he stayed the night, his rear covered by Voronov's detachment at Laloğlu. He intended next day to take possession of the Emirhane ridge which, it was evident, protected the flank of the retreating Turkish forces. The suppositions of Myshlayevski had proved correct.

On the 6th, Gabayev occupied the Emirhane ridge with four battalions of Caucasian Rifles, while Baratov, approving his subordinate's plan, sent the Caucasian Cossack Division round by the north of the ridge to Beyköy. With the appearance of the Russians on the ridge, the Turks accelerated their retreat through the Eşek-meydan Pass. But Gabayev's movement had come forty-eight hours too late, and the two Turkish divisions had time to escape encirclement: the 30th marched by the paths round Beyköy towards Kosor on the Merdenik-Oltu road, while the 31st joined the 32nd at Bardiz. Baratov's Cossacks, impeded by deep snow to the north of Emirhane, failed to establish contact in time, while the Eşek-meydan Pass was only occupied on the 7th—three days later than the date intended by Myshlayevski.

The deputy commander-in-chief now became furious with Bergmann and, on the 6th, obtained the viceroy's assent to the appointment of Yudenich in succession to Bergmann as commander in the field; but the encircling operation had failed, and nothing was left to Yudenich but the pursuit of the remnants of the retreating Turkish army.

But while Bergmann had proved so inefficient in the organization of the counter-manoeuvre from Novo-Selim, Kalitin had achieved the complete liquidation of the Turkish force which had taken Ardahan. Kalitin had been informed by Armenian fugitives that Stange's troops had not even occupied the old forts located round the town. He decided to move one of the Siberian Cossack regiments round to the north and north-west of the town, with the object not only of blocking the Akhaltzikhe road but also of threatening the road to Ardanuch which was Stange's line of retreat. The three plastun battalions were to attack along the main road from Akhalkalaki, with the other Siberian Cossack regiment covering their left. He held only one regiment of Kuban Cossacks in reserve, but he was expecting support from Istomin's brigade from the south.

The dense fog of 3 January allowed the Russians to approach the outskirts of Ardahan without being observed. A smart artillery fire from mountain batteries probably impressed Stange with the strength of the force against him. When at 6 p.m. Kalitin ordered the assault, the Turks were already trying to withdraw along the Ardanuch road. The Siberian Cossacks, coming up from the east, attacked with the sabre and made havoc of the Turkish infantry columns.[1] When the plastuns and some of Istomin's men stormed the town, they took another thousand prisoners and the four mountain guns which constituted Stange's artillery. Practically the whole of the Turkish Çoruh force was dispersed, and only a remnant succeeded in making their escape across the Yalnizçam-dağ.[2] Istomin's brigade moved on to reoccupy the Yalnizçam-dağ Pass, Ardanuch and Artvin, while Kalitin's force marched south towards Merdenik and Oltu.

Kalitin reached Merdenik on 7 January—the same day as that on which the head of the long column of the Turkish 30th Division was also debouching on to the Merdenik-Oltu road. One regiment of this division was sent to cover the Agundir Pass, bringing the Turkish force at that point up to about 3000 men. The rest of the 30th Division concentrated behind the Penek defile. Meanwhile Kalitin was joined by the Caucasian Cossack Division and two battalions of the 263rd Regiment, the force which had been sent to pursue the retreating Turks from Beyköy to Kosor. During 10 and 11 January the Turks put up a gallant resistance along the valley of the lower Bardizçay, using with some skill four mountain guns which they managed to salvage from the Agundir Pass. Their resistance, however, was broken, and the 30th Division really ceased to exist as a fighting force. On the 12th, Kalitin reoccupied Oltu; he was ordered to proceed to Nariman and farther south in order to cut the retreat of the 31st and 32nd Divisions which were withdrawing from Bardiz.

There still remained about 10,000 Turks in the Bardiz area.[3] When the 17th Turkistanskis had been withdrawn from Dovgirt's force to reinforce Przevalski at Sarikamiş, the Turks had retaken Yayla Bardiz;

[1] The Turks lost 500 killed and 300 prisoners, together with the colours of the 8th Infantry Regiment.

[2] Russian losses in the Ardahan action did not exceed 200 killed and wounded.

[3] Composed of 4000 men of the 31st Division and 6000 of the 32nd.

they were only about six miles distant from advanced elements of XI Corps who had appeared round Güllü-dağ, and Yudenich had at one moment feared combined action against his extreme right flank.

The first act of Yudenich, when he became commander-in-chief in the field (6 January), was to direct his attention to the liquidation of the Turks at Bardiz. On the 8th he ordered Baratov to concentrate a striking group in the area Kizilkilise-Çermuk. From here three separate columns had to undertake an encircling movement: Colonel Rodionov, with one battalion of the 10th Caucasian Rifles and one plastun battalion and six guns, had to move from Çermuk along the Bardiz-çay valley against Bardiz village; Gabayev, with the 11th Caucasian Rifles, was to attack the area between Yayla Bardiz and Bardiz village in conjunction with the right flank of Dovgirt's detachment at Çilhoroz; and Gulyga, with two plastun battalions and four mountain guns, had to carry out a wide outflanking movement from the north, passing over wild mountain tracks by Nisunk and descending into the Bardiz valley at the point, two miles to the east of the village of Norsin, where the *top yol* comes down to the valley of the Bardiz-çay. Baratov was keeping two battalions in reserve at Çermuk. The plan was a bold one, since the three columns were composed of six battalions (4000 men) and ten guns against a force of some 10,000 Turks with twenty guns. But there was an element of surprise in the Russian move.

On 9 January, Gulyga, with the greatest difficulty, reached Nisunk. Rodionov attacked during the same evening, but was driven back with losses and had to draw on the reserve. Gabayev, marching in snow to the belt, met with resistance at Yayla Bardiz and halted. Yudenich became anxious and asked Baratov to act with the greatest energy.

On the 10th, Gabayev captured Yayla Bardiz (200 prisoners and three guns). Connexion with Dovgirt was established and six guns were brought up to bombard Bardiz village. Gulyga, in the meantime, was meeting with extraordinary difficulties and advancing literally step by step. On the night of the 10th–11th, Gulyga made a descent into the Bardiz valley and took the village of Bardiz by surprise; 2000 prisoners of the 31st Division fell into his hands. This force proved to be a rear-guard covering the retreat of the remains of the Turkish army along the *top yol*.

The last skirmishes of the Battle of Sarikamiş took place on 15–17 January, when small groups of the 32nd, 31st and even of the 30th Divisions continued to offer resistance to units of Kalitin's force, coming up the Oltu-çay, and to the 11th Caucasian Rifles, led by the tireless Gabayev along the *top yol*. Turkish losses were about 4000 (of which the majority were prisoners) in the Bardiz fighting, and they left a further 2000 prisoners and a score of guns along the Oltu-çay. Nevertheless, the 32nd and 31st Divisions escaped complete destruction, and the remnants of X Corps reached Hasankale after sixteen days of terrible retreat, 3000 men bringing with them several guns. On 18 January Kalitin entered Id and the pursuit ceased.

In the meantime Yudenich had undertaken a counter-offensive against XI Corps. On 10 January Güllü-dağ was recaptured by two battalions of the 153rd (Bakinski) Regiment, moved up to support the completely exhausted 4th Turkistan Rifle Brigade. Yudenich realized that a frontal attack would be costly, particularly since the Turkish centre lay along the famous Zivin position. He decided to dislodge the Turks by a surprise outflanking movement. Dovgirt, with his really heroic Turkistan riflemen—now reduced to 1500 men with four guns—was ordered to move from Güllü-dağ against the rear of the Zivin position.[1] Dovgirt marched for five days in very deep snow at the rate of two to three miles a day; Yudenich considered him lost. But by the 15th this extraordinary movement, perhaps the most extraordinary in all this extraordinary Sarikamiş battle, was successfully accomplished. Abdul Kerim took the appearance of Dovgirt's small column in his rear as the signal for the retreat which he had already decided to make. On 16–18 January, XI Army Corps retired to the positions which they had occupied before the beginning of the offensive of the vice-generalissimo.

Regarded in perspective, Enver's winter offensive seems to have had no real chance of success in the conditions of topography and climate in which it was undertaken. In the early stages these conditions were already operating seriously to the disadvantage of the advancing Turks. Enver began the offensive with an army about 95,000 strong.[2] The Russians only had 65,000 available in Bergmann's and Istomin's groups and among the sparse reserves which could be brought to the front. But this ratio had been radically altered by the time the Turkish deployment had brought them to the decisive field of battle at Sarikamiş; casualties, mostly from frost and desertion, had already reached the figure of 25,000. In a week, adverse natural conditions combined with inadequate preparatory work had deprived

[1] Region of Horum-dağ.

[2] IX Corps, 25,000; X Corps, 30,000; XI Corps, 35,000; Stange's force, 5000. The 37th Division has not been included in these figures since it remained inactive to the south of the Aras.

the Turks of their superiority of force; the miscalculation of Enver became obvious.

This miscalculation had been exposed before the operation began by the experienced commander of IX corps, and the vice-generalissimo had exacted his resignation. Ahmet Fevzi Paşa had also emphasized the necessity for special preparations in equipment and commissariat before undertaking the proposed movement and he had urged the reinforcement of the Third Army by at least one more army corps from Istanbul. Enver, however, had objected that the transport of one more corps from the capital would have involved certainly a month's delay and that the Russians would have had time also to observe the Turkish preparations; the element of surprise would have disappeared—and this element had a preponderant part in Enver's plan.

Strengthened by one additional corps, the Turkish Third Army might have been able to push back the Russians as far as Kars; but, to the losses involved in the outflanking march, some 25,000–30,000 losses in battle would have been added, and four Turkish army corps, 60,000 strong, would have been incapable of pressing the invasion of the Caucasus into the Georgian valleys through the Borjom gorge and the lateral affluents of the Kura. Even with four army corps there was no real chance for the success of a grand Pan-Turanian offensive during the winter months.

As it was, the offensive resulted in a complete catastrophe for the Turkish Third Army. Its losses were appalling: about 75,000 men with most of its artillery. All that remained in mid-January were 18,000 men (of whom 15,000 belonged to XI Corps), and at the turn of the month the Turks could scarcely muster, with hasty complements from depots, 30,000 men for the defence of the Erzurum region.

Russian losses were formidable enough: 16,000 killed and wounded, and 12,000 sick, the majority from frost-bite. All the regiments of the Turkistan Rifles had suffered particularly heavily; they had each an average of 600–700 men left, out of an original strength of 2500.

The only favourable opportunity for Enver had been the dangerous chance of the presence of men so temperamentally unsuited for command as Myshlayevski and Bergmann at the head of the Caucasian army. Fortunately for the Russians the presence of General Yudenich, combined with the fine fighting spirit of the officers and men whom he had trained, stopped the magnificent *élan* of the Turkish askers who followed with such admirable courage their foolish and incompetent leader. The Russians showed that they possessed excellent

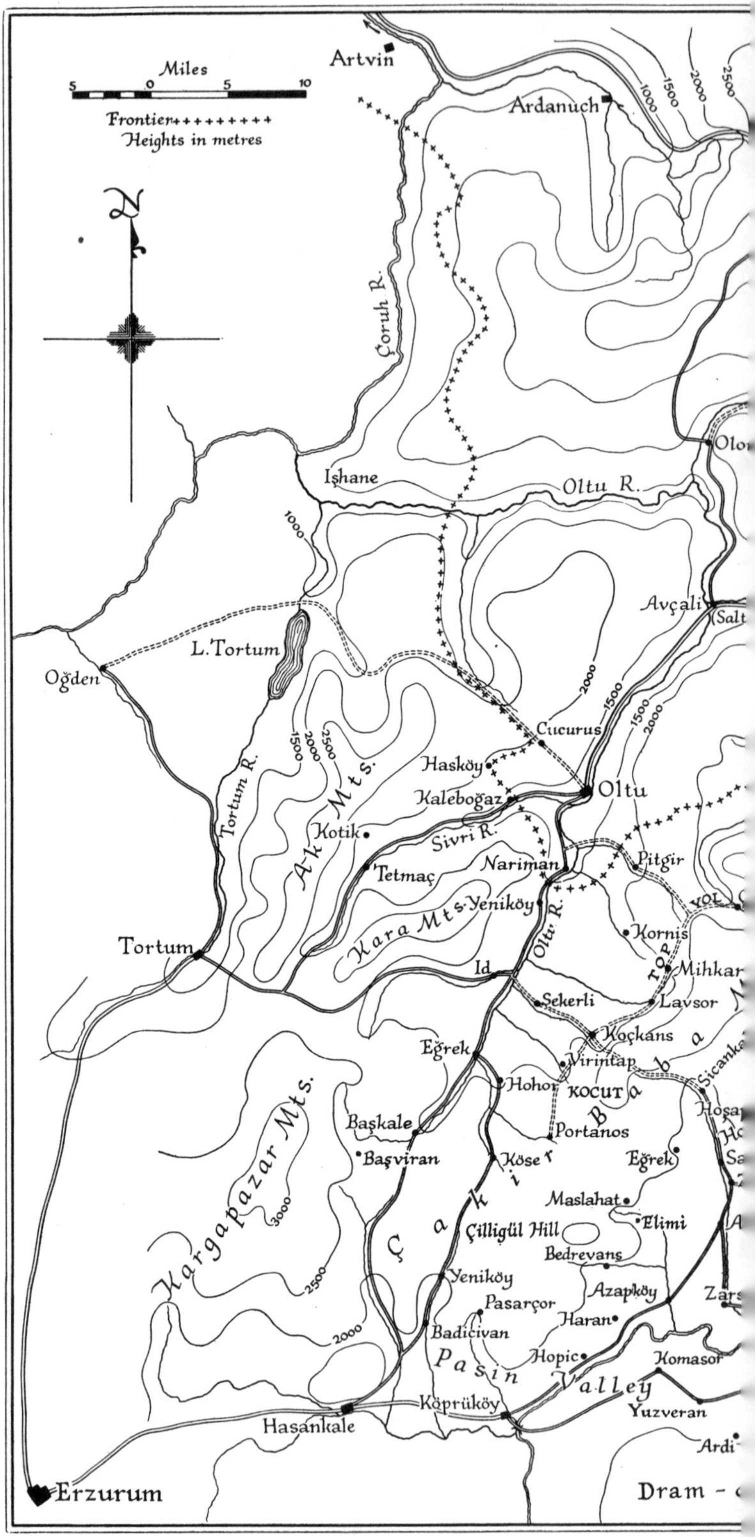

Map 19. War on the Caucasian frontier, N

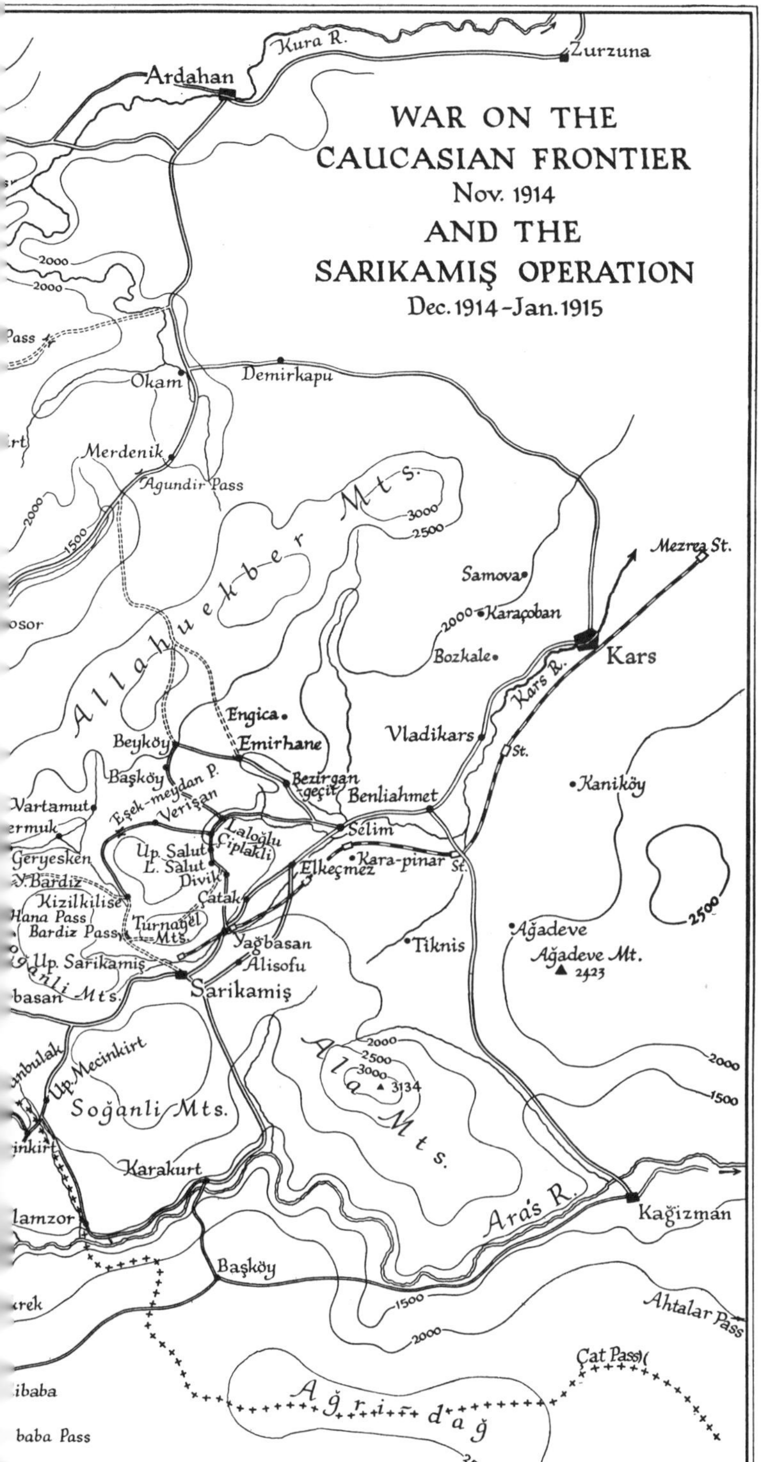

arikamiş operation, Dec. 1914–Jan. 1915

material for command among such general officers as Przevalski, Kalitin and Gabayev.[1]

The battle of Sarikamiş has an interest beyond the unusual topographical conditions in which it was fought. It was a battle of annihilation in which both Turks and Russians gave proof of incredible endurance and determination. If Alaca-dağ remains a classic feat of Russian arms, Sarikamiş must live in history as the most heroic manifestation of the spirit of the Turkish fighting man.

[1] Yudenich lost no time in appointing Kalitin to the command of the I Caucasian Army Corps and Przevalski to that of the II Turkistan Corps. The commander of the 66th Infantry Division, General Oganovski, who had ignored Myshlayevski's order to withdraw from the Eleşkirt valley, was appointed to the command of the IV Caucasian Corps, in process of formation. On the Turkish side the command of the Third Army passed to Hafiz Hakki, like Enver one of the new political soldiers of the Young Turk Revolution, but a man who had shown resolution and initiative in the most critical circumstances. Hafiz Hakki was strongly criticized by Lt.-Col. Şerif Köprülü, chief-of-staff of IX Corps at Sarikamiş, in his book *Sarikamiş Ihata Manevrasi* (The Sarikamiş Encirclement Manoeuvre), cited by Commandant Larcher in *La Guerre turque dans la guerre mondiale*, p. 391, n. 2.

CHAPTER XXII

THE CONSEQUENCES OF SARIKAMIŞ

MAPS 1, 19; cf. also 23, 38

AFTER his return to Istanbul Enver made all possible efforts to conceal the truth and to represent the catastrophe of Sarikamiş as nothing more than a local setback to the Third Army. This was not difficult in the conditions then existing in Turkey. But the arrival at the capital of a number of German officers who had taken part in the operation contributed to the revelation of what had really happened. Liman von Sanders had no illusions, and the German general staff was duly informed. It did not take the defeat of its ally very seriously. At that moment the attention of von Falkenhayn was directed to the Balkan theatre. Before pushing the Turks into active participation in the war the Germans had given a formal promise to open a direct way between the Bosporus and central Europe across Serbia and Bulgaria. These promises had been based on the knowledge that the Austro-Hungarian staff was preparing its second offensive against the Serbs. But while this offensive had failed in November–December 1914, the Turks were still insisting on a fulfilment of the German undertaking which they regarded as a condition vital to an effective Turkish war effort. The losses suffered in the Pan-Turanian offensive had to be recovered, and new Turkish armies had to be developed; reserves of men were available but arms and munitions were lacking, and those smuggled through Rumania were quite insufficient. Soon the news of British preparations for an attack on the Dardanelles made the insistent demands of the triumvirate more and more urgent.

In London the Russian victory had strengthened the position of those who favoured an attack on Turkey. The British had available powerful naval forces, excellent bases in the Mediterranean and growing numbers of Dominion and Indian troops. The Pan-Islamic offensive against Suez, so pompously announced in November, had taken very modest proportions in January–February 1915; but it was a warning all the same, and in England the necessity of assuming the strategic initiative in the Mediterranean as well as in Iraq was being increasingly appreciated.

When, somewhat impressed by the pessimism of Myshlayevski, the Russian government had asked London (3 January 1915) to divert Turkish forces from the Caucasian border by a naval or military demonstration elsewhere, British preparations had been accelerated. In a few days it became clear that the Caucasian army was scarcely in need of assistance; but the scale of the Russian victory seemed to open the way for ample action against Turkey. In the middle of January the British government determined to adopt the offensive initiative in the Straits area. France and Russia were invited to participate. Both countries appeared, however, somewhat reluctant to comply. If the French attitude were comprehensible, the position of the Russians is more difficult to explain. But it would seem that the predominance of French military doctrines[1] at Russian general headquarters, reinforced by the personal indifference of the Grand Duke Nicholas to all kinds of combined operations, and complicated by a certain political reserve on the Straits question, is the only possible explanation.[2]

Nevertheless, although sceptical with regard to an active participation of Russian naval and land forces in the operations against the Straits, Russian general headquarters was willing to make a demonstration to support the Anglo-French attack. On 19 February, Admiral Eberhardt, commanding in the Black Sea, was informed that, in the immediate future, operations by an Anglo-French naval force with an expeditionary corps were planned against the Dardanelles. The Black Sea fleet would support this action by a demonstration against the Bosporus which, in the case of the success of Russia's allies, might be developed into an occupation of the Bosporus by Russian naval forces acting in conjunction with those of her allies. In these instructions no reference was made to the participation of Russian land forces; but it was not long before Russian general headquarters appreciated the fact that if the Russian navy was to participate in the eventual occupation of the Straits area, Russian land forces must be available to share in the occupation of Constantinople. Orders were

[1] 'The way to Constantinople passes through Berlin.'

[2] For discussion of British views and the alternatives of the Dardanelles or Alexandretta, see Rt. Hon. W. S. Churchill, *The World Crisis* (1915), vol. II, chap. 5. The crisis of the Sarikamiş battle had, in fact, already passed when the Russian appeal reached London. By the end of January, the grand duke, without discouraging allied action, was careful not to commit Russian forces—particularly the Black Sea Fleet—to combined action against the Bosporus forts (see dispatch from Sir G. Buchanan to Sir E. Grey quoted by Mr Churchill, *ibid.* pp. 157–8). On 3 March the Russian foreign minister informed the British ambassador that 'the Emperor...could not in any circumstances consent to Greek co-operation in the Dardanelles'. In Mr Churchill's view, the refusal to accept the co-operation of the Greek army at this date was fatal to the ultimate success of the Dardanelles plan—'a terrible fatality'. For details, see Churchill, *ibid.* pp. 197–202.

given to press the organization of the Seventh Army in Odessa and the Crimea (at that time existing only on paper) and to prepare units for an eventual overseas expedition. This move had primarily political significance: the Russians were determined to take part in any occupation of Constantinople and the Bosporus. At the same time they did not regard as practical active operations against the Bosporus coasts. The Allies were precisely informed as to the Russian attitude; they were asked to regard the concentration of the Seventh Army in Black Sea ports as nothing more than a demonstration. And it must be said that this demonstration was not entirely without effect: the Turks were well informed of Russian concentrations; they took them quite seriously, and during March and April were anticipating Russian landing operations in the Bosporus area.

The Russians had no reserves available for the complement of the Seventh Army; so the demonstration plan had an unexpected repercussion on the Caucasian army. Yudenich had scarcely had time to form a new corps—V Caucasian (3rd Caucasian Rifle Division, 1st and 2nd Plastun Brigades)—when it was transported to the Crimea and Odessa. After some time spent in the Black Sea ports, V Caucasian Corps was moved, in May, to Galicia, where all available forces were being concentrated in the hope of stopping Mackensen's offensive; for after three months of doubts and desultory talk among the trio, Falkenhayn, Hindenburg, Conrad von Hoetzendorff, it had been decided to concentrate the maximum forces of the central powers, not against Serbia, but against the Russians in Galicia, since the situation in the Carpathians appeared more dangerous than at the Dardanelles, and Austria needed help before Turkey.

The Turkish Third Army had been reduced by the end of January to a strength of less than 20,000 men fit for the line, and reorganization could only be accomplished by drafts from the First and Second Armies.[1]

The divisions of XI Corps were first reconstituted and in February they were able to occupy the November line in the Pasin valley. X Corps was reorganized rather later, and meanwhile its units in being were employed in observation of the routes from the Oltu and Tortum regions. Hafiz Hakki was busy with this work of reorganization when, in March, he died of typhus at Erzurum.[2]

The official attitude of the vice-generalissimo was that nothing of importance had happened on the Caucasian frontier, and that the

[1] These were replaced by reservists. Thrace was teeming with men who had been called up, while Anatolian reserves were less numerous on the spot.

[2] A disease which had proved fatal to Heimann, Shelkovnikov and Loris Melikov the younger in January–February 1878.

Pan-Turanian offensive had been only temporarily deferred. Enver found a facile politico-military cover to his failure by encouraging the activity of the forces on the extreme right flank of the Third Army which had arrived too late to take part in the fatal operations against Sarikamiş. These forces were the 37th and 36th Divisions, which had reached the area of operations only in January.[1] One of these so-called Iraq Divisions (the 36th) Enver decided to move towards the Persian border in order to exploit the occupation of Tabriz by a detachment of Turkish volunteers and Kurdish irregulars, following Chernozubov's withdrawal as a result of the panic orders of Myshlayevski.

Enver chose to find in the Russian evacuation of Tabriz an important victory for the Pan-Turanian movement which might be held to offset, and indeed to outweigh, the regrettable episode at Sarikamiş. He gave the command on the borders of Persian Azerbaijan to his uncle, Halil Bey, who, in the middle of January, arrived at Diyarbekir. The 36th Division was to constitute the bulk of Halil's force, but he was given also new formations of Frontier Guards and gendarmes[2] and some half-trained Kurdish mounted units. Two new divisions (3rd *bis* and 5th *bis*, later the 51st and 52nd) were in process of formation out of regular and reservist units which were arriving from Istanbul during February and March.

The Turkish effort to create new units reached its apex during the period March–November 1915. In February the Turks had available forty infantry divisions, many of them under strength, with a total of 500,000 men. By November, despite severe losses at the Dardanelles, they had raised this strength to fifty-two divisions—total 800,000 men. It was the climax, and never again during the campaigns of 1914–16 did the Turks approach the same number of divisions and men.[3]

The Dardanelles campaign, and the Russian demonstrative concentration in the Black Sea ports, contributed strongly to this really

[1] The so-called 36th *bis* and 37th *bis* Divisions, formed during November, out of reservists, in the Murat-su valley, had lost the majority of their effectives from sickness and desertions; they were now absorbed in the regular 36th and 37th Divisions.

[2] The Frontier Guards and gendarmes, in Turkey as in Russian Caucasia, were generally formed of picked (and relatively well-paid) men with an excellent knowledge of the area in which they were operating.

[3] Distribution of Turkish divisions in March and November 1915 was as follows:

	Straits area	Caucasian front	Iraq	Syria	Arabia	Total
March	17	11	3	5	4	40
November	28	12	5	3	4	52

titanic effort, which in fact exceeded the normal capacity of the country and overreached the natural development of Turkish man-power. The German advisers readily supported this over-development of the human resources of the doomed empire. There was no lack of small arms—passed through Rumania—but the supply of artillery and shells was deficient until after the conquest of Serbia and the entry of Bulgaria into the war on the side of the central powers. All this feverish work, if it gave some appreciable results in the area of the capital and the Straits, was far from producing competent fighting forces on the more remote fronts. When the Russians again met the Turkish Third Army in the field it was far from being the same army which, in the previous winter, had gone to its devoted fate at the hands of its infatuated vice-generalissimo.

In complete contrast to the hasty, and at the same time hysterical and desultory, reorganization of the Turkish army was the systematic and very efficient work accomplished by General Yudenich and his staff. It must be said, however, that their task was much easier than that of the Turkish war leaders; for to reorganize a numerically small army in a restricted and well-prepared advance base area is quite a different proposition to the reconstitution of the armed forces of a vast, undeveloped and somewhat chaotic empire.

Yudenich had long realized the necessity of simplifying the machinery behind the Caucasian army. He definitely left Tiflis and installed at Kars his field staff of fifteen or twenty reliable officers. His first efforts were directed to filling the losses suffered during the winter battle. Since great numbers of reservists had been called up throughout Caucasia (and many had arrived from Russia) he was able to bring the divisions of I Caucasian and II Turkistan Corps up to strength by the end of February. In March, the Turkistan Corps, withdrawn to the rear for reorganization, was able to take over the Oltu region.[1]

The next step was the creation of new units. The 3rd Caucasian Rifle Division, which had done so well under Gabayev, left for European Russia together with the 1st and 2nd Plastun Brigades and the 20th Infantry Division. Yudenich thus lost forty-three battalions, but he hurried on with replacement and soon had seventeen new rifle battalions ready.[2] Ten new battalions came from the Cossack

[1] The I Caucasians had continued to hold the lines in the Pasin valley throughout the winter.

[2] The 4th Caucasian Rifle Division with four-battalion regiments; third battalions were created in the regiments of the 2nd Caucasian Rifle Brigade.

lands.[1] There remained a deficit of sixteen battalions which Yudenich partly filled by an increase in the militia formations.[2] The Cossack lands also provided numerous mounted formations of 'the third call', partly in regiments (8) and partly in single *sotni* (17). At the same time the Caucasian Cavalry Division (three dragoon regiments and one Cossack regiment) returned from the western front together with the 2nd and 3rd Transbaikal Cossack Brigades. In comparison with the 108 battalions and 117 *sotni* available in December 1914, Yudenich now had ten infantry battalions less, but the strength of his mounted arm was almost doubled. His artillery was also stronger—340 guns against an earlier 300—as new sections were created and new horsed batteries arrived.

Numerically the Caucasian army was stronger than it had been at the outbreak of hostilities, despite the fact that 50,000 men had been despatched to make up the complement of the Seventh Army. By the summer its effective field strength was 130,000 infantry and 35,000 horse. Not far short of 100,000 men had been drawn from the reservist cadres to fill losses and form new units; and 160,000 more remained available to cover future losses and to fill the rear services and the needs of internal security.

Yudenich, like other great commanders, gave the greatest attention to the welfare of his troops. Supplies, sanitary services, clothing and fuel—the last two, prime factors in the climatic conditions of the region—were greatly improved. Much work was done on roads, and the network of field telephones and telegraphs was considerably extended. Horses, mules and bullocks for transport were collected in vast numbers, and even camel transport was organized in detail.[3] A light Décauville railway was laid from Kars to Merdenik to facilitate supply to the Oltu area, and another Décauville was laid along the main road from rail-head at Sarikamiş to Karaurgan. Further, the change from narrow to normal gauge on the Archangel line allowed of the utilization of the old rails and vehicles for a line laid to unite Shahtahti station, on the Erevan-Julfa line, with the Eleşkirt valley. (Cf. Map 38.)

Yudenich and his staff worked hard to give the Caucasian army the maximum of mobility in the modern conditions of 1915; a 'railroad army' based on the newly industrialized areas of Transcaucasia

[1] These were the 4th Kuban Plastun Brigade (six battalions) and the Don Foot Brigade (four battalions).

[2] And one new Armenian *druzhina* was added to the existing four.

[3] Yudenich imported from Khurasan numbers of the hardy Bactrians with their drivers.

already stood along the ranges which covered the ultimate fastnesses of Anatolia. Yudenich had used well the months of respite afforded by the Sarikamiş victory and the British assault on the Gallipoli beaches. The general situation did not offer conditions favourable to a Russian offensive before the autumn. But already in the late spring the Caucasian army seemed strong enough to allow of limited action with a view to improving its front in preparation for the opening of a full-scale offensive.

CHAPTER XXIII

FLANK OPERATIONS ON THE CAUCASIAN FRONT, JANUARY–MARCH 1915

MAPS 1, 19; cf. 26, 23, 38

ENVER'S incursion into Russian territory had been followed by some troubles among the Muslim population of the Caucasian borderland. In the valleys of the Çoruh and the Oltu-çay there were very mixed elements: Christians (Armenians) predominated in the towns of Artvin, Ardanuch, Ardahan and Oltu, while Muslims were in the majority over the countryside; these Muslims included groups of Georgian origin, like the Laz and Acars, Turks, remnants of the old Tartar hordes, and Cherkesses who had settled, after 1864, on what was then the Turkish side of the border. Irrespective of their racial origins, all the Muslims proved more or less ready to help the Turks, particularly when they came as an invading army. Thus the Cherkesses of Upper Sarikamiş stubbornly defended their stone saklyas by the side of the Turkish askers, and the needy inhabitants of the uplands provided scanty food to the divisions of Hafiz Hakki during their desperate march across the Allahuekber Mountains.

But these 'moral' allies, if useful, could hardly justify the Pan-Turanian aspirations of the vice-generalissimo. Small scattered bands of armed men, recruited from local elements, took the field to the north of Oltu and in the district of the Posof-çay, where, in 1828, the more numerous levies of Ahmet Bey Himshiashvili had threatened Akhaltzikhe. More active were the armed bands who came from across the Turkish border: the Laz from the coastal ranges of the Pontic Alps and the wild Kurds from the Dersim who followed down the Çoruh and the Oltu-çay.

Only a few of these irregulars held their ground at Ardahan when Kalitin's Cossacks approached the town; most of them preferred to escape in time across the Yalnizçam-dağ and to leave Stange's askers to their fate. After the recapture of Ardahan, Istomin's brigade (with the three battalions of General Henike which had formed the original garrison of Ardahan) moved along the road to Artvin. The pass over the Yalnizçam-dağ was occupied after some skirmishes and Istomin slowly proceeded to Ardanuch. However, in February his brigade, together with the other two regiments of the 20th Division, was ordered to return to the rear to prepare for transport to European Russia. At the same time the three plastun battalions of Genike's brigade were included in II Turkistan Corps. In these circumstances the re-establishment of internal security in the Çoruh valley and the recapture of Artvin was entrusted to General Lyakhov, now commanding at Batum.[1]

General Lyakhov, an energetic officer known for his rather violent character, had orders at the beginning of January to clear the enemy from the Çoruh territory and to thrust them away from Batum as far as possible along the coast. Lyakhov's forces were limited to twelve battalions with eight guns and two *druzhiny* of Georgian militia.[2] The vastness of the region and its very difficult topography demanded a great effort from this slender force.

Turkish regular forces in the region were limited to about 1500 men who had survived the disaster at Ardahan, and two or three battalions of Frontier Guards. Irregular numbers, as always, fluctuated according to the situation of the moment; when Lyakhov started his operations there were 3000–4000 Acars round Artvin and Ardanuch; and about 5000 Laz, partly camouflaged by Stange as the 'Georgian Legion', were based on the port of Hopa. The Turks had only eight guns in the coastal region; and during March–April two or three regular battalions arrived there as reinforcements. Lyakhov therefore had a force of about 10,000 regulars and 5000 militia—adequate enough to deal with 10,000 of the enemy, who were mostly irregulars.

Lyakhov began an offensive movement along the coast beyond the mouth of the Çoruh during the first week of January. This region consists of a series of forest-covered bluffs with steep cliffs overhanging the sea. The bluffs are separated by deep ravines which lend themselves naturally to defence. At the same time the proximity of the sea gives a naval force the opportunity of enfilading each successive position. Admiral Eberhardt, however, refused to send naval units to Batum since he considered the port unsuitable as a base and

[1] The 1st Plastun Brigade, sent to Akhaltzikhe in January with the object of overaweing the restless Acars of the Posof-çay, was transferred to Russia, but, after Enver's defeat, quiet returned of itself to this region. In February Yudenich sent two Turkistan regiments and Cossacks to the lower Oltu-çay and the Kurds withdrew into Turkey. In March the Russians occupied the Işhane triangle between the Çoruh and the Oltu-çay while three plastun battalions covered the frontier at Olor.

[2] The 264th (Georgiyevski) Regiment; 19th Turkistan Rifle Regiment; four plastun battalions (one of the 1st Brigade and three of the 3rd); one battalion of Frontier Guards and the Tiflis and Kutaisi militia. The artillery was soon strengthened by two newly formed batteries, each of eight guns.

exposed to attack by the *Goeben* and *Breslau*.[1] It was only after 9 January, when two modern long-range 254 mm. cannon had arrived from Kronstadt, that the admiral consented to co-operate and despatched first two, and later two more, destroyers to Batum. With this small flotilla and the addition of some armed steamers on which he had mounted guns and mortars, Lyakhov undertook combined operations against the wild and forested coast round the mouth of the Çoruh. While the land forces (19th Turkistanskis and the plastun battalions) advanced over the precipitous and wooded mountains which run down to the Elysian shore of this remote land, the naval flotilla brought a flanking fire to bear upon the successive defensive positions of the Turkish and Laz riflemen. In the second half of January, the land force occupied the territory between the Çoruh and the Russian frontier. Then Makriali village was taken; and, in February, after many small engagements the Turkistanskis occupied Hopa, the small port about twenty miles from Batum which the Turks had used to supply their forces in the Çoruh valley. From Hopa, men and material could proceed by the gorge of the Içkale to Borchka on the Çoruh.

During March the land force gained another six or seven miles along the coast and reached the village of Arhavi and the river of the same name, beyond which the Turks had prepared fortified positions.

With the occupation of the Hopa-Arhavi region, General Lyakhov could regard his task along the Black Sea coast as complete for the moment. Meanwhile units of the 264th Regiment and the plastuns had cleared the district of Borchka and the Dzangul copper mines. On 27 March, Artvin was taken after a sharp fight, and the remains of the enemy's irregular bands were soon driven out of the surrounding districts. Fighting died down over the bare ridges of the Çoruh mountains and through the early-flowering forests of Lazistan, until, in the following year, a new situation on the Erzurum front opened up the possibility of further combined operations along the coast.[2]

The operations on the southern extremity of the Russian line had assumed more serious proportions. Here the panic order of Myshlayevski had involved the Russians in a very unfavourable situation.

[1] In 1910 it had been decided to disarm Batum together with some other Russian maritime fortresses, but the matter had proceeded slowly, and at the beginning of the war there were still some 6 in. guns and heavy mortars in position. On 10 December 1914, the *Goeben* had attacked Batum without much effect; cf. *supra*, p. 248, n. 2.

[2] The 19th Turkistanskis remained in occupation of the Hopa-Arhavi coastal zone; the plastun battalions and the militia held Borchka and Artvin and observed the frontier. Lyakhov lost the 264th Regiment which was incorporated in II Turkistan Corps.

General Chernozubov, taking the instructions of his superior literally, had withdrawn his troops across the Aras, evacuating Tabriz and the Lake Urmia region. The numerous Christian population of the region had fled in panic and the countryside lay at the mercy of Kurdish bands. (Cf. Map 38.)

By his disastrous order Myshlayevski had in one day destroyed all the careful work of pacification which the Russians had been pursuing for some years among the Kurdish tribes of the Turco-Persian border. The presence of detachments of Russian cavalry throughout northern Azerbaijan had not only kept the Persian Kurds in peace but had had a strong deterrent influence also on the Turkish Kurds.

The latter, with mobilization in Turkey, had been organized in four tribal (*aşiret*) cavalry divisions; but they had proved so unsatisfactory that the commander of the Third Army had decided to demobilize two of them. Generally speaking, as in the war of 1877, the Kurds had shown little enthusiasm for serious military activity. But their attitude changed when it became known that the Russians were evacuating Tabriz and the Urmia region; they saw in this temporary situation signs of Russian weakness and an opportunity for plundering the Armenian, Nestorian and Persian populations. In a few days numerous armed bands appeared all over the country round Lake Urmia: the celebrated Khan Simko, formerly friendly to the Russians, took the field; and the tribesmen were joined by Turkish volunteers (under the leadership of the fanatical Ömer Haci of Mosul) and encouraged by Turkish agents. On 14 January, Ömer Haci entered Tabriz; other irregular groups were plundering the towns round Urmia and moving towards Dilman. Even General Chernozubov was stirred to action; he sent General Nazarbekov[1] with four battalions, two Cossack regiments and eight guns to occupy Khoy, with patrols out to Dilman and Kotur.

One of Yudenich's first actions as commander-in-chief in the field was to order Chernozubov to re-establish the whole situation in Persian Azerbaijan. Leaving Nazarbekov to screen the Khoy-Dilman region, Chernozubov marched from Julfa on Tabriz.[2] During 26–28 January enemy irregulars, 4000–5000 strong, tried to resist in the Safian area but they were dispersed, mainly by artillery fire. Chernozubov entered Tabriz on 30 January.

[1] Nazarbekov was an Armenian (Nazarbekian), later commander-in-chief of the army of the Armenian republic; he was executed during the purge of 1937.

[2] His force consisted of four battalions, Antranik's Armenian *druzhina*, and two Cossack regiments with sixteen guns.

Nazarbekov, meanwhile, had not been very active in the Khoy region; he only took Dilman on 1 March. But the enemy irregulars withdrew across the frontier into the Başkale area and the situation in Persian Azerbaijan seemed to be restored.

News, however, still travelled relatively slowly in 1915, and there was room for a time-lag between events and the development of decisions based on them. Enver, as has been seen, had been impressed by the Russian withdrawal from Tabriz in January, and the raids of the Kurdish irregulars had encouraged his hopes of Pan-Turanian or Pan-Islamic action against the Russians in north-west Persia. During March, Halil Bey was very busy concentrating in Van the regular units designed for his Persian expedition. The attention of the Russians continued to be directed towards the same area.

Spring was coming and the reorganization of the Caucasian army was nearing completion. The frequent calls for drafts to Europe were a serious embarrassment to Yudenich.[1] But he knew well the defects of Turkish organization and supposed that the reorganization of the Third Army would be a slow process. The general situation in the Middle East was not unfavourable: Anglo-French forces were attacking the Dardanelles and the British were ready to extend their operations in Iraq towards Bagdad.

The strategic significance of Persian Azerbaijan had not escaped the attention of the general staff of the Caucasian army; an advance from that region towards the Tigris valley through Ruwandiz or across the Nestorian country would completely outflank all Turkish dispositions north of the Armenian Taurus.[2] The capture of Diyarbekir might

[1] Yudenich had planned to add to I Caucasian and II Turkistan Corps, IV and V Caucasian; but in March there were no units available for the constitution of V, and Oganovski's IV Corps consisted only of three regiments of the 66th Infantry Division. II Turkistan had only seven three-battalion regiments (one at Batum) and the only complete division in I Caucasian was the 39th. The general reserve was limited to the 4th Caucasian Rifle Division and some Kuban and Terek battalions. Yudenich planned, in March, to give real strength to Oganovski's IV Caucasian Corps by including in it the 2nd Caucasian Rifle Division (of three-battalion regiments) taken from Persia and to reinforce Persia with his new cavalry. He now had four Caucasian Cossack divisions: 1st (in I Caucasian corps), 2nd (in IV Corps), and the 4th and 5th in Persia (the latter developed from the Transcaspian Cossack Brigade). (The 3rd Caucasian Cossack Division and the so-called 'Wild' Division—recruited from Dağistan, Chechnia and the Tekke Turkomans of Transcaspia—was in Europe.) To these horsed divisions were added later the two Transbaikal Cossack Brigades and the Caucasian Cavalry Division.

[2] This difficult region had not proved impassable to Byzantine armies operating from Caucasian bases during the campaigns of the Emperor Heraclius (A.D. 626–7) against the Sassanian capital in Mesopotamia (Ctesiphon). For references, see *supra*, p. 8, n. 2.

threaten with collapse all the Ottoman power over Armenia, Kurdistan, Iraq and even Syria. But such vast plans demanded formidable forces based on the Tabriz region, which was an unsuitable base in view of its limited communications with Russia and Transcaucasia. An ambitious undertaking against the eastern provinces of Turkey was at any rate not to be envisaged for the year 1915. At the same time it was clearly unnecessary to attempt to outflank in depth the extreme right flank of the Turkish front when this flank was known to be only rather weakly held by the 36th and 37th Divisions. It seemed enough to Yudenich to try a direct blow by IV Caucasian Corps moving from the Eleşkirt valley towards Malazgirt and Muş, while flanking guards co-operated by advancing from Bayazit to Van and from Persian Azerbaijan on the same point. Such an operation offered two advantages: its success would shorten the extended Russian left flank; and it might promise the occupation of the important strategic points of Muş and Bitlis in the near future.

The first stage of this limited plan imposed the stabilization of the situation on the Persian border; and this seemed all the more urgent, since an enemy group 'of unknown strength' had suddenly embarked on offensive action in this area. Turkish concentrations in the region of Muş and Bitlis had been completed in March; and in April the 36th Division, led by the impatient Halil Bey, was marching round the southern shore of Lake Van in the direction of Başkale.[1] In the middle of April Halil appeared at Urmia town with an advance force of several battalions, six guns and a mob of Kurds.[2] He marched north on Dilman.

Chernozubov ordered the concentration on Dilman of the rifle battalions and Cossacks from Khoy and Julfa; but this took some time and on 29 April Halil attacked Dilman, which was held only by one Russian battalion, the Armenian volunteers and six guns. After a sharp engagement the Russian force withdrew. But within twenty-four hours General Nazarbekov had succeeded in gathering, four miles to the north of Dilman, reinforcements of six battalions and two Cossack regiments with a battery. On 1 May Halil attacked on a wide front and tried to envelop the Russians with 10,000 infantry and

[1] Halil also had with him some units of the improvised 3rd *bis* and 5th *bis* Divisions and several thousand irregular Kurdish horse.

[2] In view of their internal security duties Chernozubov's forces were rather dispersed and owing to difficulties of supply he had sent back two rifle regiments to Julfa. Three Cossack regiments were at Khoy; one infantry regiment with one battery was between Khoy and Kotur; one battalion, one Armenian *druzhina* and one battery at Dilman; two battalions, one Cossack regiment and one battery at Tabriz.

mounted Kurds. Five Turkish attacks were beaten off and Halil left nearly a thousand dead on the field. Most of the Kurds deserted and, on the night of 1–2 May, Halil anticipated a Russian counter-attack by retiring across the frontier. Nazarbekov did not pursue the initiative and seemed satisfied with the reoccupation of Dilman.

Before the incursion of Halil Bey into Azerbaijan, Yudenich had taken the first measures for a concentration of troops on his left flank, in accordance with his plan for an advance towards the basin of Lake Van and the valley of the Murat-su. Meantime the savage raids of the Kurds had put the whole country between the two great lakes into a state of anarchy. The Armenians and the Christians of the Nestorian sect were the principal sufferers; at last the exasperated population of Van, where the Armenian nationalist movement (*Dashnaksutzion*) had always been strong, rose in revolt, and after fierce street fighting took possession of the ancient city by the lake (13–14 April). The Gendarme Division (formed out of militarized police battalions) under Kâzim Bey, reinforced by several regular battalions rushed up from Bitlis, held the insurgent population under siege.

Well informed by Armenian agents of the situation at Van, Yudenich determined to take advantage of it. Since March four of the Armenian *druzhiny* had been concentrated at Bayazit.[1] Yudenich added to this force of sturdy nationalist volunteers the 2nd Transbaikal Cossack Brigade of General Trukhin, and he ordered this officer to cross the Teperiz Pass over the Ala-dağ and advance on Beğrikale and Van. Trukhin's detachment represented the extreme left column of the advancing front of IV Caucasian Army Corps. Yudenich planned this movement for the month of May, by which time he considered that the regrouping of his forces would have been completed and Persian Azerbaijan purged of irregular elements.[2]

The Caucasian Cavalry Division and the 3rd Transbaikal Cossack Brigade were also sent by rail to Tabriz. Yudenich planned to use this imposing rally of horsemen to overawe the Kurdish districts.[3] Yudenich felt that the spectacular movement of this cavalry mass would have a moral effect on the Kurds which could not be obtained by the more prolonged (and more costly) internal security operations

[1] The fifth *druzhina* was with Nazarbekov, and the sixth was still in training.

[2] Yudenich calculated that the pacification of Azerbaijan would allow of the 2nd Caucasian Rifle Brigade being moved to reinforce IV Caucasian Army Corps. The 4th Plastun Brigade was moved to Azerbaijan in April to replace the more seasoned Rifle Brigade.

[3] They represented thirty-six squadrons and *sotni* with twelve mountain and eighteen horsed guns and eighteen machine guns, a total of 6000 mounted effectives.

of small columns of infantry and Cossacks. He proved right; for, as General Charpentier led his splendid dragoon regiments and clouds of Cossacks from Tabriz to Dilman and from Dilman to Suj-bulak, Basil Nikitin, the Russian vice-consul at Urmia, was able to report that the passage of Russian cavalry had re-established Russian prestige in Persia: the movement of horse made an enormous impression; in May the Russians became the complete masters of the situation in Persian Azerbaijan.

Meanwhile the Russians moved somewhat slowly to the relief of the Armenian insurgents in Van. Trukhin did not reach Beğrikale until 24 May. Nazarbekov, who from Dilman had been ordered to support Trukhin, was at Başkale on 7 May. A few Turkish patrols were observed on Çuha-gedik leading to Hoşap; but Nazarbekov did not press a further advance until 25 May.[1] On the 29th he reached Hoşap where he still found no enemy, because Halil, deserted by his Kurds and uneasy at the advance of Trukhin's column, accelerated the withdrawal of the 36th Division. He by-passed Van, moving towards Bitlis—whither he was soon followed by Kâzim Bey's gendarme battalions. The siege of Van was thus hastily raised, and on 31 May Trukhin entered the city where he was enthusiastically acclaimed by the Armenian inhabitants. On the same day Trukhin's patrols came in contact with the advance-guard of Nazarbekov between Van and Hoşap.[2]

Yudenich was very displeased with the leisurely operations of General Nazarbekov, which had allowed the 36th Division of Halil Bey to escape without a fight. As a result Nazarbekov's command was broken up; Yudenich decided to leave the Caucasian rifle regiments to Chernozubov for the maintenance of order in Azerbaijan, while Nazarbekov's Rifle Brigade (with the exception of the 8th Regiment which remained with Chernozubov) and also the Plastun Brigade were now embodied in IV Caucasian Army Corps, and in June these units were ordered to Erciş round the northern shore of Lake Van, with a view to a later advance on Malazgirt.

Meantime, Charpentier's cavalry, having concentrated at Dilman where men and horses were rested, marched by Lake Van to Adilcevaz (26 June). The mounted units had covered more than 500 miles in the

[1] Nazarbekov at this time had under his command the 6th and 7th Caucasian Rifle Regiments, the 4th Plastun Brigade, two Cossack regiments, one Armenian *druzhina* and twelve guns.

[2] The Russians penetrated into the mountains of the Nestorians (Koçanis-Çölemerik) and brought back the patriarch, Mar Shimun, who was offered a safe residence at Khoy.

saddle, but they joined IV corps in very good condition for they had had the advantage of the late spring grazing everywhere along their route.[1]

Trukhin and his Armenians and Cossacks were engaged during June in clearing the Turks from the southern shores of Lake Van. The Armenian *druzhiny*, under such popular leaders as Antranik, Hamazasp and Dro, fought with great *élan*, and drove the Turkish gendarmerie units from Vastan on the lake and from the two villages of Çatak and Müküs on the road to Siirt. On 20 June the Armenians took Sevan, on the 25th, Sorp—both villages on the shore of the lake. The Turks, however, anxious for the security of the important town of Bitlis, brought up the bulk of the 36th Division and the Armenians retired to the line Norkev-Müküs, which they were able to hold with the help of the Transbaikal Cossacks.

[1] On 25–26 May the cavalry had had some difficulty in crossing the Jagata River, swollen by the spring rains; and at Suj-bulak, where the Russian consul had been decapitated, they had had a skirmish with the Kurds. At Ushne there was some resistance from Turkish Frontier Guards and gendarmes, but on 2 June the pacification of northern Azerbaijan was completed by the occupation of Urmia.

CHAPTER XXIV

ADVANCE OF IV CAUCASIAN ARMY CORPS AND COUNTER-OFFENSIVE OF THE TURKISH THIRD ARMY. BATTLE OF MALAZGIRT, JULY 1915

MAPS 1, 20, 21, 22; cf. 38

THE failure of Halil Bey's effort to penetrate into Persian Azerbaijan and raise there a strong Pan-Turanian (or Pan-Islamic) movement; the insurrection of the Armenians and Nestorians of the Van *vilayet*; the presence of strong Russian cavalry formations in the country between the two lakes; and the manoeuvres of General Oganovski's troops in the Murat valley—all these factors tended to make Turkish general headquarters anxious for the Bitlis-Muş region, the successful occupation of which by the enemy might outflank the stronghold of Erzurum and open the way into the interior of Anatolia.[1]

The Turks and their German advisers reacted to the danger with considerable ingenuity and skill. Halil Bey's group which had abandoned the Van region was still to the south of the lake, ready to defend the approaches to Bitlis.[2] Part of the two *bis* divisions were being concentrated at Muş, and the region of Kop was occupied by the 37th Division. Soon the 2nd (Regular) Cavalry Division arrived at the latter place from the Pasin valley and began to patrol the area between the Murat-su and Lake Van. The effectives of the Muş group were below strength, and in May there were probably rather less than 40,000 infantry and horse on the extended Turkish right

[1] Pomiankowski (*Zusammenbruch*, p. 147) states that the Armenian rising in Van *vilayet* 'influenced decisively' the failure of Halil Paşa's offensive in Azerbaijan. The rising held the Turkish gendarmerie division round Van at the critical moment when Halil was checked at Dilman. 'The connexion between the Russian offensive in April and the Armenian and Nestorian rising is manifest and it can be stated without exaggeration that the Armenian rising wrecked Enver's "Pan-Turanian offensive" in Persia.'

[2] Halil had under his command the 36th Division; elements of 3 *bis* and 5 *bis* Divisions; the Gendarme Division, Frontier Guard units and Kurd irregular cavalry. The two *bis* divisions were formed of excellent material, since they had been improvised out of available regular battalions and even regiments from the First and Second Armies; they were soon renamed the 51st and 52nd Infantry Divisions and, with the 36th and 37th, were included in the Third Army. (The two latter and older divisions, recruited in Iraq, were less reliable.)

flank. The commander of the Third Army and his chief of staff (the German Major Guse) determined to strengthen this flank and in fact to move the whole centre of gravity of the Third Army towards the south-eastern side.

The re-establishment of the three army corps decimated at Sarikamiş had proceeded fairly well during the months of February and March. XI Corps in the Pasin valley had been the first to be reorganized and brought up to strength. In April X Corps was strong enough to man fortified lines in the valley of the Tortum-çay and to occupy the highlands at the sources of the Sivri and Oltu rivers. The reconstruction of IX Corps took longer, but in May the 17th, 28th and 29th Divisions were being reconstituted and trained in the Erzurum region. The Turkish command decided to move these three divisions of IX Corps to Hinis where they were to complete their training and to be available on the flank of any advance of IV Caucasian Corps. It was a skilful (and secret) move which, as late as the middle of June, remained unknown to the staffs of Generals Yudenich and Oganovski.

When Oganovski's offensive developed, the Turks prepared to maintain an active defence of the Muş plain and to use the divisions of IX Corps to strike as strongly as possible against the advancing IV Caucasian Corps which appeared to be weak in infantry. The credit for the plan probably belonged to Major Guse (supported by Halil Bey), and Enver, as usual dreaming of a big offensive, gave his enthusiastic approval. A special active group of the Third Army was formed under the command of Abdul Kerim Paşa (officer commanding XI Corps) with Major Guse as his chief of staff: it included the 36th and 37th, 51st and 52nd, and 17th, 28th and 29th Infantry Divisions, the 2nd Cavalry Division and numerous auxiliary and irregular formations. The 17th and 28th Divisions of IX Corps were moved to the valley of the Murat; and Abdul Kerim, proceeding to Muş, took with him even one of the divisions of XI Corps from general reserve on the upper Aras. The Russians still had no knowledge that, instead of three or four enemy divisions (two of which were believed to be incomplete) they now had in front of them eight divisions with a total strength of more than 70,000 men.

This was the fundamental miscalculation of the Russian general staff, and the fault lay with Yudenich's intelligence officers, who had failed to utilize adequately the facilities for information which were available from among the numerous (and pro-Russian) Armenian elements of the area of action. The absence of IX Corps from the

Erzurum area was interpreted by the Russian intelligence as evidence that that corps had not in fact been reconstituted. At the same time Oganovski failed to use the mass of cavalry at his disposal for obtaining by adequate reconnaissance the vital information which might have saved him from disaster.

Oganovski's troops of the 66th Infantry Division and the 2nd Caucasian Cossack Division had passed the winter in the region of the Kiliç-gedik[1] (Felç) Pass and the Hamur defile where they covered the approaches from the Eleşkirt plain to the valley of the Murat. Along this most important line of the proposed Russian advance only mounted patrols and advance parties of the 37th Turkish Division had been observed. At the beginning of May the Russians occupied Tutak without encountering any resistance. Oganovski now saw an easy possibility of taking possession of the important town of Malazgirt, which he entered after a skirmish on 17 May. At the same time he pushed his left forward to Adilcevaz on the shore of Lake Van. The Turks meantime were concentrating at the town of Kop (Bulanik) and on the hills behind that town. Oganovski decided to continue his advance as soon as Charpentier's cavalry arrived at Adilcevaz where it was expected by the middle of June. However, by that date Charpentier was still at Dilman, and the slow-moving Nazarbekov, with the three regiments of the 2nd Caucasian Rifle Brigade, was in the area Başkale-Hoşap. The 4th Plastun Brigade was the only unit which was actually on the march to reinforce IV Caucasian Corps.

In the second half of June Oganovski still felt so confident in his own strength[2] that he attacked and took the town of Kop, which was held by three Turkish battalions. But with all his optimism he realized that it would be no easy task to storm the fortified hill positions beyond Kop.

These positions were naturally well adapted to defensive action. To the west of Kop and three or four miles from the stream of the Murat begin the Belican hills, about 900 m. in height.[3] This ridge, lying across the ways from Malazgirt to the plain of Muş, runs a distance of about fifteen miles to the south and comes to an end west of Lake Nazik. Between Lake Nazik and Lake Van there is a gap of low-lying country, where the land slopes gradually to the plain of

[1] T., 'Gap of the sword'.

[2] Three infantry regiments and Cossacks between Malazgirt and Lake Van and two Cossack regiments on the eastern side of the Murat.

[3] The plain of Malazgirt itself has an altitude of about 760 m. Malazgirt was the scene of the decisive defeat of the Byzantines by the Turks in 1071.

Ahlat.[1] The defensive advantages of the Belican positions were obvious, since behind them lay one of the rare bridges (the Şeyhin) connecting both banks of the Murat. Their length, however, required at least three divisions for serious defence. By 1 July the Turks probably had in the Kop-Ahlat area only the 37th Division with one of the *bis* divisions and the horse of the 2nd Cavalry Division. The gendarme division and the other *bis* division were in the region Bitlis-Tatvan-Sevan, while IX Corps was somewhere between Hinis and Muş.

Oganovski ordered Charpentier, on arrival at Adilcevaz, to march on Ahlat and secure control of this important gap. After that step had been taken, the most promising manoeuvre was clearly the out-flanking of the Belican positions by massed cavalry thrown through the Ahlat gap, to be combined with a simultaneous attack by Trukhin's group (reinforced) against the Tatvan-Bitlis line. There is little doubt that the penetration of a strong cavalry group through the Ahlat gap and the capture of Tatvan would have forced the Turks to abandon the Belican positions; and the way to Muş might have been left open to the Russians.

It seems, however, that General Oganovski had in view no such comprehensive operation as the conquest of the upland basin of Muş and the Bitlis valley. The advance of IV Caucasian Army Corps had been designed by general headquarters of the Caucasian army as a limited operation to shorten the Russian front by attaining the line Kop-Ahlat. No action against Muş and Bitlis had been planned for the summer of 1915. But this minimum programme of the attainment of the Kop-Ahlat line involved the capture of the Belican positions, since without them the Russian possession of Malazgirt remained insecure. Concerned with this minimum programme and without any wider plan, General Oganovski considered an elaborate manoeuvre unnecessary. At the same time he supposed the Turkish force on the Belican positions to be not stronger than one division—and this, not one of the best—and he relied on the superior quality of the excellent regiments of his 66th Infantry Division. He took the decision to storm the Belican positions so soon as his first infantry reserve, the

[1] The site of an important medieval city and a frequent battlefield in the perennial wars between the Armenians, the Byzantines and the Selcuk Turks. See A. Şerif, *Ahlat Kitabeleri* (The Inscriptions of Ahlat), Istanbul, 1932; and art. 'Ahlat' in *I.A.* Lynch, *Armenia: Travels and Studies*, II, pp. 280–97, gives a detailed description but fails to understand the significance of the Muslim period in the city's history. Lynch remains the essential authority on the topography of the area between the Aras and Lake Van. But see also arts. 'Ağri' in *Türkiye Kilavuzu* (Guide to Turkey), I, Ankara, 1946; and in *Inönü Ansiklopedisi*.

4th Plastun Brigade, should have arrived. As for the mass of Russian cavalry which might have been used decisively, General Charpentier was repeating bitter complaints about the shortage of feed in the dry summer heat of the Armenian uplands, and he protested that the massing of one and a half horsed divisions in the narrow space of the Ahlat gap would make the forage question particularly acute.

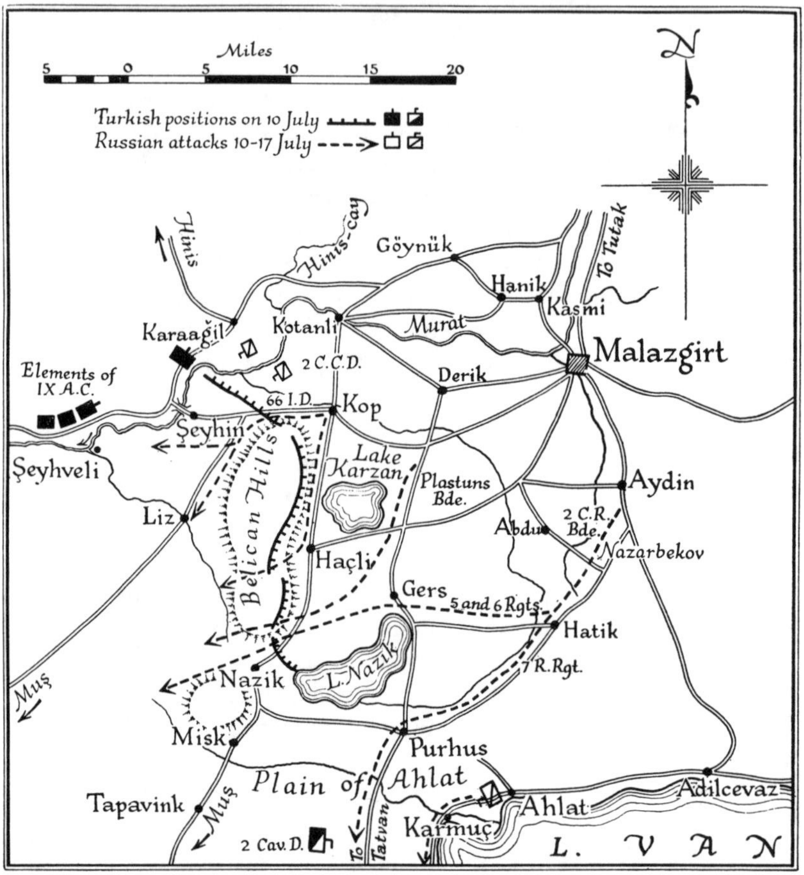

Map 20. Malazgirt: first phase, 10–16 July 1915

Thus after ten days' further delay—which permitted the two Turkish divisions to improve their entrenchments on the Belican heights—Oganovski moved his infantry to the assault (10 July).

The 262nd (Groznenski) Regiment moved out from Kop town, with the 261st (Akhulginski) Regiment advancing on their left round the little Lake Karzan; and two plastun battalions of the Terek were directed against the heights between Lakes Karzan and Nazik. The 259th (Gunibski) was held in reserve behind the centre. The fighting continued from 12 to 15 July in the face of a stubborn Turkish resistance. Already, on the second day, Oganovski had found it necessary to begin the dispersion of his cavalry mass by detaching the Transbaikal Cossack Brigade and the Cossack regiment included in the Dragoon Division to strengthen his right flank.[1]

On 16 July the infantry of the 66th Division at last succeeded in capturing the heights between the Murat and Lake Karzan. But strong Turkish reinforcements were observed moving along both banks of the river; and along the west bank the Cossacks were pressed back near Şeyhin Köprü (Sheik's Bridge).

Good news came only from General Trukhin who had passed to the offensive and retaken the village of Sevan on 30 June. His advance to the west continued and on 17 July Oganovski ordered Charpentier with two of the dragoon regiments[2] to capture Tatvan at the south-west corner of Lake Van, and to establish liaison with Trukhin to the south of the lake. Tatvan was easily captured by the dragoons, and in a few days they were joined there by the Chita Cossacks and two of Trukhin's Armenian *druzhiny*. The 7th Caucasian Rifles also followed Charpentier from Purhus to Tatvan.

Involved by his successes in serious fighting along a wide front from the Murat to the approaches to Bitlis, Oganovski lost grip of the operations in which his troops were engaged. The initial aim of his planned manoeuvre had been overreached; he had taken not only the Belican heights controlling the approaches to Kop and Malazgirt, but Tatvan as well. However, the fighting continued and in some places the reinforced Turks began to counter-attack. On 17 July, Oganovski ordered his main force[3] to pursue their advance westward. On the 20th a kind of encounter battle began near the village of Liz, while, along the western bank of the Murat, the Caucasian Cossack Division was pressed back rather dangerously by Turkish infantry attacks. Another dangerous spot developed on the left flank of the main force where the four Kuban plastun battalions were ordered to maintain

[1] Marching up from Adilcevaz, the 2nd Caucasian Rifle Brigade began to arrive at Abdu on the 11th; Oganovski used them on his left, moving the 7th Regiment to Purhus, south of Lake Nazik, and sending the 5th and 6th Regiments to reinforce the Terek plastun battalions between Lakes Nazik and Karzan.

[2] The third regiment, the Tverski Dragoons, had already been detached to the Kop sector.

[3] The three regiments of the 66th Infantry Division; the Tverski Dragoons; two regiments of Cossacks and the two Terek plastun battalions; with the 6th Caucasian Rifles in reserve.

liaison with General Nazarbekov, who was advancing with only one regiment (5th Caucasian Rifles) beyond Lake Nazik in a general direction to the south-west. Thus between 18 and 20 July the Russians were moving in two divergent directions, west and south-west, while the Turks were massing their infantry divisions on both flanks of the main force. Abdul Kerim had decided to pay no particular attention to the Russian forces between Lakes Nazik and Van and to concentrate his superior strength against the Russian group in the Kop region. Thus Oganovski's main force was attacked simultaneously by four divisions in front and on both flanks.

On 22 July the fighting became particularly severe, and on their right the Russians were definitely outflanked. Oganovski had no reserves at hand except the Don Foot Brigade which arrived on the night of 22–23 July. This brigade had had a defective training and had never before been in the line. Early on the morning of the 23rd, the Cossack infantry, attacked near the river by the Turks, fled in panic and laid open the way to Kop to the pursuing enemy. The retreat of Oganovski's main force over the hills towards the east became inevitable; and on the same day, with the 6th Caucasian Rifles, the Tverski Dragoons and the mounted Cossacks in rear-guard, the regiments of the 66th Division, having already lost 40% of their strength in the last fortnight's fighting, began their withdrawal towards Malazgirt.

In the meantime, three of the four Kuban plastun battalions, left without definite orders, had joined General Nazarbekov to the west of Lake Nazik, and only the remaining one battalion now stood in the way of the Turkish divisions (51st and 52nd) advancing between Lakes Karzan and Nazik. Strangely enough, General Nazarbekov, with six battalions of rifles and plastuns, simply remained in distant observation of the gallant fight which the one Kubanian battalion put up against the advancing Turkish division. Nazarbekov was anxious for the future of Charpentier's group at Tatvan, so he rounded Lake Nazik and retreated by its southern shore to Purhus. Thus a large gap was opened in the Russian front to the north of Lake Nazik, and the Turks, after smashing the gallant 22nd Plastun Battalion, were in a position to advance on Malazgirt in an effort to reach it before Oganovski's main force.

By the 23rd Oganovski had comprehended at last the gravity of the situation in which he was involved. He ordered Charpentier to move immediately with his two dragoon regiments to Malazgirt. Charpentier, from Tatvan, approached Malazgirt on the 25th at the moment when the Turks were approaching the town from the west, south-west and south, being with difficulty delayed by the Russian rear-guard in front of them. Charpentier decided that he was already too late; he suddenly turned back and retreated to Adilcevaz where he arrived on the night of the 25–26th.

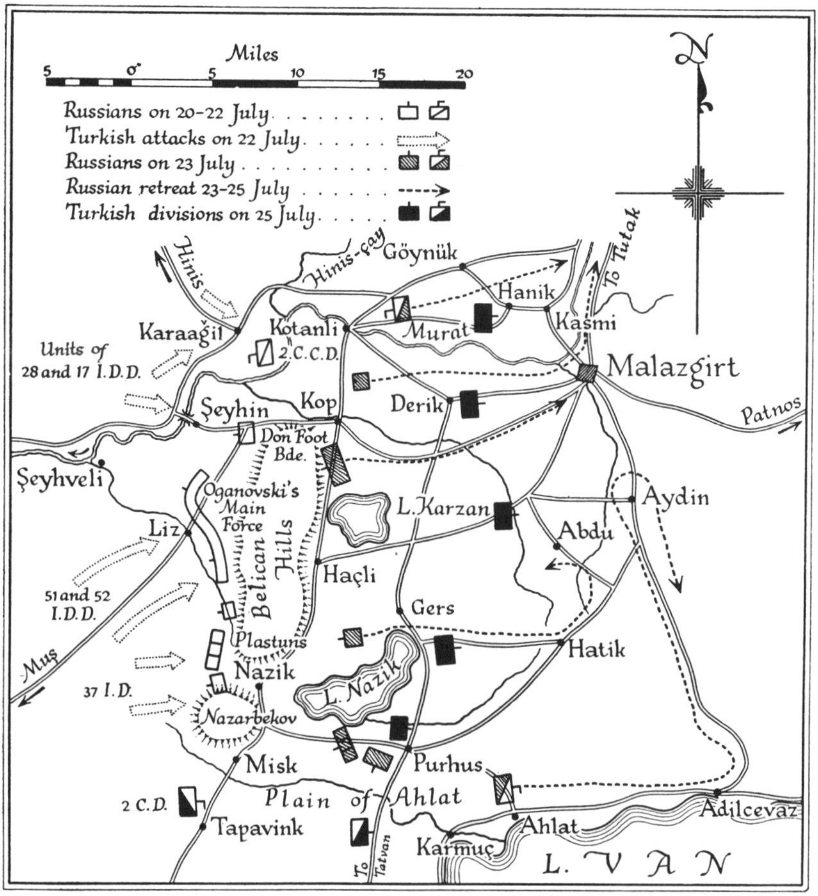

Map 21. Malazgirt: second phase, 20–26 July 1915

Nazarbekov, with the 5th Caucasian Rifles and three plastun battalions, had remained inactive at Purhus all through the 24th; next day he also retired to Adilcevaz. It was a moment of complete disorganization in the Russian command. On the left flank the 22nd Plastun Battalion alone resisted to the north of Lake Nazik and made a fighting retreat first to Abdu and then on Malazgirt. By contrast to this, Nazarbekov's troops retired in order, neither molested nor followed by the enemy. On the southern side of Lake Van, informed that Tatvan had been evacuated by Charpentier, Trukhin, with two Armenian *druzhiny*, the Nerchinsk Cossacks and one horse battery, retired to the east along the southern shore of the lake. He was not followed by the Turks; on 26 July he reached Sevan and, three days later, Vastan and Van. Van was evacuated on 4 August by Nikolayev's occupying detachment and Trukhin's column,[1] which withdrew across the Persian border to Dilman.

The whole of the Russian left flank had thus made good its own retreat, leaving Oganovski, with his centre and right, already badly mauled in a fortnight's fighting, to meet the full force of the Turkish attack on Malazgirt. Round Malazgirt Oganovski had now only fifteen battalions, and one regiment of dragoons and a few Cossack mounted units,[2] a total of perhaps 15,000 men to confront at least five Turkish infantry divisions and one cavalry division, the majority of whom were fresh troops. The obvious confusion in the command depressed even the best units and disorder spread to the rear where the Don Foot battalions were continuing their headlong flight.

On 22 July Yudenich was informed of the seriousness of the situation by his quartermaster-general, General Tomilov, whom he had sent to Malazgirt to report. General Oganovski's chief of staff, General Ryabinkin, was immediately replaced by General Mdivani and, on the 24th, Tomilov met at Eleşkirt Mdivani on his way to Malazgirt. But the battle of Malazgirt was already lost and it was too late to save the town, which was evacuated on the 26th. Tomilov and Mdivani tried to organize resistance in the Kiliç-gedik Pass and the region of Hamur. But the morale of the troops was getting worse during every day of the retreat; all the supply trains of the 66th Division were lost (although the artillery was saved) and the roads were congested by thousands of Armenian fugitives with their carts and cattle. On the 31st, Tomilov reported that there was no chance of checking the retreat of IV Caucasian Corps; the battle of Malazgirt had taken on the aspect of a serious defeat. By 2 August the Russians were in full retreat across the plain of Eleşkirt and the Turks were penetrating into it along a wide front.

[1] The Chita Cossacks joined them at Van, but the 1st and 2nd Armenian *druzhiny* continued the retreat with Nazarbekov's column.

[2] Nine battalions of the 66th Infantry Division; two Terek plastun battalions; one Kuban plastun battalion and three battalions of the 6th Caucasian Rifles.

CHAPTER XXV

ABDUL KERIM'S INVASION OF THE ELEŞKIRT VALLEY AND YUDENICH'S COUNTER-MANOEUVRE, AUGUST 1915

MAPS 1, 22; cf. also 13

THE news of the Malazgirt victory made a great impression in Istanbul. The situation in the Gallipoli peninsula had been tense during the months of June and July. The Turkish counter-offensive of May had failed, and though the Allies made little progress they had persisted in their attacks during the two following months and the losses of the Turks were growing: 58,000 from mid-May to mid-July. The Malazgirt victory raised a certain optimism in Turkish governing circles and the vice-generalissimo decided that his abortive plan for a great Pan-Turanian offensive should be revived. Perhaps the commander of the Third Army, Mahmut Kâmil Paşa, who had the reputation of being a shrewd and cautious man, was able to understand the limit of the possibilities opened up by the counter-offensive in the valley of the Murat. But the right flank group of the Third Army, united under the command of Abdul Kerim, was virtually independent of general headquarters at Erzurum and was controlled under direct orders from the War Office in Istanbul. Thus Abdul Kerim, somewhat infatuated with his victory and strongly supported by Enver, was ready not only to occupy the Eleşkirt valley but to pursue the Russians in their retreat to the Aras.

This was not, however, an easy task. In all the operational region between the shores of Lake Van on the south, Bingöl-dağ on the west and Şarian-dağ on the east, Abdul Kerim disposed of seven infantry divisions, or nine if the gendarme battalions and one division of XI Corps on the march to Hinis are included. It was a considerable force, even with divisions at a real strength of only 6000–7000. But Abdul Kerim, in planning an advance across the Ağri-dağ and an invasion of Russian territory beyond the Aras, had to secure his communications in the valley of the Murat and across the plain of Eleşkirt. He had also to cover the Van region and the ways to Hinis across the Şarian-dağ, notably the track from Tahir (in Russian hands) which crossed by the Mirgemir Pass and descended to Reşü whence it forked to Hinis and to the Murat valley.

It seems that the 36th Division with some gendarme battalions and irregular formations were left to defend the approaches to Bitlis along the northern and southern shores of Lake Van, while one division of IX Corps (29th) was moved up to the Mirgemir Pass, to be replaced at Hinis by the arriving division of XI Corps. With the five remaining infantry divisions[1] and the 2nd Cavalry Division, supported by Kurdish horse, Abdul Kerim was prepared to cover his new advance base in the Eleşkirt valley and at the same time to develop his offensive across the Ağri-dağ. Since Russian reserves were in a position to march simultaneously on Karakilise from Tahir (from the west) and from Diyadin (from the east), Abdul Kerim clearly could not move north across the Ahtalar Pass more than three infantry divisions—an insufficient strength for a big offensive, even when Oganovski's troops had suffered severe losses and had been shaken in morale by the Malazgirt battle and the long retreat which followed. The Russians could easily send 10,000–15,000 fresh troops to reinforce Oganovski and to organize a defensive line on the Ağri-dağ and along the Aras in numbers at least equal to those which Abdul Kerim could move across the Ahtalar.

It is probable that all these considerations very soon became clear to Abdul Kerim, for his offensive *élan* quickly slowed down. Karakilise was occupied by the Turks on 2 August. Next day Turkish infantry and horse followed the retreating Russian columns along the road from Karakilise to Upper Darabi, while a Turkish division moved towards the town of Eleşkirt. On 4 August Turkish advance units approached the Ahtalar Pass where Oganovski was prepared to stand on a defensive position. The Turks, however, did not attack and seemed preoccupied by the situation on both their flanks. Only small parties of Kurds penetrated across the frontier by paths in the region of Lake Balik.

Informed by Oganovski of the cautious attitude of the Turks during the 4th, Yudenich considered it possible that Abdul Kerim might be intending to take up defensive positions along the Şarian-dağ and in the defile of the Murat. He therefore gave the order to begin a counter-manoeuvre which he had been preparing since 30 July: he planned not only to defeat the Turks but to cut the retreat of all their forces which had entered the Eleşkirt plain.

Since the day on which Malazgirt had been evacuated (26 July) Yudenich had recognized that the badly dispersed forces of Oganovski would be compelled to retreat across the Eleşkirt plain. Oganovski

[1] The 37th, 51st (ex-3rd *bis*), 52nd (ex-5th *bis*), 28th and 17th.

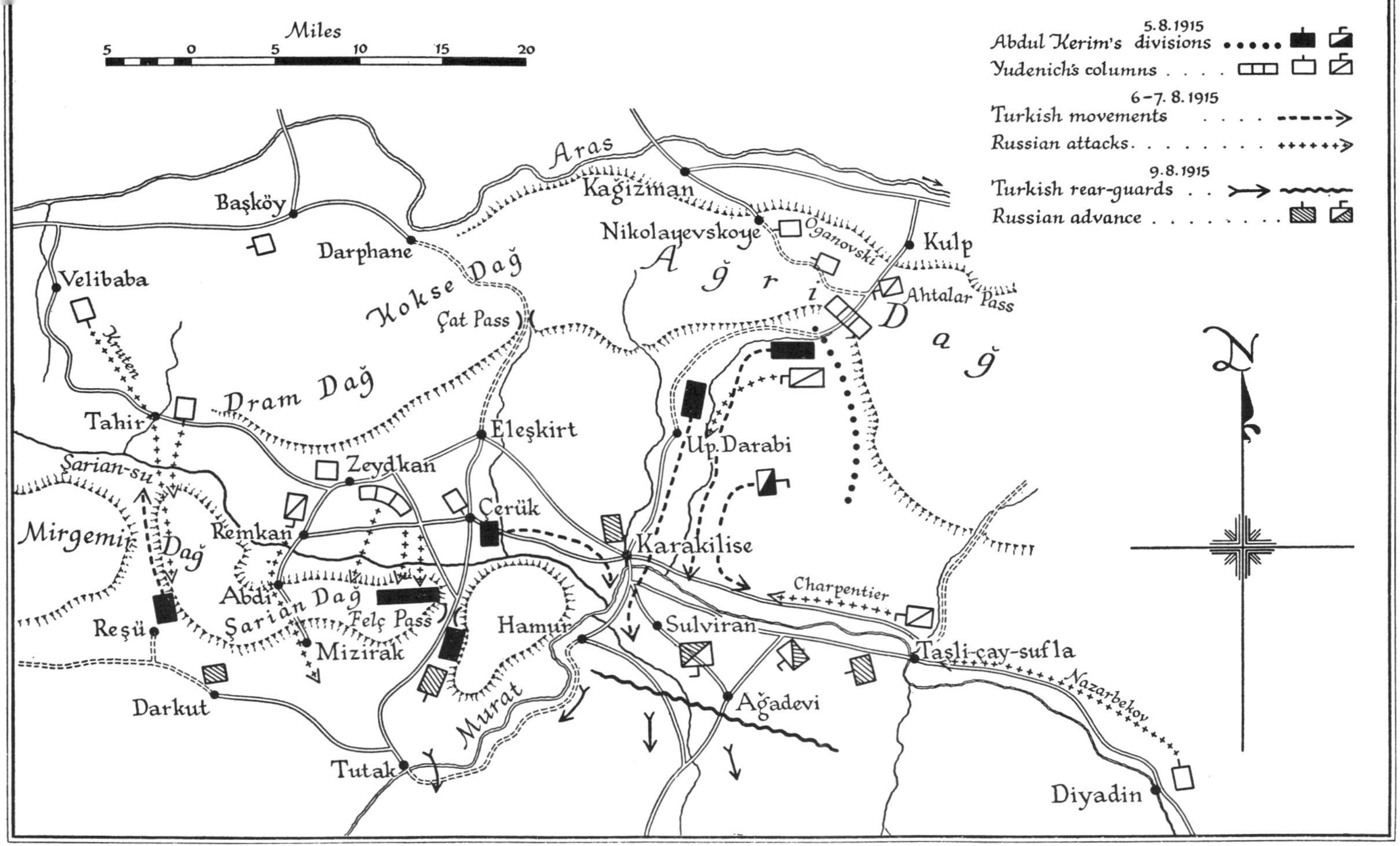

Map 22. Operations in the Eleşkirt valley, August 1915

was sending repeated requests for reinforcements but, as events developed, the commander-in-chief showed himself less and less inclined to send his unfortunate subordinate driblets of reinforcements which might simply become involved in the withdrawal and the demoralization of the retreating troops. He preferred to prepare an independent counter-manoeuvre, and concentrated a strong strategic reserve at Tahir. By the 30th, Yudenich had in this area twenty battalions of infantry, thirty-six *sotni*, and thirty-six mounted and horsed guns, all under command of General Baratov.[1]

The essentials of the counter-manoeuvre planned were kept a complete secret, communicated only to General Baratov and his chief of staff, Colonel Ern. Oganovski was not told a word about Yudenich's plan, and his requests for reinforcements were neatly refused. The commander of IV Caucasian Corps was asked only to fight a delaying action on the Ahtalar Pass, and he was told that if he found himself compelled to retreat towards the Aras he could rely upon the commander-in-chief to take some necessary measures. At the same time Yudenich insisted on the acceleration of the movements of those dispersed units of IV Corps which were retreating from Lake Van on Diyadin by way of the Teperiz Pass. Of these troops, General Charpentier had been moving with the two dragoon regiments (Severski and Nizhegorodski) since 26 July from Adilcevaz through Beğrikale and Teperiz to Diyadin where he arrived on 1 August. By Oganovski's orders he continued his march west into the plain of Eleşkirt, and on 3 August he was at Taşli-çay-sufla, twenty miles from Karakilise. Here he had to await General Nazarbekov's infantry which was proceeding more slowly by the route which the dragoons had followed. Nazarbekov only arrived at Taşli-çay on the 8th when the counter-manoeuvre of Yudenich was already in full swing.

In preparing his manoeuvre the commander-in-chief displayed perfect self-control and a capacity to ignore alarmist agitation from the rear. The retreat of Oganovski and the installation of his headquarters at the village of Nikolayevskoye near Kağizman had been the source of panic reports which circulated in the rear of the army and even in Tiflis. The Christian population of the Aras valley began to take flight to the north; nervous imaginations saw Kurdish bands not only in the Aras valley but in the vicinity of Aleksandropol. The viceroy wired to ask the commander-in-chief if Kağizman could be held and

[1] The 14th Caucasian Rifle Division (sixteen battalions reorganized from fresh *cadres*); 153rd (Bakinski) Regiment (four battalions); 1st Caucasian Cossack Division (six regiments): a total force of 18,000 foot and 4000 horse.

if Erevan were safe. Yudenich replied that he was ready not only to stop but to defeat the enemy, and old Count Vorontsov-Dashkov was sensible enough to preserve his confidence in his veteran deputy.

In concentrating his strategic reserve at Tahir, Yudenich intended to be in a position to meet all possible moves of the Turkish Third Army. If the Turkish command suddenly decided to attack the main Russian positions of II Turkistan Corps and I Caucasian Corps in the valleys of the Oltu-çay and Pasin, Baratov's group could counter-manoeuvre from Tahir on Yuzveran-Ardi and strike at the Turkish right flank in the Pasin valley. But this development seemed improbable since it had been established since 1 August that the divisions of the Turkish IX Corps were acting in the valley of the Murat. It seemed more likely that the Turkish offensive would develop in the direction of the Ahtalar Pass and the Aras valley, in the neighbourhood of Kağizman. In this event, when the better part of the Turkish offensive group had become engaged in the region between Karakilise and the Ahtalar Pass, Baratov had orders to march quickly and decisively to the Kiliç-gedik and the defile of the Murat near the village of Hamur in order to cut the enemy's line of retreat. Yudenich moved one more regiment up to Tahir (the 156th) as a liaison reserve to maintain contact between the Russian forces in the Pasin valley and Baratov when he moved east.[1]

On 1 August Yudenich himself arrived at Karakurt and Baratov got the order to be ready to move at any moment.

On 3 August Baratov's shock group concentrated at Tahir; on the same day, the 17th Turkistanskis arrived at Başköy, while the 13th Turkistanskis and the 156th Regiment were marching to Velibaba. As soon as the commander-in-chief decided the zero hour, Baratov had instructions to send one regiment of Caucasian Rifles through Zeydkan in the direction of Karakilise where it had to engage the outposts of the left flank of the Turkish main force wherever they were met. Twelve hours later, General Vorobyev, commanding the 4th Caucasian Rifle Division, had to march with the remaining three regiments of his division, through Zeydkan and Karsyan to the Kiliç-gedik which he had to occupy as rapidly as possible.[2] At the same

[1] The same consideration caused Yudenich to move two Turkistan regiments to Başköy as a general reserve replacing the 4th Caucasian Rifle Division. Yudenich had only retained on the main front (Pasin-Oltu-çay) two regiments of the 39th Division, five Turkistan regiments, the 264th (Georgiyevski) Regiment and some militia *druzhiny*.

[2] Vorobyev was assigned the 153rd Bakinskis as a reserve. The Kiliç-gedik ('Gap of the sword') is also known by the shorter name, Felç, which is used on Map 22.

time the 1st Caucasian Cossack Division was to move from Tahir on Remkan and Abdi, from where the Cossacks were to climb the mountain paths over the Şarian-dağ in the general direction of Tutak. Their task was to reach Tutak and cut the tracks leading from Karakilise to Malazgirt. Yudenich proposed to order Oganovski to throw his cavalry against the enemy along the track from Ahtalar to Karakilise simultaneously with the advance of the different units of Baratov's group, while Charpentier's dragoons and Nazarbekov's infantry engaged the right flank of the enemy from the direction of Taşli-çay-sufla.

This plan was quite reasonable and natural; its weakness lay in the lack of passable tracks or even paths in the region between Remkan-Abdi and Tutak.[1] Further, the success of the counter-manoeuvre depended on the extent to which Abdul Kerim might have concentrated his forces against the Ahtalar Pass and neglected to give adequate protection to the Kiliç-gedik and the Hamur defile. Yudenich gambled on the self-confidence of the Turkish commander who, he thought, might have concentrated almost the whole of his force for the invasion of the Aras valley.

In the event it proved that the bad organization of Turkish supplies had resulted in the gradual echeloning of the Turkish divisions instead of their being grouped in the Ahtalar-Karakilise area. It seems that by 3–4 August there were not more than two divisions to the east and north-east of Karakilise, and possibly only one division between Eleşkirt town and the Kiliç-gedik. Two other divisions were only in process of concentration between Tutak, Hamur and Karakilise, and the rest of Abdul Kerim's forces lay near Malazgirt and even more deeply to the rear. In these conditions Yudenich's manoeuvre could not achieve a complete encirclement of Abdul Kerim's army even if the movements of the Russians were carried out with rapidity and precision; and they were not.

Zero hour had been given to Baratov for the evening of 4 August.[2] The 13th Turkistanskis made an immediate start, and Baratov himself moved at dawn on the following morning. During the 5th the 13th Turkistanskis established contact with the enemy on the track between Eleşkirt and Kiliç-gedik near Çerük, and they pushed back the enemy in the direction of Karakilise. In the evening of the same day Baratov was approaching Karsyan while his Cossacks were climbing the slopes of Şarian-dağ farther to the west.

[1] Available Turkish maps gave no paths connecting Remkan and Abdi with Korun village from where there was a track to Tutak.

[2] The plan and time were being withheld from Oganovski and other commanders.

Serious fighting began on the morning of the 6th. The Caucasian Rifle Regiments made a frontal attack on the Kiliç-gedik, and the 15th managed to reach the flat wide top of the Mizirak ridge which formed the left flank of the Turkish position. The Turks, however, maintained a stubborn resistance and continued to receive reinforcements. No news came from the Cossacks on the Şarian-dağ. Oganovski and Charpentier were ordered to advance along their respective fronts, and Oganovski's Cossacks reported that the Turks were withdrawing from Karakilise. Abdul Kerim had evidently understood the threat implied in the Russian movements and alarm was already spreading among his troops.

The Turkish commander, anxious for the Kiliç-gedik and the Hamur defile, decided to reply to the Russian manoeuvre not only by a timely withdrawal of his advance troops; he had in mind also the possibility of a counter-manoeuvre of his own. He ordered the 29th Division of IX Corps which was marching on the Mirgemir Pass to force its pace and to take the offensive in that direction. The Mirgemir Pass over the western shoulder of the Şarian-dağ is only four miles to the south of Tahir, the point which served as Baratov's advance base. Without any special information, Abdul Kerim guessed correctly that a threat against the Mirgemir would be a threat to the most sensitive spot of the Russian deployment.

On the afternoon of the 6th Baratov, in front of Kiliç-gedik, got a report that strong Turkish forces were approaching the Mirgemir.[1] Baratov was so disturbed that he detached a battalion of the 153rd Regiment at the very moment when it was preparing to support the attack of the Caucasian rifles on the Kiliç-gedik positions. The general thus acted directly contrary to the idea of Yudenich, which was definitely that the shock group must not be diverted from its task by the development of the situation in its rear. To meet eventualities Yudenich was holding the necessary reserves at Başköy and Velibaba. Informed of the enemy's presence near the Mirgemir, he dispatched, on the night of 6–7 August, the 17th Turkistanskis and the 156th Elizavetpolskis via Tahir to the relief of the Frontier Guards at Mirgemir.

Before these reinforcements could arrive the Turks had occupied the pass, and during the night their advance parties had descended into the ravine of the Şarian-su within two miles of Tahir. Early in the morning of the 7th the battalion of the 153rd Regiment sent by Baratov came up, forded the Şarian-su and engaged the enemy on

[1] From a Frontier Guard battalion which had been posted at this point.

the slopes of Hayran-göl. The 17th Turkistanskis did not approach the Mirgemir until 11 p.m. of the same day. The march had been long and difficult and troops were weary; but the Colonel (Kruten) was aware that if he delayed his attack the Turks would be able to concentrate the strength of a whole division on the pass. He attacked with great boldness in the dark and succeeded in retaking the pass. Kruten's initiative changed the whole situation: on the morning of the 8th the Turkish 29th Division, encamped to the south of the pass, was so impressed by the Russian success that they refrained from further action. The Elizavetpolskis arrived from Tahir to reinforce Kruten; and on the following day (the 9th) the Turks were observed to be retreating to the south. Kruten followed them over the southern slopes of Şarian-dağ to Reşü and Darkut.

The manoeuvre of the Turkish 29th Division was well conceived but it was weakly executed; the troops of this newly reconstituted unit were scarcely fit to undertake a particularly difficult offensive operation. This Mirgemir diversion, however, helped Abdul Kerim to gain at least twenty-four hours in which to organize his retreat from Karakilise, since Baratov in front of Kiliç-gedik, having sent the 153rd to the rear, had left the Caucasian rifle regiments without fresh reserves and they had been held throughout the 7th by the obstinacy of the Turkish defending infantry. The pass was only captured on the morning of the 8th. The retreat of the Turks began to be disorderly when it became apparent that Cossacks were already in Tutak.[1]

On the fourth day of Yudenich's counter-manoeuvre the defeat of Abdul Kerim became evident; and his defeat already had all the elements of a serious debacle. Abdul Kerim's troops had been saved only from complete disaster by the slowness of their concentration in the Eleşkirt plain. At the moment of the alarm on 5 August the total of Turkish troops to the north of the Kiliç-gedik was between four and five divisions. The units of the two divisions left to hold the Kiliç-gedik suffered badly enough, but their defence during 6 and 7 August enabled their comrades to make good their retreat by the track running along the eastern bank of the Murat and by another track passing through Ağadevi, ten miles to the east of Hamur.[2]

But the retreat of Abdul Kerim was far from orderly. All stores and a part of the field artillery were abandoned at Karakilise (which

[1] The 1st Gorsko-Mozdokski Regiment.

[2] Both these tracks were bad by comparison with the road passing from Malazgirt through Tutak and Kiliç-gedik to Karakilise; but both the key points of this road were soon in the hands of the Russians.

was occupied on the 7th by the Cossacks of IV Corps advancing from the Ahtalar Pass).

On 7–8 August Abdul Kerim stood on a defensive position on the line Hamur-Ağadevi. Baratov failed to close, while neither Charpentier nor Nazarbekov came up with due speed. The Turks thus gained two further days, but as the *sotni* of the 1st and 2nd Caucasian Cossack Divisions rode after them along both banks of the Murat they were fast losing cohesion. Their losses during the advance into and retreat from the Eleşkirt valley were not less than 10,000 killed and wounded and they left 6000 prisoners in the hands of the Russians.

The result of the eastern Euphrates operation was thus a heavy blow to the Turkish Third Army. The 37th Division suffered severely, and all three divisions of IX Corps had to be reconstituted—no easy task with the continuing drain imposed on Turkish manpower by the growing losses at Gallipoli. The 51st and 52nd Divisions were in better state and remained available for dispatch, in September, to Iraq, where the threat from the Anglo-Indian invaders was becoming acute.

On the Russian side this new success of the Caucasian army was particularly welcome in August 1915, at a moment when the western armies had lost Warsaw and all Poland. Yudenich had no intention of developing his success in the direction of Muş. He had not the troops available for an offensive undertaking. IV Corps had been seriously weakened by the Malazgirt battle and the retreat to the Ağri-dağ. In September it only remained strong enough to occupy the line through Mirgemir Pass, Kiliç-gedik, Tutak, Patnos and Adilcevaz, and to reoccupy the city of Van and the town of Vastan at the south-east corner of the lake. Turkish patrols remained in possession of Malazgirt and Ahlat, and they held the villages along the southern shore of the lake.

CHAPTER XXVI

LULL ON THE CAUCASIAN FRONT: YUDENICH'S NEW PLANS. SEPTEMBER–DECEMBER 1915

MAPS 1, 19; cf. also 23, 38

IN August, in Istanbul, the collapse of Abdul Kerim's offensive was forgotten in the anxiety caused by the latest phase of the British assault on Gallipoli. The failure of the Suvla Bay operation only became clear at the end of the month. But it was known that fresh reinforcements were available to the British, while Turkish losses had been very severe and twenty-two out of a total of fifty-two divisions in the Turkish army were already engaged in the Dardanelles zone. On the Caucasian frontier, therefore, a strictly passive attitude was imposed on the Turkish Third Army. The enormous losses of the Gallipoli campaign were absorbing all available reserves,[1] and the Anatolian reservoir of man-power had been emptied. The difficulty of filling the gaps was so great that IX Corps never again attained normal strength, and by the end of the year none of its divisions contained more than 6000 men.[2]

XI and, particularly, X Corps were the only real units available to Mahmut Kâmil Paşa during the last four months of the year. On the Black Sea coast some small reinforcements had arrived before the development of the Suvla Bay offensive.[3] Between the Pontic coast and the Çoruh there were sixteen battalions with ten guns. But the rainy season did not favour any movement against Lyakhov's force in position along the right bank of the Arhavi river; there were a few small skirmishes from time to time.

Yudenich also had his man-power difficulties. The serious Russian losses in the campaign of 1915 on the western front prevented the dispatch to Caucasia of any reservists from European Russia. Normal

[1] A total of 600,000 engaged, with 200,000 killed and wounded and 100,000 sick. For varying estimates, see Larcher, *Guerre turque*, pp. 237, 242.

[2] The 36th and 37th Infantry Divisions were reduced to two regiments each.

[3] An infantry division was reconstituted on the Black Sea coast: the 9th Regiment arrived from the capital; the 8th Regiment was brought up to strength following the big losses of the Ardahan operation; and the 7th Regiment (a total loss after Ardahan) was replaced by the 17th Regiment of the 6th Infantry Division. To these must be added seven battalions of Frontier Guards and irregulars.

local reserves had been anticipated in the earlier formation of new units and by the necessity of filling losses in existing units.[1] A great number of older or untrained men were available for the militia (*opolcheniye*), and these were formed into battalions and brigades serving on lines of communication and internal security duties. Some of these militia units Yudenich found good enough for employment on the less important sectors of the front, and the better elements were put into training for field operations.

Yudenich's efforts were on the whole limited to strengthening existing regular units.[2] At the same time he wisely developed the Frontier Guard units formed of picked men with a fine tradition. Five Border Regiments, each 3–4 battalions strong, had been created by the autumn and they soon proved their worth. By October, Yudenich was holding the front with a minimum of troops but he had strong reserves stationed in depth, where he was able to provide rest and excellent living conditions for all ranks.[3]

Yudenich had always regarded Persian Azerbaijan and the Van region as an operational unity, and for action there General Chernozubov's Azerbaijan-Van force had been created. Weak in infantry but strong in horse it was essentially a mobile force[4] and, while the rest

[1] The losses of the summer campaign of 1915 were not serious, even having regard to the Malazgirt disaster: a total of less than 10,000.

[2] The complements of the three-battalion Turkistan regiments were raised to four battalions, and the 4th and 5th Brigades of II Turkistan Corps were raised to the strength of divisions. The 2nd Caucasian Rifle Brigade also became a division.

[3] The Russian order of battle on the Caucasian front during the period of lull was as follows:

Black Sea coast and Çoruh valley: 19th Turkistanskis and one plastun battalion between Arhavi and the mouth of the Çoruh; three plastun battalions and some militia *druzhiny* at Artvin.

II Turkistan Corps (*Russian right centre*): Işhane, Ak-dağ, Sivri-çay valley, Karşin, Mt Kocut: three plastun battalions; 264th (Georgiyevski) Regiment; 13th, 14th, 18th and 20th Turkistanskis; with the 15th, 16th and 17th Turkistanskis and the Siberian Cossack Brigade in reserve at Oltu.

I Caucasian Corps (*centre*): Kocut, Sanamer, Zanzak, Horsan, Tayhoça, Eşek-Elhas, Mirgemir Pass: Don Cossack Foot Brigade; eleven militia *druzhiny*; 154th (Derbentski) and 156th (Elizavetpolski) Regiments, with 155th Kubinskis and 1st Caucasian Cossack Division in reserve at Sarikamiş.

IV Caucasian Corps (*Russian left*): Mirgemir Pass, Hamur (on the Murat), Patnos: held by the 5th Caucasian Cossack Division (General Nikolayev). Tutak: 262nd, 263rd and 264th Infantry Regiments. Reserve: 2nd Caucasian Rifle Division (Karakilise) and Caucasian Cavalry Division (Charpentier) (Kağizman).

Strategic Reserve of Commander-in-chief: 153rd (Bakinski) Regiment; 4th Caucasian Rifle Division; 4th Plastun Brigade—total twenty-six battalions with 2nd Caucasian Cossack Division at Karakurt and Kars.

[4] Chernozubov had four Armenian *druzhiny* and one *opolcheniye* (militia) brigade of eight *druzhiny* as infantry; but he had also the whole of the 4th Caucasian Cossack Division and the two brigades of Transbaikal Cossacks.

of the front lay quiet, it engaged in some fruitful manoeuvres against the weakened right wing of the Turkish front. Already in August General Trukhin, with the 2nd Transbaikal Cossack Brigade and the Armenian *druzhiny*, had reoccupied the luckless city of Van. On 1 October, and again ten days later, the Turks appeared at the south-east corner of the lake. Trukhin drove them off and entered Vastan. Later, in November, he advanced westward along the southern shore of the lake, and his patrols approached to within thirty miles of Bitlis. The Turks remained passive, but the Kurds to the east and north of the lake showed fight when the autumn frosts drove them down into the plains for grazing. In the second half of November the 1st Armenian *druzhina* and a few *sotni* with two guns were sent to Erciş on the northern shore of the lake where, after some skirmishes with the Kurds, they established firm communication between Erciş and Patnos.[1]

In Persian Azerbaijan all seemed quiet during the autumn, but Persian affairs were already attracting the interest of the new viceroy and commander-in-chief in the Caucasus. The Grand Duke Nicholas had yielded his post as supreme commander-in-chief on the western front to the emperor, and on 24 September he arrived in Tiflis to replace old Count Vorontsov-Dashkov. German activity in Persia was at last forcing an active exploitation of the idea of Anglo-Russian collaboration in Persia and Iraq. The new German military attaché at Teheran, Count Kanitz, was occupying himself with the promotion of a *putsch* directed to the formation of a Persian government at Hamadan, ready to enter into an alliance with the central powers and the Young Turks. Kanitz had formed an armed force, the spear-head of which was made up of the few hundred German civilians resident in Persia and of numbers of German and Austrian prisoners who had escaped from the Russian camps in Turkistan. Some thousands of Persian militia were recruited and certain Swedish instructors in the Persian army lent themselves to the plot. All these activities were known to the British diplomatic and consular representatives and, since the brigade of Persian Cossacks, organized some years before by the Russians, seemed inadequate for the preservation of order in the capital, they asked for Russian intervention.

One of the first actions of the grand duke on arrival in the Caucasus was to tell Yudenich of the urgency of forming a special expeditionary force for Persia. This task was far from agreeable to the deputy commander-in-chief; for, while the political reasons for the move left

[1] I.e. between the Azerbaijan-Van force and the left flank of IV Caucasian Corps.

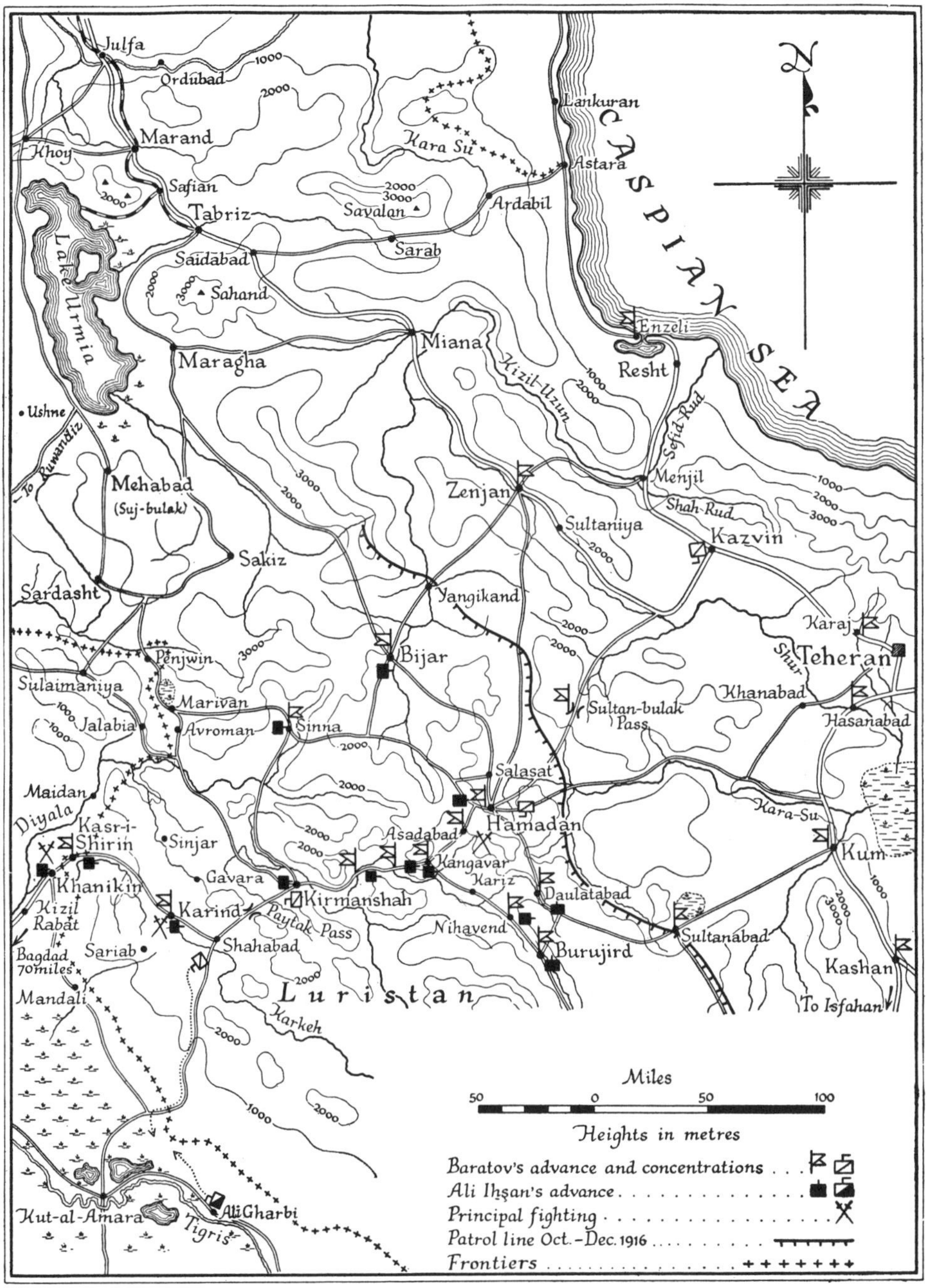

Map 23. Baratov's expeditionary force in Persia, 1915–16

him indifferent, he saw that his Caucasian army was to be weakened at the moment when he was preparing for further active operations.

Yudenich, already short of infantry, could only detach two of the newly formed Border regiments with two militia *druzhiny*. He allocated, however, the whole of the Caucasian Cavalry Division,[1] the 1st Caucasian Cossack Division and two mixed Kuban Cossack regiments built up out of single independent *sotni*. The artillery available was sixteen horsed and 12 mountain guns with two 4·8 in. howitzers—sent for their moral effect on the population. General Baratov was placed in command of a force which totalled 6000 infantry and 8000 horse.

The transport of this big cavalry concentration across the Caspian Sea was not an easy affair; but the difficulties were overcome by the crews of the Volga-Caspian steamers and barges more rapidly than had been expected. On 12 November General Baratov, with his staff and a Cossack advance-guard, landed at Enzeli (now Pahlavi); and in a few days he was followed by the two Kuban regiments and the whole of the 1st Caucasian Cossack Division. Baratov had instructions from the grand duke to advance on Kazvin and, without occupying Teheran, to isolate the capital from the possibility of penetration by German, Turkish or germanophil Persian groups. By 1 December Baratov had the 1st Caucasian Cossack Division at Kazvin, while his mixed Kuban brigade was at Karaj, twenty miles north-west by west of Teheran, with patrols cutting the roads from Hamadan and Kum. The Caucasian Cavalry Division and the infantry were still en route across the Caspian and from Enzeli to Kazvin. Baratov's landing at Enzeli precipitated events in the Persian capital: the *Putsch* was tried—and failed—while the young Shah Ahmet took refuge in the Russian embassy. As it seemed that Baratov was moving on Teheran, the so-called 'Committee for the Defence of Islam' and the pro-German 'democratic' faction of the Persian Mejlis fled, under the protection of Count Kanitz and about 3000 Persian gendarmes, to Kum. The first objective of the expedition had been attained without a shot being fired. The logical sequel was to clear the pro-German elements out of Kum and Hamadan. With the advance on these two cities, early in December, a real Persian campaign had begun, but its full development did not come until the early months of 1916.

Considerations other than the exploitation of easy successes in Persia were occupying Yudenich during November 1915. The

[1] The inefficient Charpentier had been replaced by Prince Beloselski in command of this famous dragoon division.

British assault on Gallipoli was obviously nearing liquidation, and in October the Austro-Germans, joined now by the Bulgarians, had opened their third and crushing offensive against Serbia. The direct route from central Europe to Turkey would be open within a few weeks.[1] The new situation implied that heavy Austro-German artillery would soon be available both at the Straits and on the Caucasian front. Furthermore, Yudenich appreciated that with the British evacuation of Gallipoli, twenty-two Turkish infantry divisions would become available for other fronts. Even a part of these divisions on the Caucasian front might turn the balance in favour of the Turks. Yudenich watched the situation at Gallipoli with the greatest anxiety. Fortunately for the common interest, the British only began the evacuation of the Ari Burun sector on the night of 19–20 December, and the whole operation was so skilfully prepared and executed that all Turkish forces on the peninsula were engaged there until the last day. A month at least had been gained for the preparation of Yudenich's new plan and he had not been idle during that month.

Since the end of October a new operational idea, inspired by the changed situation in the Balkans and at Gallipoli, had guided the activity of the Russian commander-in-chief. It was in complete contrast with the dilettantism of the Turkish vice-generalissimo and with the doubts which in all circumstances seemed to trouble General von Falkenhayn.

Enver and Falkenhayn met at Orsova in December; and at this small Danubian town so romantically associated with past incidents of Ottoman history, Enver made the rather startling proposal to send twenty or thirty Turkish divisions to participate with German, Austro-Hungarian and Bulgarian troops in a common decisive action somewhere in Europe, presumably beginning on the Italian front. Falkenhayn had expected the contrary: a demand for the participation of German troops in operations on the frontiers of Turkey. He was not aware that the Young Turk triumvirate, after the repulse of the British from the Dardanelles, were fearing an effective occupation of their dominions by the Germans. On the other hand, the German general's appreciation of the Turkish army, in spite of its defensive victory at Gallipoli, was low in respect of offensive operations; and he failed to see the advantages which Turkish co-operation in the west could bring. He recommended, rather reasonably, another form of co-operation: the development of action against the Russians

[1] But a year later than the date originally promised by Berlin to the Young Turk triumvirate.

in the Middle East, and more particularly against the British, since he knew them to be weak in Iraq and sensitive with regard to any move against the Suez Canal. With his usual easy-going capacity for adopting the ideas of others, Enver made von Falkenhayn's proposals his own, and by the end of 1915 was busy with the evolution of a new general plan. Wide movements of Turkish troops from one end of the empire to the other were planned for the first months of 1916, but these had hardly begun when Yudenich undertook his great winter offensive.[1]

Yudenich's idea of offensive action was simple and clear. He had anticipated that Turkish divisions from Gallipoli would be free some time in November. He estimated that three months would be necessary for the transportation of any of the Gallipoli divisions to Erzurum. Furthermore, he had no doubt that if the Turks were planning an offensive they were planning it for the spring. Their concentrations should not be complete before March–April. He planned therefore to strike a decisive blow at the Turkish Third Army by 15 January and to reduce it to such a state that, even after the arrival of reinforcements from the west, it would be in no situation to undertake any serious action against the Caucasian frontier.

Yudenich's tactical idea was also simple and was suggested by the real facts of the situation. The Turkish Third Army occupied carefully fortified positions on a front seventy-five miles in length. Both flanks were protected by high and difficult ranges: on the north the wild outliers of the Pontic Alps, and on the south the massif of Bingöldağ rising to an altitude of 3,000 m. No outflanking movement could be contemplated, particularly in the conditions of the Armenian winter. But the length of the line and the relative weakness of the Turkish defending army—nine weak divisions with an effective of scarcely more than 60,000 bayonets—suggested the feasibility of a break-through at a selected point. Such break-through seemed most difficult in the area of the Pasin valley where the Turkish field works, in three successive lines, were very strong and where excellent roads might ensure the rapid arrival of reserves from the nearby bases at

[1] In *The World Crisis* (1916–18, vol. III, pt. I, chap. 3, 'Falkenhayn's Choice'), Mr Churchill expresses the view that Falkenhayn made a fatal mistake in deciding for the offensive against Verdun rather than striking at Russia in the Ukraine and the Caucasus. 'An upward thrust of Turkish armies sustained by German troops and organized by German generals would have conquered the Caucasus.... The Austro-German front against Russia might have stretched from Riga to Astrakhan, with little more expenditure of force than was required to hold the existing Eastern line.... From the Caspian once navally commanded, Persia was a cheap and easy prey' (p. 81).

Hasankale and Erzurum. Yudenich thought a break-through more possible across a sector of the mountains dividing the Pasin plain from the valley of the Oltu-çay, where the left flank of the Turkish XI Corps joined the right of X Corps. He regarded surprise as the principal condition of success; and complete surprise demanded not only secrecy in the final concentration for attack but secrecy throughout the long and laborious activity necessary to prepare the troops for the winter offensive.

In his preparatory work, conducted with the aid of his small and efficient field staff, Yudenich was in his proper element. The preparations for the Köprüköy battle and for the subsequent assault on Erzurum may be regarded as a classical example of staff work conducted by a leader who had the rare capacity to see clearly the whole ensemble of an operation without forgetting or neglecting a single detail.

Munitions and food supplies were brought forward in the necessary quantities to the front. But, in order not to raise suspicions that an offensive was intended, all transport took place gradually and by consecutive stages. Their accumulation was officially explained by the necessity of making preparations for a quiet and well-organized winter in the inclement conditions of the Armenian highlands. A special network of field telephones was laid without attracting any curiosity. Troops were carefully equipped for the winter; but these attentions required no explanations in view of Yudenich's reputation as a paternal commander. To every soldier was issued a short fur coat (*polushubok*), trousers lined with cotton-wool, felt boots (which he put on when marching through snow or resting afterwards in bivouac), thick shirts, warm gloves and a fur cap well covering both head and ears. Since no wood was available on the bare mountain sides where it was proposed to operate, short logs from the Soğanli pine forests were cut and collected in dumps in advance; every soldier was duly issued with two of these logs to carry, so that they were able to enjoy fires in temperatures which had frozen the unfortunate Turkish askers a year earlier in their march along the *top yol*.

The more the preparations advanced the more concerned was Yudenich with the vital necessity for the maintenance of absolute secrecy. Neither troops nor officers, not even colonels and brigade commanders, had wind that anything was in the air. Yudenich took only a few of the corps and divisional commanders partly into his confidence. For the sake of maintaining a camouflage of normality, regiments, early in December, were ordered to send parties of officers and men to Tiflis to make the usual purchases in preparation for the traditional Russian Christmas feast, which, the troops were informed, would be celebrated in rare style along the front. Yudenich planned to open his offensive actually during the period of the Russian Christmas and New Year.[1] Meanwhile the rumour was intentionally spread that soon after Christmas a part of the forces on the main front would be transferred to Persia where climatic conditions were more favourable to winter operations. In order to attract the attention of Turkish agents, large quantities of corn and forage were purchased and stored in Azerbaijan in preparation for 'the troops expected there'. With the object of further interesting the enemy, a false wireless order was transmitted in clear. The *mise en scène* went so far that a few days before the offensive was opened, one of the regiments of the 4th Caucasian Rifle Division was sent by train from Sarikamiş to Kars and from Kars to Julfa with as much publicity as possible.[2] During the last five days before the attack no one was allowed under any pretext to pass between the front and the rear areas; special patrols were in observation along every road and trains from Kars to Tiflis were leaving empty.

All these measures proved effective, and on the Turkish side no one had the slightest suspicion of the feverish work which was proceeding behind the Russian lines during the second half of December and in the first days of January. Troop concentrations were carried out only during the hours of darkness; and during the day Turkish observation posts on the mountains could sometimes view Russian columns slowly moving towards the rear—whence they returned during the night. For a long time the secret was kept even from responsible senior commanding officers. General Vorobyev, in command of the 4th Caucasian Rifle Division, to whom the break-through was entrusted, was only informed of his intended task a fortnight before the battle.

By 31 December Yudenich considered that his preparations were almost complete. Only then did he go to Tiflis for the purpose of reporting to the grand duke his decision to attack the Turkish Third Army with the object of inflicting such a defeat as to effect a partial annihilation before reinforcements could arrive at Erzurum. The grand duke himself had some doubts as to the wisdom of the undertaking, but his military advisers, Generals Yanushkevich and Palitsin,

[1] 25 Dec.–1 Jan., old style; 7–14 Jan., new style.

[2] The 4th Caucasian Rifle Division were to take an important part in the coming attack; but the unlucky regiment failed to return in time to take part in the Köprüköy battle and only participated in the assault on Erzurum.

raised no objections. The matter was clinched by an opportune telegram from General Alexeyev, chief of staff to the emperor, who drew the attention of the grand duke to the recent Turkish successes against the British in Iraq, which he explained by the arrival there of reinforcements from the Caucasian front.[1] As the liquidation of the Gallipoli operations became an accomplished fact, Alexeyev was becoming anxious lest Turkish divisions should appear on one of the European fronts.[2] The Russian army on the western front had not yet recovered from the enormous losses which it had incurred during the campaign of 1915; Alexeyev felt that it was unnecessary to retain 150,000 Russian troops in Caucasia (a figure which included all the second line internal security units) when they could be used with advantage in the west. Placed between the dilemma of sending part of his Caucasian army to Poland and of authorizing Yudenich's offensive, the grand duke did not hesitate.

Yudenich had secured the grand ducal approval. On his way back to the front he invited the commanders of II Turkistan Corps (Przevalski) and of I Caucasian Corps (Kalitin), with their respective staffs, and also General Vorobyev, to meet him in Kars. Here he explained his plan of attack, but no written orders were issued to the corps commanders before 9 January. II Turkistan Corps had to begin their attack on the 10th; I Caucasian Corps and Vorobyev's shock rifle division on the 12th.

[1] The 51st and 52nd Divisions had arrived in Iraq.

[2] As proposed by Enver in the Orsova conversations with Falkenhayn. The Turkish XV Army Corps was, in fact, sent to Galicia in August 1916. The commander was Yakup Şevki Paşa, who refused to communicate with Austro-German H.Q. except in Turkish. He was replaced by Cevat Paşa. Cf. Larcher, *Guerre turque*, p. 112 and n. 3.

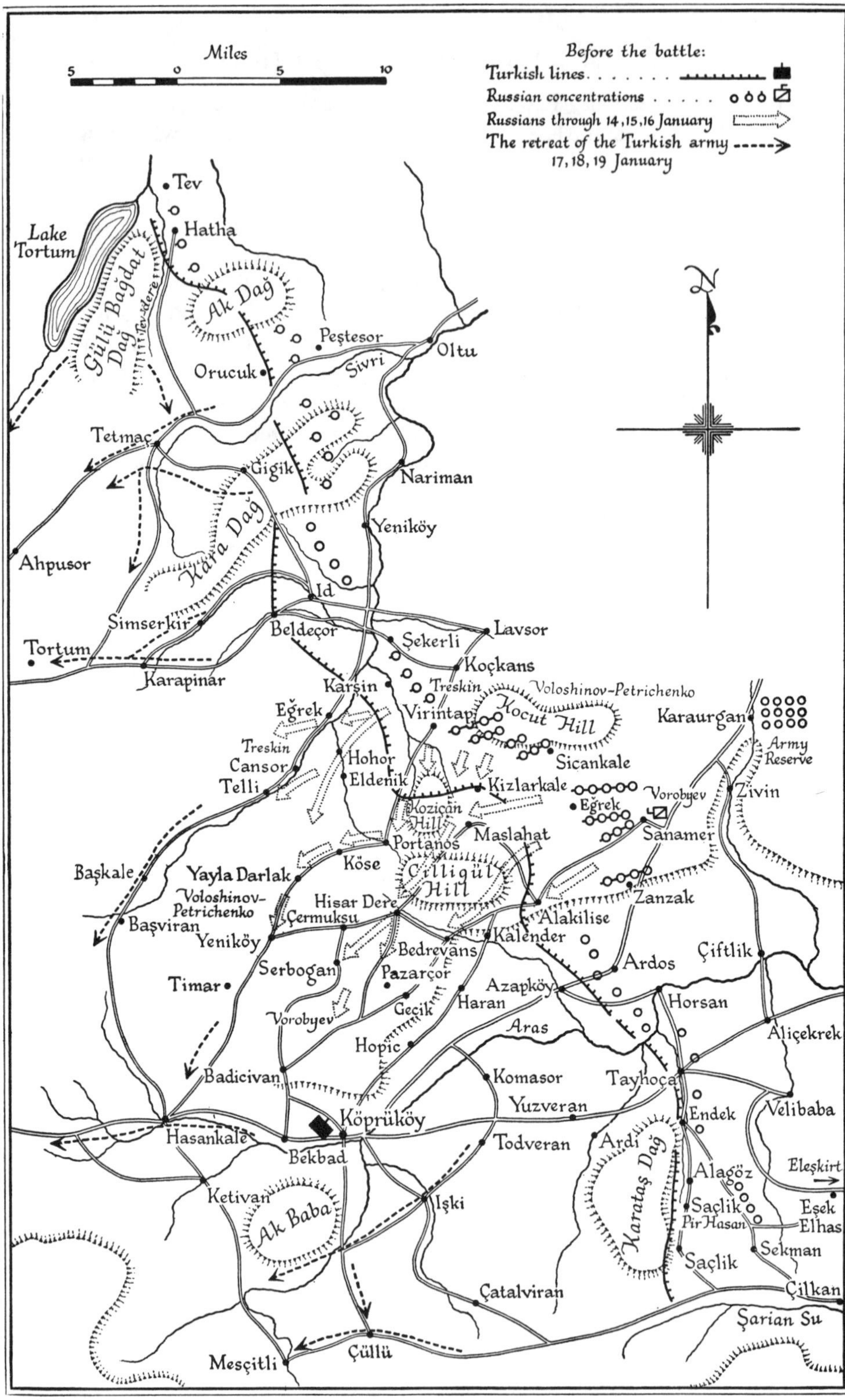

Map 24. The Çakir-baba region and Köprüköy operations, 10–19 January, 1916

CHAPTER XXVII

THE BATTLE OF KÖPRÜKÖY, 10–19 JANUARY 1916

MAP 24; cf. also 1, 19, 38

THE Russian winter offensive came as a complete surprise to the Turks. No one had envisaged such a possibility in Istanbul, and on the spot everyone, from the commander of the Third Army to the simple askers, believed that the gavurs were thinking of nothing but celebrating their Christmas week in peace and hilarity. Mahmut Kâmil Paşa himself was on leave in the capital, and his chief of staff, Major Guse, had gone to Germany to recover from the effects of typhus.

The strength of the Turkish Third Army remained somewhat below normal level. Its eleven regular divisions[1] had an average effective of between 5000 and 6000, and of these there was always a certain proportion of men in hospital. With the addition of the 2nd Cavalry Division and auxiliary and irregular formations, the total strength of the Third Army (excluding the Black Sea coast force) probably amounted in January 1916 to between 70,000 and 75,000 men with 120 guns. Yudenich's staff supposed also that in the region of Erzurum there might be about forty battalions of recruits (for complements), but his estimate seems to have been exaggerated.

These forces were hardly adequate for holding the fortified front which covered Erzurum and at the same time for protecting the strategic right flank with such important points as Hinis, Muş and Bitlis. Mahmut Kâmil had had no alternative but to leave the defence of Muş and Bitlis to the 36th Division with some cavalry and auxiliary units. All the remaining divisions (including the remnant of the 37th) were concentrated for the defence of the approaches to Erzurum. On this Erzurum front, generally called 'the fortified lines of Köprüköy', the Third Army had some 65,000 infantry with 100 guns. The right flank seemed quite safe for the winter: Russian numbers in the area of IV Caucasian corps were known to have been much reduced as a result of the dispatch of the expeditionary force to Persia; communications were lacking and the natural wealth of the Murat-su

[1] IX Corps: 17th, 28th, 29th; X Corps: 30th, 31st, 32nd; XI Corps: 18th, 33rd, 34th, also 36th, 37th. To these regular infantry divisions, 15–20 auxiliary battalions (gendarmes and Frontier Guards) must be added; also the 2nd Cavalry Division and several thousand Kurds.

valley had been ruined by the flight of the laborious Turkish and Armenian peasantry during the offensives and counter-offensives of the Russian and Turkish armies. All that region seemed empty and dead in the winter of 1915–16. The Turkish army suffered from the results of the displacement of population, since the region round Muş had always been an important base of supplies for troops in the Erzurum region. The decree dismissing all Armenians from military service was also an embarrassment to the Third Army. There were very few Armenians in the ranks among active officers, but they were numerous in the rear services and did the bulk of the office work for which the true Turks had at that time little taste or aptitude.

The front occupied by the Third Army was divided by mountain ranges and river valleys into three distinct sectors.

The left sector covered the area between Lake Tortum and the ridge of Çakir-baba which divided the valleys of the Sivri-çay and Oltu-çay from that of the Aras. The length was about thirty-five miles. The importance of this sector lay in the tracks which led to Erzurum up the ravines of the Tortum-çay and Sivri-çay and over the Gürcü-boğaz Pass.[1] The tracks up the Tortum and Sivri ravines were joined by a track from the Oltu valley which forked west from Id.

The central sector (eighteen miles) was considered to be still more important. It stretched from Çakir-baba to the Aras south of Azapköy. The Turkish line here crossed the excellent road to Erzurum which passes along the Pasin valley.

The right sector (about twenty miles) between the Aras and the Dram-dağ range seemed to the Turkish command to be immune from the danger of a Russian offensive, since it covered a difficult hilly country with few tracks. It offered also a second line of defence in the bend of the Aras backed by the Ak-baba ridge.

The disposition of the Turkish defending forces was well known to the Russians, who had studied it in detail during the long period of lull.

The left sector was occupied by the three divisions of X Corps reinforced on the right by the 34th Division of XI Corps; on the left of this sector, between Lake Tortum and the long precipitous ravine of the Çoruh, there was a flank guard of several battalions of gendarmes and Frontier Guards.

In the central sector the defence was entrusted to the 33rd and 18th Divisions of XI Corps.

[1] Literally, 'Georgian throat' or 'gate'; G. *Kartlis Qeli*. Down to the later Middle Ages the valleys of Tortum and Oltu were Georgian country and the 'gate' gave access into Georgia from Erzurum.

On the right sector were the 28th and 29th Divisions of the IX Corps and units of the 37th Division, protected on the extreme right flank by Frontier Guard battalions and Kurds.

The general strategic reserve consisted only of the 17th Infantry Division of IX Corps (6000 strong) stationed near Hasankale.

The positions of the central sector, especially in the Azapköy plain where there were two defensive lines, were carefully fortified and protected by extensive wire fields. In the mountainous region occupied by X Corps ring-form entrenchments were everywhere well adapted to the terrain. In some places trenches had been blasted out of the rock with dynamite. The Turks proved very skilful in siting machine-gun nests, which in the circumstances were likely to prove the best defensive arm. They also made good tactical use of their not very numerous artillery. Strong artillery groups were concentrated on both sides of the Aras: three on the slopes to the north of Haran (left bank) and two on Tyk-dağ near Endek (right bank). These batteries could cover with flanking fire the greater part of the Turkish positions in the Aras valley. The defect in the Turkish dispositions was the lack of an adequate general reserve which could be directed to any threatened point. Mahmut Kâmil was responsible for this; he had merely asked the commander of X Corps to detach any available troops for this purpose, if the strength of his own positions permitted. The corps commander had naturally preferred to keep his reserves under his own hand.

Yudenich and his staff had not underestimated the strength of the enemy's forces and positions. Rather to the contrary; their information placed the strength of the Turkish infantry and artillery at about 80,000 men and 180 guns. Yudenich had only a limited number of troops available. He decided therefore to concentrate on the Erzurum front eighty-five battalions, leaving in the valley of the Murat-su only twelve battalions[1] supported by the 2nd Caucasian Cossack Division. Of the troops of IV Caucasian Corps, three regiments of the 66th Division were sent secretly into strategic reserve, replacing the 4th Caucasian Rifle Division (less one regiment) which was to be used as a shock group. The 5th Caucasian Rifle Regiment was sent to reinforce the Russian left flank along the right bank of the Aras.

The Russian tactical plan was based on the following observations:

(*a*) The Turkish front consisted of three sectors which were clearly defined by the Çakir-baba ridge and by the Aras; the liaison between the northern and central sectors seemed rather weak.

[1] The 6th, 7th and 8th Caucasian Rifle Regiments.

(*b*) It was apparent that the Turks had no strong reserves and their dispositions appeared to be of a cordon character.

Yudenich therefore decided to break through the Turkish front at the point of junction of the central and northern sectors, in the region of the Çakir-baba ridge. The break-through had to be carried out by the 4th Caucasian Rifle Division strengthened by one or two regiments from the reserve. With the greatest secrecy General Vorobyev's Rifle Division had been concentrated near Sanamer and hidden with extreme precautions out of Turkish observation.

The attack was planned in three stages:

First two days. Attack by II Turkistan Corps along the front from Lake Tortum to Çakir-baba; the principal aim of this attack was to create a false impression of the intended direction of the main Russian blow.

Following two days. Attack in full strength by the Russian 39th Division along both banks of the Aras, with a view to drawing the last Turkish reserves into the conflict.

Fifth day. A break-through by the shock group in a general direction north-east to south-west, to be directed on the *sixth day* due south against Köprüköy.

The direction of the break-through had been carefully studied. General Vorobyev had orders to concentrate in the area Sanamer-Eğrek and to attack along a narrow front: Kizlarkale-Maslahat-Elimi. His action was to be secured by two flanking columns. To the north General Voloshinov-Petrichenko[1] had to advance from the region of Sicankale ('Rat castle') and Kocut-dağ and capture the Koziçan shoulder of the Çakir-baba ridge. The possession of the Koziçan-Portanos region was calculated to secure Vorobyev from any flanking attack from the Turkish northern sector. Voloshinov-Petrichenko's right flanking column had to come into action on the third day of the general offensive and to take Koziçan on the fifth day. Simultaneously a left flanking column (155th Kubinskis) had to advance from Zanzak on Alakilise and the important spur of Çilligül. As it was intended that Vorobyev, after the break-through at Maslahat-Elimi, should develop his attack across the northern slopes of Çilligül, the possession of this mountain seemed necessary to secure him against a possible attack against his left flank from the south. Vorobyev, in fact, had to open the door and the two flanking columns had to secure the door-posts, Koziçan on the north, and Çilligül to the south. This was,

[1] With the 18th Turkistan Regiment, the 264th Georgiyevskis, and the Don Foot Brigade.

properly speaking, the only real attack intended; all the rest were diversionary operations.[1]

General Przevalski, commanding II Turkistan Corps, deployed his forces before 10 January. Along the Tortum-çay between the Tevdere and the Oltu, plastuns and Cossacks watched the opposing flank guard of the Turkish X Corps. Opposite the Turkish 30th and 32nd Divisions which were massed along both banks of the Sivri-çay, the 13th and 15th Turkistanskis had been allotted only a demonstrative task. Between the Sivri-çay and Oltu-çay, the position held by the Turkish 31st Division had a defensive centre of resistance on the high and difficult Karadağ.[2]

Przevalski directed his principal action against Karadağ. He thought that its capture would endanger all the Turkish positions in the valleys of the Sivri and the Oltu and would thus have the effect of diverting, more effectively than any other stratagem, the Turkish reserves from the centre. And that was the general aim of his action, as planned by Yudenich. The task was a difficult one. Moreover, Przevalski, with sixteen battalions at his disposal, had in front of him the twenty-seven battalions of X Corps.

The attack on Karadağ was entrusted to the 14th Turkistanskis, strengthened by one battalion of the 17th. At the same time the rest of the 17th had to attract to itself the bulk of the Turkish 31st Division which was responsible for the defence of the left bank of the Oltu-çay. Przevalski retained two battalions of the 16th Turkistanskis in reserve and gave a special task to Colonel Treskin, commanding the other two battalions of this regiment and the whole of the 20th Regiment. Treskin had to concentrate in the Şekerli-Virintap area on the eastern side of the Oltu-çay and make an energetic attack in the direction of Karşin-Hohor, co-operating with the right column of Voloshinov-Petrichenko's group which had been ordered to attack Koziçan-dağ from the north (from Virintap-Koçkans). All this region between the Oltu-çay at Karşin and the Koziçan shoulder of the Çakir-baba ridge was defended by the 34th Division of the Turkish XI Corps with the bulk of its forces in the area Karşin-Eğrek-Hohor with only a few battalions at Koziçan-dağ and Portanos. This division, nine battalions strong, had to meet the combined attack of the columns of Treskin

[1] The attack of II Turkistan Corps along the northern sector from Tortum-göl to Virintap; the attack of the 153rd Bakinskis and the 156th Elizavetpolskis in the Azapköy plain; and the attack of the 154th Derbentskis and the 5th Caucasian Rifle Regiment on the right bank of the Aras.

[2] T., 'Black mountain'. It was called *Hai-dağ* by the Russians. *Hai*, in the Armenian language, means 'Armenian'.

and Voloshinov-Petrichenko, totalling eighteen battalions (and the average Russian battalion was rather stronger than its Turkish counterpart). Yudenich knew well that the Çakir-baba ridge would be the weakest point of the Turkish defensive line, and the attack against the northern slopes was to be co-ordinated with the main break-through by Vorobyev along the southern side of the ridge.

According to plan the attack of II Turkistan Corps started on 10 January while all still remained quiet on the front from Çakir-baba to the Pasin plain. The Turks were not asleep, and a strong fire met the demonstrative activity of the Russians along the front of X Corps. The Russian attack developed against Karadağ, but conditions proved desperately difficult. The Turkistanskis succeeded in ascending the eastern end of a very narrow ridge which led up to the peak; but here they met a formidable defence. Deep trenches had been blasted out of the rock and they were flanked by two 'towers' built of big stones. Here machine guns, and even two mountain guns, had been stationed, and they swept with their fire the narrow ledge of approach which fell on either side in vertical precipices. The repeated attacks of the Turkistanskis, maintained in an appalling frost, failed on both the 10th and 11th. Strangely, General Przevalski neglected to stop this useless ordeal during the two following days when the diversionary action of his corps was complete; and several hundreds more perished along the horrid ridge of Karadağ.[1]

On the second day of the Russian attack the Turks were already confident that X Corps could put up a successful resistance all along the line from Tortum-göl to the Oltu-çay; on the 11th, Abdul Kerim Paşa[2] contemplated moving his reserve up to the northern sector, but in the afternoon the order to do so was cancelled.

The only useful result of the diversionary action of II Turkistan Corps was to the east of the Oltu-çay. The Turks massed almost the whole of their 31st and 34th Divisions between Karadağ and Karşin and Colonel Treskin, while making a skilful demonstration in the Şekerli-Virintap region, was able to move a part of his force in the direction of Hohor and Eldenik where he was in touch with the 18th Turkistanskis (the right column of Voloshin-Petrichenko's group).

Not completely satisfied with Przevalski's work, Yudenich opened the second stage of the offensive at dawn on the 12th; I Caucasian Corps attacked. On the right bank of the Aras, the 154th Derbentski

[1] One battalion of the 17th Turkistanskis alone lost 300 killed or frozen.

[2] Commanding the Third Army in the absence of Mahmut Kâmil Paşa.

Regiment, their left flank protected by the 5th Caucasian Rifle Regiment, attacked across the frozen ravine of the Tayhoça towards Endek village and the height to the west of it.[1] The 28th and 29th Divisions of the Turkish IX Corps were occupying positions from the Aras south to the Karataş Mountains.[2] The 28th Division put up a stubborn resistance to the assault of the Derbentskis: they lost some trenches, then counter-attacked; not without difficulty, the Derbent men held what they had taken during the day.[3]

On the left bank of the Aras the 153rd Bakinskis attacked with an *élan* in keeping with the reputation of their regiment. Neither men nor officers had been told that their attack was only of a diversionary character; and, under the fire of the enemy's machine guns and skilfully grouped artillery, losses grew. By evening only a few Turkish advance trenches had been taken.[4]

On the right flank of the Russian 39th Infantry Division there was a serious failure on the part of the 155th Kubinski Regiment (always somewhat unlucky). Their task was to capture Çilligül-dağ, the southern door-post of the area of the planned break-through. The chief of staff of I Caucasian Corps volunteered to organize this attack personally. It was organized badly in spite of the distinguished reputation of General Lastochkin. The 155th attacked without artillery preparation, and remained without artillery support when the Turkish 33rd Division of XI Corps counter-attacked in the afternoon. The Kubinskis were thrown back with serious losses.

On 13 January, the situation must have appeared rather favourable to Abdul Kerim Paşa: stable on the front of X Corps, excellent at Çilligül-dağ, satisfactory on both banks of the Aras. Abdul Kerim was now convinced that the attack on X Corps was only a diversion, and that the furious fighting along both sides of the Aras indicated the main direction of the Russian offensive. He decided to introduce his reserve, the 17th Division of IX Corps, into the Azapköy sector. On the right bank of the Aras, units of this fresh division had to support the 28th and 29th Divisions while the 33rd Division, victorious on the previous day at Çilligül, had to counter-attack the (obviously weak) right flank of the Russians in the Pasin valley. The Turkish commander did not suspect that an overwhelming attack was preparing against the unfortunate 33rd Division; and he did not realize that the

[1] Called by the Russians *Tyk-dag*.

[2] T., 'Black-stone mountains'.

[3] At a critical moment the regimental priest, cross in hand, rallied the hesitating troops.

[4] The 18th Division of the Turkish XI Corps was defending this area.

battle was in reality already lost, since five out of nine of his available divisions were out of reach of the vital Köprüköy sector.[1]

On the morning of 13 January the Turkish 33rd Division boldly advanced towards Alakilise, their right flank covered by the 18th Division. The third division of the corps, the 17th, went into the line near Azapköy. Yudenich, somewhat disappointed but not worried by the failure of the Kubinskis on the previous day, strengthened the Azapköy sector with the 156th Elizavetpolskis from corps reserve. At the same time Vorobyev, instead of awaiting his own zero hour fixed for the following day, manoeuvred the 16th Caucasian Rifle Regiment to threaten the right flank of the Turks advancing towards Alakilise. The Turks were stopped. In the plain the Bakinskis and the Elizavetpolskis continued to make furious attacks, despite heavy losses. In this sector almost all the first line of Turkish trenches passed to the Russians. The attacking troops were now extremely tired; at the same time the questioning of prisoners from the Turkish 17th Division convinced Yudenich and his staff that the Turks had introduced their last reserves into the battle. On the left of the Russian line, along the right bank of the Aras near Endek, the 154th Regiment had beaten off seven Turkish counter-attacks by units of the 28th and 29th Divisions. Meanwhile interesting reports were coming in from Voloshinov-Petrichenko. Despite the terrible conditions on the high ridge of Çakir-baba and the snow-storm which persisted throughout the 13th, his columns were advancing, digging paths through the snow which here lay to a depth of five feet. The Don Foot battalions, incadred with the 18th Turkistanskis and the Georgiyevskis, were doing well and proving themselves different troops from those who had fled in panic near Malazgirt.

Assessing all the reports, Yudenich decided on a general attack at dawn on the following day (14 January), with the participation of all his reserves.[2] So on Russian New Year's Day, the columns of Voloshinov-Petrichenko approached the Koziçan shoulder from the north, east and south-east, while Treskin's troops were attacking along the front Hohor-Eldenik, thus cutting the communications of the Turks on Koziçan with the Oltu-çay valley. A bright winter sun broke through as Vorobyev directed the attack of the 261st Akhulginskis and

[1] The three divisions of X Corps and the 28th and 29th Divisions of IX Corps. Only XI Corps with the support of the 37th Division was ready to meet the Russian drive against Köprüköy.

[2] The 261st Akhulginskis and the 263rd Gunibskis joined Vorobyev, while the 262nd Groznenskis reinforced the Azapköy sector; all these regiments belonged to the 66th Infantry Division.

the 16th Caucasian Rifles against Alakilise and of the 263rd Gunibskis and the 14th and 15th Caucasian Rifles on the narrow front Maslahat-Elimi. Twenty-six field guns and eight howitzers supported the assault. By 11 a.m. the Maslahat-Elimi positions were taken. The Turkish 33rd Division and parts of the 18th were thrown back—in one direction on Kalender, in the other towards Çilligül and Hisar-dere.[1] Vorobyev occupied Stavut and in the evening reached Hisar-dere; it was the beginning of the break-through, since from Hisar-dere there was a path to Pazarçor and from Pazarçor to Köprüköy a track in good condition (ten miles). Just before the operation began, Yudenich had sent Vorobyev the Siberian Cossack Brigade, the only horsed force which proved really efficient in winter conditions in the high mountains. This brigade was now with the advanced infantry at Hisar-dere.

Meanwhile in the Aras plain the 153rd and 156th, supported by the 262nd, were still attacking. The Turks gradually lost ground (and a few hundred prisoners with guns), but some units proved able to counter-attack. On the right bank of the Aras the Russians were making slow progress from Endek towards Ardi.

The sixth day of the battle (15 January) proved decisive. Voloshinov-Petrichenko continued his attacks on the Turkish positions defending Koziçan-dağ, and in the evening, after desperate hours among the snow and rocks, one battalion of the 264th Georgiyevskis and two of the Don Foot captured the fortified summit. The Russians knew that across the shoulder of Koziçan, itself a part of the Çakir-baba ridge, ran a section of the *top yol* fit for the movement of troops to Portanos and Köse.[2] At the same time Treskin's column was already approaching Köse from the north.

During the day of the 15th Vorobyev's regiments marched in a long column by a path leading from Maslahat to Hisar-dere round the northern slopes of Çilligül mountain, which was still occupied by a few Turkish battalions. In the evening his advanced troops appeared within a few miles of the villages of Serbogan and Pazarçor. Both these points were in the rear of the Turkish positions on the Kalender-Haran line which represented the second line of defence in the Azapköy plain. The remnants of the Turkish 33rd Division with the 18th and units of the 17th Division continued to hold this line during the 15th against a Russian pressure from the east, which was not very heavy, since a decision was expected from Vorobyev's movement.

[1] T., 'Castle valley'. Russian sources called the place *Karabaş* (T., 'Blackhead').

[2] The first march of the route taken by the Turkish IX Corps during the Sarikamiş operation.

On the right bank of the Aras, the Turkish forces were regrouped so as to be nearer to the river. On the extreme left of the Russian line the 5th Caucasian Rifles occupied Saçlik and Alagöz, while the 154th, step by step, advanced to the west on the heights between Endek and Ardi. In this sector alone of the active front the Turks were now numerically stronger than the Russians, and their batteries in the region of Komasor gave effective help in the defence.

On the night of 15–16 January, the Siberian Cossacks were sent forward from Hisar-dere on Köprüköy with orders to blow up the historic bridge over the Aras. In the dark of a wintry night in a blizzard of snow the Cossacks lost direction and, after wandering about for hours, turned back.

On the morning of the 16th, Vorobyev's infantry advanced and occupied Çermuksu, Serbogan and Pazarçor. But Vorobyev's advance, meeting with little resistance, was very slow over mountain paths deep in snow. Even worse conditions impeded the columns of Voloshinov-Petrichenko; yet he succeeded miraculously in reaching Portanos, Köse and even Yayla Darlak.[1] In this way the communications of the Turkish X Corps and of the 34th Division with Hasankale were cut. Treskin, in the meantime, had reached Cansor and forced the 34th Division to clear out of all the region of Karşin-Eğrek, whence they retreated along the left bank of the upper Oltu-çay towards the village of Başviran. Voloshinov-Petrichenko had effected something more than the protection of Vorobyev from the north. While Vorobyev had achieved a break-through on a narrow sector of front, Voloshinov-Petrichenko and Treskin had opened a wide gap between two groups of the Turkish Third Army, X Corps and the 34th Division to the north and the other five divisions of IX and XI Corps to the south.

Already on the 16th the Turkish troops in the plain of the Aras had begun to be conscious of the presence of the enemy in their rear; their resistance slackened, they evacuated Çilligül mountain and Kalender. The artillery was hastily moved nearer to Köprüköy. Only strong rear-guards at Bedrevans and Haran kept up the battle. The Russian regiments of the 39th Division (obviously exhausted) were not very active on this fifth day of their frontal attack. On the right bank of the Aras the symptoms of Turkish retreat were clear enough. Late in the evening of the 16th, Yudenich warned the commanders on the Aras that the Turks would probably try to withdraw in the night in order to escape the threatened encirclement by Vorobyev's columns:

[1] Called also Karacli Pass.

the utmost vigilance was demanded from the troops and readiness to take up the pursuit.

However, the order of the commander-in-chief was badly executed. The Turks in fact retreated on the night of the 16–17th with great speed, and before their movement was observed by the troops of the 39th Division in front of them. During the day of the 17th, Vorobyev descended from the mountains to Badicivan, and in the late afternoon he occupied Köprüköy where his troops were joined that evening by the advance force of the 39th Division. Extremely irate, Yudenich that evening telegraphed Kalitin (commanding I Caucasian Corps) declaring that his corps had given the enemy the possibility of escaping and was at that moment marking time on the same spot. He ordered Kalitin to make his columns move; the horse were to be moved at least to Hasankale.

On the morning of the 18th the Siberian Cossacks reached Bekbad and later in the day advanced to Hasankale, meeting some resistance from Turkish rear-guards. Of the infantry, the first in the running were Voloshinov-Petrichenko's troops who, on the same day, entered Timar, some five miles to the north-east of Hasankale. Vorobyev was concentrating his men to the west of Köprüköy, the 39th Division to the east. On the right bank of the Aras the Turks were retreating in good order to the fords higher up, unmolested by their exhausted antagonists of the 154th and the 5th Caucasians.

Early on the morning of 19 January, the Siberian Cossacks attacked Hasankale where the Turkish rear-guard of four battalions was completely annihilated.[1] The Cossacks proceeded the same day to Kuruca, where their horsed battery fired the first shots against the forts of Erzurum on the slopes of the Camel's Neck.[2] That day marked the end of the Köprüköy battle.

Abdul Kerim had ordered a general retreat on the night of the 16th–17th. The commander of X Corps, however, decided to begin the withdrawal of his troops only on the night of the 17th–18th. The day of the 17th was used for movements to the rear and the destruction of irremovable stores. II Turkistan Corps was warned in time and General Przevalski had twenty-four hours longer than his colleague of I Caucasian Corps in which to organize the pursuit. But the topographical conditions and the good order in which the Turkish

[1] Their losses were 1000 sabred and 1500 prisoners and 4 guns taken.

[2] On the road several hundred more prisoners were taken from different regiments of the Turkish 18th and 33rd Divisions of XI Corps; the 17th Division of IX Corps was retiring in the greatest disorder.

X Corps was retreating did not give any opportunity to the Turkistanskis to exert a dangerous pressure. On the Turkish left the 30th and 31st Divisions tried to hold on to the passages across the Ak-dağ[1] which separated the valley of Tortum from those of Sivri and Oltu. But at Kara-pinar[2] on their right the 31st Division was outflanked, since, with characteristic egoism, the commander of X Corps had left the neighbouring division of XI Corps (the 34th) to its fate. This division, exhausted by a difficult struggle in the frozen mountains against the converging columns of Treskin and Voloshinov-Petrichenko, proved to be in a bad state, and from 16 January it was retreating in small broken groups on Telli, Başkale and Başviran. On 19 January the 31st Division evacuated Kara-pinar and passed into the Tortum valley, by which route all three divisions of X Corps were retiring, one after the other, on to the forts of Erzurum.

The battle of Köprüköy was an important victory for the Caucasian army. The fortified front covering Erzurum had been broken and now the fortress of Erzurum could be attacked directly or besieged. The Turkish Third Army had been defeated and had suffered serious losses: nearly 15,000 killed, wounded and frozen, to which must be added 5000 prisoners and about the same number of deserters. It had been 65,000 strong before the battle; only 40,000 defeated troops took refuge behind the forts of Erzurum. Material losses were not so considerable: 20–30 guns captured or abandoned. XI Corps suffered particularly, losing perhaps 70% of its effectives; but the losses of X Corps were slight. Russian losses were also considerable: 10,000 killed and wounded and 2000 in the hospitals, seriously frost-bitten.[3] The latter numbers are comparatively small when the terrible conditions, in which the men of Voloshinov-Petrichenko's group marched and fought, are recalled.[4] Yudenich's careful preparations for winter operations in the frost-bound mountains had proved efficient.

All the preparatory work of Yudenich's staff was effective. It was really an excellent example of competent team-work, particularly the arrangements which ensured complete surprise of time and place for the principal attack. The extraordinary precautions which he took gave Yudenich a well-merited success. He was somewhat stronger than the enemy in infantry,[5] and he had a score of Cossack regiments while

[1] T., 'White mountain'.

[2] T., 'Black springs'.

[3] For the first ten years after World War I, badly frostbitten men were a common sight in the streets of Caucasian towns.

[4] The 18th Turkistanskis, the 264th Georgiyevskis and the Don Foot Brigade.

[5] Infantry numbered 75,000, or 80,000 if the militia *druzhiny* are included.

the Turks had virtually no cavalry in the field. He had also a very powerful artillery: 230 guns against 100 of the enemy. But he had against him carefully fortified positions in many sectors of the front, which sometimes coincided with the most difficult topographical conditions. His cavalry was almost useless, and his superiority in artillery was neutralized by the terrain. His only chance was to concentrate in the greatest secrecy a local superiority of strength in the sector chosen for a break-through, and to attack energetically in other sectors to prevent the enemy from regrouping.

On the front selected, Yudenich concentrated between Virintap and Çilligül mountain 35,000 infantry[1] against the 13,000 men of the Turkish 33rd and 34th Divisions. He had twenty-one battalions (17,000 men) against X corps (about 20,000 strong) and eight (later twelve) strong battalions (10,000–14,000 men) against the 17th and 18th Divisions (which were reinforced by single odd battalions and attained about 15,000). He had only eight battalions and militia (9000 men) on the right bank of the Aras against the whole of the 28th and 29th Divisions with single battalions of the 37th (15,000–16,000 men). However, the unfavourable balance in the diversionary sectors was not dangerous, even on the right bank of the Aras where the Turkish counter-manoeuvre could scarcely be expected to give important results.[2]

The operation was logically planned and excellently prepared; but it did not give all the results expected. Though the Turkish Third Army had been defeated, Vorobyev's group had failed to cut off any part of it. This large force found itself unable to advance quickly enough after the break-through, because this took place on too narrow a front.[3] Vorobyev's column consisted of five regiments and a numerous artillery; it was too large a body of men to carry through a rapid march over the single path to the north of Çilligül.[4]

[1] The groups of Treskin, Vorobyev and Voloshinov-Petrichenko, with the 155th Regiment: 42 battalions in all.

[2] Since Abdul Kerim retained two divisions on the right bank of the Aras, he obviously considered this sector, together with the valley of the Sivri-çay, the most vulnerable part of his front; he regarded the plain to the north of the Aras as being strongly fortified against frontal attack and the Çakir-baba ridge, across which the Russians pressed their break-through, as inaccessible.

[3] The failure of the 155th to take the southern door-post of Çilligül made the front still narrower.

[4] It is worth noting that the 153rd and 155th Regiments, having sustained 2000 casualties each in the too enthusiastically pressed diversionary attacks of the early days of the battle, were not in a condition to follow the enemy on the night of 16–17 January.

CHAPTER XXVIII

BEFORE THE STORM OF ERZURUM, 20 JANUARY TO 10 FEBRUARY 1916

MAPS 1, 19, 31, 25

THE victory at Köprüköy raised the spirit of the Caucasian army to a high degree, despite the severe fighting, the serious losses, and the difficult conditions of mountain war in winter. A determined and coherent leadership and the excellent organization of liaison, supply and sanitary services had inspired all ranks with confidence. At the same time there was plenty of evidence of the disorder and demoralization of the retreating enemy. The askers and their officers taken prisoner were sulky and resentful; full of criticism of the command and lacking confidence in the existence of reinforcements which had been so often promised but which had never arrived.

For some days after the battle a wave of extreme optimism pervaded the Caucasian army. Some of the commanding officers, Kalitin for instance, urged an immediate assault on the Deve-boyun positions. Yudenich himself perhaps shared in the optimism but his responsibility made him more cautious, particularly when, with the grand ducal congratulations, arrived the counsel not to extend so successful an offensive and the injunction to remain satisfied with the results already obtained. Tiflis general headquarters was not too pleased that the troops had advanced as far as Hasankale; and defensive positions in the neighbourhood of Köprüköy were recommended by the military advisers round the grand duke.

Yudenich wanted fuller information on the exact situation of the Turkish Third Army. He sent his trusted collaborator, Colonel Maslovski, chief of the operational section of the field staff, to reconnoitre the region of Hasankale and the approaches to Erzurum. In addition to the Cossack patrols and the activities of Armenian agents, Maslovski now had the advantage of the observation of the fortress area by about twenty Russian planes, the first to operate in Caucasian war; he reported that Erzurum seemed unprepared to meet an attack. Trenches, ditches and roads were deep in snow; and large stores of munitions and supplies had been abandoned between Köprüköy and Hasankale. No reinforcements had recently arrived in the fortress area. It seemed doubtful whether the vast fortified camp constituted round Erzurum could be successfully defended by the strength of the defeated and numerically reduced Third Army. Maslovski particularly appreciated the importance of reports provided by patrols of Vorobyev's Rifle Division: the high and rocky ridge of Kargapazar, lying to the north of the Deve-boyun ridge and dividing the basin of the Aras from that of the Kara-su (Western Euphrates) was not occupied by Turkish troops. The Turks evidently considered these formidable mountains as an inaccessible obstacle between the plains of Pasin and Erzurum.

With the unanimous support of his subordinate commanders, Yudenich now addressed the grand duke with a formal request for authority to undertake the storm of Erzurum. He met with considerable opposition: the grand duke particularly objected to making available the last reserve of eight million rounds stored at Kars.[1] When, however, the matter was referred to the *Stavka*, Alexeyev expressed full confidence in the judgement of Yudenich. At last, on 23 January, Grand Duke Nicholas gave the necessary authority—although Yudenich took upon himself the full responsibility. His staff at Karaurgan immediately began the urgent work of preparation. The losses at Köprüköy were filled from reservists and from the younger men of the militia formations. A modest siege artillery (thirty-four pieces) was improvised from the fortress guns at Kars, and transported to the front under conditions of the greatest difficulty. The main road, through Karaurgan, Köprüköy and Hasankale, was widened, while the Décauville railway was carried forward from Sarikamiş to Karaurgan. Yudenich was able to organize several motor convoys, another innovation on the Caucasian front. The arrival of the small but well-trained Siberian air squadron proved of great help for reconnaissance.

The losses in the regiments in action at Köprüköy were soon filled, but Yudenich could secure no fresh units.[2] Thus situated, he decided to undertake action designed to prevent the Turks from bringing to Erzurum units from other sectors of their front; he ordered de Witt, who had replaced Oganovski in command of IV Caucasian Corps, Lyakhov in Lazistan, and Chernozubov in Azerbaijan, to begin offensive operations by 20 January.

[1] The bulk of the reserve, 15,000,000 rounds, had already been sent to the western front during the summer of 1915.

[2] Yudenich borrowed three battalions of the 6th Caucasian Rifles from IV Caucasian Army Corps, and recalled the 13th Caucasian Rifles from their trip to Julfa. He ordered the 4th Plastun Brigade to return from Azerbaijan, but it failed to arrive in time for the Erzurum operation.

In the Artvin area small units from Lyakhov's group crossed the frontier and by the end of January had established liaison with the plastun battalions acting as the flank guard of II Turkistan Corps in the Çoruh valley. More impressive operations in the coastal region developed somewhat later with the arrival at Batum of the battleship *Rostislav*, sent to support the land forces.

The task of IV Caucasian Corps might appear unusually difficult since, by the middle of January, available troops consisted of only two rifle regiments and one Cossack division with one Armenian *druzhina* and a militia brigade. Yudenich indicated as the first aim of the operation by IV Caucasian Corps the capture of Hinis, the important road junction through which reinforcements from the Muş-Bitlis area might move up to Erzurum. Hinis, however, is separated from the Mirgemir Pass (where lay the most advanced troops of the right flank of IV Corps) by the eastern part of Bingöl where the few tracks seemed impassable in the middle of the Armenian winter. General de Witt decided to take the longer route through Malazgirt with his main force,[1] leaving only the 7th Caucasian Rifles with the task of trying to march on Hinis from the Mirgemir. The only chance for this expedition lay in the fact that the Turkish forces left to cover the Hinis-Muş-Bitlis region were weak; and the Turkish commander took his first duty to be the defence of Muş and Bitlis at which points the majority of the troops at his disposal were concentrated.[2]

Cavalry and Kurds were met and thrown back at Malazgirt. The 8th Caucasian Rifles continued the advance to the west, and on 25 January had a sharp engagement with Turkish infantry which, with four guns, was defending a position at Kara-köprü on the Hinisçay. Cossacks and Armenians outflanked the Turks and they retired southward towards Muş. Next day (26 January) Hinis was occupied by the 8th Caucasians; the left column of de Witt's corps had accomplished its task more easily and more rapidly than could have been expected. In the meantime the right column (7th Caucasian Rifles), advancing from the Mirgemir Pass, had run into deep snow and bitter cold in the high mountains. Very few of the enemy were encountered, but the unfortunate regiment left 200–300 more or less frozen men and all the animal transport along the route. The 7th Caucasians reached Hinis several days after the 8th. IV Caucasian

[1] The 8th Caucasian Rifles, the Cossack division, the Armenian and several militia *druzhiny*.

[2] The 36th Infantry Division (two regiments only); nine battalions of gendarmes and Frontier Guards; the 2nd Cavalry Division and Kurdish formations.

Corps now held two points (Hinis and Malazgirt) from which an offensive might be launched against Muş, the junction for routes leading westward along the valley of the Murat-çay to Harput, and south through the main chain of the Taurus to Diyarbekir.

Chernozubov's Van-Azerbaijan group supported the operation against Hinis by moving round the northern and southern shores of Lake Van. The attention of the Turks was rather focused on the Lake Van region, and a considerable part of their forces had been retained round Bitlis. By 3 February Chernozubov's troops were marching on Adilcevaz to the north of the lake, while along the southern shore they had occupied Norkev. They continued to advance in the directions of Ahlat and Tatvan.

The capture of Hinis protected the left strategic flank of Yudenich's main army. Meantime Yudenich was engaged in grouping the troops of I Caucasian and II Turkistan Corps with a view to the forthcoming attack on Erzurum. He was not quite satisfied with the slow progress of General Przevalski's troops on the north of the Russian front. The advance of II Turkistan Corps had been delayed by the difficult character of the mountainous country which bars access from the upper valleys of the Sivri and Oltu into that of the Tortum-çay. Moreover, the troops of the Turkish X Corps were retiring in good order; their rear-guards were resisting, when called upon to do so, with notable stubbornness. Unfortunately for these gallant troops, the leadership of X Corps was not very skilful. The commander of X Corps was ordered by Abdul Kerim Paşa to defend the northern front of the fortress of Erzurum, namely, the forts of Kara-göbek ('Black navel') and Tafet which had been specially built to cover the northern and southern entrants of the Gürcü-boğaz Pass. This pass gives access from the upper valley of the Tortum-çay into that of the Kara-su (western Euphrates), and it is therefore the northern gate into the plain of Erzurum. The commander of X Corps interpreted his orders in rather a narrow sense and decided to concentrate his main force on positions protected by these two forts: between a shoulder of Dumlu-dağ and the northern end of Deve-boyun, where Fort Tafet stood. In the Tortum valley his rear-guards were only fighting delaying actions and even the important position to the north of Kizilkilise, a ridge between the source of the Tortum-çay and the Gürcü-boğaz defile, was occupied only by one battalion with machine guns.

Since 20 January the 4th Turkistan Rifle Division had been moving forward in small detachments along a wide front from Lake Tortum to the sources of the Sivri-çay. While the 13th and 15th Regiments penetrated into the Tortum valley to the north of Tortum town, the 14th Regiment sent its advance parties across the ridge at Liska village[1] towards the position covering Kizilkilise. On the 23rd, a battalion of this regiment skilfully outflanked the Turks and they left the way into Kizilkilise open to the Russians. On the following day the whole of the 14th Regiment concentrated there, and this important point, giving access into the Gürcü-boğaz defile, had been won almost without an action.

During the next two weeks, pursuant to the orders of Yudenich, Przevalski gradually pushed detachments of the Rifle Division forward on to the approaches up Dumlu-dağ (Kugans and Zinavur were occupied on 5 February). Yudenich had in view preparations for the passage of this high and difficult ridge when the moment came to attack Erzurum. Meantime, the 5th Turkistan Rifle Division was concentrated in the area of Kizilkilise where the men were put on to the laborious task of making tracks and runways across the snow-covered mountains for the field guns and howitzers necessary for the reduction of the Kara-göbek and Tafet forts.

On the Russian left flank, in front of the forts on Deve-boyun and Palandöken, the troops of I Caucasian Corps were deployed between Upper and Lower Toy[2] villages and Ali-baba-dağ.[1] On the extreme left flank, a special detachment under General Chikovani kept the Palandöken forts under observation.

The principal attention of Yudenich, however, was directed to the high ridge of Kargapazar[3] which lay between the Turkistan and Caucasian Corps. As has already been related, Russian patrols had discovered at an early stage that this ridge, evidently regarded by the Turkish staff as inaccessible to the enemy, was not occupied by troops nor was it included in the defences of Erzurum. On 23 January Yudenich had ordered the immediate occupation of Kargapazar by advance parties of Vorobyev's 4th Caucasian Rifle Division. Something happened to cause Vorobyev to defer the execution of this order until the 25th, and when, on that day, a rifle battalion was climbing the ridge from the east a Turkish battalion was marching up to establish possession from the west. After a sharp encounter the Russian riflemen were successful in throwing back their rivals; the Turks managed only to hold on to the western slopes. The Turkish staff still failed to understand the significance of the Russian interest

[1] Liska and Ali-baba-dağ are not marked on available maps.
[2] T., 'Bustard'. [3] T., 'Crow Bazaar'.

in the ridge; from time to time Turkish patrols appeared on the high plateau of Kargapazar and there were skirmishes almost every day. The Russians continued, however, to retain control, while their working parties set to work to build tracks up which guns might be brought. From 31 January Kargapazar was garrisoned by the whole of the 4th Caucasian Rifle Division. The cold on the plateau (about 3000 m. high) was so intense and the snow so deep (lying often to the height of a horseman) that working parties were only able to stay there for five or six hours at a time. The regiments responsible for Kargapazar had to maintain a constant rota system between the plateau and the base, and there was an uninterrupted stream of men climbing up and climbing down. When, however, the day for the storm came, the whole of the 4th Caucasian Rifle Division concentrated on the high plateau and brought most of their guns into action there.

On the western slopes of Kargapazar, Turkish troops (never stronger than two or three battalions) remained relatively inactive, inclined to hibernate in the miserable conditions of cold. In general the Turkish command in Erzurum maintained a rather passive attitude and clearly did not believe that the Russians could intend to attack before the spring, by which time the Turks were confidently expecting the arrival of large reinforcements.

Despite the defeat at Köprüköy an optimistic interpretation of events in Armenia continued to prevail in the capital. Enver, encouraged by the conversations with von Falkenhayn at Orsova, had ordered a mass movement of the divisions, now rather overcrowded in the straits area, to the Asiatic theatres of war. It had been decided to regroup thirteen divisions: two to Iraq, two to Syria, two to Alexandretta, and seven to the Caucasian front. V Corps (9th, 10th and 13th Divisions) was to be transported by rail (and partly by sea via Samsun) to Ankara, whence it had a march of between 450 and 500 miles to Erzurum. The first units of the Second Army (which were under the command of Mustafa Kemal Paşa who had distinguished himself at Gallipoli) were detailed by rail to Ulu-kişla (north-east of Pozanti and at the head of the Gülek-boğaz Pass through the Cilician Taurus) or to Ras-al-ayn (north-east of Aleppo), whence they had to cover several hundred miles to reach Palu on the way to Muş and Bitlis.[1] (Cf. Map 17.)

[1] The divisions immediately available were the 1st, 7th, 5th and 8th. About this time the vice-generalissimo ordered the creation of a further nine divisions (nos. 53–61), but artillery, machine guns and supply trains were lacking, and they were never put in the field.

The movement of these thirteen divisions began in the first half of January, but the capacity of the Turkish railways was limited, and the majority of the units of the Second Army only reached their final stations during March. The 5th Division was the first to arrive on the Russian front; after a long march it approached the region of Diyarbekir at the beginning of March, but it came too late to prevent the capture of Muş and Bitlis by the Russians.

Enver Paşa certainly could not count on any serious reinforcements for the defeated Third Army, which had taken refuge behind the forts of Erzurum, before the middle of March or even before the beginning of April. But he still remained optimistic for the fate of the fortress; and he shared this optimism with the principal personalities in the Turkish government and their German advisers. The reputation of 'the impregnable stronghold' stood high. In spite of Köprüköy, Heimann's disastrous attempt at assault (indeed after victory at Deve-boyun) might be recalled. Furthermore the Russian Caucasian army was known to lack the technical means necessary for the opening of an operation against a strong modern fortress.

In relation to modern conditions of attack, defence and access,[1] Erzurum in 1914 was a stronger fortress than in the winter of 1877–8 when the weak and defeated army of Ahmet Muhtar had managed to hold it against the victorious forces of Loris Melikov.

Soon after the Berlin Congress, and arising out of the convention concluded between the British and Turkish governments which provided a British guarantee for the Asiatic possessions of the Ottoman empire, British engineers had elaborated plans intended to convert the old fortress area into a vast modern fortified camp. These plans were in 1888 considered to be excellent by a special secret commission set up to study the Erzurum works at the headquarters staff of the (Russian) Caucasian military district. The principle of the new fortified camp had been based on the experience of the campaign of 1877; the real defence of Erzurum was found to lie along the Deve-boyun positions. As a result, the Deve-boyun ridge (with all its advance counter forts at Uzunahmet hill and elsewhere) was included in the defensive perimeter. The work on a series of new forts was pushed forward with relative speed, and as early as 1888 Russian observers were well informed on their situation and general construction. The modernization of the fortifications of Erzurum was really complete when, during the nineties, German military advisers replaced the

[1] An important factor in a region of great climatic and topographical difficulties and limited communications.

British in the confidence of the Turks. In the first ten years of the new century only two forts—those on Palandöken—had been built on German advice.

In 1914 the fortified perimeter of Erzurum consisted of a central group of eleven forts and batteries and two flank groups, each of two forts. On the northern end of the Deve-boyun ridge on the lofty mountain of Çoban-dede (2400 m.) a fort of that name had been raised; it represented the left extremity of the central fortified camp. In front of this fort and rather under it on a lower parallel shoulder of Deve-boyun stood the fort of Dalan-göz, defending the approaches to Çoban-dede and the narrow ravine of the Toy stream. On the same line to the south, defending Hamam-dere ravine, two strong batteries had been located on the table-topped hill of Uzunahmet where such sharp fighting had taken place during Heimann's attack on Deve-boyun.[1] To the south of the main road three forts in a line covered the rocky salient which was divided from the main ridge of Palandöken by the ravine of the Yağliyan-su.[2]

This line of works from Çoban-dede to the two Ortayuks formed an outer ring of forts which (with the exception of Çoban-dede itself) were built not along the main ridge of Deve-boyun but on a series of lower lateral heights. From the approaches to this line a second line of forts could be seen behind, and above, the outer ring. These were the forts along the main ridge itself: Sivişli (commanding the saddle across which ran the main road), Yildirim, called also Ağzi-açik ('Open Mouth'), Toplak, and Gez, the last on somewhat higher ground to the north of the steep Yağliyan ravine which divides the ridge of Deve-boyun from that of Palandöken. This line was three and a half miles to the west of the first line, and lay, properly speaking, behind the open right flank of the outer ring. The whole distance covered by the forts of the first and second rings was twelve miles, from Çoban-dede to Gez.

Distinct from these two rings of the central fortified group were the two flank groups at Gürcü-boğaz and Palandöken. The Gürcü-boğaz group defended the narrow defile, eight miles long, between the southern shoulder of Dumlu-dağ (itself an outlier of Gavur-dağ) and

[1] The batteries had been named Uzunahmet and Uzunahmet Karakol.

[2] Kaburga, Ortayuk and Ortayuk Ilavesi. Kaburga was built on the shoulder over the main road behind which the Erivanski Grenadiers were taking cover at the decisive moment of the battle of Deve-boyun (Nov. 1877). Ortayuk appears in forms Oyuk and Höyük in Baki, *Büyük Harpte Kavkaz Cephesi*, vol. I, text and map of Erzurum area. *Höyük* (T.) means 'mound'. *Ilave* (gen. *ilavesi*) means 'addition'; here in the sense, probably, of 'supplementary work'.

the western slopes of Kargapazar with its north-east spur of Kandildağ. In the middle of this defile on the isolated hill over the village of Kara-göbek a fort of the same name had been built. At the southern end of the defile another fort, Tafet, covered the lower slopes of Kargapazar where the ravine of the small Köşk-çay joins the upper valley of the Kara-su. Here, immediately to the south of the fort, the plain of Erzurum begins. The distance between Kara-göbek fort and Tafet was seven miles; and five miles separated Tafet from Çoban-dede. This undefended gap was assumed to be adequately defended by the 'inaccessible' high rocky ridge of Kargapazar.

The German military experts of the nineties had paid no attention to the possible significance of Kargapazar. But they had not been so sure about the natural protection provided by the Palandöken ridge which flanked the southern end of the Deve-boyun position. There were, in fact, neither tracks nor paths across Kargapazar, but a path ran across the slopes of Palandöken at the head of the Yağliyan ravine, and farther to the south a track crossed the main ridge leading from the village of Taşkesen to the plain of Erzurum. Two new forts, Palandöken East and Palandöken West, were built near the summit of the ridge (2700 m.) to close the Taşkesen track and to hold under fire the path running south-east to north-west across the head of the Yağliyan ravine and between the Palandöken forts and Fort Gez.

Thus the fortified perimeter of Erzurum was not in fact complete but only partial, covering approximately 180° of an arc, with a radius between eighteen (Kara-göbek) and eight miles (Kaburga fort). In the west there were about 60° of undefended plain, but on both flanks of the western sector were the natural obstacles of Dumlu-dağ and Palandöken. An inner circle of works, already existing in 1877, might serve as a third line of defence.[1]

The forts built after the war of 1877–8 corresponded to the technical notions of the epoch of Brialmont's constructions. The experience of 1914–16 demonstrated that Brialmont's forts (Liége, Namur, Antwerp, Bucharest) could not resist German and (particularly) Austrian heavy guns and big field mortars. But the Russian Caucasian army had not available guns approaching the weight of the modern pieces in use in the west. The forts of Erzurum (as Enver and his German advisers had surmised) might be expected to offer a successful resistance to any artillery which the Russians could transport to the Pasin valley. The armament of the Erzurum forts was planned

[1] But certainly not in theory in the modern conditions of war. These works were: Surp Nişan, Mecidiye, Aziziye, Ahali, Keremetli. (Cf. Map 15.)

PLATE VI

ERZURUM: VIEW FROM THE CITADEL LOOKING NORTH-EAST TO GÜRCÜ-BOGAZ AND DEVE-BOYUN

on a grand scale: 30–50 modern guns for each fort (or fixed battery), with a total attaining 1000. But these plans remained largely on paper. Russian sources of information gave the number of guns as between 300 and 700; the first number proved nearer the mark and only 25–30% of the pieces were of modern type.

More formidable than the fortress artillery were the field works which the Turks had constructed around and between the forts under the direction of a German specialist, Colonel von Posselt, who had been sent with a mission to Erzurum when the city seemed threatened by the consequences of Sarikamiş. Systems of trenches and wire-fields had been constructed along the ravine of the Yağliyan-çay as well as on the approaches to Çoban-dede and Dalan-göz and in the Gürcü-boğaz defile.

In view of the extensive character of the Erzurum fortified camp, considerable forces were clearly necessary for its defence: 30,000 infantry with appropriate machine guns and artillery for manning the forts of Deve-boyun and the intermediate field works; 25,000 to protect the flanks (20,000 at Gürcü-boğaz and 5000 at Palandöken); and a mobile reserve of not less than 20,000. It was a total of 75,000, far surpassing the numbers of which Mahmut Kâmil Paşa certainly could dispose when he returned to Erzurum from Istanbul. After the losses at Köprüköy, the Third Army had been reinforced by reservists and recruits at hand. But only X Corps had some 20,000 reliable infantry (three divisions with auxiliary formations); IX Corps could not muster more than 15,000 (of whom many were raw); and XI Corps was not stronger than 10,000. The total of all troops available in Erzurum area did not exceed 50,000. Field artillery was insufficient, and many machine guns had been lost in action by the regiments of IX and XI Corps. Morale was far from high among the troops who had been through Köprüköy, and complements were badly instructed. Since he was unable to establish his authority over the curious pretensions to independence of the commander of X Corps, Mahmut Kâmil was only able to put into the forts of Deve-boyun the three divisions of IX Corps, while he retained what remained of XI Corps as a general reserve.[1]

With all the handicaps of his situation, the Turkish commander might feel confident that behind the solid walls of the forts and supported by numerous (if not all modern) positional guns his troops had a serious chance of giving a successful demonstration of the famous Turkish stubbornness in defence.

[1] Mixed detachments manned the Palandöken forts and covered the right flank.

Yudenich's army, numerically stronger, was not so powerful that he could rely on weight of numbers to give him the victory. With the greatest efforts he had been unable to concentrate more than 80,000 infantry, of which at least a quarter were raw militia and recruits. His field artillery was likely to prove ineffective against the forts, and his improvised siege park was inadequate for serious siege operations.[1]

But Yudenich had no intention of attacking the fortress. His plan was to attack Erzurum as an extended fortified position and to break into the plain through weak points in the line. If his plan were successful, the fortress would be helpless.

[1] In 1888 the secret Russian commission had concluded that the storm of Erzurum fortress would be impossible without proper preparation by strong siege artillery.

CHAPTER XXIX

THE STORM OF ERZURUM, 11–15 FEBRUARY 1916

MAP 25

THE Russian plan of attack was quite simple and very similar to that which had proved so successful at Köprüköy. Yudenich proposed to break through an undefended sector of the Turkish line, undefended because it was considered impassable owing to the natural difficulty of the terrain and the ferocious winter. The sector chosen was the Kargapazar ridge; and the troops selected were again Vorobyev's 4th Caucasian Rifle Division, while again, on their right, Voloshinov-Petrichenko's Don Foot were sent in over the still more difficult terrain of Kandil-dağ.

In contrast, however, to the Köprüköy plan of battle, Vorobyev's attack was not the principal blow to be delivered. The success of Vorobyev would only be the decisive step whıcn would ease the task of the troops carrying out the main attack; this was to be accomplished by Przevalski's Turkistan troops in the Gürcü-boğaz defile. Przevalski had to capture the Kara-göbek fort; then break through the defile and take Tafet fort. Since it was anticipated that his task would prove extremely difficult, Voloshinov-Petrichenko's columns were to descend from the slopes of Kargapazar on to the heads of the enemy at Tafet. At the same time a part of Przevalski's riflemen had to try the same tactic to the west of Gürcü-boğaz; they had to cross the high ridge of Dumlu-dağ and fall on the defenders of the defile. By this combined attack of three rifle divisions (supported by Don Foot and mounted Cossacks) it was intended to open a passage on to the plain of Erzurum. This was to be the main attack which was to bring a swift victory to the Russians. Once the Russians could penetrate in force into the plain of the Kara-su, the forts of Deve-boyun would lose all meaning; a hasty evacuation would be the only resort for the Turks.

An attack, however, on Deve-boyun was necessary, if only as a diversion. It was known that Mahmut Kâmil had a mobile reserve. Yudenich supposed this reserve to be somewhere near Çoban-dede; in fact it was in station behind Sivişli Fort in the centre of the Deve-boyun group of forts. Again as at Köprüköy, to pin down the reserve, the courageous 39th Division were called upon to make a serious attack against the forts of Çoban-dede and Dalan-göz. At the same time the siege park from Kars, with the field howitzers, had to open the action by a bombardment of Dalan-göz, Uzunahmet and the field works in the Toy ravine. Less important demonstrations were to be directed against Kaburga and the two Ortayuk forts, while a special detachment was to approach the Palandöken forts and try to attract the attention of the Turkish command in that direction.

During the last days before the storm Yudenich showed signs of nervousness. There were rumours (which proved false) of the arrival of reinforcements at Erzurum. The commander-in-chief decided not to delay the assault any longer, although the second echelon of Kars guns had not yet arrived and rear preparations were not quite complete. On 8 February there was a characteristic scene at Yudenich's headquarters. He invited all the officers commanding sectors and columns to dinner to review the plan already communicated to them. After the meal Yudenich began: 'And now, gentlemen, to our business. Have you all got my order? We are attacking on the 10th.' Then silence. Someone began to explain the necessity for further delay; others were ready to start a discussion. But Yudenich cut it immediately with the abrupt remark: 'All right! We attack on the 11th.' And he took leave of his guests.[1]

At 2 p.m. on 11 February the Russian heavy artillery opened fire on the forts of Çoban-dede and Dalan-göz and on the Turkish batteries on Uzunahmet.[2] The Turks replied, the fire from the Uzunahmet

[1] Maslovski, p. 275, n. 1, based on Kvinitadze's unpublished memoirs.

[2] The Russian order of battle on 11 February was as follows:

Extreme right: 4th Turkistan Rifle Division: five battalions co-operating in the attack on Kara-göbek Fort; three battalions crossing the Dumlu-dağ in the Zinavur area; the rest, with some plastun battalions patrolling the right flank of the advance.

Right: 5th Turkistan Rifle Division: two regiments attacking Kara-göbek Fort (with one regiment held to reserve of II Turkistan Corps); four Don Foot battalions to occupy part of Kargapazar nearest to Kandil-dağ, and then to descend to Köşk hamlet in the Gürcü-boğaz defile on the approaches to Fort Tafet.

Right centre: 4th Caucasian Rifle Division: to descend from Kargapazar on a wide front (five miles) between Forts Tafet and Çoban-dede.

Left centre: 39th Infantry Division: 153rd, 154th, 156th Regiments to attack in direction of Forts Çoban-dede and Dalan-göz; to the left, two battalions of 5th Caucasian Rifle Regiment and 263rd Regiment to maintain passive attitude in front of forts from Uzunahmet to Ortayuk.

Left: Col. Dokuchayev's detachment: 155th Regiment and two battalions of 5th Caucasian Rifles to break through between Fort Gez and the Palandöken forts along the path across the upper ravine of the Yağliyan-çay.

Extreme left: General Chikovani's detachment: seven militia *druzhiny* to demonstrate against the Palandöken forts.

General reserve: three regiments of 66th Infantry Division.

Artillery: twelve Kars guns and seven howitzers, with forty-two field guns and eighteen mountain guns, on the left centre and left; forty-eight field guns and forty mountain guns with four howitzers were allocated to the right columns.

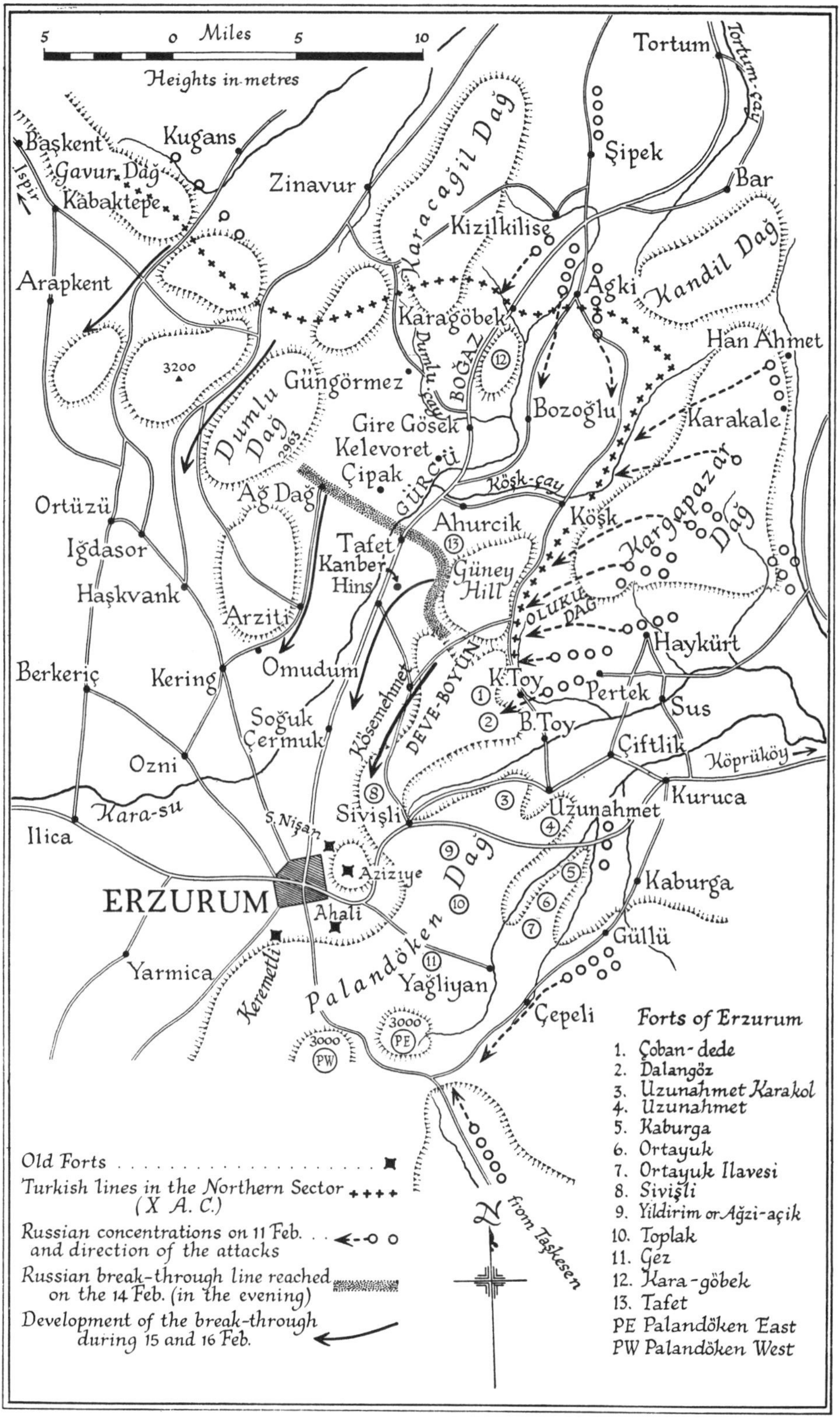

Map 25. The storm of Erzurum, 11–16 February 1916

batteries being particularly lively. This nominal artillery preparation could have only the moral effect of cheering the troops who were to go forward in the night attack. The material damage effected by the Russian guns was negligible, although the fire proved later to have been fairly accurate and the Russians had the advantage of the advice of a Turkish gunner officer who had recently deserted to them.

In the darkness and in exemplary silence one and a half battalions of the 153rd Bakinskis moved forward and succeeded in completely surprising the garrison of Dalan-göz Fort, who possibly had expected a longer artillery preparation. The Bakinskis, under their gallant Armenian Colonel Pirumov (Pirumyan), approached very near to the north angle of Dalan-göz before they were discovered. After a sharp fight, one company of the Bakinskis scaled the wall and penetrated into the fort. The rest followed. There were several hours of wild bayonet fighting inside the fort, but at dawn on 12 February Dalan-göz remained in the hands of the Russians. It was a serious success, at a cost of several hundreds killed and wounded. Another initial success had been scored by part of the 156th Elizavetpolski Regiment who penetrated up the Toy gorge and took possession of Küçüktoy height to the south-east of Fort Çoban-dede.[1] Such were the results of the night action, preliminary to the first day of the storm.

During this first day (12 February), the Turks, after concentrating their gun fire on Dalan-göz, started a series of furious counter-attacks in efforts to retake the fort. There were eight attacks one after the other. Pirumov's men inside fought under the most difficult conditions; more than once they had almost exhausted their ammunition since communication with the rest of the Bakinskis outside the fort was cut off owing to the intensive Turkish barrage. However, Pirumov held the fort until the end of the day when, out of 1400 men, he only had left 300 capable of standing to arms (many of whom were wounded). During the night the defenders of the fort were relieved by other companies of the 153rd.

Late on the 12th the 156th Elizavetpolskis started an attack on Çoban-dede. But the mountain was high and steep; rocks were slippery and climbing in the dark proved extremely difficult. During the night some of the Elizavetpolskis succeeded in crawling almost up to the walls of the fort, but with dawn they were discovered and wiped out by machine-gun fire.

[1] Here the Elizavetpolskis took 500 prisoners belonging to the 108th Regiment of the 36th Infantry Division, which was observing the slopes of Kargapazar on the left flank of the 29th Infantry Division (IX Corps) stationed in the area round Çoban-dede, Dalan-göz and Uzunahmet.

To the left of the regiments of the 39th Division,[1] parties of the 5th Caucasian Rifles and the 263rd Regiment tried to approach the Kaburga and Ortayuk Forts, but they had no success and suffered useless losses.

Again to the left, Dokuchayev's detachment had, since the morning of the 11th, been advancing along the path which led over the flank of Palandöken. But the path was lost in the darkness, and the 155th Kubinskis and two battalions of the 5th Caucasian Rifles passed the night on a high plateau swept by an icy wind. Conditions were terrible and more than 600 men (many of whom later died) were badly frostbitten. On the following morning (12th) the column proceeded against Fort Gez but became involved in a very deep ravine beyond which stood the fort. Attack was patently impractical and Dokuchayev's column was forced to retrace its steps to the 'white hell' where it had passed the previous night. During the withdrawal serious losses were suffered by gunfire from Gez and the forts on Palandöken.

On the extreme left, the Georgian General Chikovani with his militia *druzhiny* had not been so unlucky. Following the track from Taşkesen village, his column had beaten off a small Turkish attack and had approached within half a mile of Palandöken East.

During this first day the troops of Vorobyev and Voloshinov-Petrichenko were engaged in overcoming the formidable obstacles of snow, frost and rock which lay in their way, but they were slowly descending along a wide front down the slopes of Kargapazar. Their progress gave promise of decisive success on the following day. More immediate results were, however, achieved by Przevalski's Turkistanski Regiments in the evening of the first day.

The attack on the Gürcü-boğaz defile had been started early in the morning of the 12th. This attack was very well planned and, unfortunately for the Turks, it was inefficiently opposed by the obstinate and narrow-minded commander of X Corps. He thought it possible to fight his three divisions in three separate and isolated formations, each being responsible for the defence of its own region without any co-operation from its neighbours.[2] Thus, when the 32nd Division was first attacked on the 12th, it fought without the support of the other two divisions of X Corps.

General Przevalski concentrated twenty battalions for the attack on Kara-göbek and the Gürcü-boğaz defile.[3] Only eight battalions were

[1] The 154th Derbentskis were still in reserve.

[2] The 30th Division in the mountainous district of Kugans to the west of Karacağil-dağ; the 32nd in Fort Kara-göbek and Gürçü-boğaz defile; the 31st in Fort Tafet and Köşk-çay ravine.

[3] The 14th, 16th, 17th, 18th and 20th Turkistan Rifle Regiments.

operating in the Kugans area with the object of attracting the attention of the enemy towards his extreme left flank.

The main attack was planned in five columns:

Column 1. Two battalions of the 14th Regiment with field guns and four howitzers to follow the main road passing up the defile of Gürcü-boğaz through Kara-göbek village.

Column 2. Two battalions of the 14th to make an outflanking movement by the path over Karacağil-dağ leading from Kizilkilise to Güngörmez at the source of the Dumlu-çay a stream which runs behind, and to the south of, the isolated height[1] on which stood Fort Kara-göbek.

Column 3. To the west of column 2, the 16th Regiment was to carry out a deeper outflanking movement from the region of Zinavur by a path which mounted the Dumlu-dağ and passed between the two peaks of this ridge.[2] From the saddle over the summit of Dumlu-dağ there is a track giving a relatively easy incline into the plain of the Kara-su. If successful in its march, the 16th Regiment would be in a position to outflank all the northern defences of Erzurum.

Column 4. The 18th Regiment had to advance by a path to the east of the track up the Gürcü-boğaz and running parallel with it over the slopes of Kandil-dağ.

Column 5. Still deeper to the east the 17th Regiment had to penetrate into the southern end of the Gürcü-boğaz defile at a point between the junction of the Dumlu-çay and the entry of the Köşk-çay into the larger stream of the Kara-su.

The 20th Regiment was held in reserve.

All these columns met with very difficult conditions along their respective routes. The two battalions of the 14th Regiment which followed up the main track along the Gürcü-boğaz soon came under strong artillery fire from Fort Kara-göbek. The other two battalions of this regiment, advancing to the west of the defile, made rapid progress and were already descending on the village of Güngörmez in the morning of the 12th. The 16th Regiment made heavy going across Dumlu-dağ. To the east of the defile, the 18th Regiment, advancing along the slopes of Kandil-dağ, took the village of Kara-göbek at 3 o'clock of the afternoon, while parties of the 17th actually appeared to the south of the fort.

Przevalski managed to concentrate a strong artillery group in and

[1] Height, 2200 m. These names (Karacağil-dağ='Black-murmur mountain', Güngörmez='You do not see light', and Dumlu-çay='Smoky water') are indicative of the stormy and foggy character of these mountains.

[2] Heights, 3200 and 2963 m.

around Kara-göbek village and the fort was brought under a heavy fire. Before long a big explosion was observed which the Russians took to be the result of their fire; but it was apparently the preliminary to evacuation. Men of the 14th and 18th Regiments soon discovered that the fort had been abandoned by its defenders.

On the night of 12–13 February Yudenich might be well pleased with the results attained during the first twenty-four hours of the assault on Erzurum. Forts Dalan-göz and Kara-göbek had been taken; Przevalski's troops were well to the south of Kara-göbek; the Turkish reserves were pinned down by the bold attacks of the 39th Division; Vorobyev was continuing to descend the slopes of Kargapazar almost unopposed; and the Don Foot battalions were advancing despite six and a half feet of snow in their track.

On the next day (13 February) the resistance of the Turkish X Corps stiffened all along the front of the Turkistan regiments. The 13th and 15th Turkistanskis made only slow progress in the Kugans area on the extreme right of the Russian front, while the 16th was meeting with opposition in its hard climb across Dumlu-dağ. In the Gürcü-boğaz defile the Turkish 32nd Division resisted stubbornly along the ravine of the Dumlu-çay, notably round the 2410 m. level (at Kelevoret). During the afternoon the 14th Regiment took Güngörmez; but the 18th, with the support of the 20th from reserve, could make no progress down the defile. The 17th Turkistanskis were met on the Köşk-çay by units of the Turkish 31st Division—thrown fresh into action that day. The decision in the Gürcü-boğaz defile depended now on the success of Vorobyev and Voloshinov-Petrichenko.

Voloshinov-Petrichenko's Don Foot, after enormous difficulties and after losing a whole battalion frozen to death in a snow blizzard, reached Köşk village late in the evening. Vorobyev's riflemen, going along the Oluklu spur of Kargapazar, made contact with units of the 18th Division of the Turkish XI Corps, moved from reserve.

Elsewhere, along the front of the Russian 39th Division, no attempt was made to renew the attack on Çoban-dede Fort, and there was no change in the situation farther to the south. But Mahmut Kâmil still retained on the Deve-boyun positions no less than five divisions,[1] and he seemed not yet to have understood the threat on the northern sector of his front.

The day of decisive action in the Gürcü-boğaz defile was 14 February: the struggle for Fort Tafet and the southern exit from the

[1] The 17th, 28th and 29th Divisions of IX Corps and the 33rd and 34th Divisions of XI Corps.

pass. Rather late, the Turkish commander-in-chief moved up units of the 34th Division to reinforce the hard-pressed 31st and 32nd. The Turks continued to resist round Kelevoret, Çipak and Ağ-dağ, while Fort Tafet was attacked from three sides by the 17th Turkistanskis, the Don Foot and the 13th Caucasian Rifles (the first of Vorobyev's regiments to be deployed in the pass). The fort was finally taken, with 1500 prisoners and twenty guns. Meanwhile the other regiments of Vorobyev's division were advancing; at the end of the day they were descending from Güney Hill towards Hins in the Kara-su plain. It was the planned break-through. To widen it, two battalions of the 154th Derbentskis were thrown in to support the left flank of Vorobyev on Oluklu-dağ, where strong resistance was continuing. By the end of the day, along all the front from Ağ-dağ to Oluklu, the Russians stood on the rim of the Kara-su plain. On the rest of the front they were maintaining pressure without any attempt at decisive attack.

On the morning of 15 February, Russian air reconnaissance reported extraordinary movement in the streets of Erzurum and baggage trains proceeding west. At 9.40 a.m. Przevalski got a signal from Yudenich: 'Turks retreating from Deve-boyun; 4th Caucasian Rifle Division and Don Brigade are under your orders. March on Erzurum and capture it from the north.' Sometime later the Siberian Cossack Brigade reported to Przevalski with orders to advance on Ilica and cut the Turkish line of retreat.

The Turks, however, were putting up a fighting retreat. On their left the 30th Division still kept the passages leading across the Dumlu-dağ to Haşkvank. Near Kanber the 14th Turkistanskis were counter-attacked by units of the 32nd Division. The 31st and 34th Divisions of XI Corps put up a stout resistance against the Turkistan regiments, the Don Foot and the Caucasian riflemen advancing on the front through Ağ-dağ, Hins and Kösemehmet.

In the meantime the Derbentskis and the Elizavetpolskis had attacked Çoban-dede which was soon abandoned by the Turks without serious resistance.[1] By the afternoon all the forts of the outer ring had been evacuated by the Turks and occupied by the Russians. In the evening the 155th Kubinskis, still on the eastern side of the deep Yağliyan ravine, saw a big explosion in Fort Gez. It was impossible to cross the ravine in the dark, and the fort was only occupied the following morning. The Palandöken forts continued to resist throughout the 15th; Chikovani could make no serious attack on them with his militia *druzhiny*.

[1] In the fort and the surrounding batteries forty-two guns were taken.

During the night of the 15th–16th, Turkish rear-guards tried to check the Russian advance at Fort Sivişli and on field positions round it. Fighting continued throughout the night until 5 o'clock in the morning, with heavy losses to the Turks whose retreat on Erzurum became disorderly. There was no attempt to defend the old forts of 1877 nor the more ancient walls of Erzurum. At 7 a.m. on the 16th a Terek Cossack regiment entered Erzurum, followed by elements of the 263rd Regiment and of the 5th Caucasian Rifle Regiment.[1]

Although, as a result of the air reconnaissance, Yudenich had immediate information of the withdrawal of the Turks from Erzurum, the pursuit was less effective than it might have been owing to a mistake of his own commission. His order of the morning of the 15th to Przevalski to march on Erzurum and capture it from the north prevented the commander of II Turkistan Corps from moving Vorobyev on Ilica along the north bank of the Kara-su. Vorobyev reached the river to the north-east of Erzurum on the night of the 15th and asked permission of Przevalski to proceed during the night to Ilica. Przevalski was unwilling to accept the initiative, since it remained uncertain whether the Turks would try to defend the old fortifications of 1877 and the city wall. This crucial night gave the retreating Turkish columns the chance to reach Ilica unmolested. As for Przevalski's own Turkistan regiments, they were held fighting all through the day of the 16th by obstinate Turkish rear-guards on a position between Arziti and the Kara-su. The real pursuit only began on the night of 16–17 February; it continued through the following day.

In Erzurum the Russians captured 327 guns. Prisoners were not numerous: about 5000 (a greater number were taken during the pursuit). The losses of the Turks in killed and wounded were not less than 10,000. Russian casualties were 1000 killed, 4000 wounded and the same number frost-bitten.[2] Including losses at Köprüköy, the capture of Erzurum and the defeat of the Turkish Third Army had cost the Caucasian army 17,000 casualties in officers and men.[3]

[1] During the battle these troops had been in front of Kaburga, and they were therefore nearer to the city than any other unit.

[2] The after-effects of bad frost-bite are often more crippling than those of a serious wound.

[3] M. Philips Price, *War and Revolution in Asiatic Russia* (Chap. II), gives the best account in English of the battles for Erzurum. He was received by Yudenich, 'a short bullet-headed man with long moustaches,...his voice was sharp and abrupt, his manner that of one who is accustomed to command' (p. 171). Price was impressed by Przevalski: 'One of the characters of the Caucasus campaign and certainly one of the most able and useful generals in the Russian Asiatic service,...a man of deep understanding of the life and spirit of Asia' (*ibid.* pp. 200, 202).

CHAPTER XXX

CONSEQUENCES OF THE FALL OF ERZURUM. OPERATIONS ON THE STRATEGIC FLANKS TO END OF MARCH 1916

MAPS 29, 30, 31, 26

COVERED by rear-guard actions fought from the Deveboyun positions, the bulk of the five divisions of IX and XI Corps succeeded in escaping encirclement in the plain of Erzurum. During the course of 15 and 16 February, Turkish columns were retiring through Ilica, the road junction for different lines of retreat. The retreat had become disorderly by the morning of the 16th and, among the troops withdrawing from the Palandöken positions, it deteriorated into a rout.

Yudenich entered the fortress of Erzurum at midday on the 16th, and he immediately set himself to organize the pursuit. With two Cossack regiments, the 156th Elizavetpolskis and the 155th Kubinskis were moved by forced marches on Ilica, while another mounted column kept pace with them farther to the south. Ilica was taken after a sharp fight on the night of the 16th–17th; 3000 more prisoners and thirty guns were captured here and to the south of Erzurum. On the same night and during the next day (17th) a battalion of the 17th Turkistanskis and Siberian Cossacks, acting as the vanguard of Przevalski's corps, surrounded a Turkish rear-guard (remnants of the 34th Division of the XI Corps) at Kara-arz to the north of Ilica and took another 2000 prisoners with twelve guns.

Since the Turkish retreat resulted in the usual mass desertions, some 10,000—mostly in prisoners and deserters—must be added to the losses during the storm of Erzurum. Out of Mahmut Kâmil's army there remained not more than 25,000 men with thirty or forty guns. Their morale was shaken. X Corps preserved something of its spirit; but the losses of the 31st and 32nd Divisions had been severe and the strength of the corps was reduced to about 12,000. In XI Corps the 34th Division had been almost annihilated, and the other two divisions could not muster more than 5000 men fit for the line. In a state approaching disintegration the Third Army continued its retreat towards the west; and from this direction reinforcements were anxiously expected.

Between 17 and 19 February the Cossacks pressed the pursuit. IX and XI Corps were marching across the plain of the western Euphrates (Kara-su) towards Aşkale, while X Corps followed the difficult snow-bound tracks north-westward to Bayburt. Before Aşkale, on the 20th, the Cossacks ran into the Turkish rear-guard on a strong position at Abdalcik. The Elizavetpolskis and Bakinskis came up with four batteries, and after two days hard fighting took the position and Aşkale village.[1]

While the headquarters of the Third Army passed through to Erzincan, the Turkish rear-guard stood again along both sides of the Bayburt road in the area of Sapdiran and Şehir-deresi. Judging from reconnaissance reports that the country in the direction of Mamahatun was only lightly held, Yudenich formed a special flanking detachment (the bulk of it being the 154th Derbentski Regiment) which was ordered to advance along the valley of the Tuzla-çay on Mamahatun and turn the Dumanli-dağ from the south. The movement took some time, but on 13 March the flanking column reached Kumsor in the valley of the Tuzla. On the 15th there was fighting with a Turkish force in position near Parsinik. The Turks were pressed back and next day the Russians entered Mamahatun, where 800 prisoners and five guns were taken. On the 17th Yudenich's troops occupied Kotur where there is a bridge over the Kara-su. All the bend of the river to the west of Mamahatun passed to the Russians who now began their concentration for an eventual march on Erzincan. The Erzurum operation, opened by the battle of Köprüköy, had come to its logical end.

The occupation of the Erzurum plain, however, remained incomplete while the Turks continued to hold Bayburt. In this direction Przevalski's Turkistan troops met with a stubborn resistance from units of X Corps and Frontier Guard battalions. Bayburt is situated at the sharp bend of the River Çoruh,[2] and Przevalski planned an advance converging on the town from three directions.

Four plastun battalions, followed by the 13th Turkistanskis and the Siberian Cossacks, were sent up the valley of the middle Çoruh. They had in front of them only five or six battalions of Turkish Frontier Guards with some Laz irregular formations; but the frontiersmen knew the region perfectly and proved dangerous and stubborn

[1] According to the statements of prisoners, the commander of XI Corps, Abdul Kerim Paşa, was personally commanding at Aşkale where he only disposed of 3000 infantry with eight guns.

[2] The upper course of the Çoruh runs parallel to the middle course for a considerable distance, and only a ridge fifteen to twenty miles in width divides the upper from the middle stream.

adversaries. The conditions of the march towards Ispir were not bad by comparison with those which Przevalski's men had recently experienced in the highlands to the south, but the Turks fought with competent determination in every one of the numerous defiles of the middle Çoruh valley. The ancient town of Ispir[1] was taken on 23 February. The slow movement of the middle Çoruh detachment continued towards the west but, in another month, after incessant fighting and skirmishing, the Russians had only reached the district of Danzuta at the foot of the Yamanli-dağ (Ahpunuz-Hozahpur area). Bayburt was still fifteen to twenty miles to the south-west; but the Ispir column was halted, since even its modest rate of advance had been too quick for that of the two other converging columns.

After enormous difficulties the 15th and 16th Turkistanskis had penetrated from the heights of Dumlu-dağ into the deep gorge where the Çoruh finds its source. The attempt to establish a liaison with Ispir (already occupied) through Kara-hane cost the 15th Regiment the lives of several hundred frozen men.

At the same time from the south, units of Przevalski's third column (17th and 18th Turkistanskis) fought their way down the precipitous ravines of the left-bank tributaries of the upper Çoruh, the Kurt and Masat streams. In this remote region, during the first half of March, mountain war between the Turkistan riflemen and elements of the Turkish 30th and 31st Divisions continued for more than a week. By 25 March Przevalski's troops occupied positions twenty to thirty miles to the south-east of Bayburt: Vartanik on the upper Çoruh, Söğütlü and Sağirdan. On Przevalski's left flank liaison with the 39th Infantry Division was established at Dencik.

The advance to the west from Erzurum was only the final development of the Erzurum operation; it proved easier and more complete on the front of I Caucasian Corps in the Kara-su plain but more difficult and incomplete (through the failure to reach Bayburt) in the wild mountains overlooking the valley of the Çoruh. For further operations a new grouping of forces was necessary; and other factors compelled caution. With the approach of spring, large Turkish reinforcements were to be expected all along the front from Lake Van to the Black Sea. Thus, after 25 March, a certain period of lull set in on the fronts of the corps of Przevalski and Kalitin. Meanwhile the operations on both strategic flanks, planned as a whole with the Erzurum operation, had been brought to an end.

[1] Ispir and Bayburt are two of the oldest settled sites in the Armenian highland, dating back to the Urartian kingdom of the first millennium B.C.

On the extreme left of the Russian front what was left of IV Caucasian Army Corps[1] had been ordered to occupy Muş and Bitlis. All the reinforcement which Yudenich found himself able to spare from I Caucasian Corps was one battalion of the 6th Caucasian Rifles. But the troops of the Van-Azerbaijan group received orders to cooperate with IV Caucasian Corps. The advance was planned in three columns: the right column had to move from Hinis on Muş, the central column from Malazgirt on Bitlis, while the left column gave support by marching along the southern shore of Lake Van.[2]

The Turks in some force (106th and 107th Infantry Regiments and cavalry) occupied a strong position covering the two passes over which led the best tracks from Hinis to Muş: the first at Mulla-Halil, the other six miles to the north-west (north of Gümgüm).

Nazarbekov sent two battalions and one battery to engage the Turks in front of Mulla-Halil, while four battalions with the Cossack regiment and one battery had to carry through an outflanking movement in depth in the region to the south of Kara-köprü and penetrate into the valley of the Murat-su to the rear of the Turkish position. At the same time one battalion of the 7th Regiment had to surmount the precipitous heights of Bingöl-dağ and try to outflank the Turks from the west.

As the affair turned out, the left column (constituting the main force) progressed slowly and lost its way. The smaller group left in front of Mulla-Halil had to sustain three days' fighting against a superior force in an entrenched position; and the situation was only redeemed by the remarkable feat of the battalion of the 7th Regiment which succeeded in crossing the terrible slopes of Bingöl-dağ and reaching Gümgüm on the Turkish flank.

The Turks treated their position as turned and withdrew in the direction of Muş. The Russians followed, converging to Çarbuhur (10–14 February). The dispirited Turks did not now even try to defend the town of Muş which the 1st Labinski Cossack Regiment entered on 16 February, the day of the fall of Erzurum.

[1] The 7th and 8th Caucasian Rifle Regiments, some Armenian and militia *druzhiny*, and the 2nd Caucasian Cossack Division: a total of 12,000–13,000 men. The opposing Turkish force consisted of the 106th and 107th Regiments, nine battalions of Frontier Guards and gendarmes, 2nd Cavalry Division, and Kurdish formations: a total of 15,000 with twenty to thirty guns.

[2] *Right column*: 7th and 8th Caucasian Rifles; one Cossack regiment; and two field batteries (General Nazarbekov). *Centre column*: one battalion of 6th Caucasian Rifles; two Armenian and two militia *druzhiny*; three Cossack regiments and one battery (General Abatsiev; like Nazarbekov, of Armenian origin). *Left column*: two Armenian and two militia *druzhiny*; one Cossack regiment; one battery.

Meanwhile Abatsiev had been making very slow progress. The region to the south of the Murat had already become depopulated as a result of the Turkish and Russian marches and counter-marches of the previous year: the irregular bands operating on either side had begun the work which famine (resulting from comprehensive requisitions) and epidemics had completed. The remnants of the Armenian population had fled into Russian territory or had been deported by the Young Turk administration; the Muslims, scarcely more fortunate, had suffered equally from famine, epidemics and irregular slaughter, and the survivors had dispersed with what remained of their livestock into the interior of Asia Minor. Abatsiev found all the once rich valley of the Murat-su to the south of Malazgirt empty and desert, and his animal transport suffered from lack of supplies of forage. Despite the use of hardy Bactrian camels carrying more than twice the loads of horse or mule, supply columns worked badly and needed strong escorts through a country infested by bands of famished Kurds. Abatsiev had to leave almost all his mounted troops and guns in the rear, and he arrived at Tatvan on 20 February with only the battalion of the 6th Caucasian Rifles, one *druzhina* and two Cossack *sotni* with two mountain guns.

Aware that his force was inadequate to try an advance on Bitlis, Abatsiev applied to Nazarbekov for help. Reinforcements arrived both from Muş and from Van,[1] and on 27 February Abatsiev was able to move on Bitlis with five infantry battalions, five *druzhiny* and twelve guns.

The Turks prepared to resist on a fortified position on the outskirts of the town of Bitlis; it was strong enough, lying in the narrow Bitlis valley, bordered by two walls of mountain which could be neither traversed nor outflanked. Abatsiev decided on a surprise night attack by infantry without artillery preparation. He chose the night of 2–3 March. The Caucasian riflemen advanced through a blizzard of snow without firing a shot and attacked with the bayonet. After several hours of fierce hand-to-hand fighting the position was taken with all the Turkish guns (twenty) and 1000 prisoners. The Turks fled south in the direction of Siirt. On 1 March the first units of Mustafa Kemal Paşa's reinforcements for the Caucasian front had arrived at Diyarbekir; but they came too late to protect the valleys of Muş and Bitlis, the two gateways to the Taurus and the middle Euphrates.

[1] Nazarbekov sent the 8th Caucasian Rifles, two Labinski *sotni* and one battery from Muş; a Cossack regiment, two Armenian and two militia *druzhiny* and one battery joined Abatsiev from Van.

In addition to his co-operation with IV Caucasian Corps, General Chernozubov, commanding the Van-Azerbaijan group, had thought it useful to make a demonstration to the south of Lake Urmia. The 3rd Transbaikal Cossack Brigade was concentrated at Urmia town observing the frontier up to Ushne, while two Kuban Cossack regiments with the 4th Armenian *druzhina* were at Suj-bulak with patrols in the direction of the Ruwandiz Pass. It soon became clear that there was an increase of Turkish strength in the region between Mosul and Ruwandiz.[1]

On the extreme right of the Russian front General Lyakhov, on 1 February, started a diversion against the Turkish forces on the Black Sea coast. If the movement against Muş and Bitlis was hardly calculated to divert reinforcements from the Erzurum region, a vigorous action along the coast of Lazistan might well hold at Trebizond any Turkish forces sent by sea from Istanbul to Erzurum.

The littoral westward to Trebizond and Giresun very much resembles the neighbourhood of Batum where the Russians had met with such disastrous reverses in 1877–8 among the precipitous cliffs, steep ravines and dense woods of Acaristan. The experience of 1877–8 had shown that success was not to be expected in coastal operations without local command of the sea and the co-operation of naval forces, and already in the present war the action at Hopa had impressed on Lyakhov the value of enfilading fire from the sea against an enemy entrenched to defend the successive lateral valleys which are characteristic of the coast of Lazistan.

Lyakhov therefore awaited the concentration of a strong naval squadron at Batum before moving against the Turkish position beyond the Arhavi river.[2] This position, about three and a half miles to the west of the Arhavi river, well entrenched and with an excellent field of fire for machine guns, was held by seven or eight Turkish battalions with eight guns.

On the morning of 5 February the Russian squadron plastered the Turkish trenches, inflicting very heavy destruction; in the afternoon the Turks were preparing to abandon their trenches when a rash attack by the 19th Turkistanskis imposed a cessation of the bombard-

[1] Marshal von der Goltz had taken over the command in Iraq in December 1915.

[2] Admiral Eberhardt had placed at Lyakhov's disposal the battleship *Rostislav* (four 10 in. and eight 6 in. guns) and the gunboats *Kubanets* and *Donets* (each armed with two 6 in., one 120 mm. and some smaller guns). There were also four destroyers and two armed steamers. *Donets* had been sunk in the port of Odessa during the Turkish naval attack which preceded Turkish intervention; she had been later salvaged.

ment from the sea. But when the shelling was renewed on the following morning the Turks abandoned their lines, leaving 500 dead behind them, as against Russian casualties of 160—mostly unnecessary. The same sequence of events was repeated at Vice, where on 15 and 16 February, *Rostislav* shelled the Turks out of their trenches behind the Vice river.

The relatively rapid advance of the Russians along the hitherto virgin coast of Lazistan filled the Turkish command with alarm, and several battalions were moved from Trebizond by the coastal track to Rize. A defensive position for the troops retiring from Vice was chosen, skilfully enough, behind the Büyük-dere river, about ten miles to the west of Vice and four miles to the east of the small fishing port of Atina. This position was by nature very strong, since cliffs fell vertically to the sea and westward towards the valley of Büyük-dere. This valley, as its name implies, was rather spacious, and an attacking force would have to cross a wide field of fire; the ravine of the river itself was steep, and the water deep with only one passable ford.

All these conditions were considered by General Lyakhov in conference with naval officers, and he accepted their proposal to land infantry in the rear of the Turkish position, somewhere near Atina. The idea of a landing had interested the Russians since the beginning of the combined operations between the army and naval forces. There was a special type of small cargo boat engaged in the coastal traffic along the Black and Azov seas. They were called *elpidiphores* and were between 1000 and 1500 tons. Specially built for trade in shallow waters at the mouths of the Black Sea rivers and at small coastal ports where loading facilities were lacking, the elpidiphores had their engines and the crew's quarters built aft. The fore two-thirds of these rather long ships consisted of vast holds and, when these were empty, an elpidiphore lay deeper in the water aft than fore and could thus run very close inshore. The elpidiphores seemed to Lyakhov to be well suited for the transport and rapid landing of troops. A trial was successfully carried out near Batum, and it was decided to plan the operation against the Turkish position at Büyük-dere on the basis of the landing of two battalions with two mountain guns in the rear of the enemy.

On 4 March *Rostislav*, *Kubanets* and four destroyers appeared off the mouth of the Büyük-dere river and bombarded the Turkish positions for several hours; at the same time Russian infantry could be observed in movement on the eastern side of the valley. The same evening two elpidiphores took two plastun battalions on board at Batum while guns and horses were loaded in a transport. Under escort of two

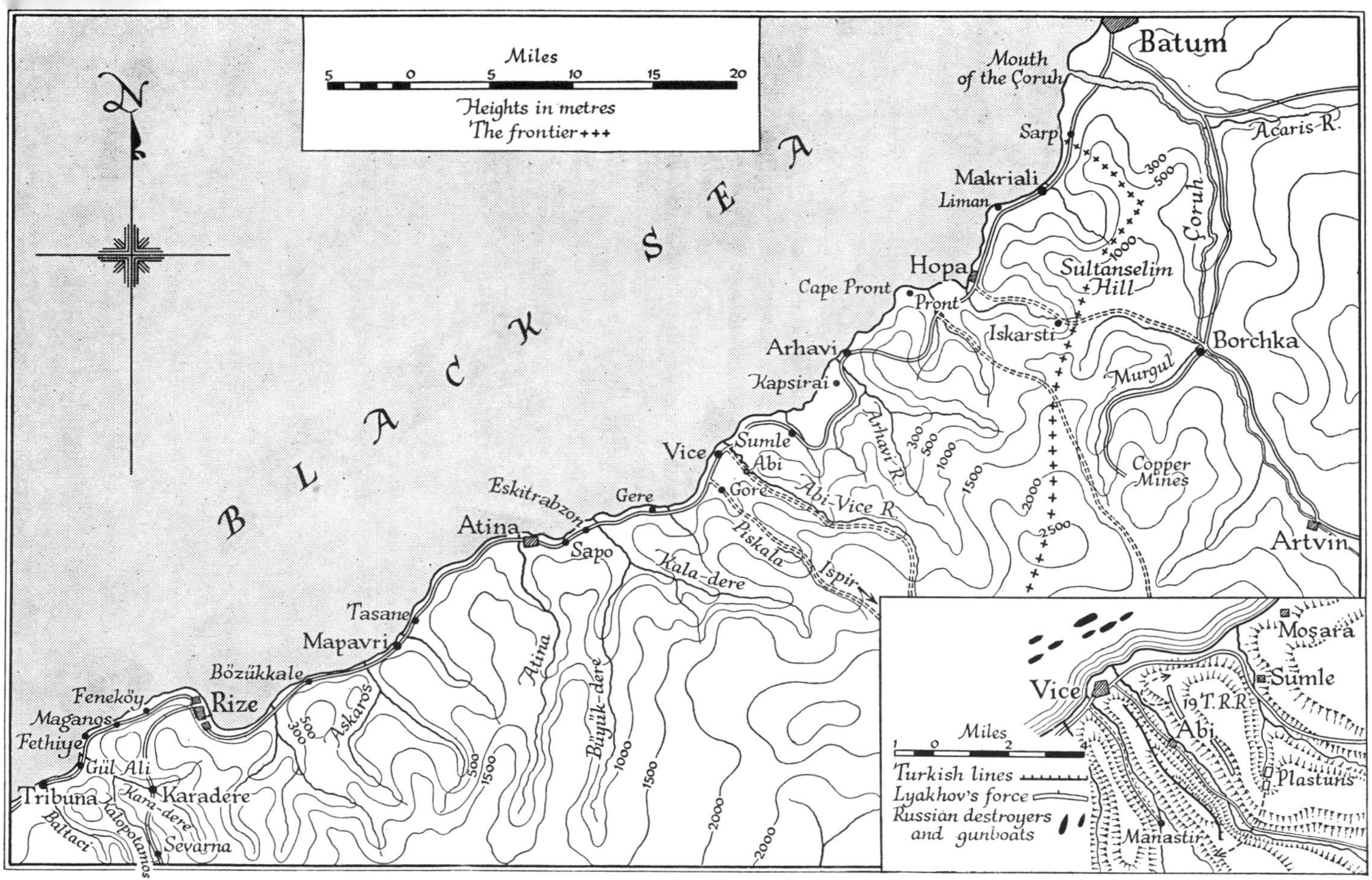

Map 26. Lazistan, spring 1916

destroyers the convoy sailed to Hopa, from where during the night it moved without lights to Atina. Atina was reached just after 5 a.m. and at a quarter to six the landing began from the first elpidiphore. It was completed in twenty-five minutes; and the second elpidiphore disembarked its cargo of Kubanian riflemen in fifteen minutes. Only when all the infantry were on shore did Turkish posts observe the enemy and open fire. One destroyer replied; and as dawn broke, the transport (with horses and guns aboard) could be seen approaching under the escort of the three other destroyers and *Rostislav*. The plastuns moved towards Atina where they made prisoner two Turkish companies. As soon as news reached them of the landing in their rear, the Turks on the Büyük-dere position fled precipitately into the mountains. The operation had been accomplished without any loss.

Lacking neither energy nor initiative, General Lyakhov decided to exploit his success without delay. Within his reach was now an important objective, Rize, the first real (if small) port to the west of Batum. The landing operation was repeated on the night of 6–7 March at Mapavri and met with only slight resistance. On 8 March the Russians occupied Rize and presently pushed their patrols forward to the River Kalopotamos to the east of the small town of Of. Here the advance of the Black Sea coast detachment was temporarily halted. At Rize General Lyakhov was only thirty miles from Trebizond. He had achieved a strategic success which could not fail to have important consequences.[1]

[1] Lyakhov had available, when he reached Rize, eight battalions of the 19th Turkistankis, two plastun battalions, and the newly created 5th Border Regiment, eight *druzhiny*, two *sotni*, a sapper battalion and sixteen guns. His losses had been less than 300 killed and wounded. When compared with the expensive and unsuccessful operations against Batum during the War of 1877–8 undertaken *without* local naval superiority, Lyakhov's success was, of course, illustrative of the significance of combined operations along this wild and difficult coast.

CHAPTER XXXI

BARATOV IN PERSIA, DECEMBER 1915 TO MARCH 1916

MAP 23

THOUGH the appearance of a Russian force not far from Teheran had been enough to dissipate the adventurous German plan for a *Putsch* in Persia, the participants had been able to make good their escape and to become a focus of trouble at Kum and Hamadan. By 1 December 1915 the whole of the 1st Caucasian Cossack Division was concentrated at Kazvin. Without awaiting the rest of his force, General Baratov sent one brigade, on 3 December, against Hamadan. The ancient city was occupied after a brief fight with Persian gendarmerie units and armed volunteers on the Sultan-bulak Pass. A week later a brigade of Cossacks moved on Kum, whence another force of 3000 Persian gendarmes, together with the 'democratic' fraction of the Persian *Mejlis*, took flight. On 20 December Kum was occupied without a fight. Baratov left one regiment in that city, to observe Kashan where the dissident politicians and their bodyguard had taken refuge. Another regiment was moved in the direction of Burujird, the centre of the warlike tribal area of Luristan; but here the population remained neutral.

Meantime, Count Kanitz, with the personnel of the German and Austrian embassies and some armed followers, had established himself in Kirmanshah. Marshal von der Goltz was now in command in Iraq, and he thought it his duty to do something to protect the friendly elements in Persia. He sent one Turkish battalion of regular troops with a mountain battery to Kirmanshah, followed shortly by three more battalions.

Baratov's forces, however, were quickly concentrating at Hamadan. Between 15 and 20 December the Caucasian Cavalry Division arrived; while battalions of two border regiments were on the march from Kazvin. On 25 December, dragoons and Cossacks were skirmishing with the enemy at Asadabad. On 13 January 1916 they marched on Kangavar, where a more serious fight took place with Count Kanitz's followers, supported by the single (and unlucky) Turkish battalion. Kanitz's bands were easily dispersed and the Turks retreated with losses. From that day Count Kanitz completely disappeared and was never more heard of; it seems probable that the adventurous diplomat was killed by his own volunteers.

Towards the end of January Baratov had 8000–10,000 men with twenty-two guns concentrated at Hamadan.[1]

The Grand Duke Nicholas sent Baratov orders to liquidate the centre of resistance which still existed at Kirmanshah where units of Turkish infantry and artillery were concentrated under command of the German, Major Bopp. The four Turkish battalions were reinforced by about 6000 Persian gendarmes and irregulars under officers of the Swedish mission. The enemy forces proved to be rather scattered: there were 3000 Persian gendarmes and Kurds at Nihavend on the Kangavar-Burujird road, and other elements at Sinna and elsewhere. The Cossacks and dragoons dispersed the Nihavend force on 8 February and occupied Daulatabad. On the 22nd the Turkish regular infantry and some bands were easily beaten at Sinna; and on the 26th Baratov entered Kirmanshah and, a few days later, Kashan. Cossacks were detached to Sinna and Bijar to observe the Turkish frontier.

By 1 March Baratov might consider his task in Persia as complete. He left infantry and Cossacks in Kirmanshah, with detachments at Daulatabad, Sultanabad and Kashan. The dragoons were distributed between Kazvin and Hamadan. All seemed quiet in Persia. But more complicated political and strategic problems were soon to involve Baratov's force in a new expedition which was to prove less fortunate than his first effort.

[1] The Caucasian Cavalry Division, two border regiments (four battalions) and two Cossack regiments; two more Cossack regiments observed Burujird and Kashan; and there were two Cossack regiments and militia at Kazvin and near Teheran.

CHAPTER XXXII

THE CAUCASIAN FRONT TO APRIL 1916. TREBIZOND OPERATION

MAPS 1, 17; cf. 28, 38

THE news of the fall of Erzurum reached Enver Paşa during his tour of inspection in Syria and Palestine in the second half of February. The vice-generalissimo decided not to proceed to Iraq, as he had planned, and he made a hurried return to the capital (8 March).

To a man of Enver's temperament dramatic measures were essential, and the first and easiest of these was the dismissal of Mahmut Kâmil. The commander of the Third Army was replaced by Vehip Paşa who had already gained an excellent reputation at the Dardanelles.[1] It was clearly urgent to strengthen the shattered Third Army, but it was not within the power of Enver to accelerate the progress of the reinforcements which were already on the way. By March only the 3rd Cavalry Division had arrived in the intermediate area between the Third and Second Armies. Other divisions which were to make up the Second Army were on the move, but they had to travel as far as Ulu-kişla by the Bagdad railway which was overloaded with the men and material of the 2nd, 6th and 4th Divisions en route for Iraq. After his personal inspection of the Taurus and Aleppo areas the generalissimo might have been expected to realize the difficulties, but he showed no appreciation of the limitations of the Turkish railway system in his new and (as usual) vast strategic plans which were born, a few days after his return to Istanbul, as a result of conversations with his German 'sympathizers', von Lossow and Bronsart von Schellendorff.

A kind of 'decisive blow' against the Russian Caucasian army was planned. The Third Army, reinforced by V Corps, was to engage the Russians along the front between the Kara-su and the Black Sea at Of. Meanwhile the principal Turkish forces were to be concentrated at Harput and Diyarbekir for an advance along the line Kiği-Oğnot-Muş-Bitlis, with the object of outflanking the main Russian army and

[1] Vehip Paşa (1877–1940) was a younger son of Mehmet Emin Effendi, mayor of Yanina, and a brother of Esat Paşa the commander at Çanakkale during the Dardanelles campaign. Vehip ended an adventurous career as chief of staff to Ras Desta on the Ogaden front during the Italo-Ethiopian War of 1935–6.

threatening the Russian rear. The task was assigned to the Second Army which had been placed under the command of Marshal Izzet Paşa.[1]

To an incorrigible amateur like Enver his plan must have seemed clever and impressive, particularly as explained in a short directive, which was illustrated by a sketch indicating only the principal points for concentration and advance. But it was hopeless except on paper. The time allowed for the concentration of the Turkish forces was miscalculated in a spirit of irresponsible optimism, and the whole conception was based on the unwarrantable assumption that the Russian Caucasian army would remain inactive on its existing positions during three of the best campaigning months of the year.[2]

Enver produced his new offensive plan a few days after the capture of Rize by the Russians. The rapid and easy success of Lyakhov's combined operations now suggested to Yudenich the idea of an immediate action against Trebizond. The possession of this port, the best of the inferior roadsteads along the northern shore of Anatolia,

[1] The 1st, 5th, 7th and 8th Divisions of the Second Army were already assembling, while the 11th, 12th and 14th and the 48th, 49th and 53rd had been assigned to it. The 1st, 8th, 11th, 12th and 14th had gained laurels at Gallipoli. Izzet Paşa, as commander of an army group, was senior to Vehip commanding the Third Army. Marshal Ahmet Izzet Paşa (1864–1937) is not to be confused with General Hasan Izzet Paşa, dismissed by Enver from command of the Third Army before Sarikamiş. A Macedonian, like so many Turks who played a leading role in the first quarter of the century, Ahmet Izzet had had a distinguished military and administrative career without committing himself too dangerously to the Young Turk triumvirate. Pomiankowski had a good opinion of him as an able and reasonable man. Later the marshal was Turkish military representative at Brest Litovsk and Bucharest and, at the end of the war, he was called upon to lead the government which signed the armistice of Mudros.

[2] The concentration for the 'decisive blow' involved the dispatch of three or four divisions by rail to Ankara and of eleven divisions (in addition to the three already en route for Iraq) to the Taurus rail-head at Ulu-kişla. Both railways were one-track and their maximum capacity was twelve trains per day on each route; in practice not more than eight trains per day got under way. In a month only two out of four divisions were carried from Haydar-Paşa to Ankara. From here they had a month's march (468 miles) to Erzincan. The concentration of four new divisions on the front of the Third Army, in fact, required three months and could not be completed before the second half of June. (Complements for the existing divisions of the Third Army were also using the Ankara railway.)

The conditions for the concentration of the Second Army were worse. The transport of a single division from the Bosporus to Ulu-kişla took thirteen days. As this line was serving also Syria and Iraq several months were necessary to complete the transport of the eleven divisions of the Second Army as far as the Taurus. From Ulu-kişla troops had to march 550 miles to the Harput-Kiği area (of which nearly 200 were over bad mountain tracks). As the first units of the Second Army were only arriving at Diyarbekir by the middle of March, a simple calculation might have convinced Enver that the new dispositions required of the Second Army could not near completion before 1 August.

was a necessity to the Russians if they wished to consolidate their hold on Erzurum.[1] At the same time the possession of Trebizond was essential to any extension of operations against Erzincan and the interior of Anatolia.

The imperial *Stavka* (the tsar's general headquarters) was well pleased with the successes of the Caucasian army, and Yudenich's reputation stood high with General Alexeyev.[2] In Russia the glory of the Caucasian army was the only brilliant relief in the gloom which followed the great retreat in Poland in 1915. But with all this, Russian general headquarters had not changed their constant view of the strategic meaning of the Caucasian front. A few lines of a letter from General Alexeyev to Admiral Eberhardt are characteristic: 'A decision in the war will depend principally on the course of events in the European theatres, i.e. the Franco-Belgian and our own western. Our European strategic front (more than 1200 kilometres long), the uncertain position of Rumania, the number and quality of the enemy forces on our western front—all this makes our military situation so serious and so complicated that we have no right at present to disperse our troops and to direct them to the accomplishment of tasks, which are secondary even if appreciable in themselves, in remote theatres of war.'[3] This letter was written on 29 January in the interval between the Köprüköy battle and the storm of Erzurum. But even the capture of Erzurum could not influence the considered appreciation of the war by *Stavka*.

Yudenich, for his part, was far from the idea of a great strategic offensive against Turkey. He wished only to consolidate advantages already attained and to remove any threat of a renewal of the Turkish offensive against Transcaucasia, an offensive which remained the only fixed objective in the erratic mind of the Turkish generalissimo. Yudenich knew that many of the Turkish divisions released from Gallipoli would be directed to the Caucasian front, and he accordingly put forward plans and made preparations designed to meet the threat.

Yudenich's own resources in Caucasia were limited and were now nearing exhaustion. He could fill his losses in the winter battles by drawing on reservists and on the *opolcheniye* (militia) formations, but

[1] The *chaussée* connecting Trebizond with Erzurum (via Ardasa, Gümüşane, Bayburt, Aşkale) had for many years been in excellent condition and, in peace-time, was the principal avenue of communications for eastern Asia Minor. From Tekke, south-east of Gümüşane, the Trebizond *chaussée* forks south-south-west to Erzincan.

[2] Yudenich had been awarded the high distinction of the St George's Cross (2nd grade) after the capture of Erzurum.

[3] Cited by Novikov, *Operatsii flota protiv berega na Chernom more v 1914–1917 gg.*, pp. 167–8.

he was not in a position to build many new units.[1] He undertook, however, a more normal distribution of units within corps.[2]

The development of communications had always preoccupied Yudenich. During the spring months the light narrow-gauge railway working as far as Karaurgan was continued to Erzurum; it was proposed to carry it forward to Mamahatun and eventually to Erzincan. The new Maku line continued under construction from Shahtahti station via Diyadin to Karakilise in the Eleşkirt valley; and a branch of the Maku line soon connected Bayazit with a point on the north-east shore of Lake Van, across which a steamboat service was working by the early autumn of 1916. The normal gauge line from Julfa to Tabriz had been running since the autumn of 1915; during the following months a branch line was constructed from Safian station to Şerif-hane on the shore of Lake Urmia; and on this water a flotilla of light steamers and barges was soon organized.[3] At the same time work was in progress on hundreds of miles of roads behind all sectors of the front. Motor-transport facilities were increased,[4] and air strength was raised to four air sections.

Preparations for a resumption of combined operations were under way along the Black Sea coast. Here some differences of view emerged between the military and naval commands. Admiral Eberhardt proved rather reluctant to lend his warships to continuous support of land operations. Batum, where there were no repair shops, was not a

[1] Thus the 5th Caucasian Rifle Division was created and, in May, sent to the Erzurum region where it was to prove almost as efficient as the 3rd and 4th. The two *opolcheniye* brigades which had taken part in the winter battles were simply renamed the 1st and 2nd Transcaucasian Rifle Brigades and incorporated as the 6th Caucasian Rifle Division. From garrison troops at Kars were formed the 1st and 2nd Kars Infantry Regiments, and the Armenian *druzhiny* were constituted into six Armenian rifle battalions. A Georgian rifle regiment (four battalions) was also formed.

The creation of artillery units proved more difficult. The new 5th Caucasian Rifle Division was allocated only one section of eighteen mountain guns. But the arrival of new 4·8 in. howitzers from Russia made possible the creation of two sections (of twelve pieces each) which were allotted to IV and V Caucasian Army Corps.

[2] The regiments taken from the 2nd Caucasian Rifle Division and the whole of the 66th Infantry Division were returned to IV Army Corps, which now again approached normal complement. The 4th Caucasian Rifle Division was posted to I Caucasian Army Corps, and the new 5th Rifle Division replaced it in general reserve. During the period of transition the Kuban battalions of the 4th Plastun Brigade were recalled from the Van-Azerbaijan area (May).

[3] For the railway construction work five special railway battalions had been brought from Manchuria.

[4] At the beginning of the war there had been only one motor-transport detachment. By the summer of 1916 there were five: two in the Erzurum area, one at Trebizond, one round Muş, and one in Persia.

suitable base for any considerable naval force, and the naturally cautious naval commander-in-chief regarded its defences as inadequate. With no bases at all along the coast of Lazistan it was difficult also to service destroyers. On the other hand, General Yudenich was unable to understand why the powerful Black Sea fleet[1] should not be in a position to secure complete mastery over enclosed waters where the enemy could only muster two new and five old battleships and one battle cruiser (*Goeben*).

In February, Yudenich drew the attention of Eberhardt to the fact that a Turkish division (the 10th of V Corps) had been carried by sea to Samsun, and he renewed his requests to the naval command that they should prevent Turkish sea transport along the Anatolian coast. Admiral Eberhardt reported to *Stavka* that it was virtually impossible to maintain a continuous naval patrol over the exit from the Bosporus, owing to the distance from base at Sevastopol. The admiral also indicated the difficulties (including danger from U-boats) of a continuous tactical support of land forces operating along the Lazistan coast. He preferred a combined operation of greater size and importance but of limited duration.[2]

These differences of opinion between the naval and the Caucasian commands only had a meaning when lack of available land forces had limited coastal operations to the gradual advance towards Rize. After the fall of Erzurum and a promise given by *Stavka* to release two plastun brigades for the Caucasus, Yudenich pressed the plan for a combined operation against Trebizond, and *Stavka* offered no objection to a project which was acceptable at once to the commanders of the Caucasian army and the Black Sea fleet.

On 20 March, Admiral Eberhardt came to Batum to confer with Yudenich. The naval commander-in-chief accepted the plan of operations and promised the support of the fleet, but he laid down that he would only co-operate for the definite and limited period of time necessary for the transport and escort of the two plastun brigades to Rize. Troops were to be embarked at Novorossiysk on 1–4 April.

During the second half of March different rumours reached General Lyakhov's headquarters at Rize, brought in by rather imaginative Greeks from the villages along the Pontic coast. It was

[1] Lately strengthened by the new *Empress Maria* and awaiting *Catherine II*.

[2] Naval headquarters at Sevastopol resented the dispersal of ships and waste of time during weeks and even months of local tactical support given by naval units to troops operating along the shore. They were not opposed to a concentration of the battle fleet for one decisive combined operation. For Russian naval and air operations against the Zonguldak coalfields, see Bibliography, (6) (*d*) (vii), pp. 550–2.

reported that strong Turkish forces (Gallipoli divisions) had already arrived at Trebizond accompanied by 5000 Germans. The fortifications of Trebizond were described as very formidable and defended by powerful guns. At the same time Russian naval intelligence had reports that 'a certain number' of German U-boats had appeared in the Bosporus and the Black Sea.

All this caused an atmosphere of some nervousness on the staffs of Admiral Eberhardt and General Lyakhov, but Yudenich did not feel any serious anxiety. He admitted the possibility of the arrival of Turkish reinforcements at Trebizond, but he found this to be all the more reason for rapid action.[1]

The Turkish land forces opposing Lyakhov in fact consisted of the heterogeneous collection of regular and irregular battalions (7000–8000 strong) which he had fought before. These were now stationed with twelve guns on heights to the west of the small coastal town of Of and partly to the east of it between two streams. Lyakhov's force, which had been slightly reinforced, was still modest and hardly sufficient for any serious action. Lyakhov had seven battalions along the Kalopotamos some six miles east of Of.[2] (Cf. Maps 26, 28.)

From Of a track, which farther up deteriorates into a path, runs across a shoulder of the Pontic Alps and ultimately leads to Bayburt. Lyakhov was preoccupied with the possibility that the Turks might attempt to turn his position from this direction, and he sent a part of his detachment up the stream of the Kalopotamos to protect his left flank. But he supposed his adversary more enterprising than he really was. No Turkish force tried the passage of the snow-bound ridge of the Pontic chain, while the force on the coast preferred to withdraw farther to the west where they took up better positions behind the Kara-dere stream which falls to the sea in two branches at the town of Sürmene. Here, with some battalions of the 10th Infantry Division brought up from Trebizond, they were able to concentrate about 12,000 men with twenty guns.

[1] In reality all these rumours were, perhaps intentionally, exaggerated. With the exception of a few coast batteries, the fortifications of Trebizond did not exist. Reinforcements from Istanbul were moving slowly: the advance-guard of V Corps (10th Division) was concentrating in the region Gümüşane-Bayburt; a few battalions only reached the threatened front on the coast. One U-boat was cruising along the Lazistan coast (U 33), but its performance was limited to the sinking of a Russian hospital ship (the French steamer *Portugal*) on 30 March. On 6 April it was discovered and rammed by the destroyer *Strogi*. (This, luckily for the Russians, was on the day before their landing operation.) The damaged U-boat escaped destruction but moved immediately towards the Bosporus. She was the only U-boat in the region but this fact was unknown to the Russians.

[2] The 19th Turkistanskis, one border regiment and one plastun battalion.

Lyakhov now considered his force not only too weak to attack but in danger of a counter-stroke from a stronger enemy, and possibly of an outflanking manoeuvre by the track down the Kalopotamos. On 28–30 March Lyakhov's infantry advanced along the coast as far as the village of Humurgan. Their commander himself became more and more nervous as rumour exaggerated the strength of the enemy on the Kara-dere position. Impatiently he pressed Yudenich for reinforcements, and the commander-in-chief in his turn asked the Black Sea fleet to accelerate the transport and landing of the two plastun brigades.

On 4–5 April, in the harbour of Novorossiysk, ten transports and twelve elpidiphores took on board the two plastun brigades with their artillery section (eighteen guns) and trains. The transports sailed on the 5th, with two steamers adapted for carrying twelve hydroplanes each, covered by two light cruisers[1] and four destroyers. On the following morning the main force of the Russian Black Sea fleet[2] met the transport flotilla at sea and accompanied it by parallel course at a certain distance to the west. Weather conditions were excellent.

Yudenich arrived at Rize early in the morning of the 7th. Here he was met by Lyakhov who described conditions in front of the Kara-dere positions as almost critical for the coastal detachment; the opening of a Turkish attack was expected within a few hours. Yudenich was rather impressed by this report. He decided to change the plan for landing the plastuns at Rize—agreed with Admiral Eberhardt—and to land at least one brigade farther to the west at Humurgan where the plastuns might give immediate support to Lyakhov's threatened troops.

Rear-Admiral Homenko, commanding the transport flotilla, had definite orders to land the troops at Rize and not at Humurgan. He refused to obey the commander-in-chief of the Caucasian army. Meanwhile Admiral Eberhardt had got the report of *Strogi*'s action against the German U-boat and a further hydroplane report that two U-boats had been observed near Trebizond. He flatly refused to change the plan and informed Yudenich that, in view of the danger from U-boats, all escorting and accompanying forces, after the landing at Rize, must return to base at Sevastopol.

Yudenich was stubborn enough; he retorted that he did not want any naval support for the landing at Humurgan. The 2nd Plastun

[1] One of the cruisers was ex-Turkish *Mecidiye*, sunk near Odessa by a mine and salvaged and put into commission by the Russians as *Prut*.

[2] Two battleships, two cruisers and six destroyers.

Brigade, which had already disembarked at Rize, was put on board the elpidiphores which worked with Lyakhov's troops. This new flotilla, covered only by an armed steamer which was at the disposal of Yudenich, sailed during the night and, without any incident, threw 5000 plastuns on to the coast near Humurgan early in the morning of 8 April. The 1st Plastun Brigade, with the artillery, marched along the coast and only reached the front line on the 10th.

All this misunderstanding between the naval and military commanders-in-chief had no sequel. Admiral Eberhardt had clearly exaggerated the U-boat danger, but Lyakhov's reports to Yudenich on the imminence of a Turkish counter-attack had been equally exaggerated. The hasty withdrawal of the Russian naval force, however, proved a lucky chance for the German *Breslau*. This light cruiser suddenly appeared off Humurgan on 10 April and sent several 6 in. shells among the Russian infantry. A day or two later a strong Russian naval squadron reappeared,[1] ready to join Lyakhov in a final combined operation against Trebizond.

By 13 April Lyakhov had available a sufficient force for an attack on the Kara-dere positions.[2] Still obsessed with the necessity of protecting his left from an attack across the main ridge of the Pontic Alps, Lyakhov detached his militia brigade with six guns at right angles to his front. The main attack, aimed at the Ahu-dağ hill, was entrusted to the 19th Turkistanskis.

Early on the morning of the 14th, *Rostislav* and *Panteleymon* opened a heavy bombardment against the Turkish positions near the coast with their 6 in. guns (the heavier calibre guns were not brought into action). During the day 1220 shells were thrown over the Turkish lines; their effect was devastating: fortifications were destroyed, gun positions smashed and troops were streaming to the rear. By 2 p.m. the Turks had practically evacuated their coastal sector and were in full retreat from Sürmene on Trebizond. The Russians on the right flank had only to await the results of the naval bombardment. But on the left the Turkistanskis had gone forward to the attack even before the naval bombardment had had time to develop. There was sharp fighting, for the Turks everywhere resisted courageously when they

[1] Battleships *Rostislav* and *Panteleymon* (formerly *Potemkin*) and four destroyers.

[2] *Right (sea) flank*: 2nd Plastun Brigade; 5th Border Regiment; one battalion of Batum volunteers.

Left flank: 19th Turkistanskis with two plastun battalions; 1st Plastun Brigade (less two battalions).

Reserve: two battalions of 3rd Plastun Brigade.

Total: twenty-one battalions (20,000 infantry) with thirty field guns.

were out of range of the naval guns. The capture of Ahu-dağ cost the Russians 1000 killed and wounded. It was an honourable but, in the circumstances, an absolutely useless performance.

Next day (15th) the Turks continued their retreat on Trebizond, covered by small rear-guards which turned to stand behind the numerous coastal streams. A few naval shells soon dislodged these delaying units. On the 16th, Lyakhov's troops reached the line through Kalafka-dere and Dorana village, eight miles to the east of Trebizond. The two battleships returned to Batum to replenish their munitions while troops were given one day's rest.

On the 18th, a deputation of the Greek population appeared in the Russian lines and informed General Lyakhov that the Turks had already evacuated the city on the night of 15th–16th. Lyakhov immediately ordered a strong column to march round to the south and cut the road to Gümüşane; on the 19th they made contact with the enemy's rear-guard at Cevizlik, fifteen miles from Trebizond on the main road to Gümüşane. To the west of Trebizond another column occupied Platana (Polathane), with its fine bay where the Russian fleet planned to organize a naval station.

CHAPTER XXXIII

ANGLO-RUSSIAN STRATEGIC CO-OPERATION. BARATOV'S ADVANCE TO KASR-I-SHIRIN, APRIL–MAY 1916

MAP 23

As has already been stated, the Russian imperial *Stavka* had not, in the spring of 1916, the slightest interest in taking the strategic initiative in the Turkish theatre of war. A great offensive was in preparation on the western front, and all the attention of General Alexeyev and his staff was directed to this supreme effort of the Russian armies. The Caucasian army, despite its brilliant victories, maintained an attitude of active defence.

It remained a constant feature of World War I that only the British appreciated the significance of the offensive initiative in the Middle East. This attitude was characteristic of the traditional maritime approach of the British to strategic problems and was in contrast to the continental conceptions of the French and Russians. If World War I be viewed as a whole, it will be seen that the Dardanelles operations (unsuccessful though they were in their immediate objective) had the effect of permanently weakening the Turkish army as a striking force;[1] and at the same time they held down the bulk of the Turkish divisions during a period when the Russians were desperately pressed in Poland. The Dardanelles campaign therefore had the effect of delaying occupation of Caucasia by the central powers for two or three years; when, in the summer of 1918, the Germans and Turks established themselves in the Caucasian isthmus and on the shores of the Caspian they found that they had arrived too late and too weak to exploit the economic resources and strategic promise of the area. Similarly, British action in Iraq checked and held off Turco-German progress through Persia to the frontier of Afghanistan; in Afghanistan, where the internal situation was not unfavourable to the central powers, the Germans might have made an excellent forward base for action against both India and Turkistan. On a long view, again, the final collapse of the central powers was promoted by the British campaigns in Palestine-Syria and in Iraq and by the Allied operations

[1] Just as the Carpathian battles of the first months of the war had shattered the Austro-Hungarian armies beyond real recovery.

in Macedonia. It stood to reason, indeed, since the main objective of German policy was the *Drang nach Osten*, that the liquidation of their prospects in Asia must act as an acid solvent on the military and Pan-German ascendancy in Germany.

During 1916 the British continued to find little response from their Allies to plans for the development of the strategic initiative in the Middle East. The French were particularly lukewarm and had been responsible for the rejection of the various proposals for action in Cilicia—action which offered the most practical prospects for effective co-operation between the British and Russian armies in Turkey.

In the spring of 1916 a new political agreement affirmed the principle of Anglo-Russian military collaboration in Persia (4 March) and gave to Russia a somewhat larger share of influence in that country than had been envisaged in the Agreement of 1907. A practical justification had been given to this revision by the successful action of Baratov's expeditionary force which, during February and March, had completely crushed the enemy elements in Persia, had captured Kirmanshah, and seemed to be in a position to guarantee Persian territory against any further Turco-German moves.[1] The presence of Baratov at Kirmanshah, only 120 miles from the Iraq border, gave promise also of immediate Anglo-Russian military collaboration, not only in Persia but also in Iraq itself.

The British campaign in Iraq had developed unfavourably since the failure of Townshend at Ctesiphon (22–5 November 1915) and his retreat to Kut-al-Amara. Since 12 December an Anglo-Indian force of rather less than 10,000 had been besieged in Kut by the Turkish Sixth Army (of an effective strength of over 25,000) under Field-Marshal von der Goltz. This new Turkish army[2] was strong enough to maintain the siege of Kut and to oppose with success the Anglo-Indian force which was moving up the Tigris to the relief of Townshend.[3] The British were reinforced but, in February, two fresh divisions (2nd and 6th) joined the Turks. Two renewed attempts by the British to break through to Kut ended in failure. Townshend's situation was becoming more difficult every week, and he informed the relieving force that he would not be able to avoid capitulation after 26 April.

The rescue of Townshend absorbed all the attention of British

[1] See pp. 322–4, 373–4 above.

[2] The 35th, 38th, 45th, 51st and 52nd Divisions: the two last had arrived from the Caucasus front after the Eleşkirt operation.

[3] Actions at Shaikh Sa'ad (7 Jan.), the 'Wady' (13–14 Jan.) and Hanna (21 Jan.); Dujaila (9 March) and Fallahiya (5–9 April).

headquarters in Iraq during March and April 1916. As a natural enough consequence the idea of collaboration with the Russians took the form of proposals for combined action to relieve Kut, and from the moment of Baratov's capture of Kirmanshah (26 February) the question became a major issue. Discussions were conducted through His Majesty's government and the imperial *Stavka*, while direct radio liaison was established between Baratov and British Relief Force headquarters as well as between Baratov and Townshend.

It must be said that the idea of Anglo-Russian collaboration in Iraq, narrowed down to participation in the relief of Kut, met with criticism at the headquarters of the Grand Duke Nicholas in Tiflis.[1] Accustomed to think in terms of the massive losses of the Polish front, the grand ducal staff was not unduly impressed by the plight of Townshend's meagre Anglo-Indian force. It was felt that the British were subordinating their plans in Iraq and Persia to what was regarded in Tiflis as a prestigial aim without regard to any wider strategic sense which they should and must have. After the failure of the second attempt at Dujaila (9 March), in which more than three Anglo-Indian infantry divisions and one cavalry brigade were involved, the Russians failed to see what effective use could be made of Baratov's force which did not exceed four infantry battalions, three dragoon regiments and a few Cossack formations. Neither could they understand the rather passive role imposed on Townshend by the epidemics which were decimating the troops under his command.[2] And as the weeks passed they became less inclined to risk Baratov's slender strength, particularly when, by 1 April, it was reported that a steady flow of reinforcements had brought Turkish numbers up to between 40,000 and 50,000 infantry with adequate artillery (and the support even of a German air squadron).

The practical difficulties of Baratov's situation were serious enough: if he advanced across the Zagros he would be leaving behind a long and inadequately protected line of communications. The supply problem at Kirmanshah was formidable, since everything had to be brought, mainly by animal transport, over routes 400 miles in length. His force consisted mainly of horse and was considered, in consequence, to be particularly mobile. But in the special conditions of war and season in Persia and Iraq, the superior mobility of horse was

[1] Baratov's expeditionary force came under the direct orders of the viceroy and not under Caucasian army headquarters (Yudenich).

[2] Cf. also A. T. Wilson's criticism of Townshend's lack of initiative during the Dujaila battle, *Loyalties*, I, p. 118.

rather in question: lack of forage was felt even in the spring, and from the beginning of the Persian summer in May cavalry could not depend on local supplies of fodder. This had to be transported in the main by animals which must consume a substantial proportion of their load en route. Even with large numbers of animals at his disposal,[1] Baratov had to limit his troops to a minimum bread ration and his horses to two and a half pounds of barley a day.[2] Further, an advance into Iraq raised the new and (for cavalry) vital problem of adequate water supply.

On 1 April Baratov was told by the chief of staff of the Caucasian army (General Bolhovitinov) that the grand duke wanted him to do what he thought possible by way of a diversion against Bagdad. The Cossack general's troops lacked all the special kit necessary to support a summer campaign on the borders of the hot Iraqi plains. A wide dispersion of forces was necessary to give the minimum of protection required to cover his long line of communications: Kirmanshah-Kangavar-Hamadan-Kazvin-Enzeli. He had to provide detachments also for Teheran, Kum, Kashan, Isfahan, Daulatabad, Sinna, Bijar.[3] He allotted two of the border battalions, four Cossack regiments and the militia formations to his line of communication, and designed for the Bagdad diversion the two remaining border battalions, the three regiments of dragoons and one of Cossacks, with four horsed guns and twelve horse-borne mountain guns: a total of about 5000 men. The two remaining Cossack regiments and twelve horsed guns were left at Kirmanshah as a mobile reserve.

Field Marshal von der Goltz had thought probable a co-operation between the British and some Russian force directed against northern Iraq. When the Turkish 6th Division began to arrive in Bagdad he moved four of its battalions with twelve guns towards the Persian frontier. An advance base was created at Khanikin,[4] and the German Major Bopp was entrusted with the organization of the defence of the border.

By 10 April Bopp had concentrated the four battalions of the 6th Division at Karind with a small reserve of two battalions at Kasr-i-

[1] Baratov had 100 lorries, 300 four-horse wagons, 1000 camels, 4000 mules, and 13,000 donkeys.

[2] Where grazing is poor, this is half the feed needed to keep mules in condition, and less than a third of that necessary for horses.

[3] Baratov had available the 2nd and 4th Border Regiments (two battalions each); the Caucasian Cavalry Division (three dragoon and one Cossack regiments); the 1st Caucasian Cossack Division (four regiments); two Kuban Cossack regiments; two militia *druzhiny*; sixteen horsed guns, twelve mountain guns and two howitzers (and four old guns with the *druzhiny*).

[4] At a distance of 130 miles by road from Kirmanshah and 80 miles from Bagdad.

Shirin, twenty miles from Khanikin; these last troops were the remnant of the force defeated by Baratov on his march to Kirmanshah. On both flanks of the Karind position and to the rear, Bopp had available about 5000 irregular horse and 1000 gendarmes. The whole was certainly not a group capable of holding Baratov's seasoned cavalry.

Difficult questions of rear organization delayed Baratov's start from one week to another. Meanwhile, in Kut, the unfortunate Townshend was becoming desperate. Strangely enough the British commander appears to have expected more from Baratov's movement than this movement could in any circumstances effect. With the failure of the third attempt at relief and the dwindling of Townshend's resources, the moment had arrived for a supreme effort and the Russians assured their allies of Baratov's serious participation.

Baratov, in the event, only proved ready to move on 20 April. With not more than 5000 men under his command, he certainly did not consider his task to be a march on Kut-al-Amara; but he hoped to reach the Iraqi border comparatively rapidly, to attract a part of the Turkish army away from Kut and thus to reduce the obstacles in front of the British Tigris Corps. If Baratov had been able to start before the end of March, his diversion might, perhaps, have proved successful: during the final British attempt to raise the siege, the Turks left only 2000 men in the lines in front of Kut and they drew off every man available to oppose General Gorringe.[1] If Baratov could have reached Khanikin by 8 April the Turks would have found themselves in a dangerous situation. As it was, Baratov, with all his goodwill, was only approaching Karind on 28 April, the day on which Townshend found himself under the necessity of capitulation. On the following day Şevket Bey's detachment was thrown back from the Karind positions. Four days later Baratov reached Kasr-i-Shirin where he got information by radio of Townshend's fate.

Baratov remained with the Caucasian Cavalry Division and one border regiment at Kasr-i-Shirin; his troops were already suffering from the unaccustomed heat and from lack of good water. Baratov decided to rest them, reorganize his supply, and attack the Turks in the position on to which they had retired at Khanikin. The decision proved more far-reaching than either Baratov or his immediate chiefs could have guessed. It implied that Anglo-Russian military co-operation was not to be restricted to its original object, the relief of Kut.

[1] Actions at Bait Isa (16–18 April) and Sannaiyat (20 April).

Important changes occurred in Iraq during the last days of the siege of Kut and soon after its fall. On 19 April Field Marshal von der Goltz died and the command of the Sixth Army passed to the most able of the political soldiers of the Committee of Union and Progress, namely, the generalissimo's thirty-five-year-old uncle, the ambitious and rather clever Halil Paşa, who was the hero of the winter advance into Azerbaijan in the first year of the war. Enver himself was on a tour of inspection in Iraq, where he arrived just in time for the laurels of Kut-al-Amara. By the middle of May, in Bagdad—where he had the advantage of the advice of the German General von Lossow—Enver Paşa had to take an important strategic decision. His first (and rather natural) intention had been to operate against the British with the whole of the Sixth Army.[1] But varying intrigues and different political and personal considerations were, as usual, influencing the decisions of the generalissimo. A campaign against the British in southern Iraq seemed more difficult and less promising than the conquest of Persia. The dynamic dreamer in Enver again prevailed over the strategist; he decided for a Turanian offensive across Iran.

The Turkish Sixth Army was divided into two army corps, XVIII and XIII,[2] while the 4th Cavalry Division was left at Mosul to cover the direction of Ruwandiz. Under the energetic command of Ali Ihsan Paşa,[3] XIII Corps had orders to defeat Baratov's small force and to invade Persia where Enver hoped to raise a national army with better success than the Germans had had.

Thus, if Russian intervention in Iraq had failed to effect the relief of Townshend, the presence of Baratov at Kasr-i-Shirin may be said to have been a factor in attracting the impressionable Enver towards a Persian expedition and imposing on him a division of the strength of the Sixth Army; and the failure of the Sixth Army to seek an immediate decision with General Maude allowed the British a period of lull in which to prepare a powerful army for the eventual march on Bagdad. To this extent the first experiment in Anglo-Russian strategic collaboration in the Middle East had proved fruitful.

[1] Enver had available the 5th, 51st, 52nd, 2nd, 45th and 6th Infantry Divisions; the 35th and 38th had practically ceased to exist after the fierce battles with the British.

[2] XVIII Army Corps: 45th, 51st and 52nd Infantry Divisions; XIII Army Corps: 2nd and 6th Infantry Divisions and one cavalry brigade.

[3] Not to be confused with Ihsan Paşa, who commanded IX Army Corps at Sarikamiş.

CHAPTER XXXIV

THE CAUCASIAN FRONT TO JUNE 1916. VEHIP PAŞA'S ABORTIVE OFFENSIVE

MAPS 1, 27, 28, 29; cf. 32

SUMMER comes slowly on the upland plains and over the bare ranges of eastern Anatolia. The spring thaw and recurrent rain storms make roads and tracks impassable; and on the passes sudden blizzards in April can be even more formidable than in February. When, early in May, the generalissimo visited the headquarters of the Third Army at Erzincan he found troops still in winter quarters and still awaiting reinforcements. V Corps had completed its concentration in April, but, in the meantime, Vehip Paşa had had to extend the front of the Third Army to meet the deployment of two Russian divisions from Trebizond.

The remnants of the coastal force and one of the divisions of V Corps occupied the small port of Tirebolu and the district along both sides of the Harşit-su as far as Ardasa, to the south of the Zigana Pass over the northern ridge of the Pontic Alps. Another division (10th) was at Ardasa and Gümüşane; and a third lay in the mountains along the Gümüşane-Bayburt road as far as the village of Balahor.[1]

The country on both sides of the Çoruh bend was the area of X Corps. Along the loop of the Kara-su and to the north of it IX Corps was in position covering Erzincan.

XI Corps (under reorganization) and the 36th and 37th Divisions (in the same state) were in reserve. In the wild Dersim Mountains the 2nd Cavalry Division and Kurdish irregulars were in observation between Erzincan and Kiği; and at Kiği they were in liaison with the 3rd Cavalry Division belonging to the Second Army.

The divisions of the Third Army averaged from 5000 (XI Corps and 36th and 37th) to 8000 (V Corps) men: a total of between 80,000 and 90,000 infantry. With only twelve guns to a division, Vehip's army was deficient in artillery. It was, however, a considerable force when compared with the strength of the Second Army in May.

Since, during March and April, priority in transport had been given to the divisions en route for Iraq, the concentration of the Second Army had been subject to frequent delays and interruptions.

[1] This area was the scene of the last battles in Paskevich's campaign of 1829.

The new XVI Corps under Mustafa Kemal Paşa was incomplete: the 5th Division was between Diyarbekir and Bitlis and the 8th between Diyarbekir and Muş; but the 7th was still in transit. Of III Corps, only the 1st Division had arrived at Harput, where Marshal Ahmet Izzet Paşa had set up his headquarters. From Kiği the 3rd Cavalry Division was patrolling along the tracks to Erzincan. Under such conditions, the Second Army was certainly not ready to undertake any large strategic operation in combination with the Third; and even the Young Turkish optimists on the marshal's staff had come to realize that the chances of an offensive in June were remote; July, or even August, seemed to be the period destined for the mounting of Enver's new 'decisive blow'.

By the end of May the total of Turkish forces on the Caucasian front (including all irregular formations and the 4th Infantry Division at Mosul) did not exceed 130,000 men with about 200 guns. The Russian Caucasian army, meanwhile, had received considerable reinforcements. During the reorganization of the Russian army in the winter of 1915–16, after the great retreat of 1915 and before Brusilov's big offensive of 1916, several scores of new divisions had been created. Some of these were to prove as good as the units of the old army and some not so good. The men were fair material, but the officers were now mostly reservists or inexperienced subalterns fresh from the military schools. To the Caucasian army were allocated the 123rd and 127th Infantry Divisions, each of them with a strength of four three-battalion regiments and a field artillery section of eighteen guns.

The successful transport of the plastun brigades for the Trebizond operation had convinced the Russian naval and military staffs of the practicability of using the sea way, and the new infantry formations were accordingly carried from Mariupol to Trebizond.[1]

[1] The transports travelled in two echelons: 127th Division sailed on 16 May (15,000 men, 4000 horses, 18 guns and 1500 wheeled vehicles); 123rd Division on 30 May (16,000 men, 4000 horses, 18 guns, 1500 vehicles). Both divisions were taken on board at Mariupol on the Azov Sea, and the Straits of Kerch were well protected during the embarcation.

Each echelon was made up of twenty-eight transports; the battle fleet accompanied the convoy with one battleship detached to watch the exit from the Bosporus. For loading, elphidores, barges and boats were used. The landing was organized in the Bay of Kovata, east of Trebizond. The operation was carried out without incident. No enemy ships were at sea at that time. *Goeben* was in repair until the end of May; while *Breslau* was at sea only between 30 May and 2 June. Of the two German U-boats in Black Sea waters, one had been damaged by *Strogi* and the other, coming from the Mediterranean, was off Trebizond on 4 June when the unloading was already complete. An attempted attack failed. The 127th Division was on shore by 21 May; the 123rd by 3 June.

Early in June V Caucasian Army Corps had been constituted on Yudenich's right flank.[1] To the left of the Caucasians was II Turkistan Corps who remained in their winter positions on both sides of the Çoruh bend.[2] On the left flank of the Turkistanskis, from a point near the main road, Erzurum-Aşkale-Bayburt, to the bend of the Kara-su, was the area of I Caucasian Corps.[3] The 5th Caucasian Cossack Division observed the extreme left flank of this corps in the direction of Kiği. A powerful strategic reserve was held in the region of Erzurum.[4] In all, opposed to Vehip Paşa's Third Army, Yudenich had a 50% superiority in infantry and double the number of guns. Such a relation of forces did not promise any chance of success for the Turks.

The Russians were weaker on their left. By the end of May, Yudenich had restored the full strength of IV Caucasian Corps which was concentrated in the Hinis-Muş-Bitlis region.[5] To the left of this corps, Chernozubov's Azerbaijan-Van force was scattered over the vast region between Muş and Lake Urmia.[6] The forces on the Russian left strategic flank were clearly insufficient for their task, but at the end of May they were as yet only opposed by the incomplete Turkish Second Army still in the first stage of their concentration between Harput-Diyarbekir and Siirt.[7]

By 1 June the Russians had concentrated along the whole of the Asiatic front about 200,000 men and 400 guns as against the 130,000 men and 220 guns at the disposal of the Turks. But the Turks were still expecting the arrival of reinforcements for the Second Army while the Russians could count on no further drafts.

[1] The 1st and 2nd Plastun Brigades; the 123rd and 127th Infantry Divisions; the coastal detachment (ten battalions): 40,000 infantry with 100 guns.

[2] Add four plastun battalions (3rd Plastun Brigade) and the Siberian Cossack Brigade: 30,000 infantry, 2000 cavalry, 74 guns.

[3] The 39th Infantry Division; Don Foot Brigade; 1st and 2nd Transcaucasian Rifle Brigades (ex-militia); 5th Caucasian Cossack Division: 30,000 infantry, 3000 cavalry, 96 guns.

[4] The 4th Caucasian Rifle Division; 5th Caucasian Rifle Division (newly formed), 1st and 2nd Kars Regiments (from Kars fortress troops), 4th Plastun Brigade (recalled from Azerbaijan): 40,000 infantry, 36 guns.

[5] The 2nd Caucasian Rifle Division, 261st, 262nd, 263rd, 264th Regiments of 66th Infantry Division, three Armenian rifle battalions, six Cossack and one horsed border regiment: 30,000 infantry, 5000 cavalry, 64 guns.

[6] The 3rd Border Regiment, Georgian Rifle Regiment, one Armenian rifle battalion, twelve militia *druzhiny*, 2nd and 3rd Transbaikal Cossack Brigades, 4th Caucasian Cossack Division: 10,000 infantry, 6000 cavalry, 30 guns.

[7] In May the 1st, 5th and 8th Turkish Infantry and the 3rd Cavalry Division were in this area, while the 4th Infantry Division was at Mosul. Total Turkish numbers in this area (excluding the 4th Infantry Division) were about 40,000 infantry (including irregulars), 11,000 cavalry, 100 guns.

The concentration of strong Turkish forces on their strategic left was known to Yudenich and his staff.[1] And on 20 May, a Cossack patrol in the region of Kiği captured a major belonging to the headquarters staff of the Turkish Second Army. This individual, of Kurdish origin, professed to nourish some grievance against his chiefs and proceeded to disclose valuable information concerning the Turkish operational plan.[2]

If the seriousness of the intended Turkish offensive was no secret to Yudenich he was also aware of the extreme slowness of the Turkish concentration. There was every indication that the Turks could not be in a position to launch their attacks before the second half of July. Yudenich therefore decided to take preventive action against one of the Turkish armies before the other was ready for the field; and he planned his action for the second half of June. But, quite unexpectedly for the Russians, Vehip Paşa, in the last days of May, seized the initiative and attacked the Russian lines in the region of Mamahatun.

The visit of the generalissimo, early in May, to Erzincan and Harput to inspect the troops of Vehip and Ahmet Izzet had not failed to stimulate the activity of these commanders. Activity took the form of reconnaissances, sometimes in force. In the middle of April the 5th Infantry Division, newly arrived from Istanbul, occupied Siirt and made contact with Russian advanced units on the tracks leading to Bitlis. There were several days of skirmishing, but Mustafa Kemal Paşa, commanding XVI Corps, had not felt himself strong enough to develop an attack on that place. In May, following the arrival of the 8th Infantry Division at Diyarbekir, Turkish reconnaissance troops made their appearance along the mountains leading over to the Muş plain. At the same time Turkish patrols were active on the front of the Third Army and their enterprise gave some results. Vehip Paşa became aware, during the second half of May, that a certain redistribution of Russian forces was in progress (in preparation for the June offensive planned by Yudenich), and he decided to take advantage of a particularly favourable moment for an attack on the front of I Caucasian Corps.

[1] The staff of the Caucasian army was more or less accurately informed of the transit situation on the Bagdad railway by their own agents, by Odessa and by the intelligence section of the imperial *Stavka*. But the most complete information, as Russian sources confirm, came through British intelligence officers in Iraq.

[2] Subsequently questioned by the British, he presented a detailed (and interesting) report on the Gallipoli campaign. He later served with the Whites during the Civil War and, after adventures in the Far East, settled in Iraq where he was still living in 1944 and had published an entertaining book of memoirs in Arabic. (W.E.D.A.)

During the last week of May the troops of the 39th Infantry Division were being withdrawn to the rear for rest in the villages of the Erzurum plain; they had suffered heavy losses in the Köprüköy and Erzurum battles and had been holding the important sector in front of Mamahatun throughout the period March–May. Meantime the sector in the bend of the Kara-su was held very lightly: the Don Foot Brigade on the right, a single regiment, the 15th Caucasian Rifles, in the centre, and on the left the 1st Transcaucasian Infantry

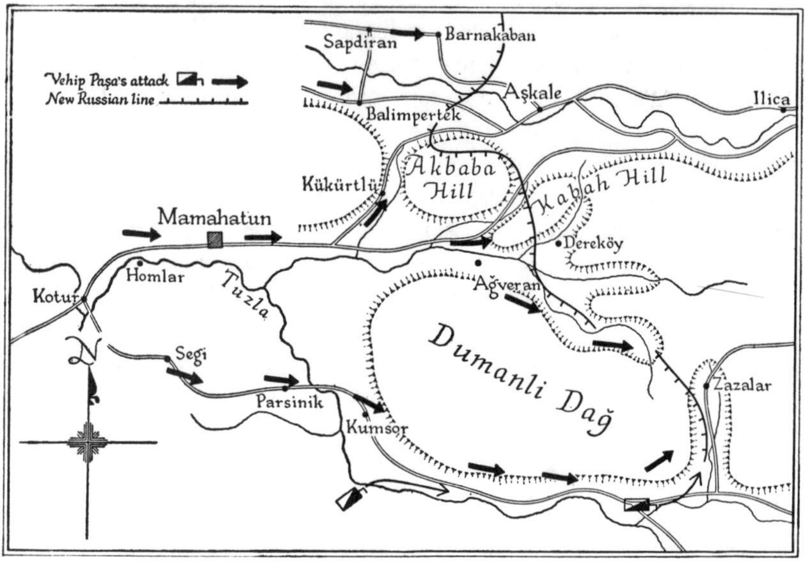

Map 27. Mamahatun, 29–31 May 1916

Brigade who were in fact militiamen. Farther to the north, between the river and the Aşkale-Bayburt road, positions were thinly held by the other improvised Transcaucasian brigade which was still awaiting reinforcement by four Kuban plastun battalions, recalled from Chernozubov's force; to the south of the Kara-su bend there were only the mounted men of the 5th Caucasian Cossack Division. It was a rather delicate moment for the front of I Caucasian Corps, where a single regular regiment was flanked by relatively raw and improvised units. The corps commander, General Kalitin, had clearly supposed that his adversary would remain passive, but Vehip Paşa, informed of the temporary weakness of the Russian front, secretly concentrated the three divisions of IX Corps, with units of XI Corps in support, for an *attaque brusque* on both sides of the Mamahatun salient.

Early on the morning of 29 May, some elements of XI Corps attacked on the right bank of the Kara-su towards Barnakaban from where the Don Foot, fighting stubbornly, retreated. More dangerous was the situation of the 15th Caucasian Rifles assailed on both sides of Mamahatun by the 29th Division of IX Corps. The lines of this regiment, which extended over twenty miles, were broken through and the Turks took the town of Mamahatun with some prisoners and two mountain guns. Farther to the south the Turkish 17th Division easily threw back the Transcaucasians and penetrated into the valley of the Tuzla-çay. During 30 and 31 May the Turkish advance continued along the Mamahatun-Erzurum road as well as up the valley of the Tuzla-çay; but, on their left, their progress beyond Barnakaban was less marked.

Yudenich took immediate steps to restore the situation. While the newly arrived Kuban plastuns of the 4th Plastun Brigade were sent to reinforce the Don Foot in retreat from Barnakaban, the veteran regiments of the 39th Division were ordered to attack along both sides of the *chaussée* to Mamahatun. At the same time two regiments of this division (the 155th Kubinskis and the 153rd Bakinskis) were directed against the right flank of the Turkish advance on the Dumanli-dağ mountains and in the valley of the Tuzla-çay. Between 1 and 3 June the Kubanians, with the Derbentski and Elizavetpolski Regiments of the 39th Division, were in hot fighting with the Turks between Barnakaban and the Mamahatun road. Here the advance of the Turks was stopped; but in the valley of the Tuzla-çay their 17th Division continued to go forward. The manoeuvre of Vehip Paşa seemed now to take the character of an outflanking movement against the left of I Caucasian Corps (which was in fact 'in the air'). Vehip Paşa was prompt in bringing his cavalry (2nd Cavalry Division) into the field while the Cossacks still remained somewhere beyond the Tuzla far from the battle. On 4 June the Kubinskis and Bakinskis attacked the Turks in the valley of the Zazalar stream—twelve miles to the north-east of Çerme. The Turkish infantry was thrown back, but at the critical moment three regiments of *süvari* made a very gallant mounted attack which held the advance of the Russian infantry. The Turks brought up units of their 28th Infantry Division which made three counter-attacks with considerable *élan*, but the Bakinski men stood firm.[1] Next day (5 June) the fighting suddenly died away all along

[1] During the fighting in the Zazalar valley the Turkish infantry and cavalry suffered very serious losses; and on the Russian side the Bakinski Regiment alone lost over 1000 officers and men.

the new front line. The Turks held their new positions from Barnakaban to the Kükürtlü Pass on the Mamahatun road, continuing along the eastern slopes of the Dumanli-dağ to the Tuzla-çay valley at Parmaksiz ('Fingerless'), through which converge several tracks from Oğnot to Erzurum over the formidable massif of Bingöl-dağ which forms the watershed between the upper Aras and affluents of the Murat-su. (Cf. Map 32.)

Both sides were for the moment exhausted; but it is clear that what really checked the offensive of Vehip Paşa between the Kara-su and the Tuzla was the news of the landing of Russian reinforcements at Trebizond during the first days of June. His attention went to his left in spite of his success on his right. He might claim a victory at Mamahatun; it was an undoubted tactical success. This tactical success might even have had serious strategic consequences since the Third Army had gained important miles on the approaches to Erzurum; they were now a few miles west of Aşkale and only thirty-five miles from the historic fortress city. On his right Vehip had advanced sufficiently to control the important tracks running from the Erzurum plain to the valley of the Endres-çay and to Oğnot; by these tracks he might establish liaison with the patrols of the Second Army which were active in the region of Kiği and Oğnot. The Cossacks had moved into the valley of the Tuzla-çay to cover the exposed flank of the 39th Division, but a wide gap of more than thirty miles divided the left flank of I Caucasian Corps from the patrols of IV Caucasian Corps on the eastern slopes of Bingöl-dağ.

The new positions gained on the right flank of the Third Army were certainly favourable to the development of the big offensive intended against Erzurum and Hasankale—an operation in which the principal part had to be played by the Second Army. But the Second Army was still far from ready for the part which it was expected to play. By mid-June Izzet Paşa could concentrate for action only six divisions,[1] 60% of his planned force. Four or five more divisions were expected to arrive during the coming six weeks.

At this relatively favourable juncture for Turkish arms, it appears that the impressionable Enver had become extremely uneasy at the loss of Trebizond and its rapid use for the landing of Russian troops; the enemy had clearly opened a new sea route which might greatly facilitate the transit of unexpected reinforcements to the Caucasian front. Vehip Paşa received orders to prevent the Russian penetration from the Black Sea coast into Anatolia; he therefore had to under-

[1] The 5th, 8th, 7th, 12th, 1st and 49th.

take a hasty offensive on the left flank without waiting for the intervention of the Second Army to consummate his success on his right.

This sudden alteration in the Turkish operational plan can hardly be justified. Taken in broad outline the situation of the two adversaries, as at 15 June 1916, was as follows:

Russians	Turks
Right group: between the sea and Aşkale-Bayburt road V Caucasian and II Turkistan Army Corps: 80,000	*Left group* V and X Army Corps and auxiliary formations: 60,000
Central group: from Aşkale to Tuzla-çay 39th Infantry Division and other formations: 40,000	*Central group* IX and parts of XI Army Corps: 35,000
Left group: from Bingöl-dağ to Bitlis IV Caucasian Corps and cavalry formations: 40,000	*Right group* Six divisions and auxiliary formations: 60,000
Strategic reserve: near Erzurum About 40,000	*Reserve*: at Erzincan 15,000

This confrontation demonstrates that the best chance for Turkish offensive action between 15 June and 1 July still lay on their right flank against the relatively weak and extended Russian left. If exploited at all, the combined plan evolved earlier in the year had to be put into action immediately and without awaiting the arrival of the four or five new divisions which were moving up to reinforce the Second Army. This Second Army, strengthened by 10,000 men from the Third, might, in a fortnight, have made the situation in the gap between the Russian I and IV Caucasian Corps extremely dangerous.

The second offensive action of Vehip Paşa, like the first, met an initial success, and for the same reason: the Turks prepared their attack by good reconnaissance work among the wild mountains of the Pontic Alps, and they were aided by local irregulars from among the Moslem population hostile to the Russians. Turkish patrols discovered the weak point in the front of the Russian right group; the point of liaison between V Caucasian and II Turkistan Corps. The area itself, north of Balahor on the Bayburt-Gümüşane road, had a certain strategic importance, since two practicable paths (partly tracks) crossed the main ridge of the Pontic Alps, one from the region of Bayburt along the valley of the Sülüklü-dere ('Leech valley') to Of, and the other, even better, from Balahor along the Kara-dere to Sürmene.

The troops of V Caucasian Corps occupied the following positions:

Between Platana and Cevizlik: 123rd Infantry Division.

Region of Cevizlik: 127th Infantry Division.

Between Cevizlik and the upper valley of the Kara-dere: 5th Border Regiment, 1st and 2nd Plastun Brigades.

Between upper streams of Kara-dere and Sülüklü-dere: 19th Turkistanskis.

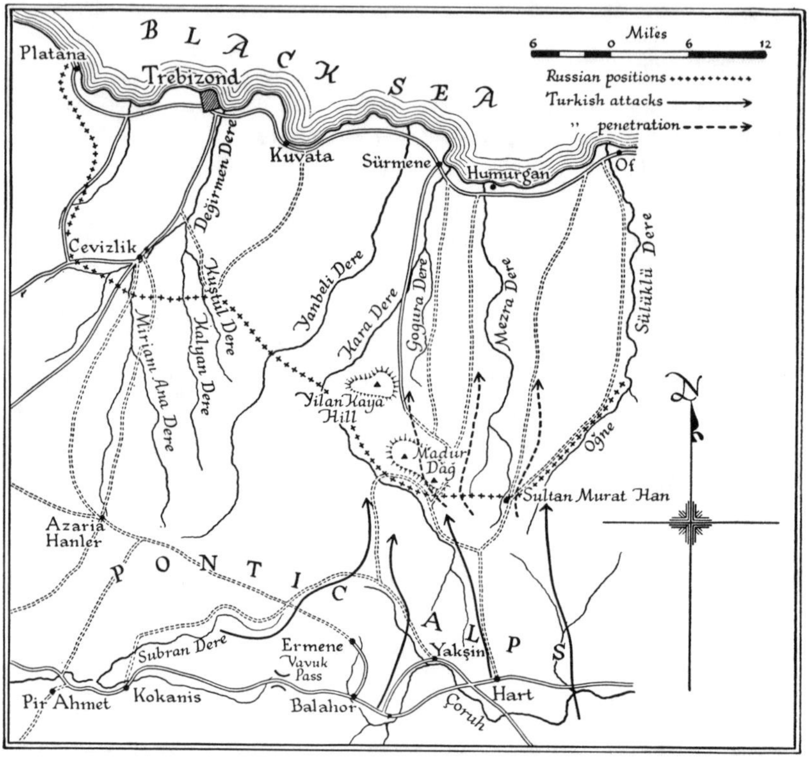

Map 28. Turkish offensive in the Pontic Alps, 26–30 June 1916

Between Sülüklü-dere and upper valley of Kalopotamos: two battalions of 3rd Plastun Brigade (in liaison with other battalions of the same brigade belonging to II Turkistan Corps on the southern side of the Kirklar-dağ ('Mountains of the Forty(?) Saints') at Orsor and Karutaş).

While two divisions of the Turkish V Corps occupied positions from Fol (Fala) village on the coast to Ardasa and from Ardasa to Gümüşane and Pir-Ahmet cross-roads, a third division was moved to Balahor where, strengthened by units of X Corps and experienced detachments of Frontier Guards, it moved secretly across the main ridge of the Pontic Alps by the two paths which led over to the valleys of the Kara-dere and the Sülüklü and fell on the positions of the 19th Turkistanski Regiment (26 June). The point of attack was well chosen. The Russian regiment was overwhelmed and next day lost their chief *point d'appui*, the great hill of Madur-dağ. Between the 27th and the 30th hasty Russian counter-attacks by Caucasian riflemen and some plastun battalions were without success. With difficulty the left flank of V Caucasian Corps retained Yilan-kaya ('Serpent rock') hill, only fifteen miles from Sürmene on the coast.

The initial stage of the new Turkish offensive was undoubtedly successful, and the surprise march of a large force across the main ridge of the Pontic chain had been brilliantly executed. But it was a victory without a morrow: the Turks had not the force available to exploit it. At the same time Yudenich had been ready since 25 June to mount his own counter-offensive on a very much grander scale. The Turkish surprise was the cause of only a slight delay in the Russian plan. The scope of the Turkish effort was accurately appraised, and on 2 July Yudenich gave orders to V Caucasian and II Turkistan Corps to start a general offensive.

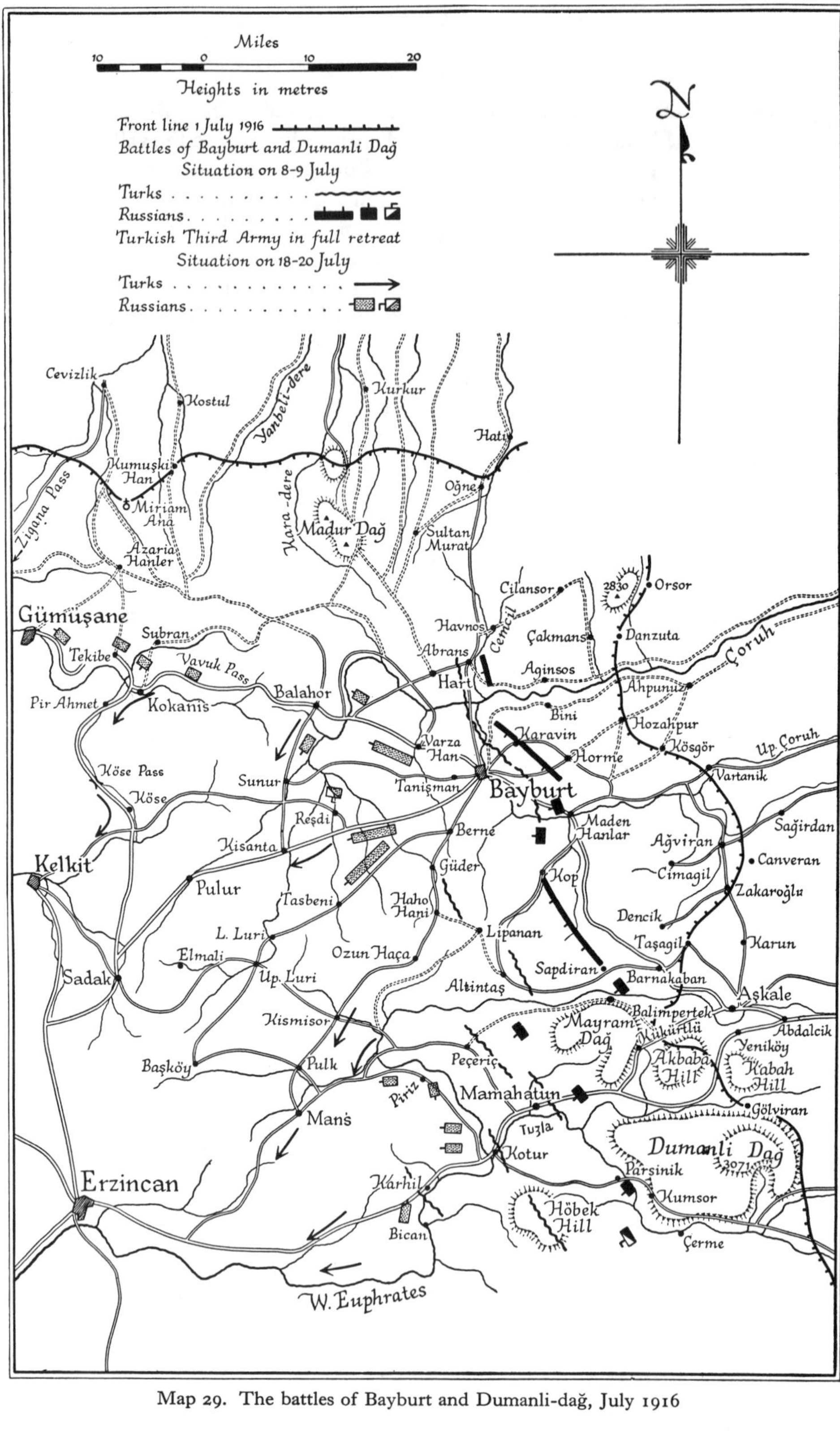

Map 29. The battles of Bayburt and Dumanli-dağ, July 1916

CHAPTER XXXV

BATTLES OF BAYBURT AND DUMANLI-DAĞ, FALL OF ERZINCAN, JULY 1916

MAPS 1, 29, 30

THE plans for a Russian offensive were complete before the launching of Vehip's daring attack across the main chain of the Pontic Alps. Yudenich directed his principal blow against Bayburt, the main cross-roads between the Anatolian coast and the uplands of Erzurum and Erzincan. An advance through Bayburt north-west to Gümüşane and west to Kelkit would have the effect of cutting in two the front of the Turkish Third Army. Again, the idea of breaking through the enemy's centre attracted Yudenich more than any other tactical manoeuvre.

The Russian commander could dispose of powerful reserves with which to carry through his operational plan. On the right the 123rd and 127th Infantry Divisions, still inexperienced in mountain war, were expected to maintain a pressure on the Turkish line between the coast and Gümüşane and along the slopes of the great Pontic ridge between the main road to Bayburt and the upper waters of the Kara-dere. Between the Kara-dere and the Sülüklü-dere the two plastun brigades with the 5th Border Regiment and the 19th Turkistanskis had to engage the enemy forces which had succeeded in penetrating the Russian lines over the wooded mountains between these two valleys. On the left flank of this group of rifle battalions the 3rd Plastun Brigade had to act as liaison with the right flank of II Turkistan Corps.

The principal blow was to come from the Turkistanskis (reinforced by three regiments from the 4th Caucasian Rifle Division) in the loop of the Çoruh. On the left bank of the middle Çoruh (which forms the northern side of the loop) four plastun battalions had to advance towards the northern angle of the river's bend. Inside the bend three Turkistan regiments, with the fourth (15th) in reserve, would mount the attack against Bayburt. To the south of the bend, along the left bank of the upper stream of the Çoruh, the front of attack was to be prolonged by the 17th and 18th Turkistanskis, usually the most successful regiments in the division, with the 20th in reserve.

On the left flank of the Turkistanskis, the three regiments of the 4th Caucasian Rifle Division, supported by units of I Caucasian Corps,[1] had to move from Dencik against the Kop-dağ Pass on the main road between Aşkale and Bayburt. Their task was to recover the hard-fought positions at Barnakaban which overlooked the plain of the Kara-su. Thus, between the upper stream of the Çoruh and the Kara-su, some thirty-four battalions were concentrated on a front of slightly over twenty miles and the density of the attack was here to be the greatest.

To prevent the Turkish command from moving reinforcements from their right to the centre the 39th Infantry Division, with the Don Foot Brigade, an Armenian battalion and two Cossack regiments in reserve, was detailed to attack along the front between the Kara-su and the Tuzla. The 1st Transcaucasian Brigade with two Cossack regiments was in observation on the extreme Russian left beyond the upper Tuzla. All these left-flank troops had already been severely exercised during the early June fighting.

In general reserve Yudenich held one regiment of the 4th Caucasian Rifle Division and the newly formed 5th Caucasian Rifle Division; the arrival of the two improvised Kars Regiments was expected. The Russian commander knew that his strategic reserve might soon be needed in the sector of IV Caucasian Corps to meet the mounting offensive of the Turkish Second Army; and his apprehensions proved right.

In the difficult country of the Çoruh bend the plastuns and the Turkistan riflemen encountered their old adversaries, the divisions of the Turkish X Corps. But this corps had already been somewhat depleted in order to reinforce the offensive of V Corps against Sürmene and Of during the last week of June. On 2 and 3 July units of the Turkish 30th Division were strongly attacked to the north of the Çoruh in their positions along the Çakmans stream by four battalions of the 3rd Plastun Brigade and to the south of the river, on the line Ahpunuz-Hozahpur, by the 16th Turkistanskis. After some sharp fighting the askers began to retreat slowly towards the west. To the south the 13th and 14th Turkistanskis made slow progress over the difficult mountains inside the loop; here the Turks seemed to intend to stand on a line nearer to Bayburt.

To the south of the upper Çoruh the Turkish 32nd Division put up a stubborn fight. Here the 17th Turkistanskis attacked at Vartanik

[1] Four battalions of the 4th Plastun Brigade and the 2nd Transcaucasian Brigade. The plastuns were the four battalions from the Kuban, the two Terek battalions having been left with Chernozubov.

along the Çoruh, and the 18th at Söğütlü and Sağirdan in the ravine of the Kurt stream. Turkish resistance proved particularly obstinate on their second line, the well-fortified Ağviran positions, which the Russians captured only on 5 July.

On the right flank of the Turkish X Corps the 31st Infantry Division held rather a long line facing south-east from Ağviran to Cimagil and the Kop-dağ Pass; continuing the line south, a division of XI Corps was standing on fortified positions between Barnakaban and the Akbaba mountain. These two Turkish divisions were attacked separately, by General Vorobyev's striking force formed of three regiments of the 4th Caucasian Rifle Division and the 20th Turkistanskis near Dencik, and at Barnakaban by the four Kuban plastun battalions under Colonel Kruten.

At Barnakaban the Turks put up a stiff resistance and, when they lost this point, they stood on their second line at Sapdiran. But in the region of Dencik two of Vorobyev's Caucasian rifle regiments broke through and by 4 July they were on the main road to Bayburt. The Turks were retiring here through Kop-hanlar[1] down the valley of the Kop-su towards the Çoruh. By the 5th the lines of the Turkish X Corps were already outflanked from the south, as Yudenich had planned; and the right flank was retreating in disorder, losing heavily in prisoners and guns. It was the beginning of a general withdrawal of X Corps on Bayburt.

By 8–9 July the plastuns had reached the line of the Cemcil-dere which enters the Çoruh from the north, less than ten miles due north of Bayburt. Inside the bend of the river the Turkistan regiments had fought their way to the line through Karavin and Mezge-tepe, while the 17th Turkistanskis had improvised a passage of the Çoruh at the blown bridge near Madenhanlar and were attacking the Mezge mountain (2300 m.) from the south. The rear-guards of the 32nd Division had fallen back on to the last ridge covering the outskirts of the town towards the east. To the south of the Çoruh all the stretch of main road between that river and the Kara-su was now occupied by Vorobyev's Rifles; and the right flank of the Turkish X Corps, thrown back into the wild mountains to the west of the main road, was in process of rapid disintegration and had ceased to exist as a fighting factor. On his left Vorobyev helped the plastuns to capture the Sapdiran position, and the defending division of the Turkish XI Corps, with its left open and exposed to serious punishment, was by 8 July in full retreat along the Kara-su.

[1] I.e. 'The inns or resthouses (*hanlar*) on the Kop Pass'.

By 8–9 July it was already clear that Vehip Paşa had lost control of the movements of his central group. Two divisions of X Corps had proved unable to defend the approaches to Bayburt against six Russian regiments and four battalions of Kuban foot.[1] On 12 July General Przevalski sent for his Siberian Cossacks and by the 14th they were round the town. Next day the 17th Turkistanskis entered Bayburt, which was already evacuated by the Turks. Patrols of the Siberians moved rapidly along the roads leading to Gümüşane and Kelkit. The fate of Bayburt had been determined not only by the fighting in the bend of the Çoruh and on both sides of it but also by Vorobyev's early break-through on to the Aşkale-Bayburt road.

The capture of Bayburt had been achieved four or five days later than Yudenich had originally estimated, owing to the great difficulty of the terrain in the Çoruh bend rather than to the resistance of the Turks. On the Russian right flank serious fighting took place only in the region of Madur-dağ, which had been so recently conquered by the Turks. Here the 1st and 2nd Plastun Brigades pressed back their antagonists, in several days' fighting, towards the main ridge of the Pontic Alps. The retreat of the Turks was accelerated by the news of the imminent Russian threat to Bayburt; and they had time to withdraw across the mountains to Balahor before the Siberians rode along the main road on Gümüşane.

As for the 123rd and 127th Infantry Divisions on the extreme Russian right, their action between the sea and the Trebizond-Gümüşane road was purely demonstrative and intended to hold down the two divisions of the Turkish V Corps in front of them.

On the left flank of Yudenich's main group the fighting had developed into a serious battle; and losses at Dumanli-dağ proved heavier than those suffered in the fighting for Bayburt. Covered on their left by the Barnakaban line and the Kara-su, the Turkish IX Corps held positions stretching south-east to the Tuzla-çay. Here on their right, the Dumanli-dağ served as a gigantic natural redoubt. The 29th Division was posted to the north of the Aşkale-Mamahatun road; the 28th on the northern salient of the Dumanli-dağ called Akbaba; and the 17th held the plateau overlooking the valley of the Tuzla with the 2nd Cavalry Division covering their right. There was a division of XI Corps in reserve at Mamahatun, while nearer to Erzincan was what passed for Vehip's strategic reserve, one division of XI Corps and units of the 36th and 37th Divisions.

As usual, the Turkish positions had been skilfully adapted to a

[1] Between 22,000 and 23,000 against less than 15,000.

terrain which favoured the defence: strong field works had been thrown up, particularly near Kükürtlü and overlooking the valley of the Tuzla; and the artillery was well placed in three main groups. For the first time the Turks had a dozen German howitzers on their front. Neither the conditions of terrain nor the numerical relation of the opposing forces could be said to favour the Russians: against twenty-seven Turkish battalions they had twenty-one battalions with nine in reserve.[1] Even if the Russian battalions were thirty per cent stronger (an average of 800:600), General Lyakhov held no advantage in numbers.

Lyakhov might have been wiser to adhere to his demonstrative role and await the result of the operation against Bayburt; but he was a tempestuous and ambitious man, impatient to fight and indifferent to losses. On the night of 5–6 July he strongly attacked, with the intention not only of holding the enemy in front of him but of defeating him as completely as possible. The 156th Elizavetpolskis with the Armenians attacked along the Mamahatun road at Kükürtlü; the 155th Kubinskis moved against Akbaba from the west, while the 153rd Derbentskis remained in observation of the Gölviran sector on the north-eastern slope of Dumanli-dağ; and the 154th Bakinskis assailed the high-terraced plateau between the summit of Dumanli (3071 m.) and the Zazalar valley. Two Cossack regiments co-operated with the Bakinskis on their left where several streams run down to form the Tuzla. Farther to the south was the Transcaucasian Brigade.

The night attack only yielded the Russians the first line of Turkish trenches along most of the front of the 29th Division. On the following day (6 July) the Turks counter-attacked energetically, especially against the Elizavetpolskis, and the Armenian battalion was badly mauled. Lyakhov moved up two battalions of the Don Foot on the right flank of the 39th Division, and one in support of the Kubinskis. During the day the Derbentskis made some progress on the slopes of Dumanli. On the night of 7–8 July Lyakhov decided to renew his night tactics with an assault directed against the outlying bastion of Akbaba; but during the preceding day Turkish counter-attacks grew in strength and the Russian commander had to engage the last battalion of his reserve. The prospects of a night attack were therefore rather desperate, but the Turkish command became suddenly depressed by the loss of the Barnakaban positions (captured during the 7th by Kruten's plastuns). The Turks feared a break-through along both banks of the Kara-su and grouped their reserves to the left.

[1] The 39th Division, Don Brigade, and one Armenian battalion.

The night assault of 7–8 July brought an important success to the Russians; the Kubinskis, supported by the Derbentskis and one Don battalion, captured the whole of the Akbaba massif, and the Turkish 27th Division was thrown back with heavy losses towards Mamahatun. Turkish resistance now weakened in front of the Bakinskis and mounted Cossacks attacked the askers of the 17th Division as they withdrew from their positions.

During the day of 8 July the troops of the Turkish IX Corps were in full retreat to the west, leaving many prisoners in the hands of the Russians. On their left the Turks were rolling back rather rapidly towards the region of Peçeriç; for the commander of IX Corps seems to have apprehended that his troops might be cut off from the only bridge across the Kara-su at Kotur. Some rear-guards tried to slow down the Russian advance on Mamahatun, but this important town was captured by the 154th Derbentski Regiment on 10 July. On the same day the 156th reached Peçeriç in the Kara-su bend. On the Russian left the 153rd reached Çerme in the Tuzla valley early on the morning of the 9th, while advance elements of the 155th appeared lower down the river.

To cover the passage across the Kara-su near Kotur the Turks took up a strong natural rear-guard position along the slopes of the Höbek mountain flanking the approaches to the bridge. The attack on this position (occupied by five regiments from IX and XI Corps) needed preparation and was only carried out by the 153rd and 155th Regiments on the night of 13–14 July. Höbek-dağ was captured in the early hours of the 14th (with several hundred prisoners). During the two following days the 39th Infantry Division came to a halt round the rim of the Mamahatun (Tercan) plain: the 156th near Peçeriç, the 157th to the north and the 155th to the south of the Kotur bridge, and the 153rd in the Girdim-su region.

The Dumanli-dağ battle had been turned into a serious defeat of the Turkish IX Corps (with one division of XI Corps). Losses were heavy; at least 12,000, of which half were prisoners. The victory cost the much-tried 39th Division about 4000 casualties; but the result of Lyakhov's spirited action was more far-reaching than Yudenich could have expected. The gain in territory was unimportant, for the Turks could hardly have avoided the evacuation of the Mamahatun plain after the loss of Bayburt. The very gallantry of the Turkish IX Corps in defence had contributed seriously to their losses and they were now a broken force in dispirited retreat towards Erzincan. The most important result of the stiff battles at Bayburt and Dumanli-dağ was that the Turkish Third Army was gravely shaken and its strength had been diminished by at least 30%. But the misfortunes of Vehip Paşa were not yet at an end.

By 15 July Yudenich might well have considered that the aim of his action had been attained if this action were of a preventive character. In the event now of an offensive by the Turkish Second Army (in process of concentration) against the Russian left strategic flank, the Turkish Third Army was no longer in a position to co-operate. Further, it was possible to move the Russian reserve to support the threatened flank; and the 5th Caucasian Rifle Division was ordered to march from Erzurum to the south-west and to cover the dangerous gap in the Russian line in the direction of Oğnot.

Meanwhile mopping-up operations proceeded along all the front from the Black Sea to the Tuzla-çay. On the coast the 123rd Division entered the valley of Fol-dere and moved up into the mountains seeking a passage into the next lateral valley, that of Harşit which debouches on the coast at the port of Tirebolu. The importance of entry into the Harşit valley was considerable, since the main Trebizond-Gümüşane road, after crossing the Zigana Pass (1983 m.),[1] descends to the Harşit at Ardasa and runs up its valley to Gümüşane.

Confusion was evident in the Turkish ranks. Rather unexpectedly, the Turkish V Corps made no serious attempt to defend Ardasa; while one division withdrew down the valley of the Harşit-su, the second retired over the mountains in the direction of Şebinkarahisar. A gap was thus opened in the Turkish front at Ardasa by the Turks themselves; and on 20 July, this important point, the gate to Gümüşane, was occupied by units of the Russian 123rd and 127th Divisions.

The 3rd Division of V Corps, together with some elements of X Corps, during these days was being pressed back up the valley of the Kara-dere by plastun battalions and the 19th Turkistanskis. The Turks tried to cross the main ridge of the Pontic Alps by paths leading to the sources of the Harşit, and from there to reach Gümüşane. Natural difficulties soon lent a disorderly character to their retreat. By 17 July the two plastun brigades and the Turkistanskis had crossed the main ridge at many points. The 1st Plastun Brigade marched direct on Gümüşane and occupied the town on the 18th, with a

[1] The Zigana Pass represents a col between the two massifs of Alaca-dağ and Kolat-dağ, themselves the bastions of a great spur which is thrown out north-westward to the sea from the main ridge of the Pontic Alps. The Vavuk, 1973 m., the second great pass on the Trebizond-Erzurum road, is a col joining Horlar-dağ with the main Pontic ridge. It constitutes the watershed between the Harşit and left affluents of the Çoruh.

big booty in stores and munitions. The 2nd Plastun Brigade, with the 19th Turkistanskis, engaged the retreating Turks ten miles to the south-east of Gümüşane at the cross-roads of Pir-Ahmet, where the main Trebizond-Erzurum *chaussée* is joined by the road which forks, farther south, to Erzincan and to the tracks down the Kelkit-su to Şebinkarahisar.

At Pir-Ahmet cross-roads there was a moment of considerable confusion: the plastuns and the 19th Turkistanskis were descending the slopes of the Pontic Alps at the moment when the 17th Turkistanskis appeared along the road from Bayburt in hot pursuit of the 30th Division of the Turkish X Corps. The Turks, now thoroughly disorganized, broke south from Pir-Ahmet by tracks over the mountains towards the Kelkit-su.

Vehip Paşa had failed to control the retreat of the troops of X Corps who had been exhausted by the strenuous defence of Bayburt. While the 30th Division had retired along the *chaussée* and over the Vavuk Pass, only to run into trouble at Pir-Ahmet, the 32nd Division had retreated by the shorter but worse track running west from Bayburt through Reşdi to the valley of the Kelkit. At the same time the men of the 31st Division were practically dispersed in the mountain country between the Bayburt-Aşkale road and the upper valley of the Balahor-çay. On the Russian side, fortunately for the Turks, General Przevalski, as after Köprüköy, had proved slow in organizing pursuit.

On 15 July, the Siberian Cossacks, patrolling ahead of the 17th Turkistanskis, marched through Bayburt. But it was only on the 19th, four days later, that the Russians came up with the Turkish rear-guard at Balahor, less than twenty miles to the north-west of Bayburt. At Balahor the Turks hardly stood nor did they turn to defend the bare and rather wide Vavuk Pass. On the 21st the Russians descended to Kokanis, from where they took part in the action at Pir-Ahmet engaged by the troops who had crossed the main ridge from the Kara-dere valley. When the Turks broke at Pir-Ahmet, the 17th Turkistanskis marched south along the *chaussée* to Erzincan where they crossed the columns of the 4th Caucasian Rifle Division slowly moving west towards the Kelkit valley.

The 18th Turkistanskis and Siberian Cossacks had followed the Turkish 32nd Division along the track leading from Bayburt into the valley of the Kelkit through Tanişman and Reşdi. At this latter place a lively rear-guard action took place on 18 July: the Turks were holding their ground against the Turkistanskis until they were taken on their right flank by units of the 4th Caucasian Rifle Division. And as the askers resumed their retreat across the open plain of Murmuş they were charged and badly sabred by the mounted Siberians; 800 prisoners were glad to surrender.

The Turkistanskis and Siberians, mixed with elements of Vorobyev's Caucasian Rifles, continued the pursuit along the two tracks leading through Reşdi and Pulur to Kelkit. This important junction of roads, leading west along the valley of the Kelkit-su to Şebinkarahisar and south to Erzincan, was taken after a short fight on 23 July. Kelkit lay at the head of one of the principal alleys of access into the interior of Anatolia.

If not quite methodically planned, the pursuit by V Caucasian and II Turkistan Corps had produced satisfactory results: Gümüşane was occupied on 18 July and Kelkit on the 23rd. It became clear to Yudenich that the advance of his right and centre might contribute profitably to the advance of his rather weaker left. Reports emphasized the hasty and disorderly character of the enemy's retreat, and Yudenich decided to exploit the opportunity. On 19 July orders were sent to the 39th Infantry Division and the 4th Plastun Brigade to continue their advance in the direction of Erzincan.

It was not an easy task to cross the Kara-su at its bend. There was only one stone bridge, at Kotur, on the historic highway from Armenia into Anatolia. It had been blown by the retreating Turks. Sappers threw up a wooden bridge which served for the passage of the 154th and 155th Regiments; the 153rd and 156th made the best use they could of the fords to the north and south of the bridge. But these, although comparatively shallow in July, were made difficult by the rapid current of the upper Euphrates. The infantry managed to cross with the water breast-high, but all guns and the divisional train had to file across the narrow improvised bridge. Under such conditions the passage of the Kara-su took three days, and another three days were occupied in the march across difficult country towards the beautiful upland plain where lay the ancient city of Erzincan.[1]

Early on the morning of 25 July advanced units of the 39th Infantry Division began to converge on Erzincan. The infantry had only mountain guns with them; the artillery was still following in a long column down the main road. But Erzincan had already been evacuated by the enemy. Vehip Paşa had seen no possibility of defending the

[1] Erzincan has been an inhabited site since Urartian times (first quarter of first millennium B.C.). The Turkish city, with many monuments of the Turkish and Armenian middle ages, was almost entirely destroyed by earthquake in 1939. The new city, built during the last decade, has become an important Turkish air base. Cf. Ali Kemali, *Erzincan tarihi* (History of Erzincan), 1932.

town with the shattered and shaken troops of IX Corps. He decided to hold only the principal approaches into the plain of Erzincan from the west and south-west with a view to slowing up further Russian penetration into Anatolia. Strong rear-guards with artillery took up positions ten to fifteen miles beyond the town: to the north-west on the mountains covering the roads to Şebinkarahisar and Zara-Sivas; and to the south-west on heights overlooking the valley of the Kara-su, down which went the difficult track by Kemah to Divriği above the great gorge of the river ninety miles west of Erzincan and covering the road junction forking to Sivas and Harput.

At midday on 25 July, Erzincan was occupied by the 154th Derbentski Regiment, while the 17th Turkistanskis were approaching the town from the north along the Bayburt road. Russian infantry continued to flow down into the Erzincan plain. The 4th Plastun Brigade soon appeared and General Lyakhov moved them against the Turkish rear-guard covering the Şebinkarahisar road.[1] At the same time the 153rd were sent to attack the Turkish position on the Kemah road. The attack by the 153rd met with a stubborn resistance by units of the Turkish 29th and 36th Divisions, the latter the only 'fresh' division left in the Third Army. The Turkish infantry were well supported by artillery on both banks of the Kara-su. Fighting continued throughout the 26th and 27th when Russian field guns and howitzers arrived. On the 28th the Turkish rear-guard was driven back, but Yudenich chose to take this success as marking the end of the Erzincan operation.

The Russian commander began to regroup his forces without delay. The four Kuban battalions of the 4th Plastun Brigade were sent south-east in the direction of Kiği. The Don Foot with the two Trans-caucasian brigades were transferred to the left bank of the Kara-su at the southern angle of the Mamahatun bend in order to maintain liaison with troops of IV Caucasian Corps in the Kiği area. Nearer Erzincan the 153rd were sent south of the river to observe the tracks running southward across the Mercan-dağ to Hozat and Harput. The 154th and 155th watched the roads to Divriği and Şebin-karahisar, with the 156th in reserve. They had to maintain contact with II Turkistan Corps concentrating in the region of Kelkit. This latter corps was now reduced in strength by the transfer of

[1] The limit of the Russian advance in this direction was the trough-like plain of Refahiye, which leads to Suşehir overlooking the middle valley of the Kelkit. Isolated by snowdrifts for four or five months a year, Refahiye remains to-day one of the most squalid and poverty-stricken towns in Anatolia. Here a squadron of Cossacks was stationed during the winter of 1916–17.

PLATE VII

THE CITEDAL OF KARS IN 1945

Vorobyev's 4th Caucasian Rifle Division which was sent to reinforce the 5th Caucasian Rifle Division in the region north of Oğnot. The 3rd Plastun Brigade kept liaison between II Turkistan Corps and V Caucasian Corps which stood on a line running from Fol on the Black Sea through Ardasa to Gümüşane. The 1st Plastun Brigade was taken into general reserve at Erzurum, together with the 17th and 18th Turkistanskis, while the 2nd Plastun Brigade was attached to Vorobyev's division en route for the Oğnot region.

After deducting the casualties suffered during the July operations (12,000) the troops remaining on the front between the Black Sea and the Kara-su numbered about 70,000 men—a force more than adequate to contain the beaten and deteriorating divisions of the Turkish Third Army.

In theory the Turkish Third Army still consisted of fifteen divisions at the beginning of August 1916.[1] But losses during the Bayburt and Dumanli-dağ battles and the retreat to Kelkit and Erzincan had been appalling: 17,000 prisoners and not less than the same number in killed and wounded. By 1 August there remained, as more or less coherent units, four or five divisions (each 3000–4000 strong) on the roads running west and north-west from Erzincan; the remnants of X Corps (5000–6000 strong) between Kelkit and Şebinkarahisar; and two divisions of V Corps in the coastal region (about 10,000). The newly formed 53rd Division was not a disposable fighting unit since its men were taken to make up the complements of some of the divisions with a strength of only 1500 or 2000 men. Vehip Paşa could put less than 100 guns into the line along all the front of the Third Army. And the process of disintegration was not checked with the end of the fighting for Erzincan: desertions continued and became one of the most difficult problems of the Turkish command, so that by 1 September the Third Army was even weaker than it had been at 1 August.

Yudenich had attained his objectives more fully and more rapidly than he had anticipated. In less than four weeks the Turkish Third Army had been eliminated as a factor in any possible Turkish offensive. The Russians had not only defeated this army, they had captured the advance bases at Bayburt and Erzincan which might have been used for spring-boards in a general offensive against Erzurum. Further, the possession of Gümüşane, Kelkit and Erzincan

[1] The twelve divisions of V, IX, X and XI Corps, with the 36th and 37th Divisions and the newly created 53rd Division which had originally been allotted to the Second Army.

now gave the Russians the necessary bases for the organization of an offensive into the interior of Anatolia. An invasion of Anatolia was certainly not an immediate possibility while the Turkish Second Army lay along the flank of the Russian strategic left, but it was a real threat for the future which converted the planned offensive of the Second Army into something like a desperate attempt on the part of the Turks to prevent the deployment of the Russian armies into the homelands of the Turkish people. Thus the gallant and abortive offensive which was now undertaken by the Second Army differed in essentials from the ambitious combined offensive operation by two powerful Turkish armies, the Third and Second, which had been planned by the generalissimo in the first months of the year 1916. Before the Second Army had been ready to move, the Third Army had been completely defeated and eliminated as a factor in a future campaign. Yudenich's July operation against the Third Army ended on 28 July. A general attack by the Turkish Second Army began on 2 August. But during the preceding fortnight Yudenich had been able to begin the movement of his reserves to his threatened left strategic flank, and after 28 July he was free to send there all the reinforcements which might be necessary.

CHAPTER XXXVI

WAR IN THE MOUNTAINS. BINGÖL AND DERSIM, JUNE–JULY 1916

MAPS 1, 31, 32

IN allotting the decisive role to the Second Army, the original Turkish plan for the offensive of 1916 had fixed the direction of the principal attack on a line through Muş, Hinis and Hasankale. There were to be supporting attacks on both flanks: in the west, through Oğnot towards Erzurum; and in the east, through Bitlis and Malazgirt into the Eleşkirt valley. But by the end of the winter, Bitlis and Muş had fallen to the Russians, and in the spring the Russian advance was developed westward along the valley of the Kara-su as well as along the Black Sea coast to Trebizond.

When, in May, Marshal Ahmet Izzet Paşa confronted the task of preparation for the great offensive he found enormous difficulties of transport and concentration. An offensive having the Muş-Hinis-Hasankale line as principal axis of advance imposed a centre of concentration at Diyarbekir and a march across the highest part of the Armenian Taurus where roads and tracks debouched on to the key towns of Muş and Bitlis, already occupied by the enemy. By comparison with Diyarbekir, Harput was more accessible, and more practicable roads connected this town through Malatya with the Bagdad railway. Harput was also much nearer than Diyarbekir to the headquarters of the Third Army at Erzincan, and Izzet therefore made it his headquarters while the Second Army was in process of concentration.

The first units of the Second Army to arrive in the region of Diyarbekir had been the 5th Infantry Division (March) and the 8th Infantry Division (April) of XVI Corps (Mustafa Kemal Paşa). In May two divisions of III Corps (1st and 49th) reached Harput, and early in June the third division of XVI Corps (7th) appeared in the district of Palu on the Murat-su between Muş and Harput. When, in May, Vehip Paşa started his operation against the Mamahatun salient, he asked Ahmet Izzet to support him by a demonstration on the front of the Second Army; and elements of the 5th Division, marching from Siirt across the Armenian Taurus, made a reconnaissance against the Russian positions to the south of Bitlis. At the same time units of the

8th Division appeared in front of Muş. Here Mustafa Kemal's troops occupied without opposition positions on Kurtik-dağ, an advance shoulder of the Hacres Mountains only five miles to the south of the town. Farther to the west, round Oğnot and Kiği, Turkish cavalry patrols were active.

The position of IV Caucasian Army Corps along this long front was somewhat precarious. Its commander could only dispose of the 2nd Caucasian Rifle Division and the 2nd Caucasian Cossack Division with some elements of Chernozubov's Van-Azerbaijan force on the extreme left flank. Yudenich had already decided to reinforce IV Caucasian Corps by the 66th Infantry Division which had for long been held in general reserve near Erzurum. By the middle of June this division was concentrated at Hinis, while one brigade of the 2nd Caucasian Rifle Division was defending Bitlis and the other Muş. With the opening of Turkish offensive operations against his right and centre, Yudenich required action from IV Caucasian Corps and ordered the capture of the positions on Kurtik-dağ which Mustafa Kemal had fortified in such dangerous proximity to Muş. The 261st Akhulginski Regiment was moved to Muş in support of the 5th and 6th Caucasian Rifles.

On 12 July, when the Russians were already within six miles of Bayburt and, on the Kara-su, had retaken Mamahatun, the assault on Kurtik-dağ began. The Turkish 8th Infantry Division held naturally strong and well-fortified positions; their resistance proved stubborn and during the fighting of the 12th and 13th the Russians made little progress. But on the night of 13th–14th a brilliant bayonet attack by the Akhulginskis (under Colonel Potto) carried a dominating sector of the enemy trenches, and in the morning the Turks began to retire along the whole line, leaving 1000 prisoners and four guns in the hands of the Russians.

This success seemed to stabilize the Russian situation in the region of Muş. A defensive position was organized on Kurtik-dağ which the 261st were left to defend, while the 5th and 6th Rifles were transferred to the right bank of the Murat-su to observe the roads from Palu where the advance-guard of the Turkish 7th Division had appeared.

Yudenich's staff had information that Turkish concentrations were taking place towards Oğnot and three regiments of the 66th Division were moved from Hinis to protect this important direction.[1]

In July 1916 it became clear that a battle or a series of battles

[1] Yudenich had now also detached from the central front and sent towards Oğnot the 5th Caucasian Rifle Division from general reserve.

between considerable Turkish and Russian forces was imminent in the region between the two branches of the great river Euphrates. Both the Kara-su and the Murat-su run, roughly speaking, from north-east to south-west and they are divided by a strip of mountainous country eighty to a hundred miles wide. Passing through Erzincan, the Kara-su (or 'Black water', called also Firat or Frat) flows westward through the gorges past Kemah until it turns sharply south-east to join the Murat-su about forty miles to the north-west of Harput.[1] The great mountainous massif of Bingöl-dağ[2] separates the sources of the Kara-su from the upper course of the Murat-su.

Under different names (Munzur, Mercan, Bağir-Paşa, Şeytan) very high and desolate ridges stretch from west to east between the two Euphrates. On the whole they approach nearer to the northern (Kara-su) than to the southern (Murat-su), but all the country between the two rivers is piled with their innumerable ramifications, counterforts and ravines. In general this mountainous country may be considered as a great westerly extension of Bingöl-dağ. Its western part (to the west of the Kiği-Palu line) is called Dersim, and its inaccessible character has made it the natural refuge against civilization of the most primitive tribes of Anatolia, the Zazas.[3]

But not only is Dersim wild and thinly populated; even the country to the east of the Kiği-Palu line is really only habitable in the valleys between the densely forested flanks of the great mountains where, from a wilderness of rocks, torrents pour down through precipitous ravines to devastate the few hamlets and cultivated fields during the wet months of spring and autumn.

Here confronting the antagonists of 1916 was a new strategic area which had never before been fought over by any but untutored bands of Kurdish mountaineers. In 1914–15 the Bingöl-dağ region had constituted for the Turks a natural and seemingly impregnable fortress protecting the southern flank of their fortified position in and around

[1] The names 'eastern' and 'western' Euphrates are explained by the fact that the 'eastern', or Murat-su, has its sources in the Ağri-dağ some distance to the east of the sources of the 'western', or Kara-su, which rises north-north-east of Erzurum. The Aras, third great river of Armenia, finds its source in Bingöl-dağ, midway between the two Euphrates.

[2] Literally 'Mountain of a Thousand Lakes', a name derived from the numerous tarns scattered over this great alpine upland. For a description of the topography of Bingöl-dağ, see Lynch, *Armenia*, vol. II. The only Englishman to visit this area since Lynch has been Mr H. A. R. Philby (1948).

[3] So it has been throughout history, and the Zaza clans in the Dersim represent probably the oldest ethnic stock in Asia Minor. The area has recently been treated in detail in a MS. work by Colonel Nazmi Sevgen, 'Zazalar ve Kizilbaşlar' (Zazas and Kizilbaşis). See also Naşit Hakki (Uluğ), *Derebeyi ve Dersim*, Ankara, 1932.

Erzurum. The great difficulty of the terrain, particularly in winter, seemed to make impractical any Russian turning movement from this direction. But in 1916, with the Russian advance through Erzurum and Mamahatun to Erzincan, the position was reversed. The Bingöl massif covered the whole of the Russian left strategic flank as far as the valley of the Murat-su and the passage-ways up through Muş and Hinis. But on the Russian left the re-entrants at Muş and Bitlis appeared to be effectively held and, to attack them, the Turks must bring considerable forces across the Armenian Taurus through the difficult Siirt gap. Diyarbekir was the essential base for an operation through Siirt, but with the principal Turkish concentration in the Harput region it remained only practical to make immediate use of a large part of the troops of the Second Army by attempting a summer passage of the Bingöl mountains.[1]

In mid-July Marshal Ahmet Izzet received repeated requests for support from Vehip Paşa, then in desperate straits. The marshal decided to open his offensive as soon as practicable and to shift the centre of gravity of his action as far to the west as possible in order to make his support of the Third Army more direct by linking his left with Vehip's right. If he did not wish to lose precious time—and that uselessly—he had to mount his offensive on the basis of the concentrations of his divisions as they stood in the middle of July.

In all his decisions the marshal was governed by the natural difficulties of the Dersim-Bingöl country where the poor quality of communications, and particularly the almost complete absence of tracks running east-west, made impracticable any changes in strategic deployment, once it had begun.

There was not much choice of operational direction for the south-to-north passage of the Dersim-Bingöl country. There were, in fact, only two alternatives. On the east, from Palu on the Murat-su (served by a good dirt road to Harput), tracks ran across the small Çapakçur plain and along the valley of the Oğnot (or Gönik). A bad track continued more steeply up the valley to the small town of Oğnot, thirty miles north of Çapakçur *ova*[2] (which was itself thirty-five miles east-north-east of Palu). Immediately to the north of the little Oğnot basin difficult paths crossed a wild mountain ridge, seven or eight miles in breadth, beyond which the valley of the Oğnot-çay widens

[1] The concentration of the Second Army was continuing slowly. In July the 11th Division of II Corps and the 47th and 48th Divisions of IV Corps arrived on the front. The 53rd Division had been diverted to fill complements in the decimated divisions of the Third Army.

[2] *Ova*, T., 'plain'.

at Boran and stretches as a grassy upland to the south-west slopes of Bingöl-dağ. A passable track continues northward and, five miles north of the ruins of Bingöl-kale, crosses a saddle and enters the valley of the Endres-çay (Peri-su)—a tributary of the Murat-su flowing west then south—which has its source on Bingöl not far from that of the Aras flowing to the north. From the sources of the Endres a track and several passes lead into the upper valley of the Tuzla-çay from where fairly good tracks go down on to the plain of Erzurum. From Çatak in the Endres valley to Parmaksiz in the region of the upper Tuzla the distance is fifteen miles, and from Parmaksiz to Erzurum about forty miles.

A march from Palu to Erzurum, even in summer, could be no easy task; but an advance in this direction might constitute a serious threat to the Russians, particularly when their main force was stretched out far to the west on the Gümüşane-Kelkit-Erzincan line.

Ahmet Izzet fully appreciated the importance of the Oğnot direction which, in addition to the threat to Erzurum, offered possibilities of an outflanking movement against Muş, since there were several tracks from Oğnot into the Muş plain and the region to the rear between Muş and Hinis. With this objective in view the 7th Division, as the left flank of XVI Corps operating against Muş and Bitlis, was stationed at Simsor with patrols out towards Haraba on the road to Muş. By 1 June, the 1st and 49th Divisions of III Corps were concentrating at Palu where the newly arrived 12th Division was sent to join them. A group of three divisions was thus available for the offensive in the Oğnot region.

It would have been logical to concentrate in the same direction the divisions reaching the front of the Second Army during the month of July, leaving a smaller force as a flank guard only in the Kiği area. But Ahmet Izzet was clearly too much impressed by the bad turn of events on the front of the Third Army; he was worried by the fall of Erzincan and he decided to move a substantial part of the forces at his disposal geographically nearer to the shattered army of Vehip. This was his principal mistake in the mounting of the last considerable offensive of the imperial Ottoman army. He dispersed his forces in three different groups in conditions where co-operation between them was practically impossible.[1] And the marshal did not sufficiently appreciate the enormous difficulties of the Kiği direction.

[1] *Muş-Bitlis group*: four divisions (5th, 8th, 7th and auxiliary units). *Oğnot group*: four divisions (1st, 49th, 12th, 11th). *Kiği group*: four divisions (47th, 48th, 14th, 3rd Cavalry).

Several practicable tracks winding across a series of mountain ridges connect the fertile and well-populated little plain of Palu on the right bank of the Murat-su with the upland region of Akmezar nearly fifty miles to the north. Akmezar is only five miles as the crow flies from the valley of the Endres-çay (Peri-su) on which stands, twelve miles north-north-west of Akmezar, the small town of Kiği. The Endres runs through a wild and precipitous defile and is joined, below the small Kiği basin, by its affluent, the Horhor-çay. The track from Akmezar crosses a high ridge into the ravine of the Horhor whence the approach to Kiği is easy. The region of Kiği, however, is a sort of blind alley which leads nowhere. To the north of the little town tower the savage heights, so appropriately named, of Şeytan-dağ.[1] The Endres penetrates to the Kiği basin after breaking through the Şeytan by an exceptionally narrow and inaccessible gorge fifteen miles long. To the north of the gorge the rapid and furious stream of the Endres flanks the northern side of the Şeytan, after descending in a general direction east-west from its source on Bingöl-dağ. The river makes its southward bend ten miles to the north of Kiği where its ravine divides the Şeytan massif from that of Bağir-Paşa. A difficult path runs up the gorge of the Endres, and other paths lead from the Kiği basin across the heights of Şeytan-dağ. Across Bağir-Paşa-dağ are a few tracks used only by the Dersim Kurds in movement with their flocks between the valleys of the Endres and the Kara-su. Even were it practicable for troops to reach the upper (east-west) valley of the Endres where going was easier, they could only find themselves in the neighbourhood of Çatak (with its tracks giving access to the Erzurum plain) which may be better approached from Oğnot.

This description of the ways of access through Oğnot and Kiği respectively will make it clear that Ahmet Izzet made a double mistake in sending an important part of his forces in the direction of Kiği; for not only was this region of secondary strategic significance, but its topography and ways of communication excluded any possibility of developing a large-scale offensive action.

From the beginning to the end of these operations Yudenich seems to have had a more accurate appreciation of the relative importance of the two possible directions for a Turkish offensive south-north across the Bingöl-Dersim country. His first measures were to protect as effectively as possible the upper valley of the Oğnot-çay and the ways leading to Oğnot. After the Kurtik-dağ action, three regiments of the 66th Division had been transferred from Muş, by Gümgüm, to

[1] 'Devil's mountains.'

the fertile and well-inhabited plain of Gündemir along the upper course of the Çarbuhur river on the southern slopes of Bingöl.[1] The Gündemir plain is connected by good tracks with the Oğnot plain, fifteen to twenty miles distant. On 25 July the three regiments concentrated in the area between Bingöl-kale and Boran, where the 5th Caucasian Rifle Division (from general reserve) was moving down from the north to join them. Yudenich ordered a general advance against Oğnot, so soon as the 5th and 6th Caucasian Rifles, in reserve at Gümgüm, had moved to Ziyaret (fifteen miles north-west of Muş) from where they had to advance through the Boğlan Pass on Boğlan village on the track to Çapakçur. This somewhat premature and adventurous offensive move may be explained by the fact that Yudenich was under the impression that Turkish concentrations were progressing even more slowly than they in fact were. Cossack patrols reported the presence of Turkish troops at Oğnot and at Boğlan. Boğlan was in fact occupied by the 7th Division (XVI Corps) and Oğnot by the 12th Division. But the Russians did not realize that the 12th Division was only the vanguard of strong Turkish columns; the 1st and 49th Divisions of III Corps were already in Çapakçur *ova* ready to march up to Oğnot.

The Russian offensive started on 29 July and the 30th was the first day of the first battle for Oğnot. The Turkish 12th Division, which had an excellent reputation earned at Gallipoli, occupied the heights on both sides of Oğnot town. Fierce fighting continued until dark and the Turks, by bold bayonet attacks, recaptured positions which they had lost earlier in the day. After regrouping during the 31st, the Russians renewed the attack on 1 August. Early in the afternoon the Turkish centre was broken; Oğnot was lost; and the 12th Division began to withdraw, leaving however only 200 prisoners and one gun in the hands of the Russians. On the 2nd the Russians followed the enemy retreating to the south; and on the following day they were near Elmali, ten miles beyond Oğnot.

Meanwhile the 5th and 6th Caucasian Rifles were advancing on the Boğlan Pass where the 7th Turkish Infantry Division had received orders to withdraw in line with the 12th Division retreating in the valley of the Oğnot-çay.

The new Russian move, together with the news from the Third Army, forced Ahmet Izzet to start his own offensive without any

[1] The fourth regiment of this division remained to hold the Kurtik-dağ position, while two regiments of the 2nd Caucasian Rifle Division were kept in corps reserve at Gümgüm.

further delay. On 1 August he ordered Mustafa Kemal to attack Muş and Bitlis. His own attack up the Oğnot-çay was to follow two or three days later after the regrouping of the retreating divisions (12th and 7th) with those advancing (1st and 49th) and the new division which was coming up (11th). The IV Turkish Corps (47th and 48th Divisions) was already concentrating near Palu for the march on Kiği.

On his side, Yudenich, in moving his main reserves to oppose the Turks from the direction of Oğnot, had not forgotten the direction of Kiği. The 1st and 2nd Transcaucasian Brigades (now combined as the 6th Caucasian Rifle Division) were marching to Melikan on the Endres-çay, thirteen miles north-east as the crow flies from Kiği. The Don Foot Brigade, with a brigade of Cossacks and the two Terek plastun battalions (recently arrived at Mamahatun from Ruwandiz) were sent to reinforce the Transcaucasians. The four Kuban battalions of the same brigade (4th Plastuns), under the enterprising Colonel Kruten, were to march towards Kiği across the Bağir-Paşa mountains by the shortest cuts they could find.

CHAPTER XXXVII

THE OFFENSIVE OF THE TURKISH SECOND ARMY, AUGUST–SEPTEMBER 1916

MAPS 31, 32

EARLY in the morning of 2 August the Russian positions in front of Muş and Bitlis were attacked by the troops of Mustafa Kemal's XVI Corps. On both sides of the narrow valley of the Bitlis- (or Basor-)çay, General Nazarbekov had 6000 infantry of the 7th and 8th Caucasian Rifles in prepared positions, supported by a section of field artillery and a battery of howitzers. One battalion of the 6th Caucasians was in reserve in the town of Bitlis. However, this usually cautious and even hesitant commander was worried and he considered his position not quite secure. Bitlis was separated from the neighbouring plain of Muş by the narrow upper gorge of the Bitlis-çay which was overhung by a rather low outlying shoulder of Nimrud-dağ; and Nazarbekov was uneasy that the enemy's auxiliary forces (which included seasoned gendarme and Frontier Guard units and Kurds) might penetrate a narrow valley to the west of the Bitlis-çay which gave access over the high ridge of the Armenian Taurus into the plain of Muş. To the east, again, he feared an infiltration up the valley of the Güzel-çay through Simek (five miles to the south-east of Bitlis) which might bring the enemy to the shores of Lake Van and even threaten his immediate rear at Tatvan.

When attacked on 2 August by the 5th Turkish Infantry Division, Nazarbekov resisted successfully; forces were fairly equally matched in numbers and fire-power. The Turks, however, renewed their attack on the following day with remarkable courage, and tension grew on both sides. On the 4th, Nazarbekov was informed that askers and Kurds were operating in the valley of the Güzel-çay. Considering that his force, on its position south of Bitlis, might become isolated, Nazarbekov, covered by rear-guards, withdrew his troops after dark to a strong position five miles north-east in front of the village of Başan in the upper defile of the Bitlis-çay (5 August).[1]

[1] North of Başan on the Tatvan road are the ruins of a medieval *han* where the celebrated Turkish traveller, Evliya Çelebi, once spent a night. The narrow gap between Nimrud-dağ and the Hacres mountains through which the road passes into

During the 6th, Nazarbekov remained unmolested by the enemy who were obviously satisfied with the reoccupation of Bitlis. But on the evening of that day news came that Muş had been lost by the 261st Regiment, and on the 7th Nazarbekov preferred to move north to Tatvan at the south-western extremity of Lake Van. Here this cautious commander still felt himself unsafe, and he proceeded to retire on Ahlat at the north-western end of the lake. Thus Mustafa Kemal's right flank had attained complete success in less than a week's

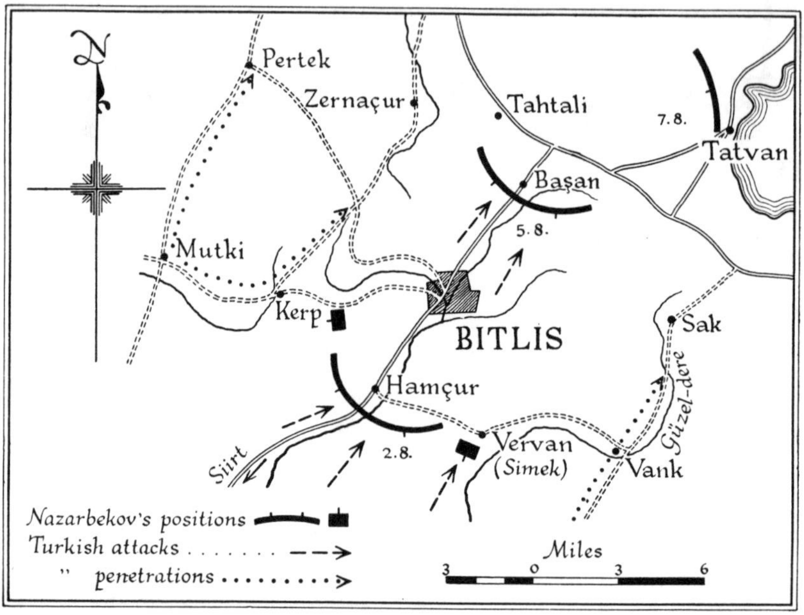

Map 31. Bitlis, 2–8 August 1916

fighting, although losses had been rather heavy. The 5th Division was no stronger than Nazarbekov's brigade, but the attached auxiliary formations, mobile and familiar with the ground, proved useful out of all proportion to their numbers. Nazarbekov had been too impressed by his isolation, and his character was not of the quality likely to respond to the special difficulties of his task.

the Muş plain is notorious for the savagery of the winter blizzards which sweep through the funnel from across Lake Van.

Bitlis is a beautiful old town, formerly of great importance, on the caravan route from Iraq to the north. Like Diyarbekir, Mardin, Antep and Aleppo, its houses are stone and built in a regional tradition which attains charm and distinction. For its history see art. 'Bitlis' in *I.A.*; for a description see Lynch, *Armenia: Travels and Studies*, vol. II, chap. VI, and Lehmann Haupt, *Armenien Einst und Jetzt*, vol. II, pp. 213 ff.

Mustafa Kemal had scored another success in the first days of the offensive of the Second Army. He deserves great credit for having been able to recreate the morale of his 8th Division after their failure to hold the Kurtik-dağ position. This division now proved capable, under his leadership, of driving the Russians from their new position on this same mountain. Here the Turks had the advantage in numbers: 5000 infantry against the 3000 men of Colonel Potto's 261st Akhulginskis. A battery of field guns and two howitzers supported the Russian infantry where a difficult approach virtually deprived the Turks of their own artillery support.

The fighting on Kurtik-dağ started on 2 August. Both sides fought with determination. The Turks continued their attacks during the 3rd and 4th without success. But on the following day, the Russian lines were broken at one point and, as there was no defence in depth, Colonel Potto, on the 6th, found himself forced to evacuate the Kurtik-dağ position (where he dropped his two howitzers and two machine guns[1]) and to withdraw from the town of Muş. The 261st Regiment had lost fully 30% of its strength in killed and wounded. The losses of the Turkish 8th Division were even more severe; its strength was reduced to 3000 men and Mustafa Kemal was not able to develop his success beyond the occupation of Muş. But the psychological effect of the 'double' (Bitlis and Muş) scored by the commander of XVI Corps was great.

The beginnings of the main Turkish offensive effort, particularly in the Oğnot direction, proved to be not so brilliant. Here, on 2 August the retreating 7th and 12th Divisions and the advancing 1st and 49th were not ready to assume the attack. During the 3rd, the Russian advance in the valley of the Oğnot-çay continued, and the Turkish 12th Division, after a rear-guard action, was again pushed back. On the 4th only the patrols of both sides were active. But on that day the commander of IV Caucasian Army Corps received bad news from Bitlis and Muş: both Nazarbekov and Potto seemed to be in difficulties. He therefore ordered the forces in the Oğnot-çay valley (three regiments of the 66th Infantry Division) and in the Boğlan region (5th and 6th Caucasian Rifles) to halt their advance and be ready for a backward movement.

On the following day, while the news from Bitlis and Muş continued to disturb the Russian command, orders came through for a general movement of the units of IV Caucasian Corps back from the Elmali-

[1] In the summer of 1945, one of these howitzers was still to be seen on the slopes to the south of Muş town. (W.E.D.A.)

Boğlan line to the eastern side of the Şerefeddin Mountains. The effect of the withdrawal was to swing back the Russian line behind the Çarbuhur river which falls to the Murat ten miles to the south-east of Gümgüm and twenty-five miles north of Muş. The Caucasian Rifles had to make a flank march through the Boğlan Pass and along the southern and eastern slopes of the Şerefeddin chain, while the three regiments of the 66th Division had to retire from the Oğnot-çay by the bad track, about thirty miles in length, along the northern side of the Şerefeddin, through Silkan to Gündemir. The artillery was dispatched by the easier but more roundabout way through Oğnot, Bingöl-kale and Gündemir.

With an active enemy both these movements, particularly that from Oğnot, might have proved dangerous. But Ahmet Izzet, even at the beginning of his big offensive, showed himself rather slow and not at all enterprising. Too late, and in leisurely fashion, the Turkish 7th Division followed the Caucasian Rifles as far as Ziyaret, where they halted to establish contact with the 8th Division at Muş. The 1st and 49th Divisions of IV Corps entered Oğnot after the last rear-guards of the 66th Infantry Division had left the town and marched eastward. It would seem that the direction of the sudden Russian withdrawal was not understood by Ahmet Izzet. His patrols only cautiously penetrated to the north of Oğnot in the region of Boran where they came in contact with those of Russian reinforcements which had reached Bingöl-kale.

The flank withdrawal of IV Caucasian Corps in the direction of the line Hinis-Gümgüm-Muş was a skilful strategic decision on the part of Yudenich, who was very worried by the Turkish successes at Bitlis and Muş. He thought probable the concentration of at least three Turkish divisions (with auxiliaries and Kurds) in the plain of Muş and he believed that the Turks might from there develop their offensive along the Murat-su or towards Hinis. He found it necessary to oppose the enemy in the Muş area with something stronger than the regiments which had already been badly shaken by the unfavourable issue of the fighting at Muş and Bitlis. The flank withdrawal to the east round the Şerefeddin Mountains opened to the Turks the road leading north from Oğnot; but Yudenich knew that it was only for a brief period. The 5th Caucasian Rifle Division was already on the upper Endres-çay, and the Siberian Cossack Brigade was moving down from Erzurum to the same region, while the 4th Caucasian Rifle Division, after five days' forced marches, reached Erzurum from Kelkit on 6 August.

Not until 12–13 August did the Turks advance in force towards Bingöl-kale. There were three or four days of fighting between the troops of the Turkish III Corps, supported by the 12th Division, and the 5th Caucasian Rifle Division; the latter, a new formation, lacked tactical experience and was thrown back with heavy losses. But the advance of the Turks was stopped by the seasoned Siberian Cossacks and the 2nd Plastun Brigade (now under Bukretov, the first defender of Sarikamiş). The Turkish divisions, rather exhausted by the difficulties of the march and the continuous fighting, took up defensive positions on both sides of the Oğnot-çay to the south of Boran, with the 11th Division in reserve near Oğnot. Meanwhile, in the wide plain of the Oğnot-çay upland, the Russians were concentrating forces adequate for a counter-blow: the 4th and 5th Caucasian Rifle Divisions, the 2nd Plastun Brigade and the Siberian Cossack Brigade —thirty-eight battalions and eighteen *sotni* against thirty-six Turkish battalions with poor irregular cavalry. By 15–16 August Marshal Izzet had missed his chance in the Oğnot direction.

The Kiği group of the Second Army had got the order to advance simultaneously with the Oğnot group. But the two divisions of IV Corps (47th and 48th) were not ready in the first days of August. Their advance only began on the 5th and 6th along the bad tracks already described. Both divisions, formed in 1915, lacked battle experience, but they had been well exercised in the region of Istanbul. Their spirit was good, but they had not been trained for mountain war; they lacked mountain guns and their field guns were almost useless in the conditions of the country round Kiği.

When, at last, the Turkish IV Corps started its offensive, the concentration of Russian forces in the Endres valley, near Melikan, had made good progress. The new 6th Caucasian Rifle Division had arrived under the efficient General Voloshinov-Petrichenko, whose own Don Foot Brigade was also there. The Cossacks of the 5th Caucasian Cossack Division were patrolling along the valley of the Endres and across the Şeytan Mountains, where they had numerous skirmishes with the horse of the Turkish 3rd Cavalry Division.

Anticipating the advancing Turkish divisions, Voloshinov-Petrichenko, who was well experienced in mountain war, occupied Kiği and the small plain to the east of the town. On the southern fringe of the plain, in the valley of the Horhor stream, fighting began on 15 August between advanced units of the Turkish IV Corps and Russian infantry and Cossacks. The Turks soon became stronger in infantry (16,000 against 12,000), but the Russians had three mountain batteries which gave very effective support. The Turks maintained their attacks for several days with great *élan*, but their progress proved negligible and their losses serious. The Transcaucasian militiamen of the newly formed 6th Caucasian Rifle Division were 'mountainy' men who fought even better than the regular troops of the 5th Division (also a new formation recruited from the normal cadres). Voloshinov-Petrichenko was aware that reinforcements of excellent quality were on their way to him. Since 3 August Colonel Kruten had been marching with his 4th Plastun Brigade from Erzincan by the shortest possible route towards Kiği. He crossed the eastern shoulder of the high Mercan ridge (2727 m.) by shepherds' paths and reached Plümür in Dersim. The Kurds of this remote region were unfriendly to the Turks, and Kruten found guides to take him across the Bağir-Paşa massif to the valley of the Endres and he joined Voloshinov-Petrichenko at Temran on 20 August, having brought his brigade over 120 miles of mountain country where scarcely a stray European traveller had passed before. He found the Russians already preparing a counter-blow in the Kiği area.

By 18 August the concentration of a strong Russian group to the north of Oğnot was complete and General Vorobyev took the offensive. His three-pronged attack was directed along both sides of the Oğnot-çay and down the Hasnavi stream (an affluent of the Çarbuhur). There followed several days' fierce fighting during which the Turks used up their reserve, the 11th Division, in a strong counter-attack with the bayonet. The Russians lost heavily on both flanks,[1] but their howitzers and mountain batteries gave them the advantage and finally Bukretov's plastuns on the Hasnavi outflanked the Turkish lines. After a week's battle the Russians had pushed the Turks down into the Oğnot plain and the four Turkish divisions on this sector were in full retreat towards Elmali.

Simultaneously with the action of Vorobyev's group, Yudenich ordered IV Caucasian Corps to resume the offensive. The 5th and 6th Caucasian Rifles attained the bridge over the Murat-su near Muş without much difficulty and the Turkish 8th Division withdrew to the hills round the town. At the same time, on the right of the Caucasians, the 66th Infantry Division on 20 August attacked the Turkish 7th Division in the Ziyaret area. The Turks tried to defend a position near Karabey (Surp Karapet monastery) covering the Boğlan Pass, but they were beaten and retired in disorder leaving several guns and 500 prisoners in the hands of the Russians.

[1] The casualties of the 4th Caucasian Rifle Division were 25 %.

The success of the 66th Division had important results. The 66th followed the retiring enemy through Boğlan and Haraba, from where their patrols gained touch with the 2nd Plastun Brigade on the extreme left of Vorobyev's group in the Oğnot-çay region. Concerned now for his right flank, Ahmet Izzet decided to withdraw his main force into the Çapakçur plain, leaving only rear-guards in the defile of the Oğnot-çay and along the mountainous barrier of Solhan-dağ. Thus by 1 September the Turkish offensive through Oğnot had failed completely, and the central group of the Second Army had been withdrawn to its starting point.

On the extreme left flank of the Russians, General Nazarbekov proved less successful. He moved forward from Ahlat to Tatvan and occupied this point at the south-west corner of Lake Van after some fighting. But, informed that the Turkish 5th Division had withdrawn to a stronger position covering Bitlis, he decided that his force was not strong enough to press an attack.[1]

In the Kiği region the arrival of Kruten's Plastun Brigade was the signal for a counter-blow. But the difficult ground in the Endres valley to the south of Kiği and along the Horhor enabled the troops of the Turkish IV Corps to maintain a courageous and determined defence. On 1 September they were reinforced by the 14th Infantry Division (diverted from III Corps) and found themselves strong enough to counter-attack. Heavy fighting continued for some days. The hard-pressed Russians were strengthened by the 156th Elizavetpolskis who had been moved up from Erzincan through Melikan. The balance of numbers was now more or less equal and the Russians had the advantage in mountain guns and better tactical training in mountain war. After 10 September the Turkish effort round Kiği seemed exhausted and the Turks remained more or less passive in their positions to the south of the Horhor stream, while the Russians maintained their lines covering Kiği town and on both sides of Temran.

In the Oğnot region there was only some sporadic fighting during September and the situation here became stabilized by the 15th, with the Russians on the line running south of Oğnot through Haraba-çay and Boğlan village. Beyond the Boğlan Pass the Russians retained a firm grip on the Ziyaret district extending as far as the right bank of the Murat-çay. Winter set in early; in the high region of Oğnot and on Kurtik-dağ, the first snow fell on 26 September. The movements of both armies were governed from now on by conditions of supply.

[1] His regiments had only 400–500 fit men per battalion.

As a first result, in the last days of September the Turks evacuated the town of Muş.

The great offensive of the Turkish Second Army had ended, if not in disaster as with the Third Army, in serious failure. Losses had been heavy: nearly 30,000 killed and wounded out of an effective strength approaching 100,000. Few prisoners, however, had been lost—only about 3000, and those in small groups. Deserters were far less numerous than in the north, where they presented a grave problem to the staff of the Third Army. The wild and inclement mountains of the Dersim and of the Armenian Taurus, sparsely inhabited by needy and hostile Kurds, did not welcome the waverer like the friendly villages of Anatolia between Erzincan and Sivas. By the end of September Ahmet Izzet found his army reduced to some 60,000 men; several of his divisions had suffered badly and could not be counted for the future.[1] Instead of four army corps only three in fact remained: XVI (5th and 8th Divisions) on the right, II (12th and 1st Divisions) in the centre, and IV (14th and 48th Divisions) on the left.

Generalissimo Enver's plan for a combined offensive by the Second and Third Armies had proved to be absolutely unrealizable in the conditions of transport and supply prevailing in Turkey. It should not have been difficult to calculate that, before the concentration of the ten divisions of the Second Army could be carried through, the Russians would have been able to attack the Third Army and then switch their reserves to oppose Ahmet Izzet. Such a probability must have been quite plain to the Turkish commanders before the end of June. After the easy capture of Trebizond the enemy was in possession of an additional and better route for supply, and it must have been evident that it would not be long before he would make an effort to clear the line Trebizond-Gümüşane-Bayburt-Aşkale-Erzurum. Thus the last moment for any kind of combined manoeuvre by the two Turkish armies had passed by the first week of July. Ahmet Izzet should have moved at least five weeks before he in fact did. At the end of June he might have achieved better results with seven divisions than he could hope to attain in August with ten; for at the earlier date he would still have had the support of the thirteen divisions of the Third Army.

Neither of the Turkish commanders seems to have had a correct appreciation of the direction of attack which could promise the most favourable results. Vehip's attempt, with less than two divisions, to threaten the Russians in the coastal region was veritably childish. He

[1] The 7th of XVI Corps, 49th of III Corps and 47th of IV Corps.

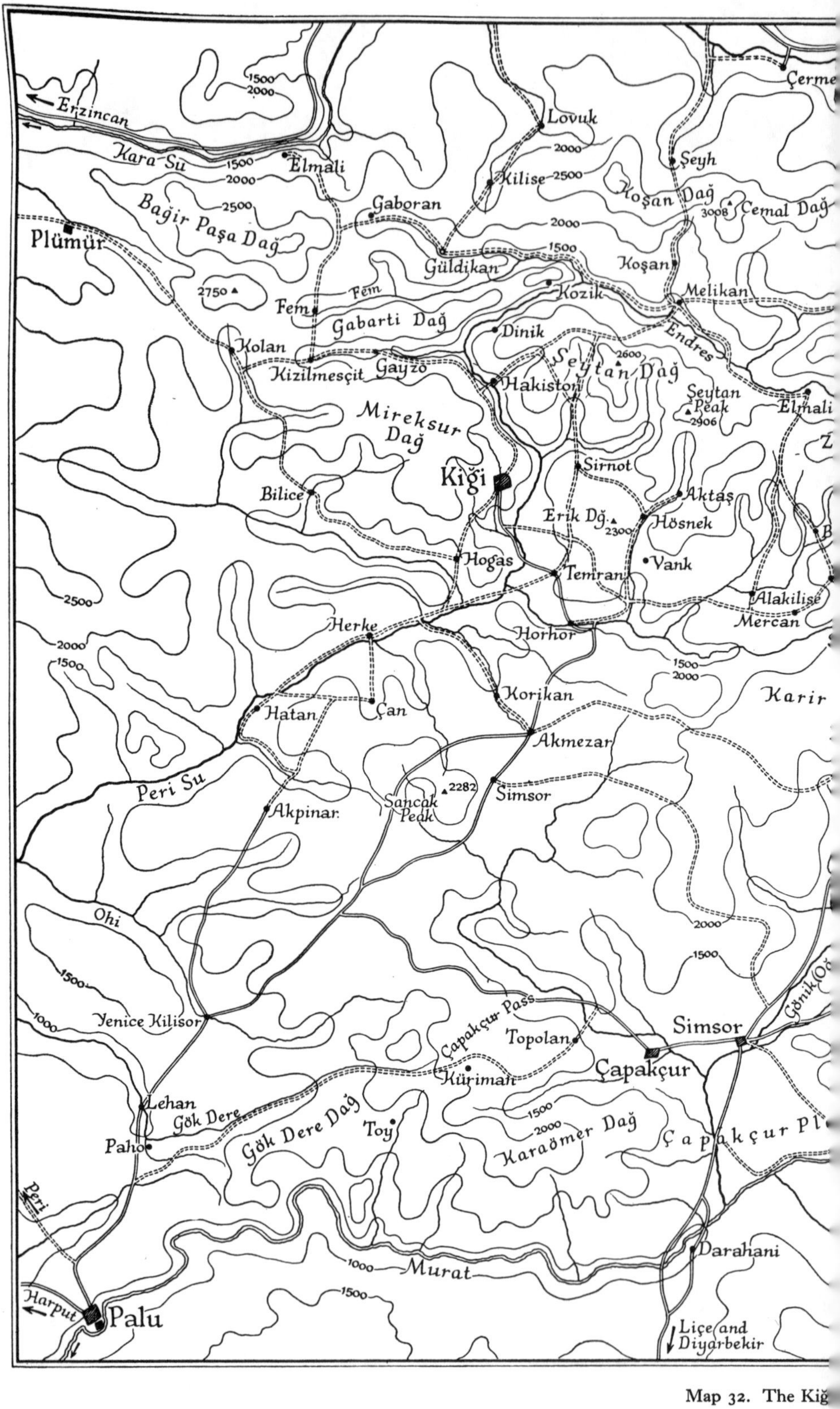

Map 32. The Kiğ

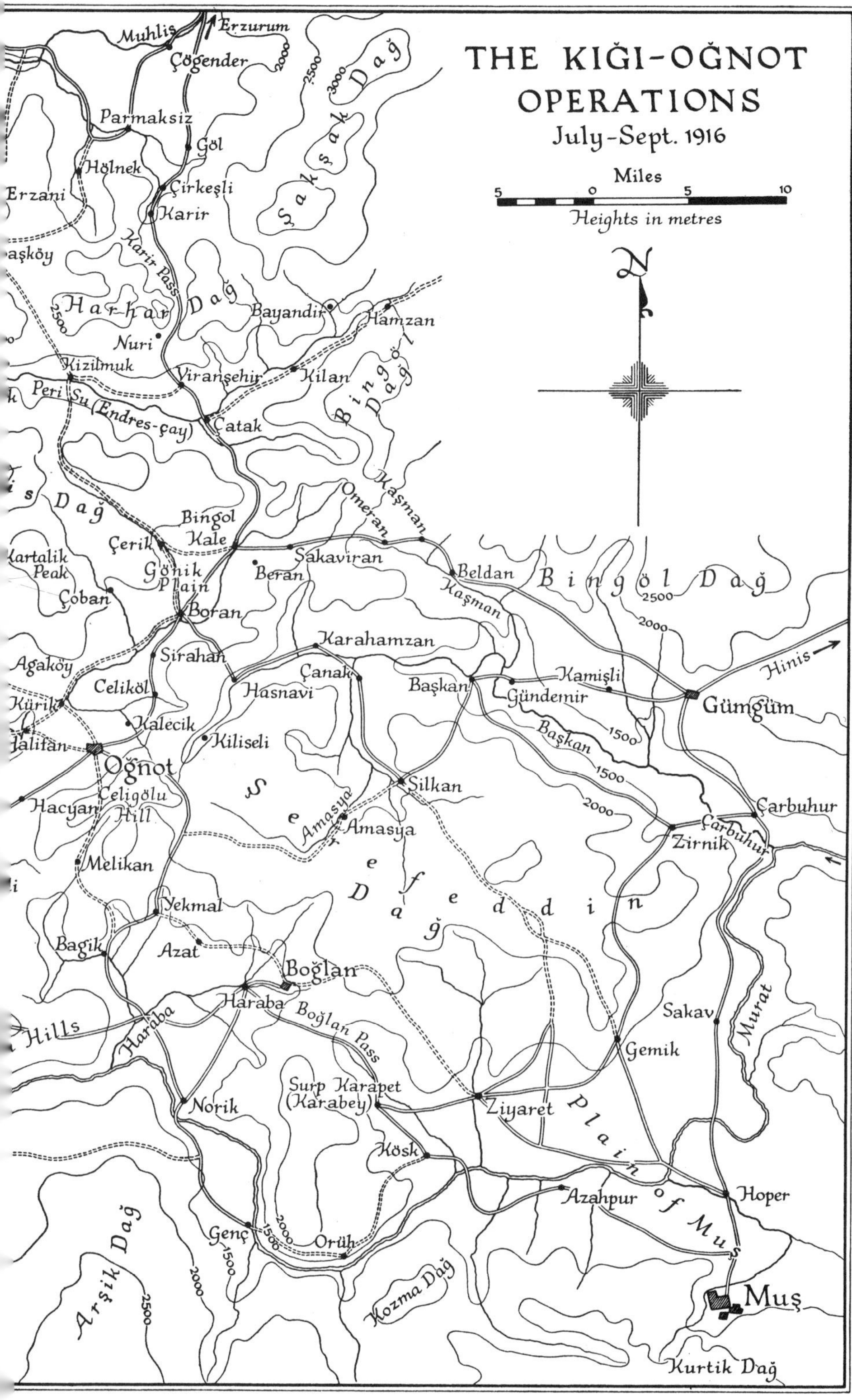

July–Sept. 1916

had just achieved a (possibly unexpected) success at Mamahatun and his troops were close to Aşkale. He could have developed this success in June by massing nine divisions on the front between Aşkale and the sources of the Tuzla. At the same moment Ahmet Izzet would have been in the position to push four divisions up the Oğnot-çay (while retaining two divisions on his right and one in reserve). Thus it would have been possible to combine the attack of thirteen Turkish divisions in the direction of Erzurum. It is likely that most of Yudenich's reserve would have been concentrated to support the 39th Division on the Dumanli-dağ front, and the seven divisions of Ahmet Izzet might well have been opposed only by the two divisions of IV Caucasian Army Corps. The Turks had, indeed, a chance of repeating the successful Malazgirt manoeuvre on a more ambitious scale.[1] Yudenich would, doubtless, have held the Turkish advance from the west and he might well have checked the advance from the south as he had done in the Eleşkirt operation of the previous year. But the effect of such an adjustment of the Turkish plan might have delayed, if it had not prevented, during 1916, the virtual collapse which took place on the front of the Third Army.

The failure of the Turkish command to adapt a preconceived plan to the immediate realities of a situation may be compared to the somewhat similar failure of the French general staff before the offensive on the Aisne in the spring of 1917. In both cases there is a curious similarity of temperamental factors in the optimism and the obstinacy of the two generalissimos, Enver and Nivelle. In both cases there was the same underlying lack of elasticity in the planning of a big offensive, and the same blind lack of imagination in the infinite capacity of the staffs concerned to ignore the known difficulties in front of them.[2]

Numbers engaged were not comparable but proportions in losses were about the same; and when it came to the launching of a desperate and hopeless adventure, the army of Ahmet Izzet was, on the whole, more fortunate than that of General Mangin. But the Turks had found in Mustafa Kemal a young commander of first quality who did not fail to impress the power of his personality even on a most unfriendly destiny.

[1] The four divisions of Vehip's V and X Corps should have been able to contain V Caucasian and II Turkistan Corps in the Pontic Alps.

[2] The French ignored the admitted inadequacy of their artillery preparation and the strengthening of the German line on the front of attack; the Turks (in spite of the lesson of Sarikamiş) ignored the insuperable difficulties of terrain in the Kiği direction. For the Aisne offensive, see (Maj.-Gen. Sir) E. L. Spears, *Prelude to Victory*.

CHAPTER XXXVIII

THE SECOND TURKISH INVASION OF PERSIA, JUNE–AUGUST 1916

MAPS 23, 33

FOLLOWING the new plan agreed between Enver, Halil and von Lossow during the conference in Bagdad at the beginning of May, a Turkish force, by the end of that month, was ready to invade Persia. The 2nd and 6th Divisions of the Sixth Army formed XIII Corps under Ali Ihsan Paşa. The infantry were strengthened by a brigade of Kurdish Hamidiye cavalry and by detachments of auxiliaries and irregulars, including Persian nationalist volunteers. The Germans had promised to provide artillery, and units of the Turkish 4th Cavalry Division from Mosul were expected to co-operate. German promises remained unfulfilled; but some 25,000 troops—and good infantry regiments among them—were available for the offensive. This was a force far stronger than any which Baratov could oppose to it.

By the end of May, Baratov had been informed by local agents and by British wireless intelligence reports[1] that a Turkish concentration was in progress in the direction of Khanikin, where the enemy was occupying a defensive position. At the same time the situation of his own troops in the Kasr-i-Shirin area was far from satisfactory; the Cossacks were suffering badly from the increasing heat of the Mesopotamian summer. A retreat, after the fall of Kut, was clearly unavoidable, even if it was only a limited withdrawal towards the healthier higher ground round Karind. But Baratov was uneasy as to the effect of a Russian retreat on the imagination of the various tribes along the 800 miles of his communications with the Caspian. He decided therefore to strike a blow at the Turks at Khanikin and to

[1] Before reaching Kasr-i-Shirin, in April, Baratov had sent Captain Ganaley with a *sotnya* of Kuban Cossacks (1st Umanski Regiment) to cross the Zagros and make a reconnaissance in the region of Zorbatiya with the object of making contact with the British at Ali Gharbi on the Tigris. This adventurous mission was accomplished with the loss of only a few men. Ganaley found the Lurs not unfriendly but cautious and neutral. The liaison with the British was, in the circumstances, nothing more than a gesture and, when Ganaley returned across the Zagros, he found his commander already retreating on Kirmanshah. Cf. A. T. Wilson's account of this incident in *Loyalties*, 1, pp. 161–2: 'A fine feat.'

start a withdrawal only after the success which he anticipated. He was unaware, at the time, of the serious offensive intentions of Turkish headquarters in Bagdad and, in consequence, did not realize that the attempt might prove rather disastrous to a modest force of cavalry thrown against two enemy infantry divisions. Fortunately for the

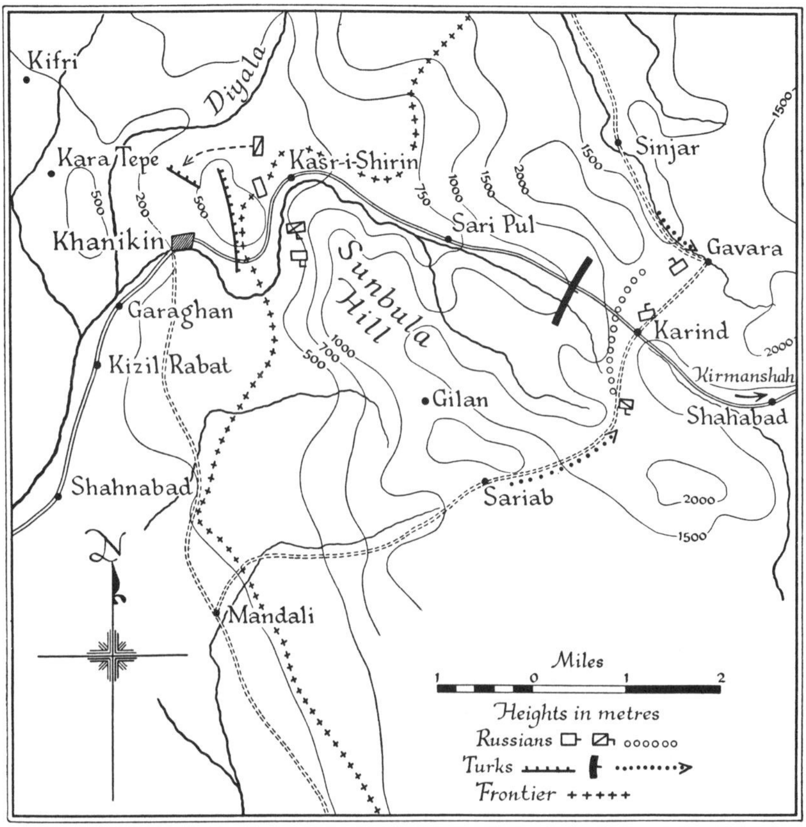

Map 33. Khanikin, June 1916

Cossack commander, Ihsan was not yet ready to begin his offensive, and when Baratov attacked Khanikin he met only the 6th Turkish Division, some units of which had already been badly mauled by the Russians in previous fights.

On 3 June, the Russian force, under command of General Prince Beloselski, approached the Khanikin position in three columns. In the centre was the Border Regiment (now of four battalions), supported by twelve guns. The Tverski Dragoons were to demonstrate on the right, while, on the left, the Severski Dragoons, with the

Khoperski Cossacks, had to outflank the enemy position in depth; the Nizhegorodski Dragoons were in reserve. Thus about 3000 infantry with 3000 horse were thrown against some 7000 Turks in a well-entrenched position. The Borderers met with strong resistance and their repeated attacks failed. The Severskis, with the Cossacks, manoeuvred skilfully and, for a short time, cut the road from Bagdad and broke up some long supply columns. The attack, however, clearly could not be developed beyond a reconnaissance in force; and the enemy's strength and will to fight proved greater than had been supposed. So, with a loss of 500 infantry, Prince Beloselski withdrew. Turkish losses were 400 men. Baratov had no alternative but to proceed with his retreat from Kasr-i-Shirin; it began on 6 June. Ihsan Paşa concentrated his corps and, two days later, crossed the Persian border.

Baratov took up a defensive position on the high plateau (1575 m.) near Karind. He called up his mobile reserve of two Cossack regiments from Kirmanshah and thus, for the defence of Karind, there were available 2500 infantry, 2000 dragoons, 2000 Cossacks with twelve horsed mountain guns and two howitzers.

Ihsan Paşa, meantime, was using his superior numbers to attempt an enveloping manoeuvre. His main force, the 2nd Infantry Division, advanced from Kasr-i-Shirin along the main road (at that time an indifferent track), while other units, with the mounted irregulars, converged on Karind from the north through Sinjar and from the south from Mandali through Kel-i-Harab. Farther to the north a flanking force of three battalions of the 4th Division, with 1000 horse and eight guns, crossed the frontier from Sulaimaniya. Turkish movement was slow, for the askers were suffering as much as the Russians from the great heat and the bad water.

Ihsan was not able to mount his attack until 28 June. During the day the meagre Russian line (3000 rifles), well supported by machine-gun and cannon fire, resisted with success. But in the afternoon, on the right near Gavara, the 1st Zaporogians were outflanked and pressed back. Baratov saw a threat to his line of retreat on Kirmanshah, and during the night the Russians withdrew. There were no defensive positions covering Kirmanshah and, on the night of 30 June, the Russians evacuated the town. Next day it was occupied by Ihsan's corps. At the same date a Turkish detachment entered Sinna and their irregulars moved on Bijar, from where the Cossack holding regiment withdrew towards Hamadan. Baratov's main force was retreating in the same direction.

Ihsan Paşa proved to be in no haste to develop his success after reaching the line Bijar-Sinna-Kirmanshah. He was clearly reluctant to become too deeply involved in the wide distances of the Iranian plateau. His troops were nowhere more than 100–120 miles from the Turco-Persian frontier, but difficulties of supply were already acute and the health of troops was rapidly deteriorating. In spite of the presence of some Germans with the Turkish XIII Corps, the organization of the rear was not so efficient, in difficult circumstances, as Baratov's had been.

The Russians themselves had many trials in the course of their retreat to Hamadan. Great numbers of horses died; and, in crossing the low-lying country on the approaches to Kangavar, hundreds of men fell out, stricken with malaria, dysentery and sun-stroke.

Baratov concentrated at Hamadan, leaving two Cossack regiments as rear-guard at Bisutun-Kangavar. Strengthened at Hamadan by the four battalions of his 2nd Infantry Regiment, he decided to fight, if attacked, on the Asadabad Pass. Turkish patrols appeared near Bisutun, but Ihsan Paşa remained inactive for five weeks. He gave different pretexts to Turkish headquarters to justify his caution. He saw no military reason for a further advance over distances which frankly frightened him. But he was overruled by the political strategy of Enver and von Lossow and, at the end of July, he received definite orders to advance from Sinna and Kirmanshah on Hamadan and from Kangavar to Burujird (where great hopes were set on an armed revival of the Persian nationalist movement).

By 1 August Baratov realized that a further enemy advance would not be long delayed. The Cossack commander aspired to cover a considerable front: Count Nirod was stationed on the Sinna road with two battalions of infantry and one Cossack regiment; Beloselski's main force of six battalions of infantry, three regiments of dragoons and one of Cossacks occupied positions in the Asadabad Pass; on the left, near Nihavend and in the Burujird region, were several Cossack *sotni* under Colonels Stopchanski and Bicherakov. Two regiments of Cossacks remained in observation round Bisutun and Kangavar. Baratov's effectives had been considerably weakened by disease; 2000 men were in hospital and several hundreds had died.

The Turkish advance began on 3 August, simultaneously with the offensive of the Turkish Second Army. There was fighting between advance elements: on the 5th at Bisutun, on the 6th at Sinna, and on the 7th Kangavar. On the Russian right the Cossacks withdrew to Salasat; and on the left they fell back from Nihavend on Daulatabad.

With less than 6000 fit men on the Asadabad Pass, Baratov knew that he could not hold Hamadan.

On the 9th, fighting was general for the possession of the ancient Persian city. An intelligent commander, Ihsan Paşa refrained from pressing costly frontal attacks and developed outflanking movements from both sides. With only one passable line of retreat Baratov had begun the evacuation of Hamadan as early as the 7th. On the night of 9–10 August the Russian troops finally withdrew from the Asadabad positions and passed through Hamadan marching north. In the forenoon of the 10th the city was occupied by Turkish advance units; on the same day the Turks appeared at Daulatabad and Bijar.

Baratov now took the only reasonable decision left open to him. He was marching towards his principal intermediary base, Kazvin (120 miles from Hamadan and eighty from Teheran). Ten days after the evacuation of Hamadan, his troops were established on the strong natural position of the Sultan-bulak Pass, seventy miles south-west of Kazvin. He had shorter and better communications behind him and his men were now enjoying a more equable climate. He could put into the line eight battalions (the Border Brigade), twenty-four squadrons and *sotni* and twenty-two guns; twenty-four further squadrons and *sotni* observed the road to Sultaniya and the tracks to Bijar; six *sotni* were in Teheran, and another six held Kum, Kashan and Isfahan. Militia and a few *sotni* guarded the *chaussée* Kazvin-Enzeli. In his report to *Stavka*, General Baratov[1] stated that, excluding militia, he had only 7000 men fit for action. It was not a large force for the protection of Persia against a Turco-German invasion.

Baratov estimated the Turkish army in Persia at 25,000. But that number was exaggerated. If Ihsan Paşa's casualties in combat had not been great, he had lost thousands of men from epidemics. His two infantry divisions (with some battalions from the 4th Division at Mosul) did not, at the end of August, exceed a strength of 12,000; and he had available perhaps 5000 horse, mostly irregular. Ihsan considered this force clearly insufficient for the conquest of Persia, with the ultimate object of opening a way to the Afghan and Indian frontiers. He therefore limited his activities to patrolling to the north of Hamadan; and he reported that his troops needed a long period of rest and reorganization before they would be in a state to move. The news from the Armenian front in August discouraged even Enver from the present pursuit of vast plans in the east. Divergencies of

[1] Baratov was now again in command of the 1st Caucasian Cavalry Corps; the style 'expeditionary force' had been dropped.

view were already developing between Turks and Germans on the subject of the future of Persia, and the gallant XIII Corps was therefore permitted a rest. But the progress of events on the British front in Iraq was to deprive them of the enjoyment of a quiet winter.[1]

One of the consequences of Baratov's retreat and of Turkish activity in the region Sulaimaniya-Marivan-Sinna was the evacuation of the strategically important point of Ruwandiz. Brilliantly captured on 15 May by Colonel Rybalchenko, the hill town had been attacked unsuccessfully in June by Kurds and units of the Turkish 4th Division. The Turks reappeared at the end of July with two regiments and twelve guns. Rybalchenko might again have resisted with success, but Baratov was already in retreat and, preoccupied by the main Caucasian front, Yudenich ordered evacuation. At the same time Levandovski's detachment was withdrawn from Neri in the Nestorian country. Yudenich had no interest in either Ruwandiz or in the general direction of Mosul. He appears not to have considered the possibilities of co-operation with the British after their recovery on the Iraq front. And the withdrawal of Baratov from the Iraq border seemed to the commander-in-chief of the Caucasian front the natural conclusion of an adventure of which he had never personally approved.[2]

[1] Sir Percy Sykes, with a mixed force of British and Indian troops and locally enlisted men, had occupied Kirman in June, Yezd in August, Isfahan in September and Shiraz in October. 'Thus, by the end of 1916, the flanks of the Mesopotamian Expeditionary Force in Persia and Arabia were no longer a source of anxiety.' Wilson, *Loyalties*, I, pp. 164–5.

[2] For details of the topography and ethnography of the Turco-Persian border, see arts. in *E.I.* under heads 'Kirmanshah' (by J. H. Kramers); 'Rawanduz' (by B. Nikitin); and 'Lak', 'Lur', 'Senna', 'Shehrizur' and 'Sulaimaniya' (all by V. Minorsky).

CHAPTER XXXIX

THE WINTER OF 1916–17

MAPS 30, 32

THE winter of 1916–17 came early and was bitterly cold. By October both the Russian and Turkish armies had passed to winter conditions. Fighting ceased everywhere and even encounters between patrols were rare.

To the Turkish Second and Third Armies this particularly severe winter proved disastrous and completed the disintegration which had begun as the result of their failures in the field. The moral collapse of Vehip's army was the result not only of the defeats at Bayburt and Dumanli-dağ but of the whole series of earlier misfortunes which had dogged the unhappy divisions of IX, X and XI Corps. The spirit of fatalism and defeat soon spread to the newly arrived V Corps. There was a lull on the Third Army front throughout August and September, but the number of deserters continued to increase. The oncome of autumn encouraged the weary askers to quit the line and to wander from one village to another where they were hidden and protected by a sympathetic population. A curious feature of the changed Turkish attitude towards the war was that the villagers in the vicinity of the front no longer abandoned their homes on the approach of the Russians: enemy occupation seemed now preferable to the conscriptions and exactions of the Young Turks which weighed so heavily on the enduring Anatolians. Deserters became a social phenomenon of the first order. Those who could not get work in the villages—and they found a less friendly welcome as the hard highland winter came down on the pauperized peasantry—took to the hills where they formed armed bands whose only livelihood lay in brigandage. In October there were at least 50,000 deserters in the rear of the Third Army. Effectives, meantime, had fallen to less than 30,000, despite the contingents of the 53rd Division drafted in to fill gaps in the different corps. Heavy snowfall at last checked desertions and many men returned to take advantage of the meagre rations which they could find with their units. Vehip's numbers again began to grow, but typhus and cold diseases offset these gains.

In fact all corps had been reduced to the strength of divisions, divisions to that of regiments, and regiments to that of battalions or even companies. A drastic reorganization of the Third Army became essential. The old V, IX, X, XI Corps disappeared and two new corps were formed which, with a gallant optimism, were designated 'Caucasian'. The two Caucasian corps were composed each of three divisions, the strength of each of which was fixed at 6000.[1] The 2nd Cavalry Division became a brigade. Artillery was redistributed and became relatively the strongest arm in the Third Army: 12–18 guns per division, with a proportion of German 150 mm. and Austrian 105 mm. mountain howitzers. The reform was a recognition of the fact that the equivalent of eight infantry divisions (60,000–70,000 men)—half of them deserters—had been lost to the Third Army during the operations of 1916. I Caucasian Corps remained in the area between Kemah on the Kara-su and Şebinkarahisar; II Caucasian Corps held the country between the Kelkit and the Black Sea coast near Tirebolu. The headquarters of the Third Army were at Şebinkarahisar. Ahmet Izzet was now in formal command of the Anatolian group of armies, and the command of the Second Army passed to Mustafa Kemal Paşa.

The agony of the Second Army had not been so acute as that of the Third, but it endured longer and the results were much the same. Ahmet Izzet had suffered a serious setback but he had not been so definitely defeated as Vehip; in October 1916 his strength stood at some 64,000 men. But the early bitter winter in the wild Dersim and Bingöl mountains imposed hardships which the Third Army did not have to endure. Natural difficulties were made worse by the fact that the Young Turk policy of deporting the Armenian population of the surrounding areas had been carried out only too effectively. At the beginning of the war Turkish supplies had been drawn from the fertile and well-cultivated upland basins of Eleşkirt, Erzurum, Muş and Mamahatun. All these had fallen into the hands of the Russians, while the Armenian cultivators had been driven off from the valley of the Murat-çay and the great plain round Harput. At the same time the Turkish peasants had, at the approach of the Russians, trekked into the interior of Anatolia.[2] The eastern vilayets had been depopulated to the extent of about seventy-five per cent of their inhabitants (and they have never since recovered their former degree of prosperity).

[1] I Caucasian Corps: 9th, 10th and 36th Caucasian Divisions; II Caucasian Corps: 5th, 11th and 37th Caucasian Divisions.

[2] It will be recalled that the relations of Paskevich, ninety years earlier, with the Muslim population of occupied territories had been excellent. World War I saw the modernization of the technique of mass terrorism as well as of the profession of arms.

While the Turks found the valley of the Murat-çay and its affluents deserted and depopulated, in the mountains they had to face the hostility of the Kurds. The Dersim Kurds had contributed their contingents of irregular horse to Enver's offensive at the beginning of the war, and they were to join later in attacking and plundering the revolutionary Russian soldiery moving back from the front in the autumn of 1917, but with the success of the Russians in the Dersim-Bingöl battles they had momentarily turned against their rulers. Turkish convoys in the Dersim were attacked and isolated units slaughtered. During September and October the Turkish rear south of Kiği was in a state of insecurity and disorder.

In October, Ahmet Izzet decided to withdraw from the higher regions of Dersim and Bingöl-dağ and to concentrate his troops nearer their bases at Harput and Diyarbekir. At the same time III Corps was dissolved and the Second Army was reorganized in three corps of two divisions each.[1] On the extreme right the 5th Division remained in occupation of Bitlis, while the 8th withdrew from Muş and the inhospitable Taurus to Diyarbekir. IV Corps held the Çapakçur plain with posts out towards Oğnot and Boğlan. II Corps was stationed at Harput and Palu with patrols in the direction of Kiği.

As the lull continued during the spring the Second Army became in effect little more than a depot for the filling of complements in the Third Army and in Iraq. In the spring of 1917 the so-called *Yildirim* Army[2] was in process of constitution in the north and it soon engulfed IV Corps. The epidemics[3] of the winter months had further reduced the strength of the Second Army and in March 1917 its six divisions did not exceed 30,000 men. The group of armies in eastern Anatolia was abolished as a formation and both Ahmet Izzet and Mustafa Kemal left the region. In the summer of 1917 only the 5th Division remained in the Bitlis-Muş area and the 12th at Palu, with the 8th divided between them.

It is clear that the outbreak of the Russian Revolution in the spring of 1917 alone saved the Turks from complete military disaster in Asia Minor.[4] The collapse of the Ottoman empire had been postponed

[1] XVI Corps (5th and 8th Divisions); II Corps (1st and 47th Divisions) and IV Corps (11th and 12th Divisions). The 7th, 48th and 49th Divisions disappeared. The 14th Infantry and 3rd Cavalry Divisions were transferred to Iraq.

[2] Cf. Larcher, *Guerre turque*, pp. 128–32. *Yildirim* is the equivalent of 'lightning' or *Blitz*.

[3] Starvation was an important factor; the normally meagre ration of the asker had been cut by two-thirds. Typhus was also prevalent. Fifty per cent of the artillery horses died of starvation.

[4] Cf. Ahmet Emin (Yalman), *Turkey in the World War*, pp. 261–5.

for eighteen months. Lieut. Field-Marshal Pomiankowski, military representative of the Austro-Hungarian empire in Turkey, wrote: 'It may be considered as certain that both Turkish armies (the Third and Second) were, by the end of the winter, in such a state that they would not have been able to resist any serious Russian attack. Fortunately for the Turks, on 9 March, the Revolution broke out in St Petersburg and soon disorganized the Russian Caucasian army and made its offensive action impossible.' Thus the military effort of the Russian Caucasian army was destined to remain without a morrow. But the results of that effort in 1916 were to prove decisive even if the effect could not be immediately observed.

Despite the very serious losses at the Dardanelles, the effective strength of the Ottoman army reached its maximum in November 1915: 800,000. By 1 March 1917 it had fallen to 400,000 and it never again exceeded this total (200,000 in March 1918). In the summer of 1916, out of a total of fifty-two Turkish divisions, twenty-six were fighting the Russians. Out of the 400,000 lost by the Turks during the period November 1915 to March 1917, not less than 300,000 must be accounted to the campaigns against the Russians (including prisoners and casualties from epidemics and desertion). If the losses at Gallipoli seriously shook the military power of the Ottoman empire and prevented, at a crucial moment, the mounting of any effective offensive against the Transcaucasian frontier, the subsequent losses against the Russians in 1916 were decisive in destroying the remaining strength of the Ottoman army and prepared the way for the British victories in Iraq in 1917–18 and in Palestine in 1918.

The operations against the Third and Second Armies during 1916 cost the Caucasian army no inconsiderable sacrifice in men. Losses for the period June–September 1916 amounted to some 50,000 (including Baratov's, which were mainly from sickness). When fighting ceased in September and the long lull on the Anatolian front began, a certain drain of man-power continued. For this the blame attaches in great degree to Yudenich's staff. Excellent in administrative and operational work, this staff always appears to have suffered from a defective intelligence section. Intelligence had proved faulty during the Malazgirt operation, before the fall of Trebizond, and before the Turkish success at Mamahatun. In the autumn of 1916 Yudenich's headquarters erred in underestimating the degree of exhaustion and wastage in the enemy armies and exaggerated the strength of the effectives which remained to oppose them. The official estimate of Turkish strength on the Caucasian front submitted to the imperial *Stavka* in October 1916 put the Turkish numbers at 344,000 men, provided with 110,000 rifles, 346 guns and 294 machine guns. As new rifles could be supplied without much difficulty by the Germans, the staff calculations implied that a force of between 150,000 and 200,000 men, with adequate equipment in cannon and machine guns, remained in front of the Caucasian army. As is now known, the Russian intelligence staff had magnified the Turkish strength to the extent of 200%.

These false estimates influenced the whole of Yudenich's dispositions for the winter months and imposed a hard and unnecessary burden on the troops. It was believed that the Second Army remained fairly strong and still capable of offensive action. It was therefore found necessary to protect the whole of the Russian front facing west from any renewal of an outflanking movement from the south. The line, Kelkit-Gümüşane-Erzincan, was strongly manned and, while the Russians watched the Kiği and Oğnot highlands only with patrols, a continuous line of trenches was maintained along the icy ridges of Bağir-Paşa and Şeytan-dağ. This line was continued to the south-east along the left side of the Oğnot-çay valley and over the southern shoulder of the Şerefeddin Mountains to the Boğlan Pass. A rifle division observed the Muş plain and maintained liaison with Nazarbekov's group north of Bitlis.[1]

Troops on the western sector of the front could rest in the comparatively clement valleys of the Kelkit and Kara-su, but those on the Dersim and on Bingöl-dağ could only shelter in the wild alpine valleys of the Endres (1500–1700 m.) and the Oğnot (1800–1900 m.). In these valleys the frost was as severe as on the upper ridges and the snow, piled in great drifts by driving gales, lay even deeper. Despite their winter equipment, troops of the 2nd and 4th Plastun Brigades and of the 6th and 4th Rifle Divisions suffered terribly; many patrols were frozen to death, and casualties from frost-bite were heavy. All these units had fought hard and suffered stiff losses in the fighting of the

[1] Winter order of battle of the Caucasian army was as follows:

Coast to Pir-Ahmet cross-roads: V Caucasian Corps (less 1st Plastun Brigade to general reserve).

Area of Pir-Ahmet, Kelkit, Erzincan plain: II Turkistan Corps with 3rd Plastun Brigade (less two Turkistan regiments to general reserve).

Area of Erzincan plain, Bağir-Paşa-dağ, Şeytan-dağ: I Caucasian Corps (with the 6th Caucasian Rifles and the 4th Plastun Brigade holding the mountain ridges).

Area of Şeytan-dağ, Oğnot-çay, Boğlan Pass: VI Caucasian Corps (newly constituted under General Abatsiev and composed of 4th and 5th Caucasian Rifle Divisions and 2nd Plastun Brigade).

Area of Boğlan Pass, Muş, Bitlis: IV Caucasian Corps.

late summer and by January their ranks were gravely thinned. Staff found it difficult to ensure the same conditions of supply which had been maintained the previous winter in front of Köprüköy. Sappers had to labour in deep snow and recurrent blizzards to improve the mountain tracks from Erzurum and the Kara-su valley, and convoys marched through snowdrifts, fogs and storms. The passage of the watershed between the upper course of the Endres and the Tuzla (Karir Pass) remained always a terrible affair for men and animals; and the northern slopes of Bingöl-dağ going down to the Erzurum plain were even worse than the southern side. There were periods when the men of the 4th Caucasian Rifle Division were receiving only half a pound of bread daily and horses only one and a half pounds of barley. There was no meat and no conserves and men were boiling soup from the flesh of donkeys, cats and dogs.[1] Only in February did headquarters find it possible to relieve this excellent unit and to bring it into reserve at Erzurum.[2] Slightly better were the conditions of the 66th Infantry Division on the southern slopes of the Şerefeddin mountains. But here, in March, typhus broke out, imported perhaps from the neighbouring town of Muş.

These winter conditions made it impossible to restore the full strength of the 2nd and 4th Plastun Brigades, of the 6th, 4th and 2nd Rifle Divisions and of the 66th Infantry Division. By 15 March 1917 their strength stood at only sixty to seventy per cent of normal. Complements, however, were gradually being found for these units and for the 39th Infantry Division and II Turkistan Corps.

Yudenich now had only limited local resources from which to try to create new formations. During the winter he had succeeded in constituting one new rifle division, the 7th, out of militia. But this last effort was not as successful as earlier ones had been. Symptoms of pre-revolutionary spirit were already manifest among these badly disciplined and badly instructed men. Two new battalions of Borderers proved more satisfactory.

[1] According to General Kvinitadze, then chief of staff of the 4th Caucasian Rifle Division; cited by Maslovski, *Mirovaya voyna*, p. 405.

[2] It was replaced by the 5th Caucasian Rifle Division.

CHAPTER XL

THE ABORTIVE ATTEMPT AT ANGLO-RUSSIAN COMBINED STRATEGY, JANUARY–MAY 1917

MAP 23; cf. also 1, 30, 32, 33, 38

BY December 1916 important British forces had been concentrated for a decisive offensive against Bagdad. The Russians were kept informed of the intended plan and their co-operation was invited.[1] The matter was raised at the inter-allied conference in St Petersburg in January 1917 and a positive decision in principle was affirmed. Details were left to British headquarters in Mesopotamia and to the grand duke's staff in Tiflis. In accordance with the predominant inter-allied view, the conference continued to regard any operation against Turkey as being of only secondary and local importance.

Already in November, the Grand Duke Nicholas had informed Yudenich of the possibility of Anglo-Russian combined action through Persia and Mesopotamia. Steps were taken to reorganize both the Russian groups in Persia: Baratov's—now named I Caucasian Cavalry Corps—and Chernozubov's, which had been transformed into VII Caucasian Corps.

Chernozubov's corps had been in particular need of infantry reinforcements, since even the two Terek battalions of the 4th Plastun Brigade had been withdrawn during the summer, leaving only the Armenian and Georgian militia formations. In September Yudenich had sent Chernozubov the 2nd Kars Regiment from general reserve;[2] later the four newly formed battalions of the 3rd Border Regiment had followed. By the end of the year Chernozubov had thrown together a whole division of infantry which went by the name of the 'Border Division' and which consisted of the two rifle regiments and the Armenian and Georgian militia formations. A specially strong mountain artillery section (twenty-four pieces) was added to the existing artillery (one field, two Cossack and one horsed mountain battery). Chernozubov was able to muster fifty *sotni* of horse—the 4th Caucasian Cossack Division and the 2nd and 3rd Transbaikal Brigades. A sapper battalion and a section of armoured cars were added. Chernozubov's total reached 13,000 foot, 8000 horse and forty-eight guns.

[1] But see Wilson, *Loyalties*, I, pp. 225–7, 247, 259.

[2] The 1st Kars Regiment remained in garrison at Trebizond.

Baratov's I Caucasian Cavalry Corps had also been reorganized and somewhat strengthened. As infantry reinforcements Baratov got only the newly formed 9th Turkistan Rifle Brigade—four two-battalion regiments of rather doubtful value, since they had been recruited and trained in the autumn of 1916. But the experienced Border Brigade (the 2nd and 4th Regiments, each of four battalions, brought up to normal strength) was a reliable unit under the command of the exacting General Yudenich junior, a nephew of the commander-in-chief. Horse and artillery remained as before: the Caucasian Cavalry Division, the 1st Caucasian Cossack Division, the Kuban Cossack Brigade. Excluding troops fit only for duty as rear details, there were available sixteen battalions (15,000), sixty squadrons and *sotni* and thirty guns.

The combined strength of the two corps amounted to 28,000 infantry, 17,000 horse and seventy-eight guns. It was not a negligible force if used at the same time and in the same direction. But simultaneous action in a common direction was made difficult by the natural conditions of the Perso-Mesopotamian theatre of war and by the particular constitution of the two corps. A distance of about 200 miles divided the axes of action of the two Russian groups: for VII Corps, Tabriz-Ruwandiz-Mosul; for I Caucasian Cavalry Corps, Kazvin-Kirmanshah-Bagdad. While the distance was not too great, it separated two areas with contrasting natural and climatic conditions. In the direction Tabriz-Ruwandiz, the bare ridges and precipitous gorges of the Zagros, deep in snow in the winter months, were scarcely passable even to seasoned Caucasian troops and the horse would be at a notable disadvantage. During the previous year the cavalry raid against Ruwandiz had proved practicable only in mid-April; and, although Chernozubov could now dispose of an infantry division, forty per cent of his available force remained mounted. In consequence, it was natural to plan offensive action by the Lake Urmia force for a date not earlier than the middle of March.

But that date was too late for synchronized action with Baratov's force. The British were expecting co-operation from Baratov in their attack on Bagdad, which was already under way in December and which was intended to reach a decisive stage in January or February. If Baratov was to undertake useful action he had to push back the Turkish XIII Corps from Hamadan to Kirmanshah during January and to reach the Mesopotamian border some time in February. Climatic conditions in the hot plains of Iraq would only permit of the Russians undertaking operations of brief duration round Bagdad.

Here again their excess of horsed troops proved a handicap. Masses of cavalry would fare badly, even if it were possible to keep them supplied, in the Iraq summer which begins to become intolerable in May. Thus Baratov's capacity for useful action would be nearly at an end about the time when the Lake Urmia force would only be beginning to get under way.

These conditions implied the necessity of planning Russian action in two stages: the first by Baratov, the second by Chernozubov. The first stage would make possible co-operation with the British in their offensive directed against Bagdad; the second would provide some support for the British if they continued their Bagdad offensive towards Mosul. Thus both Baratov and Chernozubov were limited to action of *purely local* importance. No co-operation of wide and general strategic significance seems to have been contemplated, nor even considered, by either of the two allies. The British planned to capture Bagdad and, if and when possible, to push on to Mosul; the Russians agreed to co-operate with forces already at hand in Persia after these forces had been reorganized and somewhat reinforced.

There had been almost a complete lull in Persia from August 1916 until the end of that year. The Turkish XIII Corps lay round Hamadan with patrols out along the Kazvin and Teheran roads, keeping small flank-guards at Sinna covering the direction of Bijar on the left and at Burujird-Daulatabad on the right. Some elements of the 4th Infantry Division had joined the 2nd and 6th Divisions, thus weakening the force holding Mosul without appreciably strengthening XIII Corps. This corps was gradually deteriorating owing to the ravages of epidemics, the result in part of the chronic inadequacy of the health services of the Ottoman army. Further reinforcements comprised a battery of Austrian 105 mm. howitzers and three battalions of ex-prisoners from French North African units. With not more than 10,000 fit men at his disposal Ali Ihsan dared not risk a march on Teheran, since Baratov, who might even be reinforced by part of the Lake Urmia force, would always threaten his flank from the direction of Bijar.

Ali Ihsan, with some sense, assumed the attitude that his operational objective had already been attained and he evaded all suggestions that he should move east. In December Enver Paşa was able to celebrate a political victory and the creation of a national Persian army. However, this army, gathered in the region of Burujird, never exceeded 8000 men of doubtful quality and varied origins—ex-gendarmes, volunteers and a few hundred Lur tribesmen. There was no sequel to the ambitious schemes for the penetration of Afghanistan through Persia. The Russians, for their part, provided a newly formed (10th) Turkistan Rifle Brigade and a Cossack regiment for the occupation of Meshed and for liaison with the Anglo-Indian troops in Seistan.

Baratov had begun preparations for his next campaign as early as October. The idea of co-operation with the British was not yet clear, but the necessity of establishing effective liaison with Chernozubov was obvious. The region Bijar-Sinna acquired a double interest, since contact with VII Corps could be established in this area while a move from Sinna on Kirmanshah would force the Turks to evacuate Hamadan without fighting. Baratov therefore decided to send a strong horsed column, together with the 9th Turkistan Rifle Brigade, from Kazvin via Zenjan on Bijar, a march of about 200 miles. In winter conditions this new concentration across the wild mountains of Persian Kurdistan was no easy task. The proper organization of the rear took considerable time and by January 1917 troops were progressing only gradually from Zenjan towards Bijar.

Events in Mesopotamia soon obliged Baratov to accelerate his action. With unjustified optimism Halil Paşa and his all-important nephew had for long refused to believe in the seriousness of any renewed British offensive in Mesopotamia where the Turkish XVIII Corps[1] was supposed to be adequate to hold any attack. But fighting in front of Kut began in January and soon took an unfavourable turn for the Turks. On the 21st, Ali Ihsan received a telegraphic order to move all the forces at his disposal to the support of Halil Paşa, commanding the Sixth Army. One regiment, the 44th, marched from Kirmanshah towards Bagdad. As February came in, the situation of XVIII Corps grew worse. The 45th Division was practically annihilated and the 14th Division, expected from the Second Army, had not arrived. By the middle of the month Ali Ihsan was ordered to withdraw all his troops towards Bagdad. Both British and Russian headquarters were awaiting just such a move. The Russians were in Bijar at the end of January and in February moved on Sinna from where the Turkish flank-guard was withdrawing, in part on Kirmanshah and in part on Sulaimaniya. By 20 February Baratov's main force was ready to march on Hamadan; but on the 24th the Turkish 6th Division was already retiring from Hamadan on Kirmanshah and the 2nd Division followed on the 26th.

Ali Ihsan had prepared his line of retreat in advance as he had

[1] The 51st, 52nd and 45th Divisions.

intelligently anticipated the event sooner or later. He marched with notable rapidity and covered a distance of 300 miles over mountainous country and bad tracks in three weeks. Baratov could not keep up with him. He had to establish his line of communication between Kazvin and Hamadan and then between Hamadan and Kirmanshah. The Border Brigade with six regiments of dragoons and Cossacks constituted the main column under General Radatz. It reached Kirmanshah by 10 March while the 9th Turkistan Brigade and three dragoon and Cossack regiments occupied Sinna. On the Russian left the Turks evacuated the district of Burujird and Daulatabad; Isfahan was occupied by the Cossacks.

Both the forces marching from Persia, Turks and Russians alike, were too late to participate in the battle for Bagdad. The fate of the city had been decided on 9 March, and on the 11th General Maude entered the historic capital on the Tigris. Russian co-operation at this stage had proved unnecessary. Only Colonel Bicherakov with a regiment of 'partisans'—volunteers from Cossack regiments—had time to cross the frontier and establish liaison with British cavalry at Mandali. General Maude judged the advance of his Russian allies as rather slow; the vanguard of Radatz's column only appeared at Khanikin on 2 April.

There were certain special reasons which accounted for the comparatively slow advance of the Russians and these reasons boded no good for the future. The 'February Revolution' had broken out at St Petersburg on 9 March; news of it came quickly to Kirmanshah. It was at first only known at headquarters, but it was just a matter of days before the tidings circulated widely enough among the troops. These events did not come as a complete surprise to many officers in Baratov's corps. A certain part of them were officers from the capital who had volunteered for the Persian expedition. Since the previous autumn political rumours and reports of every kind had been current in the messes. In Persia excitement had been increased by the appearance of the young Grand Duke Dmitri Pavlovich, 'exiled' by the emperor for his supposed participation in the killing of Rasputin. The grand duke got an enthusiastic reception among the majority of officers, and General Baratov, despite his official position, did not attempt to conceal his sympathies.

The other ranks had been restless for some time before the Revolution and all sorts of rumours received a ready currency. When the Revolution actually occurred, the rear details were the first to establish direct contact with revolutionary groups in the Caucasus and Russia;

newspapers and private correspondence circulated freely. The first changes introduced into the structure of the army by the provisional government under pressure from the workmen's and soldiers' soviets soon affected discipline and morale. The Cossacks and the border regiments and many long-service soldiers in the ranks remained little affected, but the crack dragoon regiments proved to contain some of the elements politically most unreliable, while the rear services were widely affected.[1]

In such circumstances every commanding officer tends to give particular attention to the living conditions of troops. Baratov took great care to organize his supply on an adequate scale over the road from Hamadan to Kirmanshah—in spite of the impatience of the British. It must be said at the same time that the disintegration of the army on the Caucasian and Persian fronts did not proceed with the alarming momentum which was in evidence on the western front. Troops still moved in response to orders and were even ready to fight when there was an order to fight.

After the fall of Bagdad the Turkish XVIII Corps retreated northward along the Tigris covering Samarra. On 15 March advance units of the Turkish XIII Corps appeared on the Diyala near Khanikin and moved south along the western bank of that stream. Halil Paşa planned to defend the line of the Diyala with XVIII Corps on his right and XIII Corps on his left with the 14th Division in reserve.[2] Halil could put about 40,000 men into the firing line and his artillery, including German and Austrian howitzers, was rather strong. He was awaiting the arrival of the 46th Division from Macedonia, and he intended to use this reinforcement on the extreme left of his Diyala line in the mountainous region of Sulaimaniya. At the end of April the Turkish commander was joined by one regiment and two or three battalions of gendarmes, with some cavalry, retreating from Sinna. After the arrival of the 46th Division, Halil had under his command more than 50,000 men with 120–130 guns.

To put against the Turkish Sixth Army, General Maude had eighty-six battalions (seventy-three of them Indian) and forty-three squadrons. But the Anglo-Indian troops had covered 120 miles in nineteen days under difficult conditions, and their effectives had been reduced by earlier losses and the fatigues and epidemics of the Tigris valley. The British retained only a slight advantage in numbers and

[1] Budenny was at this time serving as a sergeant in the Severskis.

[2] The 45th and 4th Divisions had been disbanded and their surviving elements incorporated in the other divisions.

General Maude was keen to secure Russian co-operation. But even before the liaison with the Russians was properly established the British had pushed back the Turkish XVIII Corps along the Tigris, and General Edwardes's column (which had been joined by Bicherakov's regiment of partisans) had, on 18 March, successfully attacked elements of the 6th and 4th Turkish Divisions at Bakuba. On the 25th, however, General Keary's two brigades were roughly handled by Ali Ihsan in a rear-guard action at the foot of the Jabal Hamrin. In the last week of March, Halil tried to manoeuvre against this British column with the 2nd and elements from other Turkish divisions. He could not, however, develop this, perhaps too ambitious movement; for on 2 April a strong Russian column (General Radatz) appeared at Khanikin. The Turks moved back across the Diyala and took up positions along the western bank of the river. The centre of XIII Corps was posted at Kifri with its left flank out towards Sulaimaniya and its right flank in liaison with XVIII Corps in the region of Gharfa. Although belated, the appearance of Radatz's column at Khanikin on 2 April had contributed to the failure of Halil's attempt at a counter-offensive against the British right.

At the beginning of April the British raised the question of a combined Anglo-Russian operation in the direction of Mosul. It was proposed that Radatz's column at Khanikin should participate on the British right, while another detachment from Baratov's corps should move through Sinna on Sulaimaniya. Two columns from Chernozubov's Lake Urmia force were to march from Suj-bulak, one through Bane against Sulaimaniya and the other against Ruwandiz. Late winter conditions in the mountains, however, prevented horsed movement over the high passes before the second half of April.

Meantime the consequences of the Revolution were having a serious influence on strategic prospects along the Persian border. Responding to the outcry of extremists, the provisional government replaced the commanders of VII Caucasian Corps and I Caucasian Cavalry Corps by officers supposed to be less reactionary. Chernozubov gave place to Vadbolski and Baratov to Pavlov. (Baratov was, however, rather popular with the troops, and at the end of May he was restored for a short period to his command on the demand of the soldiers' committees of his old corps.)

The provisional government was committed to the continuance of military collaboration with the western powers, and Generals Pavlov and Vadbolski accordingly received orders to take part in the summer offensive, the details of which had been elaborated during

April.[1] After some further delays the combined offensive began early in May. The task allotted to the Russians proved by no means easy. Beyond the Diyala the 2nd and 6th Turkish Divisions held well-prepared positions defended by a powerful artillery (Austro-German). There were elements of the 14th Infantry Division farther to the north, and the passes leading over to Ruwandiz were defended not only by local Kurdish irregulars but also by two regiments of the 46th Infantry Division recently arrived from Macedonia.

Between 11 and 14 May, Radatz succeeded in forcing the passage of the Diyala at three points, Dekke, Maidan and Kala'at Shirwan. On the 17th the Turks counter-attacked and recaptured two of the fords so hardly won. The heat was now severe and British operations on the Tigris had already come to a standstill. Doubtful of the readiness of his troops to sustain further privations, General Pavlov gave orders to break off the offensive and to concentrate round Kasr-i-Shirin.

Action against Sulaimaniya had proved even less decisive. Here the Russians reached Penjwin, a few miles across the Turkish frontier. On the passes leading to Ruwandiz there was some desultory fighting without any definite success. By the time that Baratov returned to his troops, fighting had ceased everywhere. The long-planned co-operation with the British had proved abortive in the circumstances created by the Revolution.

By the early summer of 1917 the Russian Caucasian army was already in process of disintegration, although the tragedy took form rather more gradually and less dramatically than on the main Russian front. The army continued into the summer to occupy the territories conquered in the previous year; there were only minor withdrawals in order to leave in the front line as few troops as possible. Muş was the first important point evacuated (in May). But even Yudenich with his constant attention to the welfare of his troops found his personal situation becoming intolerable, owing to the growing inter-

[1] The plan agreed provided for Russian co-operation on the following basis:

(*a*) Radatz, with the Border Brigade and four regiments of dragoons and Cossacks, had to force the passages across the Diyala, while Bicherakov's partisans kept liaison with the British on the Russian left.

(*b*) General Melisant, with two battalions of Turkistan rifles and a mixed division of dragoons and Cossacks was to move from Sinna on Sulaimaniya.

(*c*) In liaison with Melisant, one Border regiment and one brigade of Transbaikal Cossacks from VII Corps were to move through Bane on Sulaimaniya.

(*d*) A brigade of the 4th Caucasian Cossack Division and four Armenian battalions were to march from Suj-bulak over the Rayat and Kelishin passes in the direction of Ruwandiz.

vention of committees and delegates from Tiflis and St Petersburg. In June the greatest Caucasian commander since Paskevich resigned.[1] He was replaced by General Przevalski. The change brought no amelioration in the situation. The standstill along the front continued, while there was a depressing spread of 'incidents' and troubles in the rear. With the second phase of the Revolution (October), demobilization became spontaneous even before the negotiations at Brest Litovsk had begun.

The campaign of 1917, envisaged on the basis of an Anglo-Russian collaboration over the whole area between the Black Sea and the Persian Gulf, was never realized, but it is interesting to consider what might have been the course and the possibilities of such a campaign.

By the middle of March 1917, Turkey was in a bad and almost critical state. A considerable part of the best units of the Ottoman army had been shattered. Of the fifty-two divisions registered in September 1916, twelve had been annihilated, while the majority of those surviving had an effective strength of 6000–7000 as against the theoretical establishment of 8000–9000.[2] Not only was there no possibility of creating new divisions, but the practice became common of disbanding those divisions which had suffered severely and of embodying their remnants to bring more fortunate units up to establishment.

Excluding four (in theory five) divisions stationed in Arabia, the Turkish strength at March 1917 stood at thirty-four divisions—equivalent to the force mobilized at November 1914. Six of these divisions were dispersed on the Triple Alliance fronts in Europe, four were round Istanbul and in the region of the Straits, six in Syria and Palestine, six in Mesopotamia (or on the way there), and twelve faced the Russian Caucasian army. Roughly speaking, therefore, an equal number of divisions (twelve) were opposed to the Russians on the Caucasian front and to the British in Palestine and Iraq, while of the ten remaining divisions scarcely one could be considered as a strategic reserve. Such was the distribution of the Turkish army when, in the first months of 1917, the Allies were in a position to elaborate a combined, and potentially decisive, operation against the Ottoman empire.

As has already been indicated, in considering the strategy of the war against Turkey in general terms, the British always envisaged

[1] The Grand Duke Nicholas had resigned the posts of viceroy and commander-in-chief in the Caucasus after the March Revolution. Yudenich had replaced him in the supreme command on the Caucasian front.

[2] In some Turkish divisions effectives were reduced to 5000.

operations against Turkey in a wide strategic sense. On several occasions attempts had been made to initiate a grand strategy, but the British met with only lukewarm response from the Russians while the French were even less enthusiastic. It has been explained that the French were, and remained, prisoners of their theory that the road to Turkey lay through Berlin, and Russians of the school of the Grand Duke Nicholas were strongly impressed and influenced by it. The British, however, stubbornly adhered to a view which, in the end, gave them a complete victory in the war against the Ottoman empire. But this victory came only after a powerful effort and numerous setbacks, in the autumn of 1918. If the Russians had remained effective partners in the war and if a combined campaign had been realized for 1917, a decision against the Turks might have been achieved in the winter of 1917–18 or in the spring of 1918. And in view of the ultimate influence of the victories in the Middle East on the end of the war it is conceivable that the whole struggle might have come to an end some six months earlier.

The strategic situation of Turkey at the beginning of 1917 was determined by the following factors:

(*a*) the concentration of strong British forces in Mesopotamia and the fall of Bagdad on 11 March;

(*b*) the beginning of a British concentration against the Turks in Palestine and Syria;

(*c*) the fact that 35 % of Turkish effectives were held on the British fronts and that a further 35 % were dispersed in conditions which did not allow of their being treated as a strategic reserve;

(*d*) the control of the eastern Mediterranean by the Allies and the virtual control of the Black Sea by the Russian fleet.[1]

On the basis of these factors it is not difficult to assess the probable plan and course of events if Anglo-Russian collaboration could have been made effective for the summer of 1917. This collaboration must have taken the form of:

(*a*) direct co-operation with the British in the left flank of the Russian Caucasian front;

[1] The dreadnought *Maria* had been blown up in Sevastopol harbour as the result of a mysterious explosion in the autumn of 1917, but *Catherine II* was strong enough to contain *Goeben* while a new battleship, *Alexander III*, was nearing completion. There were, further, six old battleships which could be very useful in support of combined land- and sea-operations. New destroyers were being commissioned. Admiral Kolchak, by a skilful disposition of minefields, had effectively blocked the exit from the Bosporus into the Black Sea, thus allowing the Russian fleet an effective immunity in any operations along the northern coast of Asia Minor.

(*b*) a powerful diversion on the right flank of the Russian front combined with naval action along the coast of Asia Minor;

(*c*) co-operation by British naval (and land) forces in the Gulf of Alexandretta, at the point of junction of the Syrian and Anatolian theatres of war.

The disposition of British forces at 1 January 1917 imposed *peripheral* rather than *central* action (as in the attempt on the Dardanelles in 1915). Such a peripheral offensive, in which the British would be supporting the attack along the southern arc through Palestine and Iraq, required Russian co-operation through Armenia towards Diyarbekir and Harput which ultimately might be directed to a liaison with the British in the area Aleppo-Alexandretta. Such a combined offensive would reach a decisive phase with the capture of the principal cross-roads of the Ottoman empire at the nodal point of Kayseri.

Baratov's and Chernozubov's corps were available for co-operation in the area of junction between the British and Russian spheres of action. Here Baratov might well have moved forward from Kirmanshah on Sulaimaniya and Kirkuk, maintaining liaison on his right with Chernozubov's Lake Urmia force. An offensive by Chernozubov could have been of far greater help to the British than any action by Baratov. If his infantry strength were adequately reinforced, Chernozubov should have been able to capture Ruwandiz early in May.[1] It is highly improbable that the Turkish 46th Division—less than 10,000 Turks against 20,000 Russians—would have been able to offer an effective resistance and Chernozubov might well have entered Mosul by the end of May. The capture of Mosul would have deprived the Turkish Sixth Army of their best line of supply and retreat. No alternative would have been left to Halil Paşa other than a long and painful retreat up the valley of the Euphrates towards Aleppo.

A Russian offensive, however, would not have been limited to action against Mosul. As soon as climatic conditions became favourable, an operation against Diyarbekir might have been undertaken. This move should have proved the principal Russian offensive operation for the first period (May–July) of the campaigning year.

Yudenich would have had available for this offensive between

[1] There would have been available, as reinforcements for Chernozubov's single border brigade and Georgian and Armenian battalions, two complete plastun brigades—the 4th from the Endres valley and the 3rd from the western sector of the Caucasian front. These twelve excellent plastun battalions, together with eight border and four militia battalions and a complete Cossack division, would have been strong enough to take Ruwandiz.

50,000 and 60,000 infantry, 6000 cavalry and 178 guns.[1] On 1 March 1917, the Turkish Second Army comprised six infantry divisions (each of a reduced strength of about 6000), one weak cavalry division (2000 strong) and artillery deficient in the important arm of mountain guns. A difficult terrain could hardly have deprived the Russians of a superiority which was almost double, while the troops employed would have been all veterans of mountain warfare. And the difficulties of terrain must not be exaggerated. Diyarbekir was by no means inaccessible to an adversary penetrating the valley of the Murat-çay at Palu. For the defence of Palu and the Çapakçur plain which leads down to it, the Turks would only have been able to concentrate a maximum of four weak divisions,[2] of which one would, in effect, have been required to cover the approaches to Harput from the direction of Kiği whence the 5th Caucasian Rifle Division and the 5th Caucasian Cossack Division might have been expected to make a diversion. Three or four weak divisions, in effect no stronger than Russian brigades, could hardly have been expected to hold the offensive of three Russian infantry divisions and the Siberian Cossacks in the valley of the Oğnot-çay and the direction of Palu. There can be no doubt that by the middle of June the Russians would have been in possession of the Çapakçur and Palu plains, which are separated by only fifteen miles of mountainous country over which run several passable tracks.

From Palu it would have proved possible to organize the passage of the Taurus in two directions: into the upper valley of the western Tigris and from Darahini (now Genc) on the Murat-çay to Liçe whence good tracks lead to Diyarbekir.[3] Russian columns moving from the Murat-çay should have been able to take Diyarbekir in the first half of July. Two divisions should have been adequate for this task, while the third would have been available to observe the Turkish force at Harput. Simultaneously with the advance of VI Caucasian Corps, an offensive

[1] IV Caucasian Army Corps (66th Infantry Division, 2nd Caucasian Rifle Division, 4th Caucasian Cossack Division) and VI Caucasian Army Corps (4th and 5th Caucasian Rifle Divisions, strengthened by 1st and 2nd Plastun Brigades and the Siberian Cossack Brigade—from general reserve). Artillery included fifty-four mountain guns and twenty-eight 4·8 in. howitzers.

[2] The 11th, 12th, 1st and 47th.

[3] It is only sixty miles as the crow flies from Palu to Diyarbekir. There are twenty to twenty-five miles of difficult mountain country between Darahini and Liçe, but the col over the watershed is not higher than 1363 m. There are easy tracks over the sixty miles of plain between Liçe and Diyarbekir. For detailed topography, see Konyar, *Diyarbekir Tarihi* (History of Diyarbekir), III, pp. 266 ff.; and Cuinet, *La Turquie d'Asie*, II, pp. 465 ff.

by IV Corps would have been directed through Bitlis on Siirt, with flank columns converging from Van and Muş. To a Russian force of 25,000 infantry and 3000 cavalry on this sector of the front, the Turks would have been in a position to oppose only the weak XVI Corps —hardly more than 15,000 men fit to fight.[1] The Russian conquest of Diyarbekir must have impelled a withdrawal of the Turkish Second Army towards Harput, where after normal losses suffered in defensive actions they would scarcely have been in a position to muster more than 25,000 men.

It is interesting to consider whether it would have been possible to provide effective reinforcements for the Turkish Second Army during the period May–July 1917. Divisions were, in fact, withdrawn from the European fronts to constitute the hastily formed *Yildirim* (Seventh) Army.[2] These divisions might have been dispatched to the triangle Sivas-Kayseri-Harput. But it is known that the concentration in Palestine of the five divisions posted to *Yildirim* took six months. Even allowing for the shorter distances involved, the *Yildirim* divisions could scarcely have been available for the defence of the Harput region before 1 September.

Yudenich, in any case, would have had at hand powerful means whereby he might not only have immobilized the Turkish Third Army but also have diverted reinforcements. It would have been possible for his western sector to have undertaken an advance simultaneously with the forward movement of VI and IV Caucasian Corps. While the 39th Infantry Division and the Don Foot Brigade held Erzincan as the pivotal point of the whole Russian front, II Turkistan Corps could have marched along the valley of the Kelkit on Şebinkarahisar while V Caucasian Corps, supported by units of the Black Sea fleet, moved along the coast to Giresun. After these two points had been taken, the track uniting Giresun with Şebinkarahisar would have constituted a short supply route for II Turkistan Corps. To the two Russian army corps to the north of Erzincan— some 45,000 men with cavalry and strong artillery—the Turks could only oppose their Caucasian divisions, constituted during the previous year. These four divisions numbered round 30,000 men, in a state of rather low morale. The Russians should have reached the line through Giresun and Şebinkarahisar in June; and the subsequent threat to Sivas must have seriously shaken confidence in Istanbul, particularly since the Russians would have been in a position to threaten action

[1] The 5th and 8th Divisions, and some irregulars.

[2] See note 2, p. 438, and Larcher, *Guerre turque*, p. 274.

along the coast at Samsun, Sinop and Ereğli. The Russian threat in this area must, in fact, have had the effect of diverting the *Yildirim* divisions, otherwise available for the defence of Harput, and it would at the same time have immobilized the four divisions held in the neighbourhood of the capital.[1]

There was certainly every prospect for the success of a Russian offensive against Ruwandiz, Mosul and Diyarbekir during the first campaigning period, May–July 1917. It remains to be considered whether the offensive could have been further developed during the second period, August–October. Clearly such a development must have depended on the general situation in Europe and on the readiness or otherwise of the Germans to go to the assistance of their ally. It must also have depended on the capacity of the British to combine their offensive in Palestine with land and sea operations in the Gulf of Alexandretta. It might have proved essential also that the Russian *Stavka* should have made available one more corps from the western front to replace the losses implied in the operations of *the first period* (May–July). A suitable corps would have been III Caucasian which had earned great distinction in the west.

The fate of Harput would have been sealed by the loss of Palu and Diyarbekir. With a strength of nearly five divisions converging from Diyarbekir, Palu and Kiği, the Russians could have captured Harput

[1] It may be contended that lines of communication were inadequate to permit of a Russian offensive on the ambitious scale indicated above. For the three northern corps the problem of communication had been simplified by the occupation of the Trebizond-Gümüşane-Kelkit-Erzurum complex of roads during the previous year. The conquest of the Kelkit-Şebin-Giresun triangle would have added further and even shorter facilities.

The problem of supply for the Russian left strategic flank in an advance on Mosul and Diyarbekir would have been more complicated. The Julfa rail-head had been brought forward to Bayazit (seventy-five miles) as early as 1 July 1916, and by 1 November it had been advanced a further seventy miles to Karakilise. If work had continued at the same *tempo* the rail-head might well have been carried to Malazgirt (seventy-five miles) and perhaps to Muş (another sixty miles) by the summer of 1917. A branch line from Bayazit, through Suvalan, to Arnis on Lake Van (110 miles) was already working at the end of 1916. A harbour on the lake had been built at Arnis, and a small flotilla of steamers and barges was available by the summer of 1917. These arrangements, as they stood, greatly improved the supply situation of VI and IV Caucasian Corps. It became feasible to serve a new forward base in the Gümgüm-Muş area.

At the same time the communications of VII Corps round Lake Urmia were by no means bad. Here a branch from the main Julfa-Tabriz line had been built to the lake harbour of Şerif-hane. Navigation on Urmia was substantially in advance of that on Van. By the end of 1916 a light railway had been built from the harbour of Haydarabad (at the south-west corner of the lake) to Kala-Passova near the Turkish border on the road to Ruwandiz and Mosul. Another harbour, Gulman-hane, served the towns of Urmia and Ushne.

before any Turkish or German reinforcements would have been able to arrive (August).

The Russians would then have had the possibility of a general advance against the line of the Kizil-irmak deploying (from left to right) IV, VI and I Caucasian, II Turkistan and V Caucasian Corps, with III Caucasian (made available from the western front) in general reserve. At the same time VII Caucasian, with units of Baratov's corps, would have been holding the upper Tigris valley from Harput to Mosul, in liaison with the British advancing into northern Iraq. A Russian deployment on such a scale should have made some of the Anglo-Indian units in Iraq available for transfer to the Palestine theatre or for operation in the Gulf of Alexandretta. A new Russian offensive after the capture of Harput might have been directed against the triangle, Sivas-Kayseri-Samsun. Combined operations on a wide strategic scale would have become possible; for while the Russian fleet would have been at hand, on the extreme right, for an attack on Samsun and demonstrations against the north-western coast of Asia Minor and the Bosporus, the British would have been in a position to undertake land and sea operations in the Gulf of Alexandretta. Even if the passage of the Taurus and the Anti-Taurus had proved impracticable for the Allies before the winter of 1917–18, the Turks would have been forced to evacuate Syria a year earlier than they in fact did.

CHAPTER XLI

THE TURKISH INVASION OF TRANSCAUCASIA, 1918

MAPS 1, 19, 38; cf. 34, 35, 36, 37

SINCE May 1917 a kind of informal armistice had come into being all along the Russo-Turkish front. On the Russian side the line was only thinly held, and it became thinner as the achievements of the Revolution penetrated more deeply into the morale of the troops who had won this line at the cost of so much endurance and sacrifice. Self-demobilization started rather later than on the European front, but by the beginning of November discipline—and with it the stability of the front—had almost completely disappeared.

The Russian general headquarters which, under General Przevalski, continued to function in Erzurum viewed the situation with a gloom which was shared by the Christian minorities of the occupied provinces of Turkey, and almost equally by the non-Muslim nationalities in Transcaucasia. At the same time the western powers had reason to fear a German-Turkish advance towards the Baku oilfields at a moment when shortage of oil was tending to weaken the Austro-German war machine in Europe.

The Russian staff of the Caucasian army was not in a position to organize any effective resistance to a German-Turkish advance. The war against Turkey was almost more unpopular than that against Germany, and since April revolutionary crowds had been demonstrating against the 'Imperialists' war' and proclaiming 'We don't want the Dardanelles'. Any organized Russian resistance on the Caucasian front became impossible after the Bolshevik *coup d'état* of 7 November. As only a few hundred Russian officers were prepared to continue the defence of the Caucasian front, Russian general headquarters was compelled to rely only on the national formations developed during the period of the provisional government. These bodies were more or less legalized by the establishment of a Transcaucasian federation which was set up in response to the transfer of power in Russia to the government of the soviets. This federation comprised the nucleus of three subsequently independent states, Georgia, Armenia and Azerbaijan. But, while the existing federation did not recognize the actual (Soviet) government in Moscow, it yet hesitated to proclaim complete separation, in view of the possibility of fresh changes which might follow. The federation was not represented at the peace negotiations at Brest Litovsk, whither it had, however, been invited to send delegates by the Germans (who managed to make tactical use of the Ukrainian representation). When the soviets, on 28 December, signed an armistice with the central powers, the federation had no alternative but to follow suit, if only to retain the support of their own masses who were clamant for peace. The divergent tendencies and interests of the three national groups constituting the federation immediately became evident; for while the Tartars of Azerbaijan were prepared to base hopes for the future on Turkish friendship, the Georgians were hesitant and the Armenians dismayed and desperate. The Georgians had only an insignificant national army (not more than 10,000 men); they were prepared to enter into negotiations without delay, and their leaders relied on their own political ability to conciliate the Germans and procure protection from the Turks. The Armenians remained strongly pro-Ally—and fundamentally pro-Russian—and they tried to develop a national army with the help of Russian general headquarters on the Caucasian front and in the eventual hope of securing support from the British.

By 1 January 1918 the Armenian Corps consisted of two divisions of Armenian rifles, three brigades of Armenian volunteers, a cavalry brigade and some battalions of militia. Each of the two divisions was composed of four regiments, regimental strength being fixed at three battalions. The volunteer brigades were made up of four battalions each and the cavalry brigade was composed of two regiments each four squadrons strong. The rifle divisions were made up of men from the Armenian rifle *druzhiny* (battalions) which had seen hard and honourable fighting during the campaigns of 1914–16. Their numbers were increased by Armenians from different units of Yudenich's army who had decided to join their compatriots. The volunteers were natives of Turkish Armenia who joined the national army on the spot—in Erzincan, Erzurum, Van and the Eleşkirt valley. There was no lack of good equipment to be acquired in the rear areas of Yudenich's dissolving army, and the infantry was well provided with machine guns. The artillery might have been stronger but for the lack of trained gunners available among the Armenians. However, six batteries (each of four field guns) represented the artillery of each of the two divisions, while mountain batteries were organized for attachment to the three volunteer brigades. Units were weak in effectives: the strength of a battalion fluctuated between 400 and 600 men. Thus with twenty-four battalions of riflemen and eight battalions of volunteers, the Armenian national army did not exceed 16,000 infantry, 1000 cavalry and some 4000 militiamen. Even with the help of 10,000 Georgians (whose morale was doubtful) such a small army could not hope to hold a Turkish offensive, particularly when it is remembered that the luckless Armenians were dependent on long and disorganized lines of communication already infested by bands of hostile and angry Muslims.

The feelings of the Muslims towards the Armenian national army became clear even before the Turks began their advance. The same Dersim Kurds, who had themselves turned against the Turks in the autumn of the previous year, but had given no trouble to the Russian conquerors during the interval, became restless as they saw the Russians evacuating the regions of Erzincan, Kelkit and Bayburt and handing over to weak Armenian detachments who were concerned to protect their own compatriots. Local encounters between Armenians and Kurds immediately broke out and there were killed and wounded on both sides almost every day. In the meantime the behaviour of disbanded Russian troops on their way home was far from exemplary and gave pretext to the Turks—if they needed one—to intervene.

It would have been strange if the Young Turk government of Enver Paşa had not considered the revolution in Russia as the predestined moment for the realization of all their ambitious schemes in the direction of the Caucasus. Pan-Turanian expansion might yet compensate for the loss, or impending loss, of the Arab provinces. Even the less romantic circles in the army might at least find the moment opportune for the recovery of the traditional frontier of 1878. During the middle months of 1917 the Turks were attentively observing the course of events in Russia. By the end of the year it was decided that the moment had come to pass to action. For this action the Third Army under Vehip Paşa had been reserved; and it explains why the Turks, despite the complete lull on the Russian front and the urgent need for reinforcement on the fronts threatened by the British, kept the Third Army ear-marked for a Caucasian adventure.

The Third Army held the front between Tirebolu on the Black Sea and Kemah on the Kara-su branch of the Euphrates.[1] The Second

[1] II Caucasian (5th, 11th and 37th Caucasian Divisions) and I Caucasian (9th, 10th and 36th Caucasian Divisions) Corps, to which must be added the 5th and 12th Infantry Divisions of the Second Army, embodying remnants of the disbanded 8th Division.

Army, in the meantime, had finally disbanded, the 5th and 12th Divisions being incorporated in the Third.[1] After a rest of nearly a year all these troops were in better moral and material condition than Turkish troops on the Caucasian front had been in since the beginning of the war. But effectives remained modest with an average of only 5000 bayonets per division. With some auxiliary troops Vehip Paşa could dispose of between 45,000 and 50,000 men with 160 guns including a few Austrian and German howitzers. Such a force was more than sufficient to overcome any resistance on the part of the Georgians and Armenians.

The Transcaucasian Federation was anxious to enter into peace negotiations with Turkey at the earliest possible moment. The Turkish government expressed its willingness to negotiate, and after a month of preliminary steps a conference was proposed at Trebizond. Meantime the Transcaucasian Diet, with singular lack of sense of the realities of the situation, was busy elaborating its conception of the conditions of peace. The session opened on 23 February 1918 and on 1 March the Diet approved conditions which provided for the re-establishment of the Transcaucasian frontier of 1914 and for the right of self-determination of the peoples of eastern Anatolia with the guarantee of the autonomy of Turkish Armenia under the sovereignty of Turkey.

Some weeks before these peace proposals had been elaborated by the Transcaucasian Diet, Vehip Paşa had initiated military action against the Armenian detachments who were holding the front line following the withdrawal of Russian troops. Vehip began by sending repeated protests against the action of disbanded Russian soldiers and the massacres of Muslims committed by Armenian bands. At the beginning of February the activity of the Kurds in the Erzincan region increased and skirmishes took place with the small Armenian force still in occupation. Under the command of Colonel Morel (formerly Russian military attaché in Tokyo), the force round Erzincan consisted of three volunteer battalions (the Erzincan Regiment), one squadron and six guns with a few local irregulars—a total of 2000. Colonel Morel was virtually isolated since the nearest Armenian troops were in Erzurum (a hundred miles to the east) and his communications could easily be cut by the Kurds.

Suddenly Turkish infantry appeared in front of the Armenian post at Çardakli (on the road to Zara and Sivas) and captured it (12 February).

[1] The 1st, 47th and 11th Divisions had been moved elsewhere. The 5th was stationed at Muş, the 12th at Palu.

Another Turkish column appeared on the march from Kemah and on 13 February the Turks were within seven miles of Erzincan. Morel had no choice but to attempt an immediate evacuation and retreat on Erzurum. The retreat occupied eleven days between 14 and 24 February and was carried out with remarkable steadfastness and skill under bitter winter conditions (40–50% of troops and refugees were frost-bitten). The retreating column—as in Tergukasov's epic withdrawal from the Eleşkirt valley half a century before—had to cover thousands of panic-stricken fugitives against repeated attacks from the Dersim Kurds. The Turkish 36th and 5th Caucasian Divisions followed the retreating Armenians and occupied Erzincan and Mamahatun. Elements of the 5th Caucasian Division, in the meantime, entered Bayburt. II Caucasian Corps also moved forward against the Georgians and by the end of February had taken Trebizond and Gümüşane.[1]

These movements of Turkish troops were unopposed except for the skirmishing round Erzincan. Neither the Turkish government nor the new rulers of Transcaucasia considered themselves to be at war; and the Turks might claim that they were reoccupying their own provinces. The Armenians, however, were confronted with a very special problem: the protection of the Armenian population in the districts which the Turks were in process of reoccupying. The issue of the peace negotiations was unknown, but immediate danger threatened all Armenians living in the regions of Erzurum, Hinis, Van, Malazgirt and the Eleşkirt plain. These unfortunate people preferred not to await the coming of the Turkish army and asked for immediate evacuation beyond the Transcaucasian border. To protect them and to give them time to withdraw, a considerable part of the new Armenian national army was dispersed between the various centres populated by Armenians. Erzurum was garrisoned by a detachment under command of the famous partisan leader Antranik[2] (who had been promoted to the grade of major-general). Other detachments held Hinis and Van, while inside the old frontier two groups were concentrated at Aleksandropol and Erevan. This disposition was determined not by any strategic plan but by the protective necessities of the moment. Fear of the Tartars of Azerbaijan and Nahçivan imposed the relatively large reserve held behind the Aras.

[1] For events in Erzincan and Turkish relations with the Armenians and Kurds at this period, see Ali Kemali, *Erzincan tarihi* (History of Erzincan), 1932, pp. 117 *et seq*. Cf. also Kâzim Karabekir and Korganov in Bibliography, (6) (*d*) (vi). For a Russian account of the situation in Trebizond, see S. R. Mintslov, *Trapezondskaya epopeya*, Berlin (no date).

[2] After the war Antranik settled as a farmer in California.

The small Armenian army was thus spread over a very wide area.[1] In command was General Nazarbekov (Nazarbekian), an Armenian by origin, formerly commanding the 2nd Caucasian Rifle Division—an officer who had never been very fortunate in the field. A considerable number of imperial officers, the majority of them Armenians by origin but including also a number of Russians, made up the staff and the unit commands. About 300 young Russian officers had formed an officers' battalion which fought beside the Armenians. The Georgian army, numbering less than 10,000 men, was concentrated between Kutaisi, Akhaltzikhe and Batum. It was commanded by the Georgian General Gabayev (Gabaishvili), a capable officer, formerly commanding the 3rd Caucasian Rifle Division, who found himself unable to contend with the disorganization and lack of morale in the ranks of the Georgians. At the head of both these armies was the commander of the Transcaucasian front, the Russian General Lebedinski; but the real confidence of the Transcaucasian government was given to the Georgian, General Odishelidze.

While the Transcaucasian Diet was still awaiting the opening of the peace negotiations with the Turks at Trebizond the conclusion of the Treaty of Brest Litovsk was announced. By this instrument the Soviet government agreed to exclude from Russian territory the districts of Batum, Ardahan and Kars, where the fate of the population was to be decided by a plebiscite organized in conditions of absolute freedom. This turn of events clearly endangered the Christian minority of the area involved, since it was believed that the Turks did not take the plebiscite condition seriously and considered that the new situation simply promised a reversion to the frontier of 1877. In a state of high emotion the Transcaucasian delegation arrived in Trebizond on 8 March; but the arrival of the Turkish delegation was intentionally delayed and the conference did not open until the 14th.

By 25 February the Turkish vanguard (two regiments of the 36th Caucasian Division with Kurds) was approaching Erzurum. The Armenian patrols fell back on Ilica. By 8 March other units of the Turkish 36th Division with elements of the 5th and 11th Caucasian

[1] At Erzurum, 1st Rifle Regiment, one battalion of 4th Rifle Regiment, five battalions of Erzurum and Erzincan volunteers—3000 infantry, 400 cavalry with eight guns; at Hinis, 2nd Rifle Regiment and the Hinis and Karakilise regiments of volunteers; at Van, the 5th Rifle Regiment and two battalions of Van volunteers; at Aleksandropol, 7th and 8th Rifle Regiments; and at Erevan, 3rd and 6th Rifles and one battalion of the 4th.

Divisions had appeared along the road from Bayburt. The Armenian position at Ilica was easily outflanked by converging columns marching from Erçek in the north and from Tekedere in the south. On 10 and 11 March the Armenians were fighting round Ilica and Tekedere, but the great difference in the strength of the opposing forces was obvious and, on the night of the 11th, General Antranik gave the order to evacuate Erzurum. The city was occupied by the Turks on the morning of the 12th, when a rear-guard action at the gates raised Armenian losses at Erzurum to 600 killed and wounded. On 14 March, the day on which the Trebizond Conference opened, the Armenians retreating from Erzurum were already on the old Russian frontier at Karaurgan. On the 12th, General Nazarbekov had ordered the Armenian detachment at Hinis to evacuate that town. The 2nd Rifle Regiment joined Antranik, while the Hinis and Karakilise volunteers withdrew through Kop and Malazgirt into the Eleşkirt valley.

The first item on the agenda of the Trebizond Conference was Vehip Paşa's telegram to General Lebedinski (10 March) insisting on an immediate evacuation of the districts ceded to Turkey at Brest Litovsk. The Transcaucasian delegates protested indignantly that they did not recognize Brest Litovsk and were not bound by its conditions. Discussions continued along this line and the Transcaucasians vainly enough tried to put forward their own conditions. On 22 March a part of the delegation returned to Tiflis and came back with new proposals for partial territorial concessions. But the Turks were getting tired of the discussions and on 6 April an ultimatum from Istanbul was received demanding the complete evacuation of the disputed districts by the 10th of the month.

Vehip Paşa had been moving his troops towards the Transcaucasian frontier since the last week of March. Along the Black Sea coast the 37th Caucasian Division was marching on Batum, while the 5th Caucasian moved on Ispir in the valley of the Çoruh. The 11th Caucasian was in general reserve at Erzurum as the whole of the Turkish I Caucasian Corps (36th, 10th and 9th Divisions) prepared to cross the frontier and march on Kars. (Cf. Map 19.) On the Turkish right, IV Caucasian Corps moved forward with the 12th Infantry Division approaching Malazgirt and the 5th Infantry Division preparing to occupy Van.

On 30 March the Turks took the frontier post at Karaurgan; on the left, their patrols reached Bardiz and on the right, Karakurt. As at Erzurum, the Turks were undertaking a wide outflanking movement in order to dislodge possible Armenian resistance round Sarikamiş. Nazarbekov noted this manoeuvre. He knew that the Turks could bring three divisions against his own thirteen weak battalions (including the 3rd and 7th Regiments which had come up from Aleksandropol), and he was aware of the increased activity of Muslim irregulars on his line of communications with Kars. He feared encirclement by a superior force in the narrow Sarikamiş defile. On 5 April the Armenian lines were attacked at the point where the *chaussée* forks to Karaurgan and to Karakurt. At the same time the volunteer brigade of Colonel Morel on the Armenian right flank was attacked in the region of the Eşek-meydan Pass at Verişan and that village was lost. The danger of a Turkish outflanking movement towards Novo-Selim in the plain had increased. On the night of 5–6 April the Armenians evacuated Sarikamiş and withdrew towards Novo-Selim, where new positions were taken up in the hope of checking the Turkish advance in the direction of Kars. Among the men at the front there was no doubt that they were facing a renewal of the war with the Turks, but the newly established rulers of Transcaucasia still continued to believe in a peaceful solution and their intervention deprived the troops on the frontier of their last chance of serious resistance.

Without following any strategic plan but rather by force of circumstances, the Armenians, by 10 April, found themselves better concentrated than they had been before. They now constituted two groups: that of Kars composed of seventeen battalions in the Novo-Selim area with four battalions in reserve at Aleksandropol;[1] and that of Erevan consisting of nine battalions including the 5th Rifle Regiment and the two Van volunteer regiments who were on the march from Van to Iğdir.

Something could be done with these thirty battalions (15,000 men) even against the four or five Turkish divisions which were advancing against them (25,000–30,000 bayonets). But there was certainly little possibility of holding the Turkish advance on Kars along a line improvised to the north-east of Sarikamiş in the plain round the railway station at Novo-Selim. The thirty Turkish battalions of I Caucasian Corps with their stronger artillery and their irregular horse could easily outflank the Armenian lines.

The Armenian General Arefiev (Areshiantz) commanding at Novo-Selim evidently intended to bar the principal approaches to Kars from

[1] The Hinis and Karakilise volunteer regiments had arrived from the Eleşkirt valley.

PLATE VIII

BAYBURT : LOOKING NORTH TO THE CASTLE

the south-west—several tracks as well as the railway line and the two *chaussées* leading from Sarikamiş and from Karakurt. The Armenian line ran from Engica along a stream to Bezirgangeçit and Novo-Selim and from there to Karapinar, Tiknis, Ağadeve. Thus 9000 riflemen had to hold a front of over twenty-five miles. Three Turkish divisions were slowly concentrating against this thinly held line which, on 19 April, was attacked at two points, at Novo-Selim in the centre and at Ağadeve on the Armenians' extreme left. The Turks succeeded in capturing the mountain overlooking the village of Ağadeve, but they were thrown back as the result of a vigorous counter-attack. Another counter-attack re-established the situation at Novo-Selim but only for a short time. The Turks brought up fresh troops and pierced the Armenian line at several points on both sides of Novo-Selim. After 6 p.m. the Armenians were already in full retreat towards Benliahmet, having extricated themselves with losses which did not exceed 350 men. The Turks remained on the battlefield.

General Lebedinski (Transcaucasian commander-in-chief), together with General Nazarbekov and several staff officers among whom was Colonel Chardigny, head of the French military mission, visited the front on the day of the Novo-Selim fight and took an optimistic view of the possibilities of continued resistance. Later in the day Lebedinski elaborated this view at a meeting of the Armenian National Assembly at Aleksandropol. A majority of the Assembly voted in favour of maintaining the struggle. The attitude adopted by the Transcaucasian government and Diet was, however, somewhat different. Since 8 April the Trebizond delegates had been insisting on the acceptance of the Turkish ultimatum and on the 10th the authorization to accept had been telegraphed from Tiflis. Its arrival at Trebizond may have been delayed, since on the 12th the Turkish general commanding the troops approaching Batum sent a message demanding the surrender of that fortress and the surrounding area not later than the 13th. On receipt of this new ultimatum, the Transcaucasian Diet changed its attitude—despite the reserve of the Tartar deputies. The Diet supported the government in a sudden decision to reject the ultimatum. Thus it seemed that from 14 April the state of war had at last been legalized by the government in Tiflis. In fact, on that same day, units of the Turkish 37th Division, supported by Laz and Acar irregulars, attacked Batum. Within a few hours the commandant of the fortress, with his garrison of 3000 including 600 officers, decided to surrender. The rest of the Georgian Corps withdrew to the Muha-Evstate position, so well fought over in the Crimean

War and the War of 1877–8, between Kobuleti and Ozurgeti. This event, during the following week, produced a new change in the attitude of the Transcaucasian government. On 22 April the complete independence of the Transcaucasian Federative Republic was proclaimed by the Tiflis Diet. The head of the government, the Georgian Menshevik Gegechkori, yielded his post to another Georgian of the same party, Akaki Chenkeli, who was acting as president of the delegation at Trebizond. Thus a week after the rupture with Turkey which preceded the abortive defence of Batum, Chenkeli was entrusted with the renewal and continuance of the peace negotiations.

After the fighting at Novo-Selim the troops of both sides were moving nearer to Kars. Early on the morning of 23 April the Armenians occupied a line running through the villages of Samova, Karaçoban (Garamvartan) and Bozkale to the railway station at Vladikars. From this point the line extended east of the Kars-çay to the village of Kaniköy, six or seven miles from the town and only about three miles from the outer forts. During the morning Turkish columns deployed and skirmishing began. By 2 p.m. General Nazarbekov had received three telegrams. Chenkeli informed the general of the existence of the new independent republic. Two telegrams from the newly appointed minister of war, the Georgian General Odishelidze, ordered Nazarbekov to cease fire in agreement with the Turkish military authorities on that front. Nazarbekov was ordered to remit his government's decision to the Turkish general in front of him (Yakup Şevki Paşa) in the hope that the message might be conveyed to Vehip Paşa. Thus a complete muddle existed in the Armenian lines when the Turks informed the messenger who approached them under the white flag that they had no instructions from their army or corps commanders and that while the Armenian proposal would be communicated to Vehip Paşa they had, in the meantime, no alternative but to continue operations.

A reply from Turkish headquarters arrived on the following day accepting the proposal for an armistice on the condition that the Armenians withdrew within the line of forts round Kars. At the same time Turkish troops began to move round Kars towards Mezrea railway station thus endangering Armenian communications with Aleksandropol. During the day of the 24th, Colonel Morel was twice sent to negotiate with Colonel Kâzim Bey commanding the Turkish vanguard in front of Kars. Vehip Paşa now put forward the condition that the Armenians should surrender the forts on the left bank of the Kars-çay during the morning of the following day (the 25th) and those on the right bank during the afternoon. The Armenian withdrawal beyond the Arpa-çay was to be completed within three days. Nazarbekov found that he had no choice but to accept.

At 9 a.m. on 25 April the first Turkish units entered Kars. It was a few months over forty years (1878–1918) since the great storm of Kars by a Caucasian army under the command of the Armenian Loris Melikov.

By the evening of the 26th the retreating Armenian troops were at Başgedikler and on the 28th they crossed the Arpa-çay. All the guns of the fortress and all the war material stored there were left to the Turks.[1]

The political confusion created by events in Tiflis and by the Trebizond negotiations had prevented the Armenians from making any serious attempt to defend Kars, at least for the period necessary to permit the partial evacuation of stores and the destruction of the great reserves which had been piled up there. But if the unfortunate Armenian military leaders thought that their troubles were over after their retreat from Kars beyond the old Russian frontier of 1877 they were wrong, and new surprises awaited them. When, on 11 May, the peace conference renewed its sessions—now at Batum instead of Trebizond—Vehip Paşa declared that the old peace conditions were no longer acceptable to the Turks, since the Armenians and Georgians had replied to the original Turkish proposals by armed resistance. The territories earlier in question had been conquered by force of arms and Vehip Paşa found himself obliged to put forward new demands. These demands comprised:

(*a*) the occupation by Turkish troops of the regions of Akhaltzikhe, Akhalkalaki and Aleksandropol;

(*b*) the transfer of the control of the Aleksandropol-Echmiadzin-Nahçivan-Julfa railway, required for the transport of Turkish troops to northern Persia;

(*c*) the free use by the Turks of all Transcaucasian railways so long as the war against Great Britain continued.

These demands were an indication that *the first phase* of Turkish action in Transcaucasia was complete and that *the second phase* was now about to be undertaken. The easy capture of Batum and Kars had convinced Enver Paşa that he might revert to his Pan-Turanian offensive of 1914 and the attainment of two immediate aims seemed to have become practical politics. These aims were the conquest of

[1] Including 100 modern field guns and twelve 6 in. howitzers, 100 fortress guns of 1877 model and several hundred still older pieces—a complete artillery museum.

Baku and the annexation to the Ottoman empire of both Caucasian and Persian Azerbaijan.[1] Subsequently, if the world war and the disintegration of the Russian empire continued, even more ambitious schemes might be undertaken: the penetration of Transcaspia and Turkistan and the projection of Pan-Islamic revolt against the British into Afghanistan, southern Persia and northern India.

The two weak and newly created entities of Georgia and Armenia, across whose territories ran vital roads and railways, alone lay in the way of these intercontinental conquests. It was clearly not difficult to overcome their resistance and the forces available to Vehip Paşa were adequate for the purpose. But a wider extension of Turkish action required more troops, since it was probable that the British would not remain indifferent to the fate of Baku and northern Persia. Some reaction on the part of the Soviet Government, newly established in Moscow, was also possible. In spite of the dangerous situation of the Turkish army in Palestine, the Turkish 15th Infantry Division on its way back from the Rumanian front was halted at the Bosporus with a view to transport by sea to Batum. Even more than troops, Enver needed leaders capable of understanding and interpreting his ambitious schemes. He recalled his half-brother, Nuri Paşa, from Tripolitania (where he was organizing a rather successful resistance to Italian penetration of the hinterland) and appointed his uncle, Halil Paşa, to the command of the 'Army of Islam' on the Caucasian front.

Turkish aspirations in the eastern Caucasus soon became apparent not only to the British but also to the Germans. And the Germans did not remain indifferent to plans which might come into serious conflict with their own. In April 1918 when Enver's plan acquired definite form the Germans were already extending their occupation of Ukraine and preparing to assume control of the northern littoral of the Black Sea. The Germans reached Kharkov on the 20th and soon afterwards they were in Sevastopol and Rostov. A German representative, General von Lossow, was an active presence at the Batum Conference while the adventurous Colonel Kress von Kressenstein—not very friendly to the Turks after his own failure in Palestine—appeared at Tiflis where he established the best possible relations with the Georgian members of the Transcaucasian government. In

[1] The Tartars of Caucasian Azerbaijan and the varied Tartar and Turkoman elements who make up the majority of the population of Persian Azerbaijan speak dialects of Turkish akin to Osmanli. At the same time the culture and historical tradition of both Azerbaijans have always been Iranian rather than Anatolian or Ottoman.

the capital on the Kura the German colonel found the atmosphere very favourable to an intrigue directed to the establishment of a barrage against the Turkish designs on Baku.

The Germans were not very much interested in the success or failure of the Pan-Turanian campaign. But they were in desperate need of the oil of Baku and they were convinced that once the city were in Turkish hands there would be little oil produced even if the pipe-line were allowed to pass under the private control of Enver and all his near and distant relatives and adherents. Von Kressenstein managed to sustain a lively activity in Tiflis during the month of May and he showed himself the man for an emergency. He was aware that Georgian troops could offer no effective obstacle to the Turks, but the Turkish command might become embarrassed if Georgian detachments were covered by the German flag. For the purpose of giving a German cover to Georgian military movements, von Kressenstein mobilized all available men of German origin in and around Tiflis—from the personnel of his legation to prisoners of war and the peasants of the old German settlements like Elenendorf. Meanwhile the Georgian members of the Transcaucasian government saw in German protection the only possible salvation for their own national interest. Their small and badly organized armed force, scattered between Poti, Kutaisi and Tiflis, was incapable of preventing the Turks from occupying the line of the Transcaucasian railway. And the opinion was general among Georgians that once the Turks had been admitted to the country they would never leave it.

For the Armenians the situation was rather different. The Germans were not interested in the defence of Armenian territory and were ready rather to encourage the Turkish move across Armenia into northern Persia in order that any British move towards Baku might be checked. With only indefinite hopes of ultimate support from the British, the Armenians had no alternative but to show fight to the invader. Furthermore, the Armenians feared that the mere passage of Turkish troops through their country would encourage the Muslim elements on their southern and eastern borders to indulge in bloody and anarchic attacks on Armenian villages. Threatened with imminent invasion from the west and with the prospect of internecine racial conflict in their rear, the Armenians understood that even capitulation might only postpone the decimation—or virtual extermination—of their people. The prospect of the ultimate victory of the western powers must have seemed at least uncertain to observers in Erevan in May 1918, and it has in fact rarely fallen to the lot of a people to confront such a desperate and seemingly hopeless situation as that which threatened the Armenians in the early summer of 1918. The Turks, meantime, were in haste to fulfil their grandiose plans, and while negotiations and armistices were still the order of the day, they pursued their march beyond the Arpa-çay.

For their Transcaucasian offensive the Turks had available nine well-equipped infantry divisions—something between 55,000 and 60,000 seasoned infantry with the addition of several thousand irregulars—who in the mountainous country in which operations were to take place, were useful for diversions and for threatening the communications of a numerically inferior opponent.[1]

The wide implications of the Turkish plans, however, made these forces scarcely sufficient for the tasks which awaited them. Thus the Turks could concentrate against the Armenians (who were now in two groups based respectively on Aleksandropol and Erevan) only five out of the nine divisions allocated to the Caucasian front. The disproportion of forces, therefore, was not altogether overwhelming. Against 30,000 Turkish infantry the Armenians could rally some 20,000 rifles—units not always well disciplined nor tactically efficient, but at the same time stubborn and courageous in defence of the homes which lay at their backs.[2]

The numerical relation of the opposing armies and the particular character of the elements making up the Armenian force, who were in great part volunteers with experience of irregular war and a perfect knowledge of the terrain over which operations were to take place, might have suggested the adoption of the strategy and tactics of partisan war. Here a distinction must be made between *guerrilla war* as conducted by small groups of armed men able to disperse and gather again and *partisan war* engaged by compact mobile detachments, each 500–1500 strong and equipped with field guns and machine guns. The Armenian terrain favoured a 'partisan' strategy, since both the railways and the main roads—which the Turks were

[1] At 15 May 1918 the Turkish order of battle in Caucasia was as follows: (*a*) Batum and Oltu, 37th and 5th Caucasian Divisions of II Caucasian Army Corps; (*b*) between Kars and Aleksandropol, 36th and 9th Caucasian Divisions of I Caucasian Army Corps; (*c*) region of Ardahan, 10th Division of I Caucasian Army Corps; (*d*) in general reserve at Kars, 11th Division of II Caucasian Army Corps; round Diyadin and Bayazit with patrols in the passes of the Ağri-dağ, 12th Division of IV Army Corps; (*e*) on the march from Van towards Başkale and the Persian border, 5th Infantry Division of IV Army Corps; (*f*) awaiting transport from Istanbul to Batum, 15th Division.

[2] The Armenian command could still dispose of more than fifty field guns and 1000 cavalry organized in two regiments.

obliged to secure before they could proceed against Baku—ran along valleys which were flanked by great mountainous massifs easily defensible by small and active detachments and offering at the same time a refuge to the Armenian peasants and their livestock from the plains. For such a war the Armenians had excellent leaders—famous partisan chiefs like Antranik, Amazasp and Dro. A partisan strategy could not have prevented the Turks from taking possession of the main lines of communication, but it could have made the continuous use of these lines hazardous and difficult. A partisan strategy corresponded too to the political necessities of the Armenian situation, and primarily the need to provide refuge for the bulk of the Armenian peasant population in areas difficult to conquer pending changes in the immediate international situation. An Armenian partisan strategy, which might well have held inviolate the Armenian uplands during the summer of 1918, might have given the Armenians both heart and respite and might have constituted a serious embarrassment to the Turks in their effort to organize their advance to Baku and the Caspian. Unfortunately for the Armenians, while their newly established government continued to experiment in diplomacy, their military leaders persisted in the illusion that they were commanding a 'regular' national army which it was necessary to manoeuvre in the field according to the classic rules of 'regular war'.

The Armenians, still awaiting the results of the diplomatic permutations at Batum, were taken by surprise when the Turks suddenly initiated the second phase of their Transcaucasian offensive. On the night of 14–15 May the Turks presented the Peace Conference at Batum with an ultimatum demanding the evacuation of Aleksandropol within twenty-four hours and the withdrawal of Armenian troops to a line twenty-five miles to the east of the town. Without awaiting a reply, Turkish troops crossed the Arpa-çay at dawn on the 15th and advanced towards the Tiflis-Aleksandropol *chaussée*. In haste and disorder the Armenians began to retreat from Aleksandropol towards Delijan. As at Kars, vast stores were abandoned to the enemy.[1]

General Nazarbekov's headquarters were at Karakilise, a town thirty-five miles to the east of Aleksandropol, through which runs the railway to Tiflis and the *chaussée* to Delijan. He issued the following battle orders:

(*a*) 1st Rifle Division (twelve battalions) and the 8th Rifle Regiment

[1] Since the Russian conquest of Kars in 1877, Aleksandropol had ceased to be a fortress but it continued to be used as the principal depot for the Transcaucasian front.

(six battalions) to hold a line west of Amamli station, where a branch of the *chaussée* forks south to Erevan;

(*b*) the Erzurum, Erzincan, Hinis and Karakilise volunteer regiments (eight battalions) and the two cavalry regiments (1000 sabres) to march south along the *chaussée* to Erevan as reinforcements for the group based on Erevan;

(*c*) the Lori and Akhalkalaki volunteer regiments (four battalions) under Antranik to take up a position at Güllü-bulak where the Aleksandropol-Tiflis *chaussée* forks to Tiflis and Akhalkalaki.

As Erevan was the capital of Armenia the dispatch of reinforcements in that direction was politically unavoidable; and the main force at Amamli still had the possibility of rejoining the Erevan group by the *chaussée* Delijan-Erevan. Antranik's flank group at Güllü-bulak was a fortunate idea, but this force was too weak to be effective in regular operations.

Once in Aleksandropol the Turks had in their hands the important junction where the line from Sarikamiş and Kars joins the main Tiflis-Julfa line. It remained necessary, however, to throw back the Armenian force covering the road junction at Amamli which controlled the alternative route to Erevan and Echmiadzin round the northern and eastern flanks of the Alagöz massif. The Turks decided also to strike at Antranik's flank group across the fork of the roads to Tiflis and Akhalkalaki. The 5th Caucasian Division advanced against Antranik, while the 36th with units of the 9th were deployed for the attack on the Armenian line in front of Amamli. Here, during 21 and 22 May, the Turks, who were only slightly stronger than the Armenians in infantry, easily threw back Nazarbekov's main force. The Armenian commander decided to concentrate towards Karakilise and to stand there to the end. On the 24th one of his units even proved able to counter-attack and to drive back the Turkish vanguard to Amamli.

Meanwhile, on the 19th, the Turkish 5th Caucasian Division captured Güllü-bulak and then the Karakhach Pass on the Aleksandropol-Tiflis road. Antranik retired to Vorontsovka with the Turks on his heels. From Vorontsovka the partisan leader turned south-east towards Karakilise since he feared to be cut off from the main force under Nazarbekov. He fought a two days' rear-guard action near Jelal-oğlu, then retired to the railway station of Dseg (twenty-five miles north of Karakilise).

Between 26 and 28 May, in the region of Karakilise took place the most serious fighting in all this curious Turco-Armenian war. On this occasion the Armenian rifle regiments fought stubbornly, but the

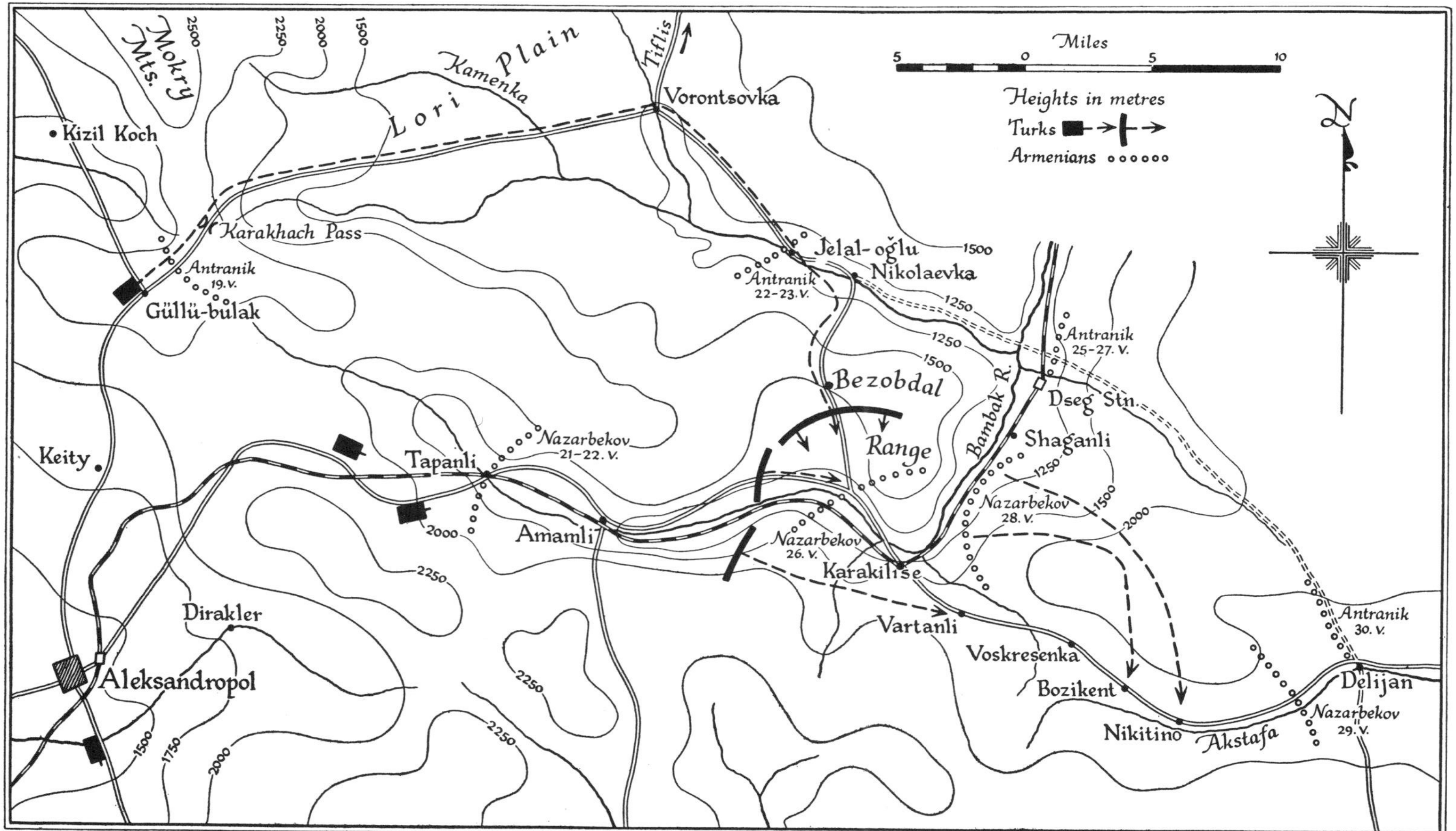

Map 34. Aleksandropol-Delijan, May 1918

Turks concentrated against the Armenian line to the west of Karakilise both the 36th Caucasian Division and units of the 5th Caucasian advancing from Jelal-oğlu. On the 27th, Nazarbekov lost the village of Bezobdal and the mountain over it, and the Armenian right flank was moved back to Shaganli station. On the following day the Turks

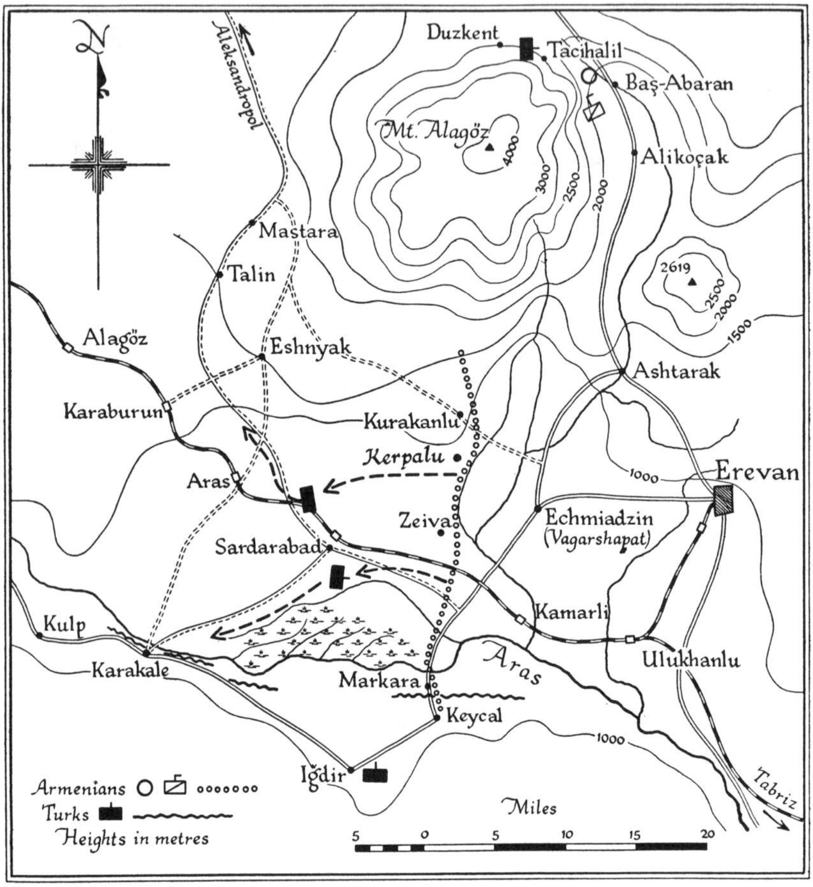

Map 35. Erevan area, May 1918

outflanked the Armenian left, captured Vartanli and cut the *chaussée* to Delijan. Only with the greatest difficulty did Nazarbekov succeed in extricating his main force. The retreating Armenian troops, marching by mountain paths, regained the *chaussée* at Bozikent and Nikitino and, on the 29th, took up a position covering Delijan. Nazarbekov's main force had now been reduced by losses, and perhaps by desertions, to 5000 fit men with a morale which was not now very high.

Next day Antranik joined Nazarbekov at Delijan, but he refused to recognize the armistice which was at this moment concluded and withdrew with a part of his men into the mountains overhanging Lake Sevan (Gökçe).

Round Erevan General Silikov (Silikian) was meanwhile fighting his own campaign. On 15 May Silikov had his main force, amounting to between 6000 and 7000 men,[1] concentrated in the area round Echmiadzin and Sardarabad. He had strong patrols out to the south of the Aras in the region Iğdir-Karakale who were in contact with pickets of the Turkish vanguard which had occupied Diyadin, Bayazit and the passes of the Ağri-dağ. On 18–19 May the Turks developed their offensive from south of the Aras. The 12th Division deployed across the Ağri-dağ with its left at Karakale and its right towards Halfali near Iğdir. At the same time Kurdish irregulars appeared on the northern bank of the Aras in the neighbourhood of Nahçivan. Silikov thus found himself under pressure from the south and east and threatened from the north following the fall of Aleksandropol and the Amamli cross-roads. In this latter direction he was covered only by two regiments of volunteers which had been detached from Nazarbekov's group and were holding the tracks over the Alagöz massif. With Amamli in Turkish hands there was a possibility that the enemy might move south by the main *chaussée* direct on Erevan (seventy-five miles from Amamli station). Silikov found that the best solution in the circumstances was to form a small but reliable force of 1000 picked riflemen under the experienced partisan leader Dro. Provided with plenty of machine guns and four field guns this group took up a position in the defile of Baş-Abaran blocking the main road thirty miles to the south of Amamli.

On 20 May units of the Turkish 12th Infantry Division occupied Iğdir. The Armenians concentrated the 1st and 2nd Van Regiments and the cavalry regiment on the northern bank of the Aras defending the bridges at Markara and Karakale. On the 21st, Silikov's main group[2] was attacked near Sardarabad by two regiments of the Turkish 11th Caucasian Division on the march from Aleksandropol. The Armenians were holding the line of villages, Kurakanlu-Kerpalu-Zeiva, a few miles to the west of their holy city and patriarchal seat of Echmiadzin. The Turks were not in greater force than the

[1] The 5th and 6th Rifle Regiments and one battalion of the 4th; the 1st and 2nd regiments of Van volunteers, one cavalry regiment and twenty-eight guns.

[2] Now reinforced by four volunteer battalions but weakened by the dispatch of Dro's group to Baş-Abaran and the provision of a garrison for Erevan.

Armenians and, using all his reserve, Silikov successfully counter-attacked on 23 and 24 May. He not only recaptured Sardarabad but pressed back the enemy as far as Ani and Mastara, some thirty miles to the north of Sardarabad. The same day Dro was fighting the 3rd Regiment of the 11th Caucasian Division which was advancing from Amamli to Baş-Abaran. Silikov reinforced him with the 2nd Cavalry Regiment and on the 25th Dro counter-attacked.[1] Fighting continued on the 26th and on the 28th Dro was reinforced by infantry sent up by Silikov. On the 29th he was able to throw back the Turks to the north of Baş-Abaran, and remained in firm possession of this important defile.

Thus both Silikov and Dro, operating on terms of numerical equality with the enemy, had each secured an important success. But the failure of Nazarbekov against the 36th and 5th Caucasian Divisions made the situation of the Erevan group still rather precarious. Nazarbekov needed reinforcements and Silikov had to move up both his Van regiments to Semenovka on the Erevan-Delijan road. He concentrated his remaining troops, with the exception of Dro's detachment, at Sardarabad and the Markara bridge. The armistice, concluded on 4 June, supervened. The Turkish 12th Infantry Division, which had remained strangely inactive during the fighting at Sardarabad, now crossed the Aras at Karakale and joined the 11th Caucasian Division. Between 7 and 9 June the Turks were still skirmishing with groups of Armenian volunteers who had refused to recognize the armistice. The Turks occupied the Julfa railway at the Sardarabad, Echmiadzin and Ulukhanlu stations. They refrained from taking possession either of the branch line to Erevan or of the Armenian capital itself. General Silikov and the partisan leader, Dro, had gained the only real—if modest—Armenian successes in this strange Turco-Armenian war which, after a confused preliminary period, had lasted just three weeks. Four Turkish divisions had been diverted to these operations, and it could scarcely be claimed that the Armenian resistance had seriously delayed the development of Turkish plans in Caucasia. A far more serious delay was, in fact, produced not by the courageous decision of the Armenians to fight but by the political manoeuvres of the Georgian leaders.

Within a week of the ultimatum to the Armenians (14–15 May), the Turks presented a further ultimatum to the Transcaucasian government in Tiflis demanding the immediate transfer of the line of the Transcaucasian railway running from Batum through Tiflis to

[1] On this occasion the Armenian horse executed a successful mounted attack.

Baku. The Transcaucasians played for delay on the ground of discussing the details of a peaceful arrangement and on the 27th a *coup de théâtre*, prepared by Colonel Kress von Kressenstein, took place in Tiflis. The Georgian members of the Transcaucasian government proclaimed Georgia a republic independent of the Transcaucasian Federation. The new republic, furthermore, accepted a German protectorate. Von Kressenstein and von Schulenburg themselves announced this protectorate from the window of Tiflis town hall. The Armenian and Tartar members of the Transcaucasian government became the guests of the now completely independent Georgian Republic.

The impression in Batum was great: Halil and Vehip were furious, and the latter warned the Georgians that the Turkish ultimatum remained in force and that on 30 May Turkish troops would move on Poti and Kutaisi. Von Lossow demonstratively left Batum on his way to Germany to report. The first consequence of this new development was that on 4 June the Turks signed a 'treaty of peace and friendship' with the Armenians. At the same time Vehip found himself unable to realize his threat to invade Georgia when he understood how well the three Prussian officers had prepared their ground.

Immediately following the proclamation of the Georgian Republic, the main Transcaucasian railway line was occupied by mixed Germano-Georgian detachments. Georgian and German flags were flying at all the stations. And the situation became something more than a joke when, on 3 June, two German battalions, sent from the Crimea, landed at Poti. Companies of German troops soon made their appearance in the ports of Sukhum and Poti, along the railway and roads leading from Tiflis to Aleksandropol, at the frontier station between Georgia and Azerbaijan and in Tiflis itself. Berlin recalled the German detachments from the Syrian front and the 217th Infantry Division was ordered to proceed from Ukraine to Georgia. Something like a real conflict had arisen between Turks and Germans. On 5 June, Enver Paşa, accompanied by General von Seeckt, sailed from Istanbul to Batum in search of a decision satisfactory to both sides.

These exalted personages came to Batum at the moment when the Turco-German conflict had reached its height. An obstinate man, Vehip Paşa had ordered his troops in the region of Aleksandropol[1] to march on Tiflis. On 10 June advancing units of the Turkish 9th Caucasian Infantry Division came into contact with

[1] Where the 9th Caucasian Division had replaced the 11th, which had been moved towards Erevan.

Germano-Georgian detachments at Vorontsovka on the main road leading to Tiflis across the valley of the Khrami and along the railway which follows northward down the valley of the Borchalu. Two German companies were in the line with some of Kress's volunteers and a number of Georgian militiamen. After brief skirmishes the Turks threw back the Germano-Georgians and took a considerable number of prisoners. On the following day this 'scandalous incident' provoked a telegram from German general headquarters, threatening the withdrawal of all German troops and officials from Turkey. The Turks were summoned to halt their advance into Georgia and to release immediately all prisoners taken in the frontier actions.

Some drastic change was clearly necessary. In a few days Enver found a new form for his Caucasian plans. Vehip Paşa, now no longer *persona grata* to Turkey's allies, was recalled to Istanbul. A new operational force, the Ninth Army, was constituted under Yakup Şevki Paşa (commander of I Caucasian Corps) with its base in the region of Aleksandropol and the Aleksandropol-Julfa railway. The object of the Ninth Army was stated to be active resistance to the British and Bolshevik threat to Caucasia and Persia.[1] In giving his Ninth Army the operational direction of northern Persia, Enver pretended to satisfy German general headquarters which remained very critical of Turkish Caucasian and Pan-Turanian plans and was interested only in creating difficulties for the British command in the East.

However, Enver had by no means renounced his plans and he now hoped to realize them with the aid of the 'Army of Islam'. Under this title it had been decided to organize within the territories of the 'Republic of Azerbaijan', now no longer a part of the Transcaucasian Federation, an armed force of Azerbaijan Tartars. A base was chosen at the ancient Muslim town of Ganja (R. Elizavetpol) lying on the Tiflis-Baku railway where the northern foothills of the Shah-dağ massif slope down to the valley of the Kura.

To form a disciplined nucleus for the Army of Islam, the Turkish 5th Caucasian Division, which was then round Delijan, had to cross Armenian territory (with or without the consent of the Armenians) and reach Akstafa station on the Tiflis-Baku railway, which station

[1] The Ninth Army was made up of the 5th, 9th, 36th and 11th 'Caucasian' Divisions and 12th and 5th Infantry Divisions. The 37th and 10th 'Caucasian' Divisions, under Esat Paşa, continued to occupy the region Batum-Ardahan-Kars while the 15th Division, on arrival from Rumania, was treated as a strategic reserve for the Turkish forces in Caucasia.

was some thirty miles south-east of the frontier of the newly established Georgian and Azerbaijan Republics. There was a good *chaussée* between Delijan and Akstafa and a fair road from Akstafa to Ganja. The Turks had only 100–120 miles to cover between Delijan and the headquarters of the Army of Islam. The 5th Caucasian Division was able to enter Ganja on 20 June. A few days later Enver's half-brother, Nuri, who had travelled by a circuitous route through northern Persia, arrived to take over command of the Army of Islam, which now numbered about 6000 Turkish regulars and 10,000–12,000 Tartar volunteers and militiamen.

Thus was consummated a skilfully conceived political move to counter the Germans: the Turks were now in a position to develop their Caucasian plan, directed first of all to the capture of Baku; but they had evaded German objections since they were using neither Georgian territory nor Georgian railways and were basing their operations on the temporarily occupied part of Armenia with the good *chaussée* from Aleksandropol via Delijan to Akstafa station.

With the last week of June the contest for Baku may be considered to have started. It was a complicated affair since there were many competitors for the possession of the oil city.

(*a*) *The Turks* based their hopes on being the strongest military force in the immediate vicinity of the Caspian. The 5th and 36th Caucasian Divisions (and if necessary the 15th Infantry Division) were available to reinforce the 9th. The Azeri militia might be developed to a strength of between 20,000 and 30,000, and Turkish agents planned to raise some 10,000–15,000 irregulars among the tribes of Dağistan. Enver not only hoped to conquer Baku in four to six weeks, but he already saw his brother's green banner on the Terek and the lower Volga and even in Transcaspia. A British move through northern Persia was anticipated, but the Ninth Army of Yakup Şevki Paşa could oppose it with four or five divisions.

(*b*) *The Germans* did not have at hand in Georgia any force sufficient to move against Baku. They continued their fight, therefore, only on the diplomatic plane. The first German move was an invitation by the 'Allies' (i.e. Germans, Austrians and Turks) to the representatives of the Caucasian peoples to attend consultations in Istanbul. During July these consultations took the form of a series of conversations and intrigues in the Ottoman capital. The Caucasians were tempted in turn by Germans and Turks, but the Georgians remained firmly pro-German and the Tartars equally pro-Turkish. The Armenians were clearly in a helpless situation and the representatives of the mountaineers, who included some wealthy *bourgeois* from the Grozny oilfields, were happy to be courted by so many imposing statesmen and generals. But in studying these Caucasian guests the Germans came to a definite conclusion: the key to the Baku problem lay not in Transcaucasia but in Moscow. The Germans now proposed to invite Soviet delegates to join the Constantinople conference, but the Turks expressed themselves as definitely opposed to the idea. The Germans thereupon entered into secret negotiations with the Soviets, and the first sign of this *rapprochement* between Berlin and Moscow was the recognition by the Soviets of a Georgian independent state (20 August).[1]

Anti-German sentiment grew rapidly in Turkish political circles, and the rumour spread that the Germans were negotiating with the Don Ataman Krasnov (who was received by the Kaiser at Spa), with the object of securing Cossack co-operation against both the Turks and the British at Baku.[2] By the end of August the Turks received news that a Germano-Soviet agreement on Baku was practically concluded.[3] A storm of indignation shook the Turkish press. Enver sent an order to his brother to take Baku immediately. This order was only accomplished on 14 September, on the eve of a series of events which were destined to bring about the complete collapse of the Ottoman empire.

(*c*) *The Soviets* were practically in the same position as the Germans in relation to Baku. The Red army had not the armed forces on the lower Volga nor in the northern Caucasus adequate to prevent the occupation of Baku by the Turks. The Soviet point of view was, however, categorically expressed: Baku must remain within the Russian Soviet Republic, since the oilfields were absolutely necessary to the economy of Russia. The Soviets had never recognized the claims of either the Transcaucasian Federation or of the Azerbaijan Republic to Baku. The Soviets were ready to resist Turkish pretensions with arms, but by force of circumstances the limited number of Red Guards available were compelled to co-operate with such uneasy

[1] Enver countered this move by landing a Turkish contingent at Sukhum. Georgian officials were expelled and the union of Abkhazia with the north Caucasian 'Mountain Republic' (*Gorskaya Respublika*) was proclaimed. When the German and Soviet governments protested, Enver denied the landing of Turkish troops and attributed events at Sukhum to the spontaneous action of the Abkhazians (Pomiankowski, *Zusammenbruch*, p. 373).

[2] The Kaiser had also received the Ukrainian, Hetman Skoropadsky, at Spa and, about this time, a Kalmyk prince.

[3] For Germano-Soviet relations at this time, see Kazemzadeh, *The Struggle for Transcaucasia*, pp. 135, 142, 150.

allies as the Armenian national militia and the Cossacks of Colonel Bicherakov, which latter force had been supported and largely organized by the British. British help was at the same time suspect as 'capitalist and imperialist intervention'. In the negotiations with Germany the Soviets went no further than the promise of economic concessions. But after the capture of Baku by the Turks and in the light of the changed situation on the western front the Soviets, on 25 September, suddenly denounced the Treaty of Brest Litovsk with Turkey and required the immediate withdrawal of Turkish troops from Transcaucasia.

(*d*) The pretensions of the *Azeri Tartars* to Baku were natural enough, since the great majority of the population of the surrounding country districts were Tartar by blood and Muslim by faith. But neither the Transcaucasian nor (later) the Azerbaijan states were recognized by the Soviets, and the Revolutionary Committee at Baku, supported by the Red Guards and, more or less, by the levies of the Armenian national organization, was the only actual power in the city and the oilfields area. While the railway to Tiflis was cut by Tartar bands the Revolutionary Committee still retained a connexion with central Russia by the Caspian and Astrakhan. At the time when the Transcaucasian Federation was set up there were leaders who genuinely believed in the possibility of collaboration between the three Transcaucasian nationalities; but this state of mind was only of brief duration, and an insignificant pretext was sufficient to start terrible street-fighting between the Armenian and Tartar sections of the population of Baku (30 March to 1 April 1918). As the Armenians found support among the Reds (who regarded the Tartars as a counter-revolutionary element) the fighting soon became a massacre of the Tartar population. Several thousands of Tartars were killed and almost half the Muslim population of Baku fled the city. The 'March events' were a very bad beginning for the Armenians since they held promise of a bloody Muslim revenge in the future. Throughout Azerbaijan the Tartars took up arms. They began to organize something like a regular force at Ganja out of ex-regiments of the so-called 'Savage Division'[1] and demobilized soldiers (with a few officers mostly of Tartar blood). The efficiency of these 'national' troops was impaired by a lack of discipline even more marked than that existing among the Armenian regulars who had fought at Amamli

[1] Recruited largely from among the Muslim peoples of Dağistan and Azerbaijan, who had distinguished themselves by their *élan* in numerous battles on the Russian western front.

and Sardarabad. Nuri Paşa found the Army of Islam neither numerous nor of any appreciable military value. And the contact of this improvised force with the askers of the Turkish 5th Caucasian Division was not calculated to improve the fighting morale of the latter.

(*e*) *The Armenians*, who in practice were relatively more heavily committed to the struggle for Baku than any of the other competitors, were actually the least interested in the fate of the oilfields. But at Baku the Armenians faced in major form the tragic problem which had tried them in the lesser towns of Turkish and Russian Armenia: the defence of the Armenian civil population which was threatened with the prospect of massacre and deportation and even with extermination. The problem existed also in the smaller centres of eastern Transcaucasia where the Armenians lived in dangerous proximity to their hereditary neighbours and enemies the Muslim Tartars. Shemakha, Shusha, Nukha and other ancient towns of the Kura basin had seen many phases of the conflict between the two races. The Armenians early began to form 'national' battalions in these places and at Kazakh near Akstafa station on the vital line of Turkish communications between Delijan and Ganja.

At Baku the situation had at first seemed particularly favourable to the Armenians. Between January and March 1918 several thousand soldiers of Armenian origin, demobilized from the Russian front, had gathered in the city. These troops, with several scores of officers, unable to proceed further to their homes, seemed to be ideal material for the formation of an Armenian national corps. The Armenian Colonel Bagratuni, with the protection of the National Committee and without opposition from the Red soviet of soldiers' and sailors' delegates, began the work of organizing his compatriots. The Turco-Tartar danger was appreciated in equal degree by both the national and Red elements and co-operation was furthered by the fact that Stepan Shaumian, the Red chief in Baku and, indeed, one of the foremost leaders of the revolutionary movement in Russia, was an Armenian. But the formation of the Baku force was only in process when Tartar volunteers moved on the city along the Ganja-Baku railway. They were stopped at Hajikabul station (April), seventy-five miles from the city, by the partisans of Amazasp, veterans of the mountain warfare round Lake Van in the years 1914–16.

(*f*) *The British* seemed to the Armenians to represent the only possibility of an ultimate salvation. During the spring and summer of 1917 the British had continued to rely on the assurances of the

Russian command in Caucasia, and the front in Armenia and northern Persia had retained the appearance, if not the substance, of normality. Even in September and October Baratov's troops were still responsible for security between Kirmanshah and the Caspian and VII Caucasian Corps was occupying the region between Tabriz and Urmia. It was only after the November Revolution and the opening of the negotiations at Brest Litovsk that the British began to concern themselves seriously with the political and military future of the area between the Black Sea and the Caspian. Self-determination was then the fashion and it was natural, with the not too remote prospect of the liquidation of the Austro-Hungarian and Ottoman empires, that the crisis of the Romanov empire should have promoted the consideration of all kinds of possibilities in connexion with the minority nationalities which now found themselves without a centre of attraction. British action in Caucasia seemed to meet the strategic requirements of the situation in the Middle East and at the same time to respond to the centrifugal phenomena which had produced, first the Transcaucasian Federation and then, within a few months, the republics of Georgia, Armenia and Azerbaijan. The negotiations undertaken by the new Soviet government at Brest Litovsk were a serious embarrassment to the governments of the western powers and the refusal of the Transcaucasian Federation to send delegates to Brest Litovsk naturally attracted their attention and invited their support. Somewhat exaggerated views were held, even at Russian general headquarters on the Caucasian front, of the prospects of holding the line conquered during the last three years with the support of national formations and some confidence was placed in the numerous bands recruited from Armenian and Assyrian (Nestorian) refugees which were ready to oppose the Turks in the region round Lake Van.

Early in January the skeleton of 'Dunsterforce' (under Major-General Dunsterville of the Indian Army) was constituted at Bagdad. It was at first intended that Dunsterville, with a numerous staff of officers and non-commissioned officers and considerable funds, should make his way across Persia to the Caspian and establish direct contact with the Transcaucasian authorities in Baku and Tiflis. It was anticipated that the Georgians and Armenians, with British support in trained cadres and funds, might be able to sustain the weight of a Turkish attack while the Azeri Tartars might be deterred from actual participation on the side of the Turks. Strangely enough no substantial British forces were moved up into northern Persia to support Dunsterville's mission—possibly in view of the difficulty of organizing the movement of large bodies of troops in the winter conditions of the Iranian plateau.[1]

With all his energy Dunsterville was only able to reach Enzeli on 17 February, and only then because the continuous stream of demobilized Russian troops returning to their homes gave him cover against the turcophil, Kuchik Khan, who was holding, with bands of Jangalis, the Manjil Pass on the road from Kazvin to Enzeli. Dunsterville was already too late: Enzeli was in the hands of Red committees, and the Reds at that moment were chiefly interested in the conclusion of the Brest Litovsk negotiations. Passage was refused to the British mission and Dunsterville narrowly escaped a Jangali ambush on his return to Hamadan. In Hamadan, Dunsterville, who found the situation daily deteriorating and the remnants of Baratov's corps quite unable to maintain internal security, began to prepare for the occupation of northern Persia by British troops. He succeeded in taking into British service Bicherakov's volunteer detachment of partisans which had done so well on the advance into Iraq during the previous year. Thus 1200 seasoned Cossacks were added to the 200 British officers and other ranks who arrived in Kirmanshah at the end of May; the two groups represented the only armed forces at the disposal of the British general. In June some serious reinforcements began to arrive: the 14th Hussars, a battalion of British infantry and another of Gurkhas; one battery, four armoured cars and five hundred lorries. But in June the Turkish offensive by several divisions against Tabriz and the Urmia region was already in full swing, and by 1 July the first movements of the Army of Islam against Baku became known.

Developments in the Caucasus certainly did not correspond with the hopes of continued resistance on the part of the Transcaucasian nations on which the original conception of Dunsterforce had been based. The Georgians had gone over to the Germans and the Armenian fight against the Turks was clearly proving ineffectual. With the Germans in Tiflis and the Turks in Aleksandropol it was obvious that either one or the other or both would not lose time in moving against Baku. Dunsterville received instructions to help the defenders of Baku—whoever they might be—in any possible way. Without awaiting the meagre reinforcements which were expected during the month of June, he set up headquarters at Kazvin. He decided to force a passage through the Jangali country and reach Enzeli. With only Bicherakov's Cossacks, their four horse-drawn guns, one squadron of hussars and a few armoured cars, Dunsterville succeeded in dis-

[1] For detailed discussion, see A. T. Wilson, *Loyalties*, vol. II, chap. II.

persing the three or four thousand irregulars which Kuchik Khan could put in his way (8 June)[1] and in a few days entered Enzeli. During July the arrival of his reinforcements allowed Dunsterville to garrison Resht and to protect the Kazvin-Enzeli road while his patrols were thrown out along the roads from Kazvin to Bijar and Tabriz. Fortunately for the British, the Turks (who were already in Tabriz) remained passive.

In Enzeli, Dunsterville found the Reds adopting an attitude very different from that which they had taken up in February; but, with the exception of Bicherakov's Cossacks, the British commander had no troops available for immediate action across the Caspian. It was decided to allow Bicherakov to co-operate temporarily and tactically with the Reds and to secure their aid in transporting the Cossacks by sea to Baku. However Bicherakov himself planned to land, not at Baku, but at Alyat, a point some forty miles to the south of Baku where the railway from Ganja approaches the Caspian coast and turns sharply north to the oil city. Bicherakov's intention was to preserve his independence of the numerous conflicting political authorities in Baku; at the same time he meant, by a bold raid along the line of the railway, to reach and destroy the important bridge over the Kura at Evlakh, thus seriously slowing any advance of heavy Turkish forces towards Baku. A shrewd adventurer and leader of men, Bicherakov was thus pursuing the twin aims of winning his transport by contributing to the campaign against the Turks and, at the same time, of bringing his men appreciably nearer to their homes on the Terek and the Kuban.

During the first week in July Bicherakov landed at Alyat, but the situation in Baku was less favourable to his plans than he had anticipated. Meantime, on the shores of the Caspian, Dunsterville was impatiently awaiting the reinforcements which had been promised him and which only came up during the first ten days of August. By the 9th he had available for immediate operational purposes only five battalions and four batteries, one cavalry regiment, a few armoured cars and one squadron of aircraft; five further battalions and another five batteries were strung out along his line of communications through Persia. 'Dunsterforce' came too late and proved too small. Its commander, now better informed than he had been six months before, had no illusions as to the potentialities of the local armed groups in Baku. Dunsterville's expedition, in fact, had no

[1] The Jangalis had been well supplied with German equipment and were commanded by a German major.

chance of success, and as it set sail across the Caspian—the only large stretch of water in the world which had never before floated British armed forces—events in France and Palestine had taken a decisive turn which made the risks involved in the weak attempt to succour Baku absolutely pointless.

At the beginning of June the forces available for the defence of Baku amounted to thirty battalions. The strength of each battalion was between 400 and 500 men and units were of a most heterogeneous character: Russian workers from the oilfields (mobilized by the Reds), Armenian ex-soldiers and levies from the Armenian civilian population, Armenian refugees from as near as Karabağ and from as far as Erzincan and Van. The majority of these formations were ill-disciplined and inefficient, but some proved not too bad and able and willing to fight. Plenty of rifles were available but there was a shortage of machine guns in working order. Fifty-four cannon were in condition, organized in two batteries of howitzers, six of field and one of mountain guns. The mounted arm was represented by two squadrons. As a defensive force it was not quite negligible, even with its uncertain and varied politics and its lack of officers and non-commissioned officers.

Curiously enough this essentially defensive force had begun its operational career with an offensive manoeuvre. This did not arise from strategic considerations: here, as elsewhere, the first task of a force largely Armenian proved to be the protection of an Armenian civil population. The Armenians, scattered in considerable numbers throughout Azerbaijan, were not, as a whole, taking refuge in Baku. There was a large group in the ancient Muslim town of Shemakha (sixty miles to the west of Baku), others at Salyany near the mouth of the Kura and in other rural centres. All these Armenians were in danger; their local armed volunteers might defend them against their Tartar neighbours but certainly could not prove effective against the Turks who had at this time already reached Delijan and who might be expected to move farther east. It was therefore decided to occupy as much of the country to the west of Baku as possible before the armed groups of Azeris were reinforced by regular Turkish troops. The Baku command decided to occupy the line from Shemakha through Aksu (on the *chaussée* from Kurdamir station to Shemakha), Kurdamir station (on the Tiflis-Ganja-Baku railway) and Petro-pavlovka village (on the Kura near the point of junction of the Aras with that river). If feasible, it was intended later to aim at the occupation of a line farther to the west running from the town of Nukha southward to the Evlakh bridge over the Kura and then along the *chaussée* to Shusha.[1] The plan was not a bad one, but it required mobile and experienced forces able to act rapidly.

By 5 June an offensive force had been concentrated at Hajikabul station adequate to push back the advanced groups of Azeris westward along the railway line. The Armenian offensive was developed by three columns: on the right four battalions moved on Aksu with the objective of opening the road to Shemakha; in the centre nine battalions advanced along the railway on Kurdamir; and on the left four

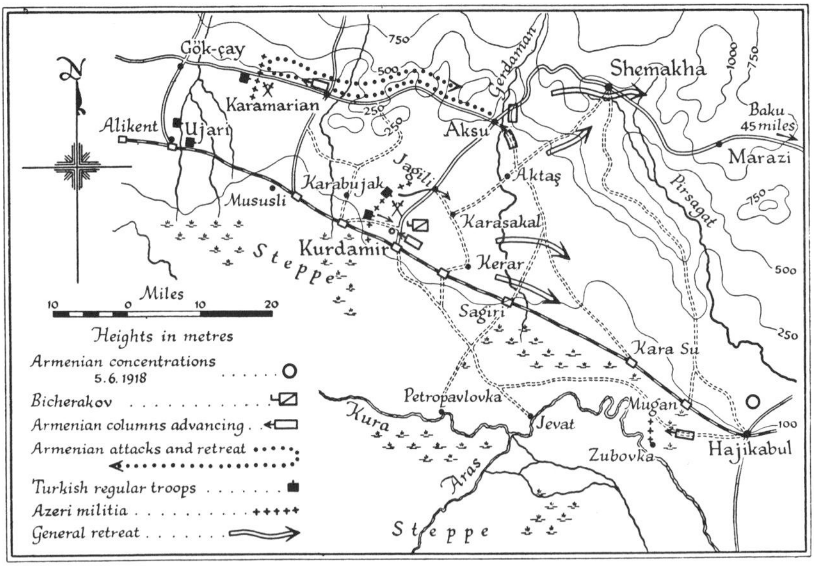

Map 36. Operations along the lower Kura, June 1918

battalions moved along the Kura. By 15 June the line Aksu-Kurdamir-Petropavlovka had been reached except on the left. Only a few skirmishes had taken place, but strong groups of Azeris were known to be concentrating at Mususli station, 20 miles to the west of Kurdamir, and at Gök-çay on the road running north of and parallel to the railway to Evlakh.[2]

The right column met with resistance at Karamarian where fighting took place on 16 and 17 June. The Armenians captured Karamarian and maintained their position but were unable to make further progress. The stronger centre column, which had been

[1] There were importan Armenian minorities in both Nukha and Shusha.

[2] The main road north of the Kura from Shemakha to the Alazani valley via Nukha and Zakatali forks, near the village of Khanapat, south to Shusha in Karabağ and south-west by west to Ganja.

awaiting results from the advance on the right—conceived as an outflanking movement—remained inactive at Kurdamir. The left was held up at Zubovka by commandos of mounted Azeri irregulars and failed to reach Petropavlovka. The Armenians remained on this line from 17 to 26 June but made no attempt to fortify it.

On the 26th, the Armenian command received information of the march of Turkish regular infantry[1] from Ganja to Evlakh and farther to Ujari station. Boldly but rather senselessly the Armenians decided to repeat their 'manoeuvre by the right', a movement which had failed against the Azeri irregulars and which was now to be pressed against seasoned Turkish infantry. On 27 June the Armenian right moved against Gök-çay and reached the heights to the east of that village. Here, on 29 and 30 June, the Armenians were attacked by Azeri militia supported by one Turkish infantry regiment with artillery. The Armenians fought stubbornly, but on 1 July they were in retreat to Karamarian and, next day, fell back on Aksu. Prior to the 27th the right column had been strengthened by four battalions from the centre, but the fighting of the last week of June had cost them 800 men killed and wounded. Furthermore, the whole force was ravaged by dysentery so that effectives at Aksu (inclusive of a volunteer formation from Shemakha) were reduced to some 2000 men. The Turks, however, remained not very active as they slowly concentrated in the region of Gök-çay and Ujari station. A few days later, on 5 July, Bicherakov landed with 1200 Cossacks and six guns at Alyat. In Baku there was at once a wave of optimism. The Reds and the Armenian Nationals agreed to put the distinguished Cossack colonel at the head of the 'active group' on the Aksu-Kurdamir line.

On 8 July Bicherakov arrived at Kurdamir. He worked out a plan of action whereby the Armenians at Aksu and Kurdamir were to hold on to their position while, with his Cossacks and 1000 picked men of the Baku force, he struck from Karasakal (on the Kurdamir-Aksu road) *between* the Turkish columns on the march from Gök-çay to Aksu and from Ujari to Kurdamir. There was, however, no time to put this plan into operation since, on the 10th, two Turkish regiments, supported by Azeri militia and mounted commandos, attacked with considerable energy on the front Kurdamir-Karasakal. The Armenian detachment at Kurdamir withdrew on Kerar station. Bicherakov had to conform to this move in order to avoid encirclement at Karasakal while the Armenian right swung back from Aksu. At the same time

[1] The 5th Caucasian Division.

the roads became crowded with swarms of Armenian fugitives from Shemakha and the surrounding countryside in flight towards Baku.

By the 15th Bicherakov had come to the conclusion that the 'active group' was unable to resist Turkish regular troops with artillery in the open field. He ordered the right column to retreat on Balajari station (a few miles to the north of Baku city) while the centre fell back on Hajikabul, in which direction they were still able to use rail transport. The Cossacks covered the retreat—no very difficult task since the Turks did not move in force beyond the line Aksu-Kurdamir.[1] By 30–31 July the remnants of the Baku 'active group' were concentrated between Balajari and Bibi-Eybat stations on a radius of four to five miles from the centre of Baku. Bicherakov's detachment lay on the right along the railway line to Derbent.

The slow *tempo* of the Turkish advance was proof that Nuri Paşa had his own difficulties. The terrible summer heat of the lower Kura basin, the lack of good drinking water, a long line of supply would all have tended to slow up a force better organized than the Army of Islam. The dysentery epidemic which had decimated the Armenians now struck the askers of the 5th Caucasian Division; and when Nuri reached Kurdamir he had only 4000 men fit for the firing line. The Azeri militiamen were deserting and there remained hardly 8000 men available for the conquest of Baku, half of whom were irregulars of doubtful value. As early as 15 July Nuri had been pressing for reinforcements, and the 36th Caucasian Division was moved from Delijan to Ganja.[2]

During the first days of August several changes occurred in the situation at Baku. The failure of the offensive operation, with the loss of 3000 in killed, wounded and sick, so impressed the Red committee that it decided to abdicate and to surrender the city. But the National Armenian organization, supported by *bourgeois* and moderate elements carried out a *coup d'état* on 1 August and, under the name of the Centro-Caspian Dictatorship, decided to continue the struggle and to seek help from the British at Enzeli. The Red units of Shaumian (Armenian) and Petrov (Russian) ceased fighting and were disarmed. The number of the potential defenders of Baku was thus reduced by a further 3000.

On 2 August, at the moment of maximum confusion in the city as

[1] Shemakha was occupied by Azeri militia.

[2] The 37th Caucasian Division had been dispatched as reinforcements to Palestine via Batum and Istanbul; it was replaced by the 15th Infantry Division (newly arrived from Rumania). The 9th and 10th Caucasian Divisions remained to hold the region Ardahan-Kars-Aleksandropol-Karakilise.

a result of the political crisis, the vanguard of the Army of Islam approached the defensive line between Balajari and Eybat stations. If the enemy had pressed an attack this thinly held line must have been broken and the further defence of the city would have been impossible. But the timely appearance of Bicherakov's Cossacks on the Turkish left and rear suddenly produced a panic and the Turco-Tartars withdrew several miles to the west in a disorder which was interpreted in Baku as a complete victory. After some pursuit of the beaten enemy the jubilant Baku levies returned to the joys of the revolutionary city—no one took the trouble to occupy and fortify the high ground to the west of the railway line.

Another surprise followed: Bicherakov, without any explanations, withdrew from the Balajari region and marched north along the railway to Derbent with his Cossacks and a further 1000 men of the Baku garrison who preferred to link their fate with that of the only efficient commander in the disorganized and threatened city. Fortunately, the impression of Bicherakov's departure was alleviated by the appearance on the same day (4 August) of the first contingent of British troops. Seventy infantrymen and a few officers were landed: a handful of men but sufficient to prove, at last, that British aid was coming, and the facile enthusiasm of the crowd was stirred by stories that 20,000 or 30,000 British troops were on their way to the Centro-Caspian capital.

This enthusiasm proved rather helpful next morning (5 August) when the Turks launched their first attack against the city. They were trying to break through from Eybat station along the main road which penetrates a gap known as Wolf's Gate in the line of cliffs separating the town itself from the valley along which the railway runs north from Eybat to Balajari. The assailants were met by artillery and strong rifle fire. At some places they reached the upper line of the cliff wall, but Wolf's Gate was held and about noon the defenders counter-attacked and the Turco-Tartars were thrown back everywhere to the line of the railway. They managed to retain, however, the heights to the west of Railway Valley. On that day (5 August) the Baku garrison proved to have put up their best performance, at a cost of some 600 men and twenty officers killed and wounded. The Turco-Tartars suffered serious losses and left sixteen machine guns on the field.

A new stage in the fight for Baku now began. The Turks were markedly impressed by the resistance met on 5 August and, perhaps, yet further impressed by the rumours of the arrival of considerable

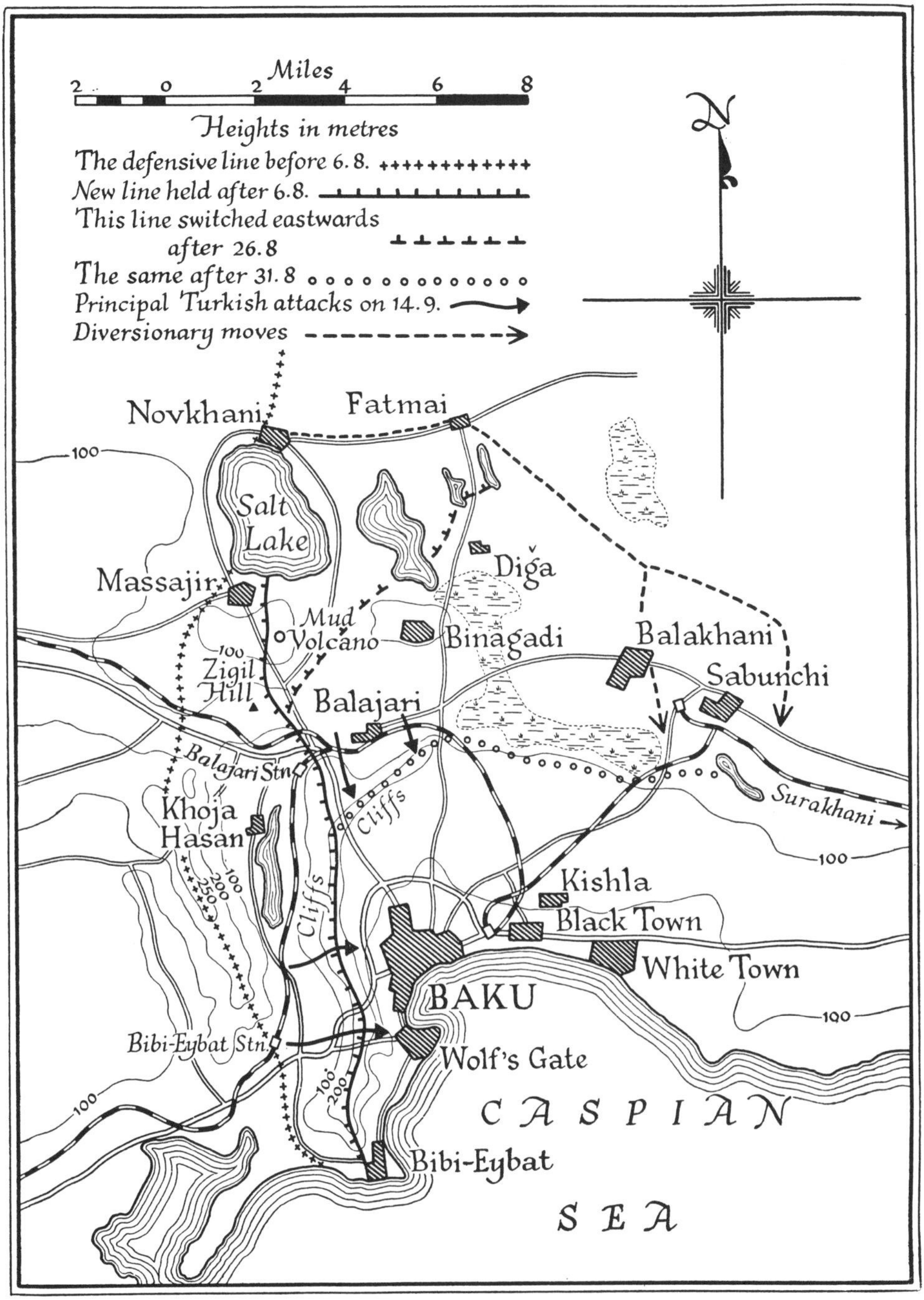

Map 37. Baku in September 1918

numbers of British troops. The commander of the 5th Caucasian Division, with his effectives reduced to some 3000 rather dispirited askers, decided to stand on his positions and await reinforcements. The 36th Caucasian Division was already on the move to join the Army of Islam. The reports received from the Baku front now decided Turkish general headquarters to move even the 15th Infantry Division from Batum to the region of Aleksandropol, while the 9th Caucasian Division was in transport to Julfa. The Ninth Army was ordered to act with more energy in northern Persia and to cut the road Hamadan-Kazvin-Enzeli, by which route British troops directed to Baku were supposed to be on the move.

In fact the hopes of the Armenians and the apprehensions of the Turks proved to be very much exaggerated. Between 4 and 17 August only one incomplete British brigade (the 39th) arrived in Baku—three battalions, a field battery and three armoured cars—under the command of Major-General Dunsterville.[1]

Dunsterville might bitterly regret Bicherakov's decision to leave the defenders of Baku in the lurch, but he understood the reason of the Cossack commander's conduct before his arrival in the oil city and even better after his own experience there. Bicherakov was certainly not a type capable of collaborating with political committees and with troops who held meetings before they determined whether or not to fight; he had only compromised with the Reds in order to secure transport for his men to the Caucasian shore of the Caspian. After his effort to hold the Turks at Kurdamir he had foreseen that his Cossacks might become involved in a siege of Baku and cut off from their route back to their homes in the northern Caucasus. So when he had news of Turkish activity in Dağistan and of bands of mountaineers in occupation of the road to Derbent and of that town itself, he found it better to march north and to capture Derbent, the gate to northern Caucasia. Accordingly by forced marches he reached and quickly captured Derbent (12 August) and remained in station there.

Meanwhile in Baku, General Dunsterville was becoming more pessimistic every day. The defensive organization of the city was bad; the Centro-Caspian troops lacked discipline and most of them were in poor fighting spirit. The Russian General Dokuchayev, now commander-in-chief, was not the man to handle revolutionary volunteers, while the Armenian Colonel Bagratuni, an efficient organizer, was an invalid who was helpless in the tide of growing

[1] The infantry were drawn from the North Staffords, Warwicks and Worcesters.

disorder. British intervention had been planned on the basis of sending limited cadres of technicians to help the Caucasian national forces to organize their own resistance while the Armenians took the view that the British had come to fight for them and to assume the burden of the defence of the oil city.

The three battalions from the English midlands could do little in the situation in which they found themselves on the shores of the Caspian Sea. Baku was well covered from the west by a line of almost perpendicular, but not very high, cliffs. At the southern end of the cliff wall the main road zigzagged to the gap at Wolf's Gate. The line of cliffs might be defended easily enough even by the half-trained Armenian militiamen; one English battalion seemed sufficient support in the Wolf's Gate area. But four or five miles to the north of the Gate, the cliffs turn sharply north-east and then disappear in flats. Balajari station lies at the point where the railway, after following a line parallel with the cliffs, turns east, then south, and approaches Baku from the north-west through the area where the most important oilfields and refineries are situated. It was natural to assume that it would not be long before the Turks turned the Balajari region from the north and, moving east by south along the railway, tried to penetrate into the city taking the whole line of defence in rear. To prevent such an attempt it was necessary to extend the defensive line northward from Balajari through the points named Mud Volcano and Salt Lake to the northern shore of the Apsheron peninsula. Balajari was about equidistant (ten miles) from the northern and southern shores of the peninsula. And the forces in Baku were certainly inadequate to defend a front of seventeen miles (excluding the three miles occupied by the Salt Lake). Furthermore, the ground stretching to the north of Balajari was completely flat except for two small hills (one of which was Mud Volcano). The only course open to Dunsterville under all the circumstances was to place the Warwicks and the Worcesters in the line to the north of Balajari, with two or three local battalions in support.

From the middle of August bands of Azeris began to penetrate into the northern part of the Apsheron peninsula, where defensive patrols had extended the line east from Balajari to the Balakhani and Sabunchi oilfields. Towards the end of the month the enemy became more active. The advance troops of the 36th Division had reached the Baku front and the Turks were extending their line to the east. On 26 August a company of the North Staffordshires in position on Mud Volcano was attacked by two Turkish battalions supported by light and heavy artillery. The company lost all its officers and eighty men. Two more English companies were thrown in as reinforcements, but the two Armenian battalions supposed to be somewhere in the region of Balajari failed to appear. Dunsterville could only muster three English companies to hold the new line established between Balajari and Binagadi hill (to the east of Mud Volcano). The Turks attacked this line on 31 August, and again the English county troops were overwhelmed by superior strength in infantry and artillery. Nor were the Armenian reserve battalions willing to go into the battle. The precarious defence line was withdrawn still further to the east, between Balajari village and Diğa.

The circumstances of the two actions of the 20th and 31st caused General Dunsterville to lose patience. He addressed his reproaches to the Baku 'dictators' on the conduct of the Centro-Caspian troops and concluded that the further defence of the oil city was a waste of time and life. After several days' discussions Dunsterville declared his intention to evacuate his troops since no power on earth could save Baku from the Turks. It was, however, impractical for the British commander to attempt evacuation without agreement with the Centro-Caspian dictatorship, and since the dictators refused to approve his proposals to surrender the city there was no alternative but to await the final attack of the enemy.

This final attack was only delayed by the slow concentration of the 36th Caucasian Division. By 12 September the Army of Islam was able to muster some 7000 or 8000 Turkish troops and 6000 or 7000 Azeri irregulars. In the city British and Centro-Caspian effectives able (or willing) to fight were less than 8000. On the 14th, 3000 Turks and Azeris appeared on the extreme right of the defensive line in the region of Surakhani. This move proved to be no more than a feint; for, on the following day, the principal attack was launched by eight or ten Turkish battalions, with strong artillery support, on both sides of Wolf's Gate. This important point was captured early in the day, but British infantry with field artillery and six armoured cars (three British and three Russian) checked further penetration. With their homes immediately threatened, some hundreds of Armenians fought with a desperate resolution.[1]

By 8 a.m. the Turkish attack at Wolf's Gate and at the angle of the line of cliffs near Balajari station seemed to have been held. A strong counter-attack might have thrown back the enemy to his original

[1] These men belonged to the detachment of 1000 Baku Armenians who had joined Bicherakov on 4 August; 500 of them had returned by sea from Derbent.

positions, but two or three unskilful attempts by the Baku volunteers failed. The Turks, meantime, began to shell the town and harbour with considerable effect on the morale of the population. By 11 a.m. the situation was clearly deteriorating and General Dunsterville gave orders for his shipping to be ready to leave. By 6 p.m. the last effort to organize a counter-attack in force had failed. An hour later the British troops began to withdraw towards the port with the North Staffords acting as rear-guard. By 10 p.m. all British were on board. The evacuation of the Baku troops and civil population continued throughout the night and by 6 a.m. of the 15th some 8000 Armenian troops and civilians had left the port. The Turks did not move forward from the line of cliffs until 5 a.m. of the 16th, when they entered the city to put an end to the massacre, pillage and incendiarism which had been started by the Tartar section of the population during the night of the 14–15th and which continued all through the day and night of the 15th. Recent estimates have placed the number of Armenians slaughtered at just under 9000.[1]

The Turkish 5th Caucasian Division remained in Baku while the 36th moved northward along the railway on Derbent. Bicherakov had already evacuated this ancient gateway of the Caucasus and retired north to Petrovsk which he reached on 2 September. Despite the desperate situation of Turkey, the 36th Caucasian Division (now only 3000 strong) continued its triumphal Pan-Turanian march along the shores of the Caspian and in October occupied Petrovsk, from where Bicherakov had already moved to the Terek. The occupation of Petrovsk was the last and curiously useless gesture of the Army of Islam. The events of September on the Syrian front had already compelled the Turkish general staff to concentrate their 10th Caucasian and 15th Infantry Divisions at Batum with a view to transportation to the Bosporus.

It remains to make reference to events in northern Persia in connexion with their bearing on the Caucasian campaign. Early in the summer of 1918 some 10,000 Assyrian and Armenian fugitives from Turkish territory had been gathered in the region of Lake Urmia. Provided with the remnants of Russian armaments and stores abandoned in this area and under command of some Russian officers (Colonel Kuzmin and others), the Assyrians and Armenians put up

[1] Kazemzadeh (*The Struggle for Transcaucasia*, pp. 143–4) gives a total of 8988, based on the findings of a special commission of the Armenian National Council. The figure may be compared with the estimate of 12,000 Azeri Muslims massacred by the Bolsheviks and Armenians during the March events in Baku; cf. p. 481 above and Kazemzadeh, pp. 72 ff.

a successful fight against the Turkish 6th Infantry Division advancing from the direction of Ruwandiz. Soon, however, the 5th Infantry Division appeared, followed by the 12th (both of the same IV Corps of the Ninth Army). Meantime, in July, the 11th Caucasian Division had occupied Tabriz while the 9th Caucasian Division reached Julfa in support. Despite the very slight British forces covering the road from Tabriz to Kazvin the Turks remained astonishingly passive. After repeated orders from Turkish general headquarters the 11th Caucasian Division undertook a slow advance at the end of August and by 6 September was at Miana. On the 12th, its patrols reached Zenjan, while the 12th and 5th Infantry Divisions, after the evacuation of the Urmia region by the Assyrians, penetrated to Bijar and Sinna.[1] Then the events of September in Syria put an end to this episode of the Pan-Turanian offensive.

[1] For details of the Assyrian (Nestorian) resistance, see art. 'Nestorians' by B. Nikitin in *E.I.* See also M. H. Donohoe, *With the Persian Expedition* (London, 1918); W. A. Wigram, *Our Smallest Ally* (London, 1920); and R. S. Stafford, *The Tragedy of the Assyrians* (London, 1935).

EPILOGUE

'THE GOVERNMENT OF SOUTH-WEST CAUCASIA' AND THE DAĞISTAN REVOLT

MAPS 1, 38, 39

IN Caucasia, at the end of 1918, the centrifugal forces released by the Russian Revolution seemed to have undone a hundred years of history. The fantasy of Enver Paşa took form in sturdy movements of revolt among the mixed Turkish and Muslim peoples who constituted not less than two-fifths of the population of the Caucasian isthmus. By a strange twist of history it fell to the maritime powers—which had given some support to the Circassians, had fought the Crimean War and had resisted the incorporation of 'the three sanjaks'[1] by Russia at the Congress of Berlin—to check the Turkish and Muslim revival in Caucasia during the crisis of the Russian state.

1. THE GOVERNMENT OF SOUTH-WEST CAUCASIA

Under the terms of the armistice of Mudros (30 October 1918) the Turkish army was obliged to withdraw to the west of the 1914 frontier. Evacuating Azerbaijan and all eastern Transcaucasia, the Turks reached the 1877 frontier at the Arpa-çay on 4 December but, in spite of protests from Major-General Thomson in Baku, Yakup Şevki Paşa managed to delay the abandonment of Kars for a further two months. The delay enabled the Turks of Kars to set up a provisional government, under Fahrettin Pirioğlu, on the basis of the Fourteen Points of President Wilson. The government, composed of local landowners, lawyers and school-teachers, reinforced by some Muslim officers of the former imperial Russian army, professed to be the administrative organ of the Kars province with the addition of parts of the adjoining territories of Aleksandropol (Gümrü), Akhalkalaki, Akhaltzikhe and Batum, where there were Turkish-speaking or (as in Batum province) Muslim majorities. A congress of delegates elected from all these districts was convened at Kars where it proceeded to form a national council of twelve members.

In the country to the east of the Arpa-çay and the Aras, the National Council took measures to support the resistance of local Turco-Tartar bands to the troops of the *de facto* government of Armenia (Erevan)

[1] Kars, Çildir (Ardahan) and Batum.

which now enjoyed the prestige of the support of the victorious Allies. Fighting soon developed between the guerrillas of the National Council and Armenian troops along the Arpa-çay and the Aras. The bands of Server Bey and Dikanli Hafiz Bey, about 500 strong, took over Akhaltzikhe from the retiring troops of Halit Paşa and drove a Georgian force up the Borjom defile beyond Atskhur, the Turkish frontier fort of 1828. Strongly supported in the Kağizman district, the Kars plain, Oltu and the lower Çoruh country, the National Council was soon in a position to muster some 8000 men, armed from the abandoned Russian dumps taken over by the Turks at Kars and other points.

Following abortive negotiations with a British mission and representatives of the Armenian government, a second national congress was convened in Kars and, on the night of 17–18 January 1919, the National Council was reconstituted as the Provisional National Government of South-west Caucasia, claiming authority over all the Turkish or Muslim areas between Batum and Nahçivan, that is, the territories, other than Azerbaijan, which had been evacuated by the Turkish army under the armistice of Mudros.

The British authorities in Transcaucasia refused to recognize these proceedings. At the beginning of March, Georgian forces under General Kvinitadze captured Atskhur and occupied Akhaltzikhe and Akhalkalaki. At the same time a British contingent was drafted to Kars. On 19 April, British troops surrounded the 'Parliament' (the building adjoining the present Halkevi) and arrested the leaders of the 'government', who were then dispatched to Batum and subsequently transported to Malta.

During May 1919, General Osebyan (Osebov) assumed command in Kars on behalf of the Armenian government in Erevan. The Turkish community (including members of the families of the provisional government) became the victims of persecution at the hands of ill-disciplined Armenian troops and many excesses were committed in the Turkish villages of the Kars plain. Fahrettin Pirioğlu, Ali Bey of Kağizman and other Turkish leaders had taken refuge in Erzurum; and guerrillas in the Akbaba massif, in the Allahuekber mountains, in Tausker and Oltu, received aid from the Kemalists whose movement in Anatolia was gaining strength.

The policy of supporting the political pretensions of the national minorities in Asia Minor, sponsored by President Wilson and Mr Lloyd George and embodied in the Treaty of Sèvres (10 August 1920), was creating serious difficulties which tended to impede the establishment of normal conditions in the Middle East. At the same time, a solution of the status of the Caucasian nationalities, based on the conception of three independent states in Transcaucasia, proved inacceptable to either of the major contending parties in the Russian civil war. The rout of the White army in the Kuban in February 1920 was followed by the Soviet occupation of Baku and the suppression of the Tartar Republic of Azerbaijan (28 April 1920).[1] Against the revolt of the Tartars of Ganja and Shamkhor (Şamhur) the Red army took drastic repressive measures, but it soon became apparent that the Kemalists were not prepared to support the Azeris and that they had abandoned the Turanist aspirations of the Young Turks. A substantial degree of Turco-Soviet co-operation soon, indeed, became evident in the diplomatic and material support which Ankara received from Moscow in the course of the struggle against the Sèvres Treaty.

With the withdrawal of British forces from Transcaucasia and the liquidation of the Azerbaijan Republic, the position of the Armenians became untenable. The economic situation of Erevan was desperate, and the government found continuous difficulty in controlling a Turco-Tartar minority which within the territory of the republic almost equalled in numbers the Armenian majority. The political initiative in the Turco-Soviet attack on the Armenians came from the Soviets. At the end of 1920 the Soviets had liquidated the White Russian army of General Wrangel in the Crimea and, although they still had to contend with the Poles, their forces in Transcaucasia were adequate to pursue a policy of reintegration of former imperial lands. But while the Kemalists had found themselves unable to support the Azerbaijan Republic, it appears that the Soviets regarded the retrocession of Kars and Ardahan as the necessary basis of a Turco-Soviet understanding in the Middle East. The sacrifices in traditional policy made by each party were at least a measure of the pragmatism of their revolutionary leadership.

In September 1920, the Turkish forces in Erzurum under Kâzim Karabekir Paşa again crossed the 1914 frontier. Sarikamiş was occupied on 29 September; and at the end of October the Turks entered Kars and reached the line of the Arpa-çay. While a Soviet regime had already been proclaimed in the remaining territory of the Armenian republic (28 November), the last act of the Dashnak

[1] By a decision of the Supreme Council at Versailles, British troops had been withdrawn from Baku and all points in Transcaucasia, except Batum, in August 1919. Batum was formally transferred to Georgia on 9 July 1920 and the British and French contingents in occupation were evacuated.

government was to sign the Treaty of Aleksandropol (2 December 1920) whereby Kars province again passed to the Turks.

In May 1920, in the crisis of the Polish war, the Soviets had concluded a peace treaty with the Georgian government.[1] On 12 February 1921, without the formality of a declaration of war, the Red Army launched an attack on Georgia—the last survivor of the three independent republics of Transcaucasia. Tiflis fell on 25 February, and on 18 March the Georgians finally capitulated. On 7 March the Turks occupied Ardahan, probably with the connivance of the Georgian command. Turkish troops then marched on Artvin and Borchka and there was something of a race for Batum which, after an exchange of shots with the Turkish advance guard, was effectively occupied by the Red army at the end of the month.[2]

The new Soviet-Turkish frontier was confirmed by two instruments: the Treaty of Moscow (7 March 1921) and the Treaty of Kars (13 October 1921). The Russians accepted the return of Kars, Ardahan and Artvin to Turkey. At the same time it was agreed that Batum should become an autonomous province and a free port; and the Soviets subsequently recognized the special character of the majority of the inhabitants of the Batum province by creating within the structure of the Georgian Soviet Republic the Autonomous Soviet Socialist Republic of Adzharistan.[3] In this area the frontier, less favourable to the Turks than that of 1877, ran from the village of Sarp, on the Black Sea a few miles south of Batum, east to the crossing of the Çoruh near the village of Maradidi. The frontier then followed the northern boundaries of the former Russian circuits of Artvin and Ardahan to the line of the Arpa-çay. Along the south-eastern sector of the frontier, the Turks received some compensation for their forfeiture of the Batum region in the districts of Tuzluca and Iğdir which had been under Russian sovereignty since 1828. The Aras, instead of the Ağri-dağ, thus became the common bound between the two states, as it remained farther east between the Union of Soviet Socialist Republics and Persia.[4]

[1] The Poles took Kiev on 6 May. The Soviet-Georgian treaty was signed in Moscow on 7 May. The First Cavalry Army (Gen. Budenny) was transferred from Caucasia to mount the counter-offensive against the Poles; cf. *B.S.E.*, art. 'Grazhdanskaya voyna v Rossii, 1917–21'.

[2] For detail of these events, see Kazemzadeh, *Struggle*; also cf. Bibliography, (7) (*b*).

[3] G. *Achara*, T. *Acaristan*.

[4] It is clear that the retrocession of the 1877 frontier with modifications distinct from those of the Brest Litovsk Treaty, is based, not on the decisions at Brest Litovsk, but on the two later treaties of Moscow and Kars. Nor can it be suggested that the Russians entered into the 1921 treaties under duress. The day before the

2. The Dağistan Revolt (1917–21)[1]

The insurrection of the mountaineers of Dağistan, successively against the White and Red regimes of the Russian civil war, constituted a somewhat tragic epilogue to the long Murid wars of the nineteenth century. At the same time the insurrection was a tail-piece to the Pan-Turanian and Pan-Islamic plans of the Young Turks. And in military history the campaign of 1920–1 remains a classic example of the potentialities of partisan war in the mountains.

The campaign has been described in detail by Comrade Todorski, commander of the 32nd Soviet Rifle Division, who was the chief of the Red forces in the field from November 1920 to mid-April 1921; and much interesting light has been thrown on the political background of obscure events by the books of Comrades Samurski and Takho-Godi, both of whom were natives of Dağistan.

When the Russian Revolution broke out in March 1917, less than seventy years had elapsed since the final conquest of the eastern Caucasus by the Russians. The traditions and legends of the epic struggle of Shamyl still lived in the minds of his countrymen, and the practice of revolt had been kept alive during the Turkish war of 1877–8 and during the period of revolutionary ferment in 1905. The Pan-Islamic propaganda of Sultan Abdul Hamit had not been without effect on the fanatical mullahs who were accustomed to visit Turkey and the Holy Places and who continued to use Arabic as the cultured language of Dağistan. More recently the Pan-Turanian propaganda of the Young Turks had attracted a few intellectuals, although the Dağistanlis themselves could hardly be considered as Turanians. It was in fact a religious fever which was smouldering in Dağistan, rather than any modern form of nationalism, when news came through to the mountain auls[2] that the tsar's army in Transcaucasia was breaking up and that the Turks were everywhere advancing.

signature of the Treaty of Kars, a preliminary treaty had been signed with Poland which brought an end to Polish-Soviet hostilities. On the other hand the Turks were passing through a difficult phase of the Greek war which was destined to continue for another year.

[1] This section is based on extensive notes given to me by the late J. F. Baddeley (W.E.D.A.).

Note on spelling. As many names in Dağistan are not of Turkish origin, the Russian spelling of east Caucasian names has been adopted, with the exception of common Turkish forms such as Dağistan, Haci.

[2] *Aul* is the native name for 'village' in mountain Dağistan. Very often an *aul* contained a considerable population, and might be, in fact, a market town for a wide stretch of country.

The priestly class of mullahs, which constituted no less than four per cent of the population of Dağistan, was the dominating political factor in the mountains.

As a matter of fact [wrote Samurski], there is no opposition on the part of the clerical intelligentsia of Dağistan to the Soviet power as the bearer of Communism. On the contrary, Muridism, which has of late made great strides in Dağistan, willingly identifies itself with Communism and it is not without reason that present day Murids call themselves Communists. In their teachings there are, indeed, traits of Communism, but it is a religious, ascetic, primitive communism, answering to that of the Christian communities of the early centuries of our era. The clerical intelligentsia of Dağistan looks upon the Soviet power not as Communistic but as atheistic and as the bearer of western civilization—'the sinful, the accursed'.... 'All European civilization is the invention of the Devil, whether it takes the form of capitalism or of communism.'... and hostility to European civilization is a phenomenon more complicated than any mere religiosity and one far more difficult to deal with. (Samurski, p. 131.)

The two chief figures who emerged in Dağistan at the end of 1917 as the leaders of the Muslim clericals were Najmudin Gotsinski, a local landed proprietor and a large sheep farmer, who had already been proclaimed Imam of Dağistan and Chechnia by his immediate followers (in traditional succession to Shamyl), and Sheikh Uzun Haci, who was called Emir of Chechnia by his friends.

It was Uzun Haci who, according to Samurski, 'more completely than any other leader gave expression to the spirit and aims of the Dağistan "clericals"'. 'He it was who most accurately expounded the teaching bequeathed by Shamyl....' Internally he sought to set up a petty 'Caliphate of the Caucasus', a theocracy based upon a democratic equality of all true believers. In his foreign policy he was inspired by an irreconcilable hatred of everything Russian and a passionate striving after complete independence from the infidels and a union with Turkey, which country he envisaged as destined to hold hegemony over all Muslim lands.

In practice, the clerical leaders found themselves compelled by circumstances to collaborate with more wordly men, such as Prince Tarkovski, an important Muslim landowner, Colonel Chermoyev, an oil magnate of Grozny—who were not unsympathetic to the Russian Whites—and Ali Haci of Akushà, a liberal intellectual, who later became associated with the Bolskevik defence committee against General Denikin.[1]

[1] Photographs of Uzun Haci, Najmudin Gotsinski, Tarkovski, Chermoyev and other leaders of the mountaineers are reproduced by Takho-Godi in *Revolyutsiya i kontr-revolyutsiya v Dagestane*.

But following the Bolshevik *coup d'état* in St Petersburg, in the late autumn of 1917 the elements in Dağistan sympathetic to one or other of the Russian *bourgeois* parties found themselves isolated, and the extremist groups rapidly gained in influence. A national committee was set up in Temir-Khan-Shura, while a revolutionary committee came into existence in the industrial port of Petrovsk on the Caspian Sea.

In order to understand the somewhat complicated military operations which followed, it is necessary to give some consideration to the detail of the geography of Dağistan. The Autonomous Soviet Republic of Dağistan up to 1926 was bounded on the north by the Andi ridge and the lower course of the Sulak, on the east by the Caspian, on the south by the Samur, and on the west by the main chain of the Caucasus, the watershed of which divides the Dağistan tribes from the mountain Georgians—Tushes, Pshavs and Khevsurs. The whole country occupied an area a little over 17,000 square miles.[1]

The name 'Dağistan' itself signifies in Turkish the 'Mountain Country', but the high mountains, comprising about two-thirds of the area, are distinguished as 'Nagorny' or 'mountain' Dağistan, and the coastlands as 'Primorski' or 'maritime' Dağistan. (Cf. Map 1.)

Mountain Dağistan consists of a compact labyrinth of interlocking mountain ridges, jagged and precipitous and (except in rare patches) bare of the forests which abound over the mountains of Chechnia and Georgia. These ridges represent offshoots of the main chain of the Caucasus, and of the Andi ridge, which is itself a minor branch of that chain. The whole, rising at Diklos-mta[2] and the Bogos[3] ridge to heights of over 4000 m., is intersected by the precipitous ravines of four streams (*köysu*), the Andi, Avar, Kara and Kazi-Kumukh, which flow in a north-easterly direction into the Sulak. The Gimri ridge may be said to constitute the division between mountain Dağistan and the maritime districts. The western flanks of the ridge are very steep and jagged, but the eastern slopes fall more gradually to the sandy foothills which run down to the Caspian coast. It is this narrow foreshore

[1] Todorski does not include in his sketch map of Dağistan the districts of Terek province (Khasav-Yurt and Kizlyar, with Achikulak) which were united with the Autonomous Soviet Socialist Republic of Dağistan as a result of the adjustments of administrative frontiers in 1926. Todorski (p. 12) gives an area of 46,036 square versts, which is incorrect, and appears to be a confusion with the area for the whole, including the newly added districts, for which Samurski (p. 13) gives an area of 46,123 square versts. The Dağistan Yearbook for 1901 gives an area of 25,888 square versts (17,258 square miles) for Dağistan as far as the Sulak.

[2] For a fine photograph of Diklos-mta, see Merzbacher, *Aus dem Hochregionen des Kaukasus*, II, frontispiece; and Dechy, *Kaukasus*, II, p. 16.

[3] For a panorama of the Bogos, see Dechy, *op. cit.* II, opp. p. 221.

of the Caspian which gives access from the steppe lands north of the main chain of the Caucasus to the flat country of Azerbaijan, and the coastal railway which runs through the ports of Petrovsk and Derbent links the oilfields of Baku and the Transcaucasian railway with the main Russian railway system. The warlike tribes of Dağistan, remote in their highland auls, are thus brought into close and dangerous contact with the most vital artery of modern Russian economic life; for it is no exaggeration to say that the stability of Russian economics is largely dependent on the continuous exploitation of the Baku petroleum area. Petrovsk must remain a nodal point in Caspian-Caucasian politics, and its importance explains the anxiety of the contending Russian parties and of the Turks to control developments in Dağistan during the period 1918–20.

The principal tribes of mountain Dağistan are the Didos and Andis,[1] occupying the upper valley of the Andi-köysu; the Avars, covering all the country watered by the Avar and the Kara and the middle valleys of the Andi and the Sulak; and the Kazi-Kumukhs in the valley of the Kazi-Kumukh-köysu. Of these the Avars are at once the most numerous and the most formidable, and it was they who gave strength to the long resistance of Shamyl to the Russian conquest in the nineteenth century. During the seventeenth and eighteenth centuries they were an important element in Transcaucasian politics, and their leaders—who do not lack the qualities of oriental statesmanship—have a long tradition of alliance with the Turks, against Russians, Persians and Georgians.

While the mountain clans were, generally speaking, free village communities with a democratic tradition, and a very specialized economy based partly on stock-raising, partly on sericulture and gardening, the inhabitants of maritime Dağistan were a peasantry holding their lands from feudal beys, whose interests attached them, rather, to the imperial regime. Azeri Turkish is the common language of the lowlands, and the people there are mostly of Tartar blood, with the exception of the Tats of Persian stock in the valley of the Samur. The population of the whole of Dağistan was estimated, in 1914, to be about a million and a quarter; of these, the warlike tribes of the inner highlands represented not more than a fifth; but all political initiative lay with them, while the agricultural populations of the maritime districts would remain either passive or sympathetic.

[1] For the language of these two tribes see A. M. Dirr, *Kratki grammaticheski ocherk andiyskago yazyka* (Tiflis, 1906) and *Materialy dlya izucheniya yazykov i narechiy ando-didoyskoy gruppy* (Tiflis, 1909).

The only towns of importance are Petrovsk (now renamed Makhach-Kala)[1] and Derbent[2] on the coast, and Temir-Khan-Shura[3] (Buinaksk) a large market town on the edge of the mountain country. A small industrial proletariat was developing in these towns, and there were certain elements in the villages who had come into contact with modern industrial conditions through working in the oilfields at Baku and Grozny.

To the north of Dağistan, and beyond the Andi ridge, was the territory of the Chechens, who occupied the highland valleys of the north-eastern slopes of the Caucasus, falling to the Terek. The Chechens are Mussulmans, and they were, after the Avars, the principal protagonists of the war under Shamyl. After the Revolution, efforts were made to co-ordinate the political action of the Chechens, as also of the Kabardans to the west of the Daryal Pass, with that of the Dağistan tribes; but the political focus of the northern tribes was Vladikavkaz and Grozny rather than Shura and Petrovsk, and the Chechens had the Cossacks of the Terek for immediate and dangerous enemies while they did not benefit from the proximity of the Turks and Azeri Tartars as allies. Both the 'Chechen Emirate' of Uzun Haci and the so-called Gorskaya Republic (Mountain Republic) of the Grozny millionaire, Tapa Chermoyev, were stillborn, and first the Whites, later the Reds had no great difficulty in extending their authority as far as the Andi ridge and the Sulak. The Chechens failed to play the great part which they had fulfilled during the Murid War, and effective Mussulman insurrection in the north-eastern Caucasus during the period 1917–21 was really limited to the heart of Dağistan.

The lack of communications in mountain Dağistan favoured the highlanders at the moment when Russian authority ceased to be effective. A light railway connects Petrovsk on the coast with Temir-Khan-Shura. From here a metalled road leads south-east round the flanks of the Gimri ridge to Khodjal-Makhi, in the valley of the Kazi-Kumukh-köysu. Second-class roads (*kolesnaya*) connect this point with other towns on the coast—via Deshlagar with Buinak, Kayakent and Berekei. From Khodjal-Makhi a road, at that time metalled only part of the way, leads south to Kumukh, the chief place of the Kazi-Kumukh.

[1] Modern Petrovsk (named after Peter the Great) was built just north of the site of old Tarku. An imaginative painting of Peter at Tarku is reproduced in Potto's *Istoricheski ocherk kavkazskikh voyn*.

[2] For Derbent, see art. by Bartold in *E.I.*; also E. I. Kozubski's detailed and very well produced *Istoriya goroda Derbenta* (Temir-Khan-Shura, 1906).

[3] For an account of Temir-Khan-Shura, see Baddeley, *Rugged Flanks*, II, pp. 2–4; also *Sbor. Mat.* XIX, pt. 1, pp. 1–77, 'Ocherki istorii goroda Temir-Khan-Shury'.

North-westward from Khodjal-Makhi, the main Shura *chaussée* went as far as Khunzakh, the Russian fortified place commanding the Avar-köysu, and from here a second-class road led via Tlokh up the valley of the Andi-köysu to the fort of Botlikh. From Botlikh, it continued north-westward to Veden and Grozny in the Chechnian country.

From the Salti bridge, over the Kara-köysu between Khodjal-Makhi and Karadağ, the main *chaussée* was also connected by a short metalled track with the fortress of Gunib, commanding the valley of the Kara-köysu.

The so-called Avaro-Kakhetinski road led from Karadağ into Georgia, but it had never been completed, and only the forty versts between Karadağ and Ghidatl were fit for wheeled traffic. The only other routes of communication with Georgia, which, until March 1921, remained independent and therefore friendly to the insurgents, were over the high passes of the main chain, which were unsuitable for the passage of baggage-animals even in the summer months.

The routes converging at Khodjal-Makhi constituted the principal lines of access into mountain Dağistan, and upon the Botlikh road depended the provisioning of the Russian forts in mountain Dağistan and the maintenance of Russian authority among the tribes. In the winter months heavy falls of snow frequently made this road impassable. The only other avenues of communication by which Shura might maintain contact with mountain Dağistan were by the difficult paths across the Gimri mountains. The principal were Karanai-Gimri, Erpeli-Gimri and Shura-Arakani through the precipitous defile of that name.

During the last months of 1917, hostilities began between the supporters of Najmudin Gotsinski in Shura and the Soviet authorities in Petrovsk, who had at that time armed about 500 men. However, it was not until March 1918, that the Dağistan Nationalists, reinforced by fragments of the celebrated Savage Division who had returned from the Russian front, were strong enough to occupy Petrovsk, whereupon the members of the Soviet fled, some to Astrakhan and some to Baku. The Nationalists then detached a force to assist their Tartar friends (Mussavatists) in Baku, but the Muslim rising in that city on 17 March was suppressed, and the Dağistanlis were repulsed without much difficulty by the Red guard.

The geographical position of Dağistan [wrote Todorski], between Moscow and Astrakhan on one side and Baku on the other, compelled the interest of all three places in whatever form of government might be established there, especially in Petrovsk, which was situated both on the land-route (railroad) and the sea-route concerned. (Todorski, p. 39.)

Soviet contingents from Baku and Astrakhan, the latter supported by ships reinforced by local revolutionary fragments, were able to occupy Petrovsk and Shura some time between March and May 1918. The bands of Najmudin Gotsinski and Uzun Haci, however, maintained a strong resistance in the immediate neighbourhood of Shura, and during June their prospects were improved by the arrival of Turkish adherents who had penetrated from Transcaucasia where the Turkish army was rapidly advancing. On 12 August, Colonel Bicherakov, with a Cossack 'commando' which had been co-operating with Dunsterforce and Russo-Armenian units in the defence of Baku against the Turks, occupied Derbent with the intention of covering his own retreat to the northern Caucasus. On 2 September he captured Petrovsk, and on the following day the Soviet in Shura, threatened by bands of Chechens who had come down from the north, also collapsed. Prince Tarkovski[1] established himself as dictator in Shura, while Bicherakov declared Petrovsk a free town and constituted it as a White rallying point opening communication with the Terek Cossacks.

On 16 September the Army of Islam under Nuri Paşa occupied Baku, and set up the Tartar government of Khan Khoiski. During October the Turks moved north along the Caspian coast and occupied Derbent and Petrovsk, from which town Bicherakov had withdrawn.

The occupation of Baku by the British and the withdrawal of the Turkish forces from Transcaucasia after the armistice, was a serious blow to the Muslim separatists both in Dağistan and in Azerbaijan. The Russian Whites adopted an aggressive policy in Transcaucasia, and the British authorities, who held the balance between the imperialists and the *de facto* nationalist governments, were placed in an awkward position. In spite of the existence of the Azerbaijan government, Colonel Bicherakov actually appeared in Baku with his Cossacks during December, and General Przevalski started recruiting troops there on behalf of General Denikin. In Dağistan, the Whites occupied Derbent and Petrovsk, and the British effort to establish a neutral zone round the latter town failed. The *de facto* administration of Tarkovski collapsed, and some of its heads seem to have welcomed the re-establishment of a Russian central authority. The Tartar General Khalilov became Denikin's representative at Shura,

[1] Prince Nuh-Bek Tarkovski was the representative of the ancient house of the Shamkhals of Tarku who, for some centuries until the Russian conquest, had ruled maritime Dağistan from the mouth of the Sulak as far south as the district of Kayakent.

and Colonel Kaitmas Alikhanov, an influential native of Avaria, commanded the Cossack garrison of Khunzakh where in the administration of the region he had the support of Najmudin Gotsinski. Uzun Haci, however, appears to have held out against the Whites in Chechnia, where actually he co-operated with the Bolshevik partisan leader, Tikalo (according to Samurski); and the 'liberal' Sheikh Ali Haci of Akushà joined the popular movement which presently developed in southern Dağistan under Bolshevik auspices.

By the end of August 1919, the Whites had suppressed the partisan warfare which had broken out against them in the districts of Akushà, Tabassaran and Karakaitagh, and in the highlands of Kazi-Kumukh, and the defeat of the partisans ended with the destruction of a number of revolutionary auls (Kodor, Durgheli, and part of Great Jengutai). But the extension of White authority into mountain Dağistan, and more particularly, the attempt to recruit men for service in the volunteer army, provoked insurrection of a more serious character round Gunib and in the districts of Kiurin and the Samur, and in parts of Avaria.[1]

In September and October 1919, the Whites suffered a number of important defeats, notably in the neighbourhood of Deshlagar, Kasumkent and the Salti bridge, as a result of which they lost several thousands in killed and prisoners, together with quantities of light artillery and machine guns and many thousands of rifles.

The Bolshevik elements attempted to bring the whole movement, which appears to have been largely popular in character, under the direction of a defence committee which was established at Levashi in the Akushà district. A curious episode followed. A number of Turkish officers had been participating in the fighting, principal among whom were Kâzim Bey and Nuri Paşa, the latter of whom was rivalling, in Dağistan, the adventures of his more famous half-brother in Turkistan. In January 1920, a conflict took place between the Turkish officers and the communists over the leadership of the insurrection. Kâzim Bey surrounded the committee of defence in Levashi, and arrested the executive, at the head of which were the well-known communists, Korkmazov and Kazbakov. The latter was shot on his way to Kâzim's headquarters. However, rival bands

[1] Dechy (*Kaukasus*, II, chaps. XVIII–XX) is the only European traveller who has described in some detail the remote districts of Kumukh and Kiurin and the approaches to the upper valley of the Samur. He made the first ascent of Bazarjusi (4487 m.), the highest peak of the Shah-dağ group. For the obscure dialects of the region, see A. M. Dirr, *Archinski yazyk* (Tiflis, 1908) and *Rutulski yazyk* (Tiflis, 1911). For a view of the middle Samur, see Dechy, *op. cit.* II, p. 272.

under Comrade Bogatyrev and Sheikh Ali Haci of Akushà delivered ultimatums to the Turks, and secured the release of the communist leaders. Nuri Paşa and his immediate following soon afterwards left the country.

Meanwhile the defection of large numbers of soldiers from the ranks of the Whites, following the news of the defeat of Denikin's main forces in the north, was strengthening the position of the communist elements within the ranks of the insurgents. During the last ten days of March 1920, the resistance of the Whites finally collapsed, and the insurgents occupied Shura, Petrovsk and Derbent. A month later Soviet troops entered Baku, and with the 'liquidation' of the Tartar Republic of Azerbaijan, the insurrectionary movement of the Muslim peoples of the Caucasus, which two years before had inspired so many hopes, seemed to be at an end.

Todorski is at pains to explain the outbreak of the insurrection of 1920–1, which was begun within six months of the defeat of the Whites by the forces of the professedly Bolshevik committee of defence. The truth, of course, is that the movement against Denikin was fundamentally anti-Russian to a far greater extent than it was pro-Bolshevik, and the hostility of the mountaineers to all things Russian did not disappear with the substitution of Red commissars for White colonels in the forts of highland Dağistan.

> The number of the [native] communist leaders in Dağistan [wrote Todorski] was small in relation to the work that had to be done, while the Russian communists who were sent there as [political] workers could only be of use in the towns. This being so, the Soviet power was unable to plant itself morally in all the recesses of the country, so that in distant parts those elements re-acquired influence in whose hands it had previously been. The endeavour to disintegrate the mountain peasantry through the creation of 'committees of the poor' resulted, indeed, in some intensification of class antagonism, but this did not suffice to plant the youthful authority firmly on its feet. Moreover, the more backward peasantry remained as before in the hands of the 'clericals' with all the resulting consequences. In this way the attempt to profit by the favourable attitude of the people towards the Soviet to some extent failed.
>
> ...One asset for the counter-revolutionary agitation was the want of any well-considered policy in relation to the mountaineers on the part of the communist leaders. The greater part of the Soviet workers applied, mechanically, the Russian method of going to work, which frequently led to precisely the opposite effect to what was intended.
>
> ...The revolutionary propaganda suffered from many very serious defects. No account was taken either of the cultural level of the mountaineer, nor of the special conditions of his existence; the propaganda was

of an abstract nature, incomprehensible and remote, as far as concerned those for whom it was meant.

...The 'poor' did not now realize any material advantage [from the Revolution]. There was hardly any land to divide. The [few] rich landowners had driven away their flocks into Georgia, and the population itself, frightened by the former power of the beys and the propaganda of the 'clericals', took no willing part in the confiscation of property and, indeed, abstained from it, in spite of the encouragement of the revolutionary bodies.[1]

...So it was, then, that in places there was no Soviet authority but only its outward show, under which continued to exist in full the former social relations, the former conditions of life. Only in the towns, in thickly populated places, and, in part, in the lowland auls was there a beginning of real progress in the revolutionary sense. (Pp. 43–5.)

Meanwhile the civil war in Russia had, with the defeat of General Denikin, merely assumed new forms. The Polish campaign, which reached an intense phase during the summer of 1920, absorbed the military strength of the Soviets; the Whites, under General Wrangel, were resuming the initiative in the Crimea, and among the Cossacks of the northern Caucasus a formidable movement, under General Fostikov, had broken out.

In Transcaucasia, during May 1920, the Red army had been threatening the Georgian frontier at various points, but the Soviet government now showed its desire to relieve itself of embarrassments on its southern frontier by the conclusion of a treaty again recognizing the independence of the Georgian Republic (7 May). About the same time, the revolt of the Mussulman Tartars of Ganja and Shamkhor, bloodily suppressed by the Soviet authorities, demonstrated the strength of the counter-revolutionary feeling which still smouldered throughout the eastern Caucasus.

Many of the leaders of the counter-revolutionary factions in Dağistan had taken refuge in Tiflis. Here counsels were divided. A group of the former members of the 'Mountain Government' were working for the establishment of an independent 'Republic of the Mountaineers'. It is probable that these had the sympathy—if not the active support, as Todorski suggests—of members of the Georgian government, for, with the Soviets in Dağistan, the Transcaucasian governments could ultimately continue to exist only on sufferance. Another organization, with the somewhat unwieldy title of the 'Committee to help the mountaineers and Terek Cossacks to

[1] The Soviets experienced the same reluctance on the part of Muslim peasants to participate in confiscations of land, both in Azerbaijan and in Turkistan.

freedom from the Bolshevik yoke', favoured collaboration with General Wrangel and the establishment of a general directorate to co-ordinate the various reviving anti-Soviet movements.

When the insurrection was finally set on foot in September 1920, it was given the character, at once, of a movement for national independence and of a holy war against the Russians. A young officer in the Turkish army, Said Bey, a grandson of Shamyl, was brought from Turkey for his name's sake only, and placed at the nominal head of the movement. He rode always accompanied by a personal escort of twenty-five sabres, all his own blood relations from Gimri, and the supreme power over the *Shariat* troops[1] of the mountain tribes lay with the council of sheikhs and Said Bey, to whom the military command was responsible. At the same time most of the principal military leaders had been closely associated with the Whites, and to this extent the movement was suspect among the more fanatical tribesmen.

The political head of the rising was Najmudin Gotsinski, who is characterized by Todorski as one of the most learned Arabists of Dağistan but as selfish, ambitious and avaricious; Todorski quotes against him the punning words of Uzun Haci, one of his chief supporters, when near his death: 'I wanted to make an Imam of him, but he turned out to be an Ivan'—referring to Gotsinski's leaning to the White generals. Todorski also quotes *The Free Mountaineer*, the organ of the Dağistanli democrats published in Tiflis (no. 65, 6 December 1920) to the effect that 'the rising might have embraced the whole of Dağistan had it not been compromised in the eyes of the population from the first by the fact that it was headed by Najmudin Gotsinski and Kaitmas Alikhanov'.

Colonel Kaitmas Alikhanov, of Khunzakh, was a brother of the more famous General Alikhanov-Avarski, who had been the chief agent of the imperial government's repression of the Georgian insurrection in 1905. He had fought on the Russian side in a previous small rising in Dağistan, and had served in the Japanese War. He had taken the side of the Whites against the Bolsheviks, and, as has been mentioned, had held command for General Denikin in Avaria.

The chief of the insurgent forces in the field was a native of Dağistan, Colonel Jafarov, born at Kudal in the Gunib *okrug*.[2] He

[1] I.e. troops fighting for the maintenance of the holy laws (*Shariat*).

[2] The *okrug* (circuit) was the unit of administration in areas under military government (such as the Cossack provinces and Dağistan), as opposed to the *uyezd* (district) in areas under civil administration.

had served as an officer in the Dağistan Horse, had commanded a mountain regiment in the White armies, and had fought under General Wrangel.

The insurgent bands were formed in 'Menshevik' Georgia out of counter-revolutionary refugees from Dağistan, and, early in September 1920, they began to dribble quietly over the main chain into the upper valleys of the Andi and the Avar, where their ranks were rapidly augmented by contingents from the majority of the auls of the Avar, Andi and Gunib okrugs.

The original bands, numbering about 600, were soon swollen to a force of some 3000, and, as a result of their early victories, they managed to secure an adequate supply of small arms and ammunition; but they had no artillery, with the exception of a light battery captured from the Red army at the end of October, and this deficiency was to prove fatal to them as they advanced into the maritime districts.

The tribesmen were all excellent shots, brave to excess, and hardy as only mountaineers can be. Best of all, they knew every inch of the ground in a country as broken and difficult as almost any in the world. The great majority were on foot, but each band had its attendant troop or troops of horsemen.

The forces at the disposal of the Dağistan provincial war commissariat, which dealt with the rising until November 1920, consisted of the following:

(*a*) The 1st Dağistan Rifle Regiment, belonging to the separate Dağistan Brigade, and composed of Russian regulars, except the 9th Company, which was native. This regiment had been formed from the Soviet forces which had hung on Denikin's rear at Derbent towards the end of 1919, augmented by 150 deserters and prisoners of war from Denikin's own Dağistan Regiment. The officers were mostly of the old army; the regiment numbered 720 rank and file, with two Maxim and twelve Lewis guns, and was well disciplined and in all respects ready for the field.

(*b*) The 38th Rifle Battalion of the *Vokhra* (= *Voyska vnutrenniya okhrany*, 'Internal security force').

(*c*) The reserve squadron of the provincial militia.

(*d*) The Frontier Guard Regiment, which was worth very little as a fighting force, and local frontier-guard squadrons.

(*e*) A battalion of the 176th Rifle Regiment, transferred hurriedly from Azerbaijan.

Early in September news was received in Shura that insurgent bands were forming in the frontier districts bordering on Georgia.

As the main line of attack came down the Avar and the Andi, the Soviet forces made corresponding dispositions, and from the commencement of the struggle were based, as to the right flank on Botlikh,[1] as to the centre on Khunzakh,[2] and as to the left flank on Gunib.[3] On 8 September the 2nd Battalion of the 1st Dağistan Regiment (200 bayonets, one Maxim, five Lewis guns) reached Khunzakh, which was normally garrisoned by a squadron of mountaineer irregulars, under Atayev, war commissar of the Avar okrug. News of the invasion being confirmed, the remainder of the regiment, with the exception of the 3rd Battalion at Petrovsk, was also ordered to Khunzakh and, on 11 September, to the number of 250 bayonets, with one Maxim and five Lewis guns, started from Shura and, threading the Arakani defile, reached its destination two days later, detaching on the way one company (eighty-five bayonets and two Lewis guns) to Gunib, where it arrived on 14 September. The commander of the Dağistan Brigade was appointed chief of the mountain war area, including the Gunib, Avar and Andi okrugs.

The 1st Battalion of the 1st Dağistan Regiment was sent on from Khunzakh to Botlikh, where, on 18 September, having left fifteen bayonets to guard the bridge over the Andi-köysu at Tlokh,[4] it was received with music and dancing by some, at least, of the inhabitants of that important aul. Then, under order from the provincial commissar, the battalion, leaving forty bayonets and one Lewis gun in Botlikh to reinforce the garrison of 100 native partisans, on the 20th started up the river through Khushdada to Echeditl and on through Saantla to Kvarshi (Atelko) with the object of occupying the aul Shiitl on the western slope of the Bogos range, and at the same time of keeping watch over the auls Sagada, Gako and Shildi, near the upper waters of the Andi-köysu. A glance at the map will give an idea of the magnitude of the task thrown thus upon a very small force.[5]

[1] For a description of Botlikh, see Baddeley, *Rugged Flanks*, II, pp. 68–75.

[2] For Khunzakh, see Baddeley, *ibid.* II, p. 55; and for a panoramic view of the Khunzakh plateau, see Dechy, *Kaukasus*, II, opp. p. 188.

[3] For Gunib and the heights of Kegher, see Baddeley, *Rugged Flanks*, II, pp. 46–9; for a panoramic view of Kegher heights, see Dechy, *Kaukasus*, II, opp. p. 248, and for Gunib, *ibid.* p. 240; for ravine of Kara-köysu, *ibid.* p. 246.

[4] For a reference to Tlokh, see Baddeley, *Rugged Flanks*, II, p. 66.

[5] This region of the middle Andi-köysu was visited by Dechy and Merzbacher at the beginning of the century, since when it has known no other European travellers. Dechy (*Kaukasus*, II, pp. 186 and 235) reproduces views of the valley of the Andi-köysu. He was in Echeditl (Etschida) and describes the inhabitants as the roughest he had met in Dağistan (there is a fine group of some of these men, opp. p. 232). For a view of Shiitl (Schaiitl) see Merzbacher, *Hochregionen*, II, p. 468. Sagada was visited by Merzbacher (*ibid.* II, pp. 395 and 431). He reports

At the same time, from Khunzakh, 250 bayonets of the 2nd Battalion of the Dağistan Regiment (including a detachment of forty-five men from Gunib) with ten Lewis guns and 100 sabres of the Avar local Frontier Guards, marched up the Avar-köysu through the auls, Ghidatl and Erta, to the Antsukh guard-house to the west of the aul Khobok.[1]

Then came a series of Soviet disasters due, as admitted by Todorski, to military blunders, but, first and foremost, to gross underestimation of the strength of the enemy. The Botlikh command, leaving thirty-five bayonets at the hostile aul Khushdada (Khusad) and five bayonets with twenty sabres at the bridge of Echeditl, on 23 September, took possession, unopposed, of Kvarshi, but being reduced now to seventy bayonets and seventy sabres (native partisans) could get no farther, the slopes of the Bogos being occupied by about 300 of the enemy. A reconnoitring force lost a company commander and seven Russian soldiers, 'fate unknown'. Meantime, behind, the enemy took command at Echeditl, Sumada and Khushdada, whereupon, pressed front and rear, the small force of Russians began their retreat, during which their native partisans melted away until there were but twenty left, then, after a desperate forcing of the Echeditl bridge, only five. The forlorn remnant, hearing a rumour that Botlikh was now in the hands of the enemy, endeavoured to reach Khunzakh, but was surrounded at Karata and taken. Small detachments which had been left to guard the line of advance or retreat had already been destroyed. Tlokh with its bridge had been captured, the small Russian guard escaping to Khunzakh. To restore the situation, twenty native riders and fifty Russian infantry were sent towards Tlokh; but at Kharakhi they were surrounded, taken prisoners and killed. The Botlikh, or western command, no longer existed.

In the centre the main, Khunzakh, expeditionary force had made its way up the Avar-köysu to four versts north-east of Ghidatl, and had there found the enemy established in some strength on the mountainous banks of the river. A gallant attack by a small detachment gave the Russians illusory possession of Ghidatl, Rosnu and Erta, but further advance was found to be impossible, and now the enemy, numbering 1500 men, succeeded in surrounding the main force. With the utmost difficulty and by dint of hand-to-hand

that the people rose against the Russians in 1877 when their village was destroyed. Dechy (*op. cit.* II, pp. 222 and 230) gives views of the Icho valley and the bridge at Kvarshi. See also Merzbacher (*op. cit.* II, pp. 567, 569) for views of the wild country round Kvarshi. The Gako valley is illustrated by Dechy (*op. cit.* II, p. 234).

[1] For a view of the Avar-köysu from Khunzakh see Dechy, *op. cit.* II, p. 181.

fighting the Soviet troops made their way back to Rosnu, after losing in killed and wounded four officers and forty-nine men. On 25 September, hard pressed by the enemy, they continued their retreat to Khunzakh.

On the same day the provincial war commissar had issued a tardy order to maintain on the right flank the line Sagada-Echeditl-Kvarshi, in the centre the line Akhvakh-Samoda on each side of the Avar-köysu, and on the left flank the auls Tlerosh and Ilib (the first twenty, the second thirty-five, versts south-west of Gunib).

On 1 October the 3rd Battalion of the 1st Dağistan Regiment (100 bayonets and two machine guns) reached Khunzakh from Petrovsk and, on the 3rd, relieved the 2nd Battalion on the Khunzakh Pass, seven versts south-west of the aul. On the 5th, the insurgents took possession of Greater Gotsatl, fifteen versts south-east of Khunzakh, threatening thence the *chaussée* between that fortress and Karadağ.[1] On the 7th, a company of the 176th Regiment, sent to guard the Karadağ bridge, was partly captured and partly destroyed; on the 9th, the insurgents made an abortive attempt on Gunib and captured the important Salti bridge over the Kara-köysu, seven versts north of Gunib.[2]

> In this way [writes Todorski], at the very commencement of military operations the enemy had destroyed the Botlikh force, besieged the garrisons of Khunzakh and Gunib, and cut communications between the latter and Shura. The greater part of mountain Dağistan was in his hands, and soon afterwards he established his authority in Arakani, interrupting the shortest route between Khunzakh and Shura. The only road in mountain Dağistan still available to the Red army was the *chaussée* Shura-Khodjal-Makhi-Salti; and the enemy, having taken possession of the Andi, Avar and Gunib okrugs, had a good base whereon to gather and equip his forces which soon reached a total of 3000 fighting men. (P. 67.)

It was now realized that the suppression of the Dağistan revolt was no easy matter, but the provincial war commissar went on blundering. The rest of the 176th Regiment came to reinforce him from Azerbaijan (one battalion of ninety-four bayonets and four machine guns, another of 350 bayonets and ten machine guns). Together with the 38th *Vokhra* Battalion, a reserve squadron of local

[1] For Karadağ defile, see Baddeley, *Rugged Flanks*, II, pp. 50, 52; Dechy, *op. cit.* II, p. 190 (view).

[2] For the importance of Salti bridge and road junction, see Baddeley, *Rugged Flanks*, II, p. 45. It was captured by Kazi-Kumukh insurgents during the abortive rising against the Russians in the summer of 1877. For views of the ravine of the Kara-köysu near Gunib, see Dechy, *op. cit.* II, pp. 191, 192.

militia and a battery of light artillery, these troops were sent towards Gunib with orders to occupy Arakani and the bridge over the Avar-köysu and also to secure control of the Arakani heights (9 October).[1] With the Arakani force was a supply column for besieged Khunzakh.

From the Eleventh Army at Baku (commanded by Comrade Hekker), was also detached the 95th Rifle Brigade (of the 32nd Rifle Division—Todorski's) consisting of the 283rd and 284th Rifle Regiments, and the 3rd Light Artillery Group. These troops reached Shura on 13 October. The commander of the 95th Brigade was ordered to hand over the 283rd Regiment and one battery to the garrison commander in Shura, and to proceed himself with the 284th Regiment to Khodjal-Makhi, there to take command of all the forces in mountain Dağistan—now called the Gunib command.

The first efforts of the Gunib command met with some success. Having joined forces, the partisan commandos of the native Comrades Bogatyrev and Karayev, took possession of Tchokh, ten versts south-east of Gunib; the 284th Regiment attained the aul of Salti, eight versts north-east of Gunib, and the 176th Regiment reached Salti post-station on the Kara-köysu. On 17 October the aul Kegher, five versts east of Gunib, was taken after fierce street fighting, and the same day the bridge at Gunib and the aul of Kharta, nine versts north-west of Gunib, were likewise occupied. Gunib was relieved, and the next problem was the relief of Khunzakh, but the advance proved difficult, and only on 2 and 4 November were communications opened with the beleaguered garrison, from Golotl and Usdal-loso respectively. The garrison had suffered considerable privations during a thirty days' siege, and it seems to have deserved the credit lavishly bestowed upon it. It was, unfortunately, characteristic of the disunited mountaineers that the Russians in Khunzakh fort were able to hire spies in the neighbouring aul of the same name, at the price of a few yards of cotton cloth or a few pounds of sugar, and learn all that was going on in the world outside.

In the direction of the relief of Gunib and Khunzakh success had attended the efforts of the Red army, but meanwhile a serious disaster had overtaken the Arakani field force. This force had left Shura on 9 October. Several days had been spent in partial approaches to Arakani by way of the famous defile of the same name when the provincial war commissar in Shura gave orders from that distance that one battalion with a troop of light artillery should watch the aul,

[1] For a view and description of Arakani (Arakanee), see Baddeley, *Rugged Flanks*, II, pp. 36–8.

while the rest of the force should march by the shortest route to Gherghebil, and from there, circumventing Arakani, attack it from the south-west. On 20 October, one battalion of the 183rd Regiment was surrounded by the insurgents at Kodutl and thrown back with heavy losses. This put an end to the circumventing movement. On 23 October, the insurgents took the offensive and beat back the Russians to Arkaz. Under fresh attacks, while the artillery fought well, the infantry lost nerve and failed to support it. As a result of the fighting the Arakani field force was reduced to inaction in a hollow between mountain ranges. On the 28th the enemy again attacked, but was once more kept at bay by the artillery and the force was slightly strengthened by the arrival of 100 local communist levies from Shura.

At dawn on 30 October the mountaineers made a frontal attack, sending at the same time a detachment to outflank the Russian left, and then quite unexpectedly attacked them in the rear as well, driving off the guard of the baggage-train, occupying Arkaz, and creating a panic. The whole Arakani field force was surrounded, the greater portion of it destroyed and the rest taken prisoner. In this fight, 700 men of the Red army were killed, and the insurgents took four guns, twenty-four machine guns and a large quantity of ammunition. The representative of the Dağistan revolutionary committee, Comrade Dudarov, and nearly all the officers—political and military—perished.

This reverse compelled the commander of the Eleventh Army in Baku, Comrade Hekker, to give serious attention to the situation in Dağistan, and the decision was taken to send further reinforcements and to transfer the whole business of the suppression of the rebellion from the provincial war commissariat to the commanders in the field. On 3 November, Comrade Todorski, commanding the 32nd Rifle Division, arrived in Shura from Baku with his divisional staff. The forces in Dağistan were reinforced by the 282nd and 285th Rifle Regiments belonging to the division, and by a cavalry brigade and a mounted mountain battery.

The inhabitants of Gimri[1] had adopted a professedly neutral attitude, which the Red army had refused to recognize as such; but, as this aul was almost inaccessible, it was merely kept under observation from the neighbouring north-easterly heights (Karanai)[2] until Christmas Day 1920, when the Red artillery commenced a siege, described as the most difficult and onerous operation undertaken by that arm during the whole campaign. It lasted until 9 February 1921, when again

[1] For a description of Gimri, see Baddeley, *Rugged Flanks*, II, pp. 19–25.

[2] For Karanai as an observation post, see *ibid.* II, p. 16.

for a time the birthplace of Shamyl was ordered to be kept under observation.

The whole of the Soviet troops in mountain Dağistan (formerly the Gunib command) became now the Botlikh relief force, with orders to act from Khunzakh against (*a*) the auls, Tchalda, Mogokh, Arakani, and Butsro, Shakhada, Moksok, to the east and north; (*b*) in the direction Tlokh-Botlikh in the centre; (*c*) along the Avaro-Kakhetinski road (up the valley of the Avar-köysu) to the south-west; and (*d*) from Gunib against the auls, Rugja and Tliakh. A good deal of obstinate fighting took place in the second half of November and the Russian objectives in the region of Khunzakh were mostly attained. At the same time the Khunzakh garrison made some little progress in the direction of Botlikh, occupying the post-station Matlas[1] and other points. The Avaro-Kakhetinski force also advanced, and the 176th Regiment, acting from Gunib, had taken the auls, Kula, Batsada, Unti (north-east of Batsada) and Shulani, but could make no further progress to the south-west as intended.

Further and notable successes of the mountaineers now followed. On 9 November, the 1st Model Revolutionary Discipline Rifle Regiment left Grozny in Chechnia with the object of co-operating in the relief of Botlikh from the north. The regiment, with a strength of three battalions and a machine-gun command, marched without mounted scouts. At Veden they left 100 effectives, either shoeless or sick, and after a rest of two days moved in the direction of Botlikh. South of the aul Khorochoi, they were fired upon by a band of 150 insurgents, but beat them off, and next day reached Khoi, where one company was left in reserve. On the 16th the regiment entered Botlikh[2] without opposition, and, on the 18th, two companies with the regimental commissar marched to neighbouring Muni with the object of recapturing two guns, said to be inadequately guarded by the enemy. After a brief exchange of shots the enemy retired to Orta-kolo, taking the guns with him.

The inhabitants of Muni were put under contribution for provisions and forage, hostages were taken and disarmament carried out; all this, added

[1] For the situation of Matlas, see Baddeley, *Rugged Flanks*, II, p. 58.

[2] The road from Botlikh through Khoi and over the Kerket Pass (2220 m.) to Khorochoi, is described by Dechy, *Kaukasus*, II, pp. 196–7. East of the Khoi-Kerket stretch lies Esen-am, the only considerable sheet of water on the northern border of Dağistan. The lake is 3 km. long and 1 km. at its greatest width with a depth of 60–80 m. Dechy gives a map of the lake (p. 199). To the south-east of Esen-am is a second lake, Ardshi-am, only 600 m. in length. Baddeley (*Rugged Flanks*, II, pp. 84–7) describes the alternative road from Botlikh through Andi. For a view of Andi aul see Dechy, *op. cit.* II, p. 184.

to cases of arbitrary requisition by the Red army soldiers of eatables and small articles of value, created a hostile attitude on the part of the villagers. (Todorski, p. 93.)

The expeditionary force, numbering 250 bayonets with three machine guns, next marched to Orta-kolo in pursuit of the same two guns and, having dispersed the defenders and occupied the aul, proceeded to make search for the guns, without taking any reasonable precautions. Suddenly the enemy made his appearance at the Russian rear and, aided by the exasperated villagers of Muni, intercepted the line of retreat to Botlikh. The force failed to cut its way through and was totally destroyed. This event had a depressing effect on the rest of the troops in Botlikh, and the enemy, aware of this fact, took the offensive and surrounded the place.

The officers of the 1st Model Revolutionary Discipline Rifle Regiment, including the commander, entered into negotiations with the enemy's leaders, who demanded the unconditional surrender of all weapons, in return for a promise of personal safety and an open road to Grozny; the officers persuaded the soldiers to give up their arms, though the latter, fearing treachery, protested for a long time. Their fear was justified. The insurgents, bursting into Botlikh, stripped the unarmed Russians and massacred them. A few individuals only managed to escape, and made their way naked to Veden, with frost-bitten extremities. There had been in Botlikh not less than 600–700 men of the Red army, with nine machine guns and a substantial supply of ammunition. This success very naturally raised the spirits and hopes of the mountaineers and added materially to their rather scanty supply of munitions, while at the same time it had a serious effect on the morale of the Russians in other parts of Dağistan, more especially since they had received news, as late as 23 November, from the administration of the Terek province, of a 'successful' advance on Botlikh and Muni.

In the main area of fighting the insurgents continued to secure some remarkable successes. On 29–30 November, the 282nd and 284th Regiments made attempts to occupy the country to the north-east of the Karadağ-Tlokh road; the first of these regiments, finding its way barred, entrenched itself on the line Gotso-Mogokh and the Mogokh bridge. One battalion of the 284th occupied Tsatanikh, while the other, caught in a blizzard, lost its way and, having been taken by the enemy in flanks and rear at Moksok, was shattered. Of 148 fighting men, only forty-seven rank and file and three officers rejoined the 1st Battalion after its retreat from Tsatanikh.

About the same time a series of reconnaissances in force was made all along the front, but the enemy was quick to take the initiative. A company of the 280th Regiment (forty-eight bayonets), with the battalion commander, Comrade Chernishev, was surrounded in Erkachi and wiped out. The rest of the regiment was surprised at Gotso and driven back to Greater Gotsatl with heavy losses; likewise the 284th Regiment from Shakhada to Ghenechutl (between Butsro and Khunzakh). An insurgent attack on Matlas was repulsed, but in nearby Mochok they surrounded and killed the native communist, Bogatyrev, and the whole of his partisan commando.

The commander in the field, Todorski, was compelled to reduce his objectives to the defence of Khunzakh and Gunib, the former of which counted as a fortress, the latter as a fort only. At dawn on 3 December, the enemy in force attacked and drove the 282nd Regiment out from Greater Gotsatl. By the second week in December both Khunzakh and Gunib were again isolated and besieged, the mountaineers having taken possession of Gherghebil and Aimaki. Pursuing their advantage, the mountaineers next moved against Khodjal-Makhi (10 December), and, after hand-to-hand fighting, drove out a battalion of the 176th Regiment, a squadron of cavalry and a company of the 283rd Regiment. Then, 1000 bayonets strong, they entered Levashi on the heels of the retreating Russians, inflicting on them a further loss of 150 killed. The Red troops took refuge at Urma, but the too rapid advance of the mountaineers could not be maintained and, retreating in turn, they allowed the Russians to re-enter Levashi.

The capture of Aimaki threatened the main Shura *chaussée*, and presently the insurgents occupied the auls, Oukhli (Oğli) and Kuletsma, and the road itself in their neighbourhood. But, chiefly through the use of armoured cars, which could operate along the *chaussée*, the Russians succeeded in driving the insurgents out of both auls, and they retired with some losses to Aimaki.

The advance of the insurgents towards Shura encouraged the tribesmen in the Kazi-Kumukh okrug where, under Ibrahim Haci, they occupied Tchokh (18 December) and, through Mukar, the aul of Kumukh itself (24 December).

On 28 December, the 283rd Regiment (which had been brought up to a strength of 356 bayonets and twenty-three sabres with nine machine guns) reached Levashi from Shura, together with a troop of light artillery and the 281st Regiment, numbering 300 bayonets and twelve machine guns, newly arrived from Baku.

The inadequate provisioning and ammunition supply of Gunib and

Khunzakh compelled Todorski to make every effort to press forward the relief of these two places. With this object a striking group was formed at Levashi of the 281st and 283rd Regiments, two batteries of light artillery and two armoured cars. At dawn (20 December) this force, under Comrade Krivonosov, set out for Khodjal-Makhi, while a small detachment by way of support took up a position on the Aimaki heights, and thence shelled that aul. The mountaineers allowed the Russians to approach Khodjal-Makhi unopposed, and then opened a hurricane fire, sallied forth and drove them back helter-skelter to Levashi, where the latter maintained themselves only owing to the help of the armoured cars. This failure was due to the Russian advance along the high road through the defile being made without adequate covering from the heights on either side. The Russians lost a large proportion of officers and about 160 rank and file; the insurgent losses were chiefly due to the fire of the armoured cars. The insurgents proceeded to occupy Tchauuli and resisted successfully two attempts to dislodge them (25 and 26 December). During the following three days they made desperate attempts on Levashi, but were driven back with considerable losses—again, admittedly, by the fire of the artillery and armoured cars. It seems evident that along this main road, where both artillery and armoured cars could operate—as was not often the case in mountain Dağistan—the attempts to dislodge and compel the further retreat of the Red army were really hopeless; and in all probability their failure had much to do with the subsequent collapse of the rising. Todorski himself writes that 'the last days of 1920 mark the turning point of the insurrection in Dağistan—from insurgent success to insurgent failure'.

Meantime fresh troops from the reserve of the 'Caucasian front'[1] had been detailed to reinforce the much tried 32nd Division. The 52nd Rifle Brigade occupied Deshlagar (154th Regiment) and Madjalis (156th Regiment) in southern Dağistan, the staff of the brigade being stationed at Derbent until 20 February, when the brigade was transferred hurriedly by rail to Azerbaijan (for the invasion of the territory of the Georgian Republic in the last week of the month). The Eleventh Army command also placed at Todorski's disposal the 14th Rifle Division (minus its cavalry brigade), of which the 40th Brigade reached Dağistan on 26 December, the 41st on 1 January 1921, and the 42nd on the following day. To the north, in the Terek province,

[1] The term 'Caucasian front' is not quite clear in the text of Todorski's book. It apparently comprised the Eleventh Army (Baku) and the Ninth (or Kuban) Army in the northern Caucasus.

the 97th Rifle Brigade was concentrated at Veden (26 December). Finally, the 32nd Rifle Division was directly reinforced by the 2nd Moscow Brigade of *kursanty*,[1] consisting of two rifle regiments and one cavalry regiment, a reserve rifle battalion, and one light (four-gun) and one heavy (two-gun) battery; but in view of the valuable nature of the *kursanty* element, the army commander forbade their use in any serious action.

These reinforcements enabled Todorski to take energetic measures at last for the relief of Gunib, the attack being made along three lines, Gimri, Aimaki and Khodjal-Makhi. The Gimri movement resulted in nothing more than an artillery bombardment; the Aimaki movement was subsidiary only; but the main attack along the *chaussée* to Khodjal-Makhi was fully successful, thanks to artillery and armoured cars. On the morning of 2 January this important aul, together with 140 prisoners of war, fell finally into the hands of the Red army. This was the first serious loss to the insurgents in the whole campaign and marked the beginning of the end, although heroic efforts delayed final defeat into the month of May. Communications were opened with Gunib, and that garrison was relieved.

The relief—for the second time—of Khunzakh was now urgent, and with the increased strength of the Red forces it had become possible; but the disposition of the insurgent troops made it indispensable to capture first their commanding position at Gherghebil. The heroic and successful defence of this aul in 1847 against Prince Vorontsov was, perhaps, the most stirring event in all the thirty years of the Murid War,[2] and the Russian losses on that occasion might well be remembered by the opposing forces, with exultation on the one side, and on the other with foreboding. Until Gherghebil was taken the other operations of the army in the field could only be subsidiary: Gimri was to be watched from the heights above it, Arakani to be reconnoitred, the country round Gunib and down the Kara-köysu to be held as far as the post-station at Salti, and the high road to Khunzakh if possible cleared of the enemy.

The first attack was made on Gherghebil and Aimaki by the 41st Brigade (121st, 122nd, 123rd Regiments, numbering 2686 effectives and thirty-six machine guns) on 8 January, in three columns, right, centre and left. Of these, the right made for Aimaki and, after a preliminary success, was driven off with the loss of the heights between Akhkent and Oukhli; the centre column drove the enemy to the east

[1] A corps of selected men undergoing an officers' training course (*kurs*).

[2] See Baddeley, *Conquest*, pp. 429–33.

of Gherghebil; but the left column, neglecting to crown the heights as it forced its way along the Kara-köysu, was entrapped and driven back in disorder to Kuppa, held by the 122nd Regiment, with the loss of 292 killed and wounded, including 21 officers.

In a renewed attack on 9 January the 121st and 281st Regiments, after heavy fighting, were driven off, and occupied some heights two versts to the north-east of the auls, while two battalions of the 122nd Regiment and one battalion of the 123rd entrenched themselves on another height (1663 m.). Their losses were seventy killed and wounded.

On 10, 11 and 12 January, after obstinate fighting, the 122nd and 123rd Regiments failed to make any progress against Gherghebil from the east or south; the 118th Regiment, supported by artillery, made a demonstration against Kharta; the 119th Regiment made an abortive attempt to capture the Salti bridge, and the 281st and 121st Regiments retreated on the 12th to Oukhli and Akhkent. The losses on 11 and 12 January were 120 killed and wounded.

Similar attempts, accompanied by obstinate fighting, continued up to 20 January with considerable improvement in the Russian position, since certain of the heights in close proximity to Gherghebil and Aimaki were taken. Then came lengthy secret instructions from the commander of the Ninth Army at Baku (Comrade Hekker) to Todorski, insisting that his objective should be changed from Gherghebil and Aimaki to Karadağ. Todorski, in reply, gave his very decided reasons for rejecting the proposed plan of operations, based chiefly on the impracticability of negotiating the existing roads and paths, all of which led through narrow defiles or along perpendicular river banks which were absolutely impassable to an advance in force. Hekker only gave way after a visit in person to Todorski, and an inspection, with him, of the Salti gorge and bridge, which, stripped of its roadway by the enemy, seemed little adapted for use in an attack on Karadağ. As to the heavy losses occasioned by the repeated unsuccessful attacks on important points such as Gherghebil, Todorski acutely remarks that, in the long run, these attacks were more than justified, because in mountain warfare it was just such places, as in the case of Khodjal-Makhi and others later on, which gave the most valuable results, and for the very reason that they had been so difficult to take: the localities became known to the last and least footpath; the weakest points in the defence were found out; the natives, when at last beaten, were profoundly discouraged; and whole regions gave in.

The army commander gave way to Todorski in all except his insistence that the attack on the Salti bridge should be something more than a demonstration. The operations initiated by Todorski were carried on with varying success up to 24 January, when the bridge was captured by a partisan commando but retaken the same day by the insurgents.

Gherghebil was now closely beleaguered on all sides and, against modern artillery and high explosive, defence was no longer possible. On the 25th the final assault took place; but desperate fighting failed to give actual possession of the aul, and, as the losses by day were considerable, the troops were ordered to maintain their positions and wait for night to make a final rush with alternate battalions, the others standing to arms in reserve. At 2 a.m. on the 26th rockets gave the signal for attack, and immediately from three sides (east, south-east and south) the Russians entered the aul, the hand-grenadiers leading. The defenders, taken by surprise, fell into confusion. Some escaped in the direction of Kodutl, others kept on firing from their stone houses. One group occupied the mosque and adjoining houses. The majority of the stone houses were taken by force, chiefly by the aid of hand-grenades which were dropped down the chimneys from the roofs and burst in the dwelling rooms. The insurgents occupying the mosque held out obstinately until 10 a.m., and in the attempt to dislodge them the attackers lost fifteen killed and a number wounded, after which orders were given to set fire to the mosque and other buildings, in which more than a hundred of the defenders were killed or burned to death. The 'liquidation' (a sinister term in general use by the Red army) of the enemy in the aul of Gherghebil lasted until 6 p.m. All the women and children belonging to the aul, together with all property, had been evacuated by the enemy beforehand to the auls on the left bank of the Avar-köysu.

During the final attack on the aul the losses of the Red army had been less than 100, and the enemy had several times that number killed. Nine machine guns were taken. Kharta and Kikuni were occupied the same day, and Aimaki, no longer tenable, on the 27th. Koroda, the Salti bridge, Murada, Darada, Maali and Karadağ fell into the hands of the Red troops without further fighting. On the 28th, a mounted detachment rode in from Khunzakh, and the siege of that vitally important position was for the second time raised. The fortress had now been isolated for seven weeks—since 7 December 1920. The garrison numbered 2000 men; the attacks had been half-hearted, and indeed, in such operations, the mountaineers, without siege guns and lacking the training and discipline necessary for formal fighting of this character, were at their worst. The real danger to the garrison had arisen from its scanty stocks of both provisions and munitions of war.

The main objectives were now Arakani and Gimri. So long as the former place held out, the latter could only be approached down the 5000 feet of precipitous rock between it and Karanai—the eastern bank, so to speak, of the three united köysus which, with the addition of the Andi-köysu here, just below Gimri become the Sulak. The defence of Arakani in this, its final, stage depended mainly on holding the far-away auls of Kuiada and Tchalda, the second of which changed hands five times in the five days from 31 January to 4 February. On the 8th, both were finally captured and the Red army turned its attention to the heights surrounding Arakani itself, the aul Kodutl being taken on 10 February. The final assault took place on the 13th, and on the morning of the 14th Arakani fell.

In accordance with Todorski's dictum, cited above, other positions were surrendered or abandoned one after the other. The Russians were now masters of all the more accessible portions of north-east Dağistan with the exception of Gimri, and on 18 February battered Gimri was occupied, without fighting, by way of Untsukul,[1] a large and turbulent aul, frequently hostile in the past to its more celebrated neighbour and at this period more revolutionary (Red), perhaps, than any other place in the whole of Dağistan. The Red troops had no sooner entered Untsukul than a Soviet partisan commando, 400 bayonets strong, was formed from its inhabitants under the communist Hasanov. In the Gunib district the recalcitrant aul of Rugja had been taken on 8 February, Sogratl on the 14th, and many others on both sides of the Avar-köysu with little delay.

The mountaineers still held out north-west of Khunzakh in the region between Ikhali and Tlokh on the Andi-köysu, with advance posts on the line through Mochok, Matlas post-station and Siyuk (twelve versts north-west of Khunzakh). They were also still in possession of all western Dağistan.

Nearer at hand, Ashiltà[2] was occupied by the *kursanty* the day that Gimri fell; but these cherished troops, who were on no account to be sacrificed, fell into a trap: 125 men of their 5th Regiment entered

[1] For a description of Untsukul, see Baddeley, *Rugged Flanks*, II, p. 35. The inhabitants were noted for their inlay work and, according to Baddeley, men from Untsukul attended the Glasgow Exhibition of 1911 to sell their wares.

[2] For description of Ashiltà, 'an earthly paradise', see *ibid.* II, pp. 26–8.

Ashiltà unopposed and piled arms; during the night insurgents from without joined the villagers within, and between them the *kursanty* were destroyed to a man.

On 19 February the position at Ashiltà was restored by the 2nd battalion of the same regiment, various detachments hurried up from Gimri and a battalion of the 126th regiment from Untsukul. In Ashiltà were found 52 mutilated bodies of the *kursanty*. (Todorski, p. 140.)

So far as can be judged, there is no reason to doubt Todorski's figures and statements, but it would be interesting to know what roused this unusual exasperation in the inhabitants of Ashiltà. The other side of the story, or most of it, as in all native wars, remains for ever untold.

The remaining auls of the region—Kakhab-roso, Betl, Tsatanikh, Ikhali, Chirket—were all taken by 11 March, and Artlukh, Danukh, Arguani and Gadari on the 13th, the victors thus gaining possession of north-east Dağistan up to the borders of Chechnia.

Farther west the enemy continued to resist through March, although resistance was now futile. Auls and dominating heights fell in rapid succession into the hands of the Red troops; Botlikh itself surrendered without a shot fired on 5 March, and Andi and other places on the 9th, giving possession of all the northern part of Dağistan up to the Chechen border.

Meantime, an event of supreme importance had occurred, one that of itself made all further resistance hopeless. Tiflis was taken by the Red army on 25 February and Batum on 17 March, and the free and independent Social Democratic Republic of Georgia became, perforce, a Socialist Soviet Republic. Several small Red columns were now organized in Georgia and sent north-east over the main chain of the Caucasus, and the last heroic remnants of the insurgent bands, under leaders not unworthy to be called the successors of Shamyl, were surrounded on all sides by overwhelming forces. The wild and mountainous nature of the border country in the west and the remaining winter snows enabled small bands of the insurgents to keep up a desultory resistance—chiefly in the Dido country—up to the fourth week in May, but then the end came.

The 'insurrection' or 'rebellion' in Dağistan was 'liquidated'. The remaining insurgents dispersed, to regain their auls, if at all, as individual fugitives.

Colonel Kaitmas Alikhanov and his three sons were taken prisoners and all four killed. Later, Colonel Jafarov surrendered voluntarily at Shura. Colonel Ömer Piralov was likewise killed, later on. Lieutenant Hasan

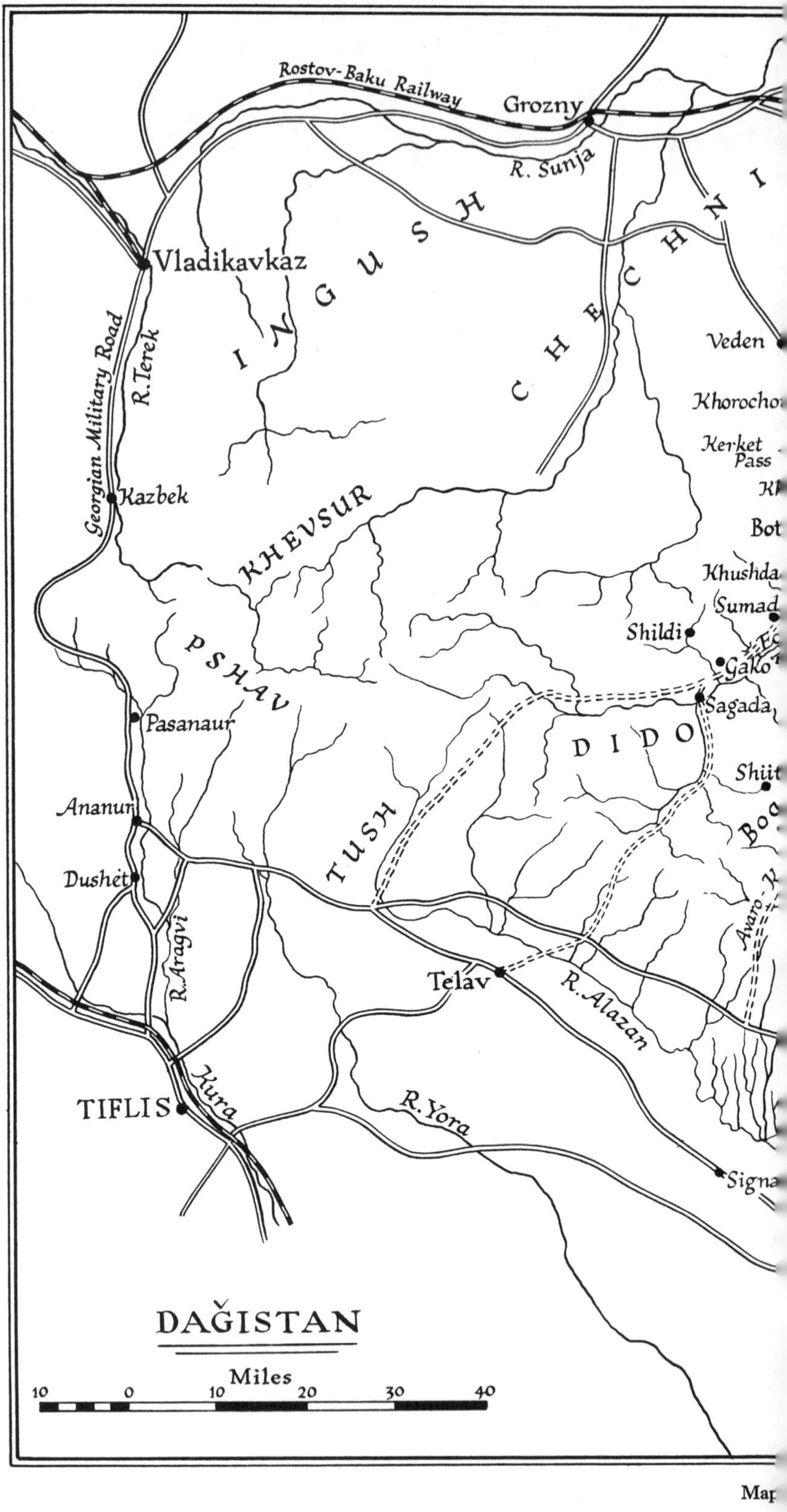
Rostov-Baku Railway
Grozny
R. Sunja
INGUSH
CHECHNI
Vladikavkaz
Georgian Military Road
R. Terek
Veden
Kerket Pass
Kazbek
KHEVSUR
Khushda
Shildi
Gako
Sagada
PSHAV
Pasanaur
DIDO
TUSH
Ananur
Dushet
R. Aragvi
Telav
R. Alazan
Kura
TIFLIS
R. Yora
DAĞISTAN
Miles
10
0
10
20
30
40

Map

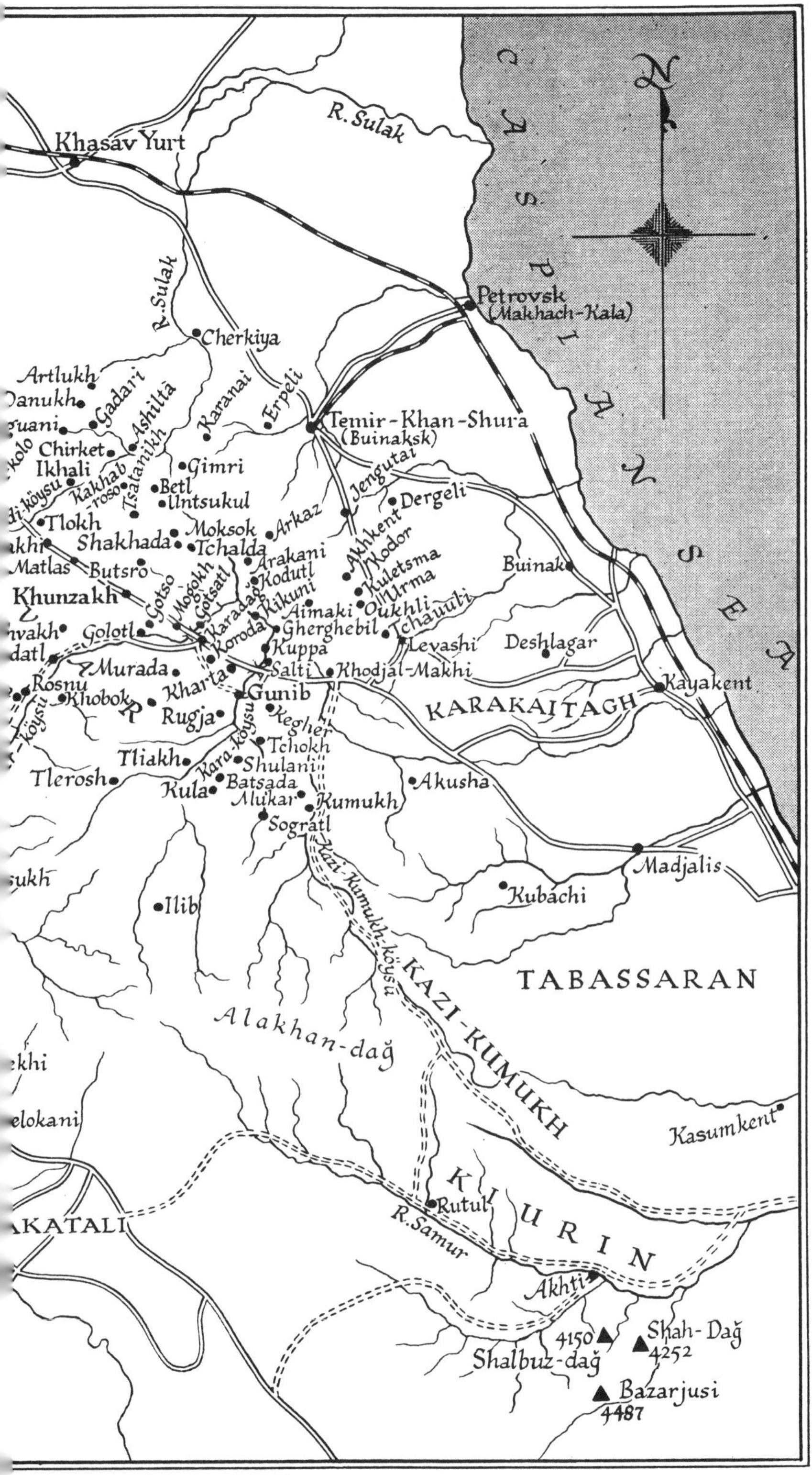
CASPIAN SEA
Khasav Yurt
R. Sulak
R. Sulak
Petrovsk
(Makhach-Kala)
Cherkiya
Artlukh
Gadari
Ashilta
Karanai
Erpeli
Temir-Khan-Shura
(Buinaksk)
Chirket
Ikhali
Tsatanikh
Gimri
Kakhab-roso
Betl
Untsukul
Jengutai
Dergeli
Tlokh
Moksok
Arkaz
Akhkent
Kodor
Shakhada
Tchalda
Arakani
Kuletsma
Buinak
Matlas
Butsro
Kodutl
Urma
Khunzakh
Gotso
Mogokh
Gotsatl
Aimaki
Oukhli
Tchaudi
Golotl
Karadağ
Kikuni
Gherghebil
Levashi
Deshlagar
Koroda
Kuppa
Murada
Salti
Khodjal-Makhi
Kayakent
Rosnu
Khobok
Kharta
Gunib
KARAKAITAGH
Rugja
Kegher
Kara-köysu
Tchokh
Tliakh
Shulani
Tlerosh
Kula
Batsada
Akusha
Alukar
Kumukh
Sogratl
Madjalis
Kazi-Kumukh-köysu
Kubachi
Ilib
TABASSARAN
Alakhan-dağ
KAZI-KUMUKH
Kasumkent
KIURIN
Rutul
R. Samur
Akhti
4150
Shah-Dağ
4252
Shalbuz-dağ
Bazarjusi
4487

917–22

Abakarov, taken prisoner, threw himself off the Ghidatl bridge into the Avar-köysu and was drowned. The head of the movement, Najmudin Gotsinski, went into hiding in Dağistan. Said Bey, Shamyl's grandson, fled to Turkey. (Todorski, p. 158.)

Todorski explains that, with recently sovietized Azerbaijan on the one side (where the whole of XI (Red) Army was stationed), with Menshevik Georgia striving to the utmost 'to oppose the advancing wave of proletarian revolution' on another, with expectant Muslim Chechnia on a third, and with the remaining (north-western) side barred by impracticable mountain ranges held by the Tushes and Khevsurs, Dağistan constituted a position of singular importance, especially as the only railway connecting Russia with the Baku oilfields ran through it.

Local conditions—the mountainous character of the country, the severity of the climate, the absence of roads, the prevailing poverty, and the warlike qualities and training of the inhabitants—made the Red army's task a heavy one; and the initial failure was the result, primarily, of underestimating the fighting powers of its opponents.

As to the Red army itself, the nine months' obstinate fighting, in which it lost 5000 men, proved sufficiently its qualities, good and bad. (P. 159.)

Politically, the movement of 1920–1 depended, even for the limited object of its indefinite prolongation, on the maintenance of the Georgian frontier in the rear, and for any remote chance of ultimate success on a triumphant outcome of Wrangel's adventure. In fact, the movement can only be considered to have been at all justifiable in the military-political sense if it be considered in relation both to the position in Transcaucasia, where conditions were in a state of flux, and in Poland and the Crimea, where in the early autumn of 1920 important elements continued to refuse to recognize the success of the Soviets against Denikin as final.

Tactically, the leaders of the insurrection, who showed considerable ability, made the cardinal blunder of using their slender forces against the Shura *chaussée*. They thus came down to fight on Russian ground rather than waiting for renewed Russian attacks on their own ground. Had they chosen, instead of taking the (abortive) offensive at Levashi, to await a Russian advance into the mountains, and had they succeeded in inflicting on the Reds (as they might well have done) the type of disaster which overtook Prince Vorontsov at Dargo in 1845, the results of such a victory might well have helped to alter the position in Transcaucasia in the spring of 1921.

BIBLIOGRAPHICAL AND SUPPLEMENTARY NOTES

(1)

GEOGRAPHY, TOPOGRAPHY AND MILITARY HISTORY: GENERAL

Documentary sources for the military history of Caucasia and eastern Anatolia are in the Tsentral'ny Arkhiv Department in Tiflis and in the archives of the Turkish Ministry of War. The principal Russian printed sources have been based on the Russian archives, but so far only a beginning has been made of the collation and printing of Turkish sources. Special studies of individual operations and unit histories have been printed from time to time in Turkish service magazines, but these, always difficult to obtain, have already become bibliographical rarities.

Works which the authors were unable to consult are given for purposes of record and are marked with an asterisk in the present bibliography.

(*a*) Bibliographies and Encyclopaedias

(*Abbreviations used for reference in text are in square brackets.*)

Miansarov, *Bibliographia Caucasica et Transcaucasica*, vol. 1, sections 1 and 2, St Petersburg, 1874–6. [Miansarov.]

Contains sections on military history, geography and topography.

*Gizetti, A. L., *Bibliograficheski ukazatel' pechatannym na russkom yazyke sochineniyam i statyam o voyennykh deystviyakh russkikh voysk na Kavkaze*, St Petersburg, 1901.

Encyclopaedia Britannica (14th ed.). [*E.B.*]

Encyclopaedia of Islam. [*E.I.*]

Islam Ansiklopedisi. [*I.A.*]

The Turkish edition, of which some fifty-one parts have appeared to 1951, contains numerous articles by Turkish and Caucasian authors on east Anatolian and Caucasian subjects which are important.

Bol'shaya Sovetskaya Entsiklopediya. [*B.S.E.*]

Complete in sixty-five volumes (1929–47). Appeared in odd numbers at irregular intervals (e.g. vol. LXV was published in 1933, vol. XXXV in 1947). The first edition appears to have been withdrawn from the market and a new and revised edition is in course of publication. A double volume covering U.S.S.R. was issued in 1948. The first edition contains articles, including useful bibliographies, on Caucasian subjects by leading Caucasian and Russian writers.

Dictionary of National Biography. [*D.N.B.*]

Russki Biograficheski Slovar'. [*R.B.S.*]

Published by the Russian Historical Society, twenty-five volumes to 1918, after which date the series was not continued. The volumes did not appear in strict alphabetical sequence and there are several gaps. Nevertheless, the series is a valuable source for Russian military history to the end of the nineteenth century.

Türk Meşhurlari Ansiklopedisi (Encyclopaedia of Famous Turks), 1 vol., Istanbul, 1948. [*T.M.A.*]

Contains popular biographies, often with portraits, of Turkish notables but lacks any selective balance: while numerous notables of the nineteenth century are omitted or given cursory treatment, many nonentities, still living, receive long and fulsome biographies.

(*b*) Documents and Periodical Publications

Akty sobranniye kavkazskoyu arkheograficheskoyu kommissiyeyu, Arkhiv upravleniya Namestnika Kavkazskago, 12 vols., Tiflis, 1866–1906. [*Akty*]

Contains a mass of original documents relating to the Russian occupation of Georgia and the Caucasus down to the surrender of Shamyl (1864); it includes a series of interesting portraits of Russian viceroys and also members of the old Georgian ruling house of Bagrationi.

Izvestiya kavkazskago otdela Imp. russkago geograficheskago obshchestva, Tiflis, from 1872. [*Iz. Kav. Ot. I.R.G.O.*]

Contains valuable papers on natural and political geography, topography and anthropology.

Kavkazski Kalendar', for the years 1854–1916 (Tipografiya Kantselyarii Namestnika, Tiflis). [*Kav. Kal.*]

Kavkazski Sbornik, 32 vols. to 1912. [*Kav. Sbor.*]

Published under the auspices of Grand Duke Michael Nikolayevich and of the commander-in-chief, Caucasian Army. It contains many important articles on geography and anthropology of Caucasia.

Novy Vostok. [*N.V.*]

Published by the Scientific Association of Eastern Studies of U.S.S.R., Moscow, 28 numbers in 21 vols., 1922–8.

Russkaya Starina, St Petersburg, 170 volumes. [*R.S.*]

Covers the years 1870–1918, when this valuable periodical ceased publication. *R.S.* contains many important articles on the military history of the Caucasus.

Sbornik materyalov dlya opisaniya plemen i mestnostey Kavkaza, vols. I–XLIV, Tiflis, 1881–1915. Vols. XLV and XLVI published after 1920 in Makhach-Kala. [*Sbor. Mat.*]

Sbornik svedeniy o kavkazskikh gortsakh, 10 vols., Tiflis, 1867–76. [*Sbor. Gorts.*]

Published under the auspices of the commander-in-chief, Caucasian Army. Important for the study of the tribes of Dağistan.

Sbornik svedeniy o Kavkaze, edited by N. Zeidlitz, 9 vols., Tiflis, 1871–85. [*Sbor. Sved.*]

Contains numerous articles on topography and demography.

Zapiski kavkazskago otdela Imp. russkago geograficheskago obshchestva, vols. I–XVIII, Tiflis, 1872–96. [*Zap. Kav. Ot. I.R.G.O.*]
Further volumes were published after 1896.

Georgica: a Journal of Georgian and Caucasian Studies, vols. I–V, 1935–7.

See also articles in *Kavkazski Vestnik*, *Voyenny Sbornik* and in *Kavkaz*, collected in *Sbornik gazet Kavkaza*, 14 vols.

Most of the above sets, as well as the varied series of the Russian Academy of Sciences, are accessible in England in the libraries of the British Museum, the Royal Geographical Society, the School of Oriental and African Studies, the Bodleian Library (Wardrop bequest) and the Taylorian Institute (Morfill bequest). The Caucasian collections at the Royal Geographical Society and the London Library have been enriched by the gifts of the late Mr John Frederick Baddeley, who also bequeathed to the London Library his important MS. *Index Caucasica*.

(*c*) Geographical: Special Studies

B.S.E., sections 'Kavkaz', 'Chernoye more', 'Kaspiyskoye more'.

Some of the best special studies of the geography and topography of Caucasia and eastern Anatolia are in English:

Baddeley, J. F., *The Rugged Flanks of Caucasus*, 2 vols. Oxford University Press, 1942. [Baddeley, *Rugged Flanks*.]
Based on personal observation and on a detailed study of Russian periodical sources. Vol. II contains a valuable bibliography.

Freshfield, Douglas, *The Exploration of the Caucasus*, 2 vols., London, 1888.
Remains capital for the topography of the main chain of the Caucasus.

Lynch, H. F. B., *Armenia: Travels and Studies*, 2 vols., London, 1904. [Lynch, *Armenia*.]
The prime source for the Armenian highland.

The work of the great Frenchman, Marie-Felicité Brosset, is the foundation of modern Caucasian studies. It was published in five volumes:

1. *Histoire de la Géorgie depuis l'antiquité jusqu'au XIXe siècle; I-ère partie, Histoire ancienne, jusqu'en 1469*, St Petersburg, 1849.
2. *Additions et éclaircissements à l'histoire de la Géorgie*, St Petersburg, 1851.
3. *IIe partie, I livraison, Histoire moderne*, St Petersburg, 1856.
4. *IIe partie, II livraison, Histoire moderne*, St Petersburg, 1857.
5. *Introduction et Table des Matières*, St Petersburg, 1858.

Brosset's prodigious researches have been summarized by his son, Laurent Brosset, in *Bibliographie analytique des ouvrages de Monsieur Marie-Felicité Brosset* (St Petersburg, 1887); a work which is at once a model bibliography and a monument of filial piety. M.-F. Brosset's *Rapports sur un voyage archéologique dans la Géorgie et dans l'Arménie* (with the rare atlas of 18 lithographed plates), St Petersburg, 1849, is important for the topography of Caucasia in the nineteenth century.

Other books in west European languages useful for the study of the geography of Caucasia and Anatolia are:

Dubois de Montpéreux, Frédéric, *Voyage autour du Caucase chez les Tcherkesses et les Abkhazes, en Colchide, en Géorgie, en Arménie et en*

Crimée: avec un atlas géographique, pittoresque, archéologique, géologique, etc., 6 vols. and 2 vols. (atlas), Paris and Neuchâtel, 1839–43. [Dubois, *Voyage.*]

Tchihatcheff, P. de, *L'Asie Mineure*, 6 vols. and 3 vols. (atlas), Paris, 1866–9.
A companion volume covers Constantinople and the Bosporus.

Erckert, R. von, *Der Kaukasus und seine Völker*, Leipzig, 1887.

Cuinet, Vital, *La Turquie d'Asie: Géographie administrative, statistique, descriptive et raisonnée de l'Asie Mineure*, Paris, 4 vols., 1890.

Merzbacher, Gottfried, *Aus den Hochregionen des Kaukasus*, 2 vols., Leipzig, 1901.

Dechy, Moriz von, *Kaukasus: Reisen und Forschungen im kaukasischen Hochgebirge*, 2 vols., Berlin, 1905–6.

Lehmann Haupt, C. F., *Armenien: Einst und Jetzt*, 2 vols., Berlin, 1910–26.
The index is published separately, undated.

The fundamental work for the study of the topography of Georgia was the Geography of Prince Wakhusht (of Kartli). The Georgian text with French translation was published by Brosset, entitled *Description Géographique de la Géorgie, par le Tsarévitch Wakhoucht*, St Petersburg, 1842. A Russian translation of the Georgian text with introduction and notes was published by M. G. Janashvili in *Zap. Kav. Ot. I.R.G.O.* XXIV, pt. 5.

Among modern works, Ivane Javakhishvili's magistral *Kartvel Eris Istoria* (History of the Georgian Nation) is important for an understanding of the detail of Georgian and Caucasian topography; vol. I, 2nd ed., Tiflis, 1928; vol. II, 1st ed., Tiflis, 1914; vol. IV, pt. 1, 1st ed., Tiflis, 1924; other parts not accessible. I. Javakhishvili is also the author of *Sakartvelos Ekonomiuri Istoria* (Economic History of Georgia) of which vol. II appeared in 1934, and of a special study on the frontiers of Georgia, *Sakartvelos Sazghvrebi*, Tiflis, 1919.

Reference may also be made to the following works on the geography of Georgia:

Khelidze, M., *Sakartvelos Chedarebithi Geographia* (A Comparative Geography of Georgia), Tiflis, 1919.

Janashvili, G., *Sakartvelos Geographia* (Geography of Georgia), Tiflis, 1920.

Javakhishvili, Alexander, *Sakartvelos Geographia: Tome I. Geomorphologia* (Geography of Georgia: vol. I, Geomorphology), Tiflis, 1926; *Geographiuli Atlasi* (Geographical Atlas), Tiflis, 1927.

These volumes, although badly printed, contain useful maps and graphs.

The following have special reference to the geography and historical topography of Caucasia:

Marr, N. Y., *Plemennoy sostav naseleniya Kavkaza: Klassifikatsiya narodov Kavkaza*, Petrograd, 1920 (published as no. 3 of Trudy komissii po izucheniyu plemennogo sostava naseleniya Rossii).

Melikset-bekov., *Vvedeniye v istoriyu gosudarstvennykh obrazovaniy Yugo-Kavkaza*, Tiflis, 1924.

Kuchayev, *Polozheniya o krestyanakh i poselyanakh Zakavkazya*, 3 vols., Tiflis, 1886.

Veidenbaum, E., *Putevoditel' po Kavkazu*, Tiflis, 1888.

Although published more than sixty years ago, these two latter works contain much interesting material.

Zakavkazye: Statistiko-ekonomicheski sbornik, published by Vys. Ekon. Sovyet Z.S.F.S.R., 1925.

Lyaister, A. F. and Chursin, G. F., *Geografiya Kavkaza: Priroda i naseleniye*, Tiflis, 1924.

Denisimov, S., *Kavkazski Kray: Putevoditel'*, Tiflis, 1928.

Godabrelidze, S. A., *Mineral'niye resursy S.S.R. Gruzii*, Tiflis, 1933.

This comprehensive work of 1128 pages contains numerous plans and graphs.

Matveyev, S. N., *Turtsiya: Aziatskaya chast'. Anatoliya: Fiziko-geograficheskoye opisaniye* (Academy of Sciences of U.S.S.R., Geographical Institute), Moscow, 1946.

Is a recent summary of the geography of Anatolia.

Otchety kavkazskago gornago upravleniya (Tipografiya Kantselyarii Namestnika, Tiflis).

Contains much interesting topographical and economic information. [Years 1909–14 were available to the authors.]

(*d*) Military History

There are no general histories in Russian or other languages covering all the Caucasian wars. The following works are useful for the history of the conquest during the nineteenth century:

Dubrovin, Gen. N., *Istoriya voyn i vladychestva russkikh na Kavkaze*, 6 vols., St Petersburg, 1871–88.

Potto, G. M., *Istoricheski ocherk kavkazskikh voyn ot ikh nachala do prisoyedineniya Gruzii*, published by Voyenno-istor. otdel. Shtaba kavkazskago voyennago okruga, Tiflis, 1899.

Baddeley, J. F., *The Russian Conquest of the Caucasus*, Longmans, 1908. [Baddeley, *Conquest*.]

The definitive history of the conquest of the eastern Caucasus down to the surrender of Shamyl in 1859; based on *Akty*, *Kav. Sbor.*, etc.

*Pokrovski, M. N., 'Zavoyevaniye Kavkaza' (in *Diplomatiya i voyny Tsarskoy Rossii v XIX v.*), Moscow, 1923; also article 'Kavkazskiya voyny', in *B.S.E.* xxx, cols. 483–505.

Cevdet Paşa, *Tarihi* (History), 12 vols., Istanbul, 1271–1301 (1854–83); in Arabic script. Covers period 1774–1826 only.

The numerous references to Caucasian affairs are cited in Mirza Bala's article 'Çerkesler' in *I.A.*

A collection of Russian regimental histories (to which the authors did not have access) existed in Paris before the war of 1939–45.

The following Turkish reviews contain unit histories and special studies which are of value:

Askeri Mecmua, Tarih Kismi (Military Review, Historical Section), Askeri Matbaa (Army Press), Istanbul, 66 parts to 1945.

Süvari Mecmuasi (Cavalry Review), Deniz Basimevi (Navy Press), Istanbul, 128 parts to 1946.

Occasional material may also be found in:

Topçu Mecmuasi (Artillery Review), Topçu ve Topçu Atiş Basimevi (Artillery and Gunnery Press), Polatli, 132 parts to 1945.

Fen Kit'alari Mecmuasi (Technical Arms Review), Harp Okulu Basimevi (Military College Press), Ankara, 119 parts to 1945.

Levazim Mecmuasi (Commissariat Review), Harp Akad. Matbaasi (Military Academy Press), Ankara, 54 parts to 1945.

Deniz Mecmuasi (Review of the Sea), Deniz Basimevi, 375 parts to 1945.

(*e*) Regional: Special Studies

The following regional studies are important:

Kazbeg, Prince Dmitri, 'Tri mesyatsa v turetskoy Gruzii', in *Zap. Kav. Ot. I.R.G.O.*, vol. x, pt. 1, 1876.

Fundamental for the topography of Achara.

Bakradze, D., *Arkheologicheskoye puteshestviye po Gurii i Adchare*, with atlas, St Petersburg, 1878.

Marr, Prof. N. Y., *Zhitiye Sv. Grigoriya Khantztiyskago*, published as vol. VII of *Teksty i razyskaniya po armyano-gruzinskoy filologii*, 1911, and containing 'Dnevnik poyezdki v Shavshiyu i Klarzhiyu'.

The 'Dnevnik' is important for the study of the historical geography and topography of Shavsheti and the lower Çoruh valley.

Uvarova, Countess, *Kavkaz: Puteviya zametki*, pt. 2, 'Abkhaziya, Adzhariya, Shavshetiya, Poskhovski Uchastok', Moscow, 1891.

The topographical studies of Wakhusht, Kazbeg and Marr were summarized by W. E. D. Allen in an article in the *Geographical Journal*, vol. LXXXIV, no. 2, August 1929, 'The Marchlands of Georgia'.

The following volumes published by the topographical office of the Caucasian section of the (Imperial) Russian General Staff give a detailed description of the frontier region:

Lisovski, *Chorokhski kray*, 2 vols.: *Karsskaya oblast'* and *Erivanskaya guberniya*.

Apostolov, L. Y., 'Geograficheski ocherk kubanskoy oblasti' (published as pt. I of *Sbor. Mat.* XXIII, pp. 1–305), remains the capital work for the topography of north-west Caucasia.

See also: *Sbor. Mat.* I, 1, pp. 1–54, 'Gorod Erivan', by S. Zelinski; II, 1, pp. 109–42, 'Nakhichevan i nakhichevanski uyezd', by K. A. Nikitin; III, 1, pp. 315–50, 'Kratkaya zametka o karsskoy oblasti', by K. Sadovski; V, 1, pp. 1–65, 'Poskhovski uchastok ardaganskago okruga karsskoy oblasti', by K. Sadovski; XXXIV, 1, pp. 1(*bis*)–181, 'Ocherki karsskoy oblasti', by F. C. Yanovich; *ibid.* pp. 182–217, 'Zangizurski uyezd', by N. Shirakuni; XXXVIII, 1, 1(*c*), 'Iz poyezdok po karsskoy oblasti', by V. I. Divitski.

The following regional studies in Turkish are useful:

Nusret, M., *Tarihçe-i Erzurum* (Short history of Erzurum), Istanbul, 1922.

Beygu, A. S., *Erzurum: tarihi, anitlari, kitabeleri* (Erzurum: Its history, monuments and inscriptions), Istanbul, 1936.

Kocagüney, Gen. Vehbi, *Erzurum kalesi ve savaşlari* (The castle and battles of Erzurum), Istanbul, 1942.

Çelik, F. and others, *Kars*, Istanbul, 1943.

Atak, S., *Doğunun kapisi: Kars* (Kars: the gate of the east), Istanbul, 1946.

Konyar, B., *Diyarbekir tarihi* (History of Diyarbekir), Ulus Basimevi, 1936.

Türkiye kilavuzu (Guide to Turkey), vol. I (1946), covering vilayets A–Ç, has articles on Ağri, Bingöl, Bitlis.

Among many articles in *E.I.* (Turkish ed. *I.A.*) the following are important: 'Abkhaz', by W. Bartold (*E.I.*); 'Aras', by Besim Darkot and A. Z. V. Togan (*I.A.*); 'Baku', by W. Bartold (*E.I.*) and by Mirza Bala (*I.A.*); 'Čerkesses', by A. Dirr (*E.I.*), and 'Çerkesler', by M. Bala (*I.A.*); 'Daghestan', by W. Bartold (*E.I.*), and 'Dağistan', by M. Bala (*I.A.*); 'Derbend', by W. Bartold (*E.I.*); 'Kuban' (*E.I.*) and 'Tiflis' (*E.I.*), by V. Minorsky.

The short items by Streck on 'Adharbaidjan' and 'Bitlis' in *E.I.* should be compared with the fuller studies in *I.A.* under 'Azerbaycan' (A. Z. V. Togan) and 'Azeri' (M. F. Köprülü) and 'Bitlis' (M. H. Yinanç). Compare also Huart's short paragraph 'Diar Bakr' (*E.I.*) with Yinanç's essay under 'Diyarbekir' (*I.A.*); Hartmann's single column on Erzurum (*E.I.*) with H. Inalcik's 'Erzurum' (*I.A.*); Hartmann's slight note on 'Eriwan' (*E.I.*) with M. Bala's 'Erivan' (*I.A.*). In *E.I.* the sections on 'Turks' by W. Bartold, M. F. Köprülü, etc., and on 'Kurds' by V. Minorsky, published nearly twenty years ago, remain capital.

Reference may be made to *B.S.E.* for the Kabardans ('Kabardino-Balkarskaya A.S.S.R.'), Chechens ('Chechenskaya A.O.'), etc. *B.S.E.* also describes Baku (1926), Erevan, with plan of the city (1933), Tiflis (under 'Tbilisi') (1946).

(2) Book I, Chapter II

THE RUSSO-TURKISH CAMPAIGN OF 1828–9

Akty, VII, *Kavkaz i Zakavkazye za vremya upravleniya Generala Grafa Ivana Feodorovicha Paskevicha-Erivanskago.*

Contains substantial documentation of events in Caucasia during the years 1827–31. Documents 752–815 cover Paskevich's dispatches from the seat of war.

Dubrovin, *Istoriya Voyn*; Baddeley, *Conquest.*

Ushakov, Col. (attached to the staff of F.-M. Paskevich), *Istoriya voennykh deystviy v aziatskoy Turtsii v 1828–1829 gg.*, with atlas of maps and plans, 2 vols., St Petersburg, 1839.

*Shcherbatov, Prince, *F.M. Knyaz' Paskevich: ego zhizn i deyatelnost' po neizdannym istochnikam*, St Petersburg (7 vols., 1888–1904) (in Sumner bequest to Bodleian).

Ahmet Muhtar Paşa, *1828–1829 Türkiye-Rusya seferi ve Edirne muahedesi* (The Russo-Turkish campaigns of 1828–9 and the treaty of Adrianople), 2 vols., Istanbul, 1928.

Lukyanovich, *Opisanye turetskoy voyny 1828–1829 gg.*, 4 parts, with maps and 35 plans, St Petersburg, 1844–7.

Chesney, Col. F. R., *The Russo-Turkish Campaigns of 1828 and 1829*, London, 1854.

A popular but useful account published at the time of the Crimean War.

Monteith, Lt.-Gen. W., *Kars and Erzeroum: with the Campaigns of Prince Paskevich in 1828 and 1829*, London, 1856.

Gen. Monteith was British representative in Tiflis charged with the supervision of the payment of the Persian indemnity to Russia under the Treaty of Turkmençay. He states that the historical sections of his book are based on Fonton, *La Russie dans l'Asie Mineure*. According to Miansarov, Fonton's work, published in 1840, is an almost literal translation of Ushakov (see above). But Monteith's simply written book contains many details and personal reminiscences which give it a value of its own.

Witzleben, F. A. von, *Darstellung des russisch-türkischen Feldzuges, im Jahre 1828, in Europa und Asien*, Magdeburg, 1829–31, 2 vols., with map.

Much interesting information on conditions in the Russian and Turkish armies is to be found in the works of officers who described the campaign in the Balkan theatre. See:

Moltke, Baron von, *The Russians in Bulgaria and Roumelia in 1828 and 1829*, English ed., London, 1854.

The introductory chapter deals with the organization of the Russian and Turkish armies.

Alexander, Capt. J. E., *Travels to the Seat of War in the East through Russia and the Crimea in 1829*, 2 vols., London, 1830.

Contains many vivid details of contemporary life in the Russian army with an interesting account of the Black Sea fleet, then under the command of Admiral A. S. Greig.

Slade, Lt. Adolphus, R.N., *Records of Travels in Turkey, Greece, etc., and of a Cruise in the Black Sea with the Capitan Pasha in the years 1829, 1830 and 1831*, 2 vols., London, 1833.

D.N.B., for biographies of Chesney, Francis Rawdon; Monteith, William; Greig, Alexis Samuilovich; and *R.B.S.* for Paskevich, Graf Ivan Fedorovich Erivanski, Knyaz' Varshavski; Chavchavadze, Knyaz' Alexander Garsevanovich; Osten-Sacken, Baron (Count, 1855) D. E.

Chavchavadze was the son of the last *Mandaturt-Ukutsesi* (Chancellor) of the Georgian kingdom, and a godson of the Empress Catherine II. He commanded the Erevan column which took Bayazit and occupied the Eleşkirt valley. He was a gifted poet and one of the leaders of the Georgian literary revival of the nineteenth century. With equal facility he translated Hafiz, Voltaire and Pushkin into his own tongue.

Osten-Sacken died in 1881 at the age of 92, after 76 years service in the Russian army. He had taken part in fifteen campaigns and ninety engagements. For a short interval after the fall of Sevastopol, he was acting commander-in-chief of the Russian army in the Crimea. He was an active writer on military and biographical subjects.

For the visit of the poet, Alexander Pushkin, to Erzurum in the summer of 1829, see Veidenbaum, *Kavkazskiye etyudy*, I, pp. 233–60, 'O prebyvanii Pushkina na Kavkaze v 1829 g.' Pushkin's *Puteshestviye v Arzroum* was published in the first number of the afterwards celebrated literary journal *Sovremennik* (1836). It was somewhat acidly reviewed by Bulganin and other critics. Like his English contemporary Byron, Pushkin had fallen under the then fashionable allure of the Orient. After travels in the Caucasus and the Crimea, he had written his *Prisoner of the Caucasus* (1821) and his rather slight *Fountain of Bagcheserai* (1824).

Resplendent in Circassian costume, the poet arrived in Tiflis in the summer of 1829 on a visit to his brother Lev and other friends serving in the army of Paskevich. He joined the army at Kotanli on 13 June and was greeted 'in gay spirit' by the commander-in-chief. After visiting Burtsov's position, Pushkin attached himself to Muravyev's column. He saw the fighting along the Hani ravine and was present when Hakki Paşa capitulated at Meliduz. Later, the poet stayed three weeks in Erzurum as the guest of Paskevich in the *konak* of the seraskier. Paskevich was a member of the cultivated circle which included his wife's brother, Griboyedov, and the Georgian poet, Alexander Chavchavadze (whose daughter married Griboyedov). But the company of the erratic Pushkin seems to have irritated the veteran field-marshal, preoccupied with military detail; finally he induced the poet to leave Erzurum (see *R.B.S.* under 'A. S. Pushkin', p. 249).

The Kurds. In 1830 risings began throughout Kurdistan: in the country between the middle Euphrates and the upper Tigris under Bedr Khan; and in the Zagros under Muhammad Paşa of Ruwandiz. Helmuth von Moltke took part in the Turkish security operations (see his *Briefe über Zustände und Begebenheiten in der Türkei aus den Jahren 1835–1839*, Berlin, 1841).

The Egyptian defeat of the Turks at Nizip in 1839 encouraged further risings under Bedr Khan and Nurullah Beg of Hakâri (1843–7). (For this period see W. F. Ainsworth, *Travels in Asia Minor, etc.* (London, 1842), 2 vols.; and A. H. Layard, *Nineveh and its Remains* (London, 1849), vol. I, chaps. VII and IX.)

During the Crimean War a rising broke out in Bohtan under a nephew of Bedr Khan; and, after the Turkish defeats in 1877–8, the Kurds of Bohtan and Bahdinan rose under the sons of Bedr Khan. Later (1880), the more formidable movement of Sheikh Ubaydullah of the Nakshbandi order spread to Persian Kurdistan.

The Kurdish movement, at first feudal and conservative and only really nationalist after the Young Turkish Revolution of 1908, was perhaps stimulated by the deft policy of Paskevich. The important Russian works by Averianov and Mayevski, cited by Minorsky (art. 'Kurds' in *E.I.* III, p. 1148) have not been available to the authors. In his MS. work, 'Zazalar ve Kizilbaşlar' (Zazas and Kizilbaşis), Colonel Nazmi Sevgen records details of the relations of the Dersim Kurds with the Russian consul in Erzurum before the Crimean War and the War of 1877–8.

Abdul Hamit II, reversing the policy of his predecessors, tried to conciliate the tribal leaders and to direct their activity against the Armenians. A military aspect of his policy was the formation of the Hamidiye units, intended to militarize the Kurdish tribesmen on the pattern of the Russian Cossacks. (For this period, see Lynch, *Armenia*, vol. II; and Sir Mark Sykes, *The Caliph's Last Heritage* (London, 1915). Sykes published a useful classification of the Kurdish tribes, with a map—now out of date.)

(3) Book I, Chapter III

RUSSIAN PACIFICATION OF THE MAIN CHAIN OF THE CAUCASUS. DEVELOPMENT OF BRITISH INTEREST IN CAUCASIA, 1830–52

Akty, VII, VIII, IX; *Sbor. Gorts.*, all vols., particularly for east Caucasus; *E.I.*, art. 'Abkhaz'; *I.A.*, arts. 'Çerkesler', 'Dağistan'; *B.S.E.* art. 'Kavkazskiye voyny'; Dubois, *Voyage*; Baddeley, *Conquest*; Pokrovski, *Zavoyevaniye*.

Esadze, S., *Istoricheskaya zapiska ob upravlenii Kavkazom*, 2 vols., Tiflis, 1907.
An excellent source for Circassia and Georgia.

Idem, *Pokoroniye zapadnago Kavkaza i okonchaniye kavkazskoy voyny*, Tiflis, 1914.

Ponka, Ivan, *Chernomorskiye Kazaki v ikh grazhdanskom i voyennom smysle i sluzhbe*, 2 parts in 1 vol., St Petersburg, 1858.

Kolenati, *Die Bereisung Circassiens*, Dresden, 1859.

*Züssermann, A. L., *Dvatsat' pyat' let na Kavkaze, 1842–1867*, St Petersburg, 1879.

Olshevski, Gen. M. Y., 'Zapiski', in *R.S.* LXXXIII and LXXXIV, 1894–5, for a description of the western Caucasus towards the end of the period of Circassian resistance (1854–60).

Stal (Stahl), Gen. Baron K. T., 'Etnograficheski ocherk cherkesskago naroda', in *Kav. Sbor.* XXI, pt. 2, pp. 53–173, for a detailed survey of the Cherkess tribes just before their dispersion in the sixties.

Berzhe, A. P., 'Vyseleniye gortsev s Kavkaza', in *R.S.* XXXIII and XXXIV, for a description of the expulsion of the Cherkesses after their capitulation in 1864.

Khadzhimukov, Prince, 'Narody zapadnago Kavkaza', in *Kav. Sbor.* XXX (2), pp. 1–50.
An interesting survey by a Bzhedukh noble.

For the later operations in western Caucasus (1861–4), see I. Drozdov, 'Poslednyaya borba s gortsami na zapadnom Kavkaze', in *Kav. Sbor.* II, pp. 387–457; V. Soltan, 'Voyenniya deystviya v kubanskoy oblasti s 1861-go po 1864 god', *ibid.* V, pp. 345–470; and K. Geins (Heinz), 'Pshekhshki otryad', *ibid.* VIII, pp. 398–509 (also published in *Voyenny Sbornik*, 1866).

British policy in Caucasia during the middle decades of the nineteenth century has not been examined in detail by a modern English historian. The first volume of Harold Temperley's *Crimean War*—so rich in material for other borders of the Ottoman empire—does not cover Caucasia. The subject seems to have been treated in a work not accessible to the authors (Dr L. Widerzal's *British Policy in the Western Caucasus, 1833–42*, Warsaw, 1933). There is a useful article by G. H. Bolsover in the *Slavonic and East European Review* (Aug. 1934): 'Lord Ponsonby and the Eastern Question, 1833–39.

British travellers to Circassia during the thirties of the nineteenth century published the following works:

Spencer, Edmund, *Travels in Circassia, Krim Tartary, etc.*, 2 vols., London, 1838.

Bell, J. Stanislas, *Journal of a Residence in Circassia during the years 1837, 1838 and 1839*, 2 vols., London, 1840.

Longworth, J. A., *A Year among the Circassians*, London, 1840.

See also *D.N.B.*, art. 'Urquhart, David'.

Istoriya S.S.S.R., vol. II, *Rossiya v XIX veke* (with folder of maps). Gives interesting and relatively objective chapters on Caucasia in the nineteenth century. See particularly chap. XI, 'Narody Kavkaza v pervoy polovine XIX v.: borba gorskikh narodov pod rukovodstvom Shamilya'. Recent items of interest cited in the bibliography are:

Tahir, Mahomed, 'Tri imama', in *Sbor. Mat.* (vol. XLV, Makhach-Kala, 1926). An edition of Mahomed Tahir's work was published by the Academy of Sciences in Moscow (1941) under the title, *Khronika Mohammeda Takhira al-Karakhi o dagestanskikh voynakh v period Shamilya*, with introduction by I. Y. Krachkovski.

Pokrovski, N. I., 'Obzor istochnikov po istorii imamata', in *Problemy istochnikovedeniya, Sbornik II*, Moscow-Leningrad, 1936, pp. 187–234.

Idem, 'Myuridizm u vlasti ('Teokraticheskaya derzhava' Shamilya)', in *Istorik Marksist*, 1934, no. 2, pp. 30–75.

Bushyev, C. K., *Borba gortsev za nezavisimost' pod rukovodstvom Shamilya*, Moscow-Leningrad, 1939.

The new *Istoriya S.S.S.R.*, vol. II, and other books and articles published as late as 1950 were the subject of a violent attack in *Voprosy Istorii* (1950, fasc. 11, pp. 101–25, E. Adamov and L. Kutakov, 'Iz istorii proiskov inostrannoy agentury vo vremya kavkazskikh voyn'). In their introductory paragraph the authors state that the entire character of the Murid and Shamyl's movement is misconstrued if it is presented as a progressive democratic phenomenon tending towards national liberation. Such an estimate of Shamyl and Muridism is anti-Marxist, contradicts historical facts, and radically distorts the real meaning of this movement, which was reactionary, nationalist, and was in the pay of English capitalism and the Turkish sultan. The authors reprint a number of documents from the Russian Foreign Office archives, including copies of some of Bell's letters, letters addressed by Shamyl to Sultan Mahmut II and to Ibrahim Paşa and correspondence from the Russian Ambassadors in Constantinople and London to Nesselrode.

Note on Lermontov and Tolstoy in the Caucasus. The epic of the Caucasian wars inspired three great writers of the Russian nineteenth century. In Russia it was the fashion for members of the most aristocratic families to serve in the Caucasian army; and the system of exiling to the provinces persons involved in the Decabrist Revolt (1825) and, later, those suspected of radical views, produced in the garrison towns along the Caucasian line a lively and intelligent society.

In 1837 Mikhail Yuryevich Lermontov (1814–41) was posted to the Caucasus following the publication of a poem on the death of Pushkin which the emperor regarded as impertinent. Mikhail Yuryevich had already spent some years of his boyhood in the northern Caucasus and, with brief intervals, he was to pass the last four years of his life in the Caucasian service. In watering places behind the line, Lermontov met many of the survivors of the Decabrist movement together with men like Belinski and Satin, the friend of Herzen and Ogarev. (For a recent account of the Decabrists in the Caucasus, see M. V. Nechkina, *Griboyedov i Dekabristy*, Moscow, 1951.)

So great was his love of life in the Caucasus that he evaded a transfer to the capital (which had been secured from the emperor on the intervention of his friend, Benckendorf). He came to boast in a letter to the younger Raevski that he had ridden along the whole length of the line from Kizlyar to Taman, had crossed the mountains and had been in Shusha, in Kuba, in Shemakhi and Kakheti; 'dressed as a Cherkess with a gun on his back', he had slept in the open, eaten 'chürek' and drunk 'Kakhetinski' (*R.B.S.*, under 'M. Y. Lermontov', p. 389). For an account of Lermontov's death, see M. F. Fedorov, 'Pokhodniya Zapiski', in *Kav. Sbor.* III (1), pp. 193–4.

Lermontov wrote a number of charming Caucasian tales ('Ismail Bek', 'Bella', 'Haji Abrek' and others). *The Demon*, one of his longer poems, has a Caucasian setting, while his novel, *A Hero of our Times*—in which the character of Pechorin is autobiographical—is a unique and brilliant mirror of the time.

Lev Nikolayevich Tolstoy (1828–1910) studied as a youth at Kazan University where he took some interest in Turkish and Arabic. At the age of eighteen, while on a visit to his brother on the Caucasian line, he rented a cottage in the mountains near Pyatigorsk where he settled for some months. At Tiflis he passed his examinations for the artillery and the next years were spent in the Caucasus. On the outbreak of the Crimean war he volunteered for service on the Danube and took part in the siege of Silistria. He served in the defence of Sevastopol, and his *Tales from Sevastopol* brought him literary fame at the age of twenty-seven. *The Cossacks* (1863), in which he developed the theme that culture is an enemy of happiness, reflected his Caucasian experiences. His moving sketch, *Haji Murad*, written in later life (1896–1904), recalls the old philosopher's youthful years as a subaltern in Dağistan.

(4) Book II, Chapters IV, V, VI

THE CAUCASIAN THEATRE DURING THE CRIMEAN WAR

(a) The Political, Diplomatic and Military Background of the Crimean War

For the political and diplomatic background of the Crimean War, see: *The Cambridge Modern History*, vol. XI, chap. XI; *The Cambridge History of British Foreign Policy*, vol. II; *Histoire de la Diplomatie* (French translation of a Soviet symposium published under the editorship of M. Potemkin), vol. I (Paris, no date), chaps. VIII and IX; Temperley, *The Crimean War: England and the Near East*, London, 1936; Tarlé, *Krymskaya voyna*, Moscow-Leningrad, vol. I, 1944; *Istoriya S.S.S.R.*, vol. II (1949), chaps. X–XIV.

The best critical sources are undoubtedly Temperley and Tarlé. Unfortunately, Harold Temperley died after his first volume had been published—a brilliant and delightful study. Only vol. I of Tarlé's work has been available to the authors. It is a reasonably objective presentation of the Russian side and contains useful bibliographical material. See also F. H. Geffcken, *Zur Geschichte des orientalischen Krieges, 1853–6*, Berlin, 1881; and E. Bapst, *Les origines de la guerre de Crimée*, Paris, 1912.

On the political and military history of Caucasia during the Crimean War, a select Russian documentation was published in *Akty*, X, 1853–4, and XI, 1854–6. These two volumes amount to nearly 2000 (beautifully printed) folio pages. Numerous papers covering personal recollections of participants in the war were published in *R.S.*, *Kav. Kal.* and *Kav. Sbor. Voyenny Sbornik* and Russian periodicals not available to the authors also contain much material. *R.B.S.* and *Akty* are invaluable for careers of individual officers.

The Caucasian operations are covered in the following general works by Russian historians of the Crimean War:

Bogdanovich, M. T., *Vostochnaya voyna 1853–56 gg.*, St Petersburg, 1876.

Dubrovin, N. F., *Vostochnaya voyna: Istoriya Krymskoy voyny i oborony Sevastopolya*, St Petersburg, 1900.

*Zayonchkovski, A. M., *Vostochnaya voyna 1853–1856 v svyazi s sovremennoy ey politicheskoy obstanovkoy*, St Petersburg (2 vols. only, published 1908–13).
This work is praised by recent Soviet historians for its technical excellence, see *Istoriya S.S.S.R.*, vol. II, p. 348.

The English military historians, Kinglake and Fortescue, in their works covering the Crimean War, have ignored the operations in Caucasia. A recent popular account by Gen. Sir George MacMunn (*The Crimea in Perspective*, London, 1935) has a chapter on the Caucasian conflict which reveals some misconceptions.

*Turkish source: *Türkiye ve Kirim harbi* (Turkey and the Crimean War), Istanbul, Askeri Basimevi, no date.

Personal accounts and recollections:

Muravyev, Gen. Count N. N., *Voyna za Kavkazom v 1855 g.*, pts. 1 and 2, St Petersburg, 1876.

Likhutin, *Russkiye v aziatskoy Turtsii v 1854–1855 gg.*, St Petersburg, 1863.

Olshevski, M. Y., 'Russko-turetskaya voyna na Kavkaze v 1853 i 1854 gg.', in *R.S.* 1884, Oct. pp. 171–86; Nov. pp. 417–32; Dec. pp. 497–514.

Potto, V., 'Vospominaniya o zakavkazskom pokhode 1855 g.', in *Kav. Sbor.* XXV, pt. 1, pp. 1–242; and 'Aleksey Petrovich Kulgachev', in *Kav. Sbor.* XXX, pt. 3, pp. 1–88.

Amilakhvari, Gen. Prince, 'Iz zapisok', in *Kav. Sbor.* XXVI, pt. 1, pp. 1–195, with continuation in following volume not available to authors.

Karal, Enver Ziya, 'Zarif Paşa'nin hatirati' (Memoirs of Zarif Paşa), in *Belleten, Tarih ve Dil Kurumu* (Bulletin of the Society of History and Language).
(The learned editor has made the best of a disappointing document.)

Potto, Gen. G. M., 'Blokada i shturm Karsa v 1855 g.: zapiski Y. A. Baklanova, razskazy K. A. Ermolova, Baron Maydila, i d.', in *R.S.* II, 1870.

Oreus, I. N., 'Shturm, blokada i vzyatiye Karsa v 1855 g.', in *R.S.* XIX, 1877.

Duncan, Charles, *A Campaign with the Turks in Asia*, 2 vols., London, 1856.

Lake, Col. A., *Kars and our Captivity in Russia*, London, 1856.
(The British officers, taken at Kars, were hospitably entertained and even lionized—in Tiflis and St Petersburg.)

Idem, *The Defence of Kars*, London, 1857.

Sandwith, Humphrey, *A Narrative of the Siege of Kars, etc.*, London, 1856.

Oliphant, Laurence, *The Transcaucasian Campaign of the Turkish Army under Omer Pasha*, London, 1856.

(*b*) The Kars Controversy

The fall of Kars was the subject of bitter criticism in England and of some acrid debates in the House of Commons. See 'Papers relative to Military Affairs in Asiatic Turkey and the Defence and Capitulation of Kars, presented to both Houses of Parliament by command of Her Majesty, 1856' (Nos. 2032, 2071), and *Hansard Debates*, May 1856. Details of the controversy may be found in the lives of contemporary statesmen:

Lane Poole, S., *Life of Lord Stratford de Redcliffe*, London, 1890. chap. XXIV.

Skene, J. H., *With Lord Stratford in the Crimean War*, London, 1883.

Martineau, J., *Life of Henry Pelham, Fifth Duke of Newcastle*, London, 1908.
See chap. IX for the Duke's brief visit to the Mingrelian coast.

See also:

Kmety, George, *A Narrative of the Defence of Kars on the 20th of December, 1855*, London, 1856.

Marx, Karl, *The Eastern Question.*
(A reprint of articles written while the author was the correspondent of an American newspaper in London. Papers CX–CXIII deal with the fall of Kars. While the author was never nearer the seat of war than London, he wrote with an insight into military events and his reportage is one of the best on the war.)

(*c*) The Black Sea

Publications of the Navy Records Society:
Vol. LXXXIII, *Russian War, 1854: Baltic and Black Sea*, London, 1943.
Vol. LXXXIV, *Black Sea, 1855*, London, 1946.

Sergeyev-Tsenski, *Sinopski boy*, Moscow, 1944.

Slade, Rear-Admiral Sir Adolphus, K.C.B. (Mushaver Paşa), *Turkey and the Crimean War: A Narrative of Historical Events.*
Chapters XV, XVII and XVIII cover the abortive attempts to establish liaison with the Cherkesses, and chapters XVIII–XXX record the events leading to Ömer Paşa's invasion of Mingrelia. The Admiral, a strong partisan of the Turks, was a supporter of the strategic conceptions of Ömer Paşa.

In *Voprosy Istorii*, IX (1951), pp. 76–96, A. Fadeyev has published a study of the political background of the Russian war against the Cherkesses ('Myuridism kak orudiye agressivnoy politiki Turtsii i Anglii'). The writer uses important documentary sources drawn from Soviet archives (*Tsentralny gos. voyenno-istoricheski arkhiv* and *Tsentralny gos. istoricheski arkhiv Gruzinskoy S.S.R.*), and he cites periodical series not available to the present authors (*Kubanski Sbornik*, up to vol. X; and *Kuban*, up to vol. VII (1949)). However, Mr Fadeyev is responsible for inaccuracies, misinterpretations of sources and omissions surprising in a historian who is clearly so painstaking. 'S. Canning' is described as British ambassador in Constantinople in 1831, and an active support of the Cherkesses is attributed to his initiative (p. 81). In fact, Stratford Canning did not occupy the post of ambassador in Turkey between the years 1828 and 1842.

Less pardonable is the writer's description (p. 92) of the adventurers Mlodetsky and Pichinichini as English agents in Circassia in 1852, when he cites as his source I. Drozdov (*Kav. Sbor.* XI, p. 467) who describes them specifically as 'emissaries of the Sultan himself' (*poslanny samim sultanom*). Mr Fadeyev's reference to the Turkish invasion of Mingrelia in 1855 is equally unscientific. The battle of the Ingur is not mentioned at all, and the withdrawal of the Turks is ascribed to the action of the Russian army and of 'the Georgian *opolcheniye* standing in the defence of their fatherland' (p. 94).

Mr Fadeyev makes the interesting point that Admiral Nakhimov's attack on Turkish ships at Sinop was intended to forestall a Turkish descent on the Caucasian coast.

The complex history of the Circassian struggle awaits a scholar with access to the vast documentation available in Soviet archives, and it is to be hoped that in the future the subject may be approached in the objective spirit which the writing of history demands.

(*d*) GEORGIA

A new history of Georgia has been published in Tiflis in Georgian and Russian editions under the joint authorship of N. Berdzenishvili, I. Janashia and I. Javakhishvili (*Istoriya Gruzii* (1946), I, pp. 466; review by I. N. Borozdin, in *Vestnik drevney istorii* (1948), I, pp. 110–15). This volume covers the period to the beginning of the nineteenth century. The history of Georgia in the nineteenth century has been the subject of a number of books and articles published in Tiflis during the last decade. In his excellent study, *K istorii Gruzii pervoy poloviny XIX veka*, G. V. Khachapuridze cites the following works on Georgian history up to the period of the Crimean War:

*Javakhov (Javakhishvili), I., *Politicheskoye i sotsialnoye dvizheniye v Gruzii v XIX v.*, St Petersburg, 1906.

*Makharadze, P., *Gruziya v XIX stoletii*, Tiflis, 1938.

*Chkhetiya, *Tbilisi v XIX stoletii*, Tiflis, 1942.

*Burchuladze, E. E., 'Gruziya v vostochnoy voyne 1853–1856 gg.', in *Trudy Tbilisskogo gos. pedagogicheskogo instituta*, VII (1949).

*Kakabadze, S., 'Krestyanski vopros i krestyanskiye dvizheniya v vostochnoy Gruzii v XVIII v.', in *Sbornik istorii klassovoy borby v Zakavkazye*, I, pp. 153–94, Tiflis, 1930.

*Pirtzkhalaishvili, A. G., 'Imeretiya i Guriya v period 1808–1840 gg.', in *Materialy po istorii Gruzii i Kavkaza*, I, pp. 1–120, Tiflis, 1942.

*Makharadze, N. B., 'Vosstaniye v Imeretii 1819–20 gg.', *ibid.* III.

See also M. K. Rozhkova, 'Iz istorii ekonomicheskoy politiki rossiyskogo tsarizma v Zakavkazye', in *Istoricheskiye zapiski*, XVIII, pp. 169–200.

(*e*) Biographical

D.N.B., for biographies of Ballard, John Archibald; Lake, Sir Henry Atwell; Oliphant, Laurence; Sandwith, Humphrey; Slade, Sir Adolphus; Teesdale, Sir Christopher Charles; Thompson, Henry Langhorne; Urquhart, David.

R.B.S., for biographies of Andronikov, Knyaz' Ivan Malkhazovich (also *Kav. Kal.* XXXI (1876); he was a nephew of Solomon II, last king of Imereti); Argutinski Dolgorukov, Knyaz' Moisey Zakharovich; Bebutov, Knyaz' David Osipovich (also detailed memoir in *Kav. Sbor.* XXIII, pp. 1 (*bis*)–130); Chavchavadze, Knyaz' Yason Ivanovich.

N. N. Muravyev is the subject of several memoirs: *R.S.* VIII (1873), critical article by A. P. Berzhe; *ibid.* X (1874), 'N. N. Muravyev v 1854–6', by A. N. Muravyev (reply to Berzhe); *ibid.* XI (1874), 'N. N. Muravyev i Knyaz' Vorontsov', by Shcherbanin. Gen. Y. P. Baklanov in his 'Memoirs', edited by Gen. Potto (*ibid.* II, 1870), is also critical of Muravyev. See further *R.S.* VI (1872) and L, LI (1886).

(5) Book III, Chapters VII–XV

THE CAUCASIAN CAMPAIGN OF 1877–8

(*a*) General Background

For the political history of the Near Eastern crisis of 1876–8:

Sumner, Humphrey, *Russia and the Balkans, 1870–80* (Clarendon Press, Oxford, 1937).
A valuable study which makes liberal use of Russian sources.

Medlicott, W. N., *The Congress of Berlin and After* (Methuen, London, 1938).
Covers the discussions over Batum and the Russo-Turkish frontier.

(*b*) General Military Studies

For the relation of the campaign in Asia to the war in the Balkans:

Greene, F. V. (Military Attaché to the U.S. Legation in St Petersburg), *The Russian Army and its Campaign in Turkey in 1877–78*, London, 1880.
The final chapter gives a brief account of the operations in Asia. In 1903 Greene's book was reprinted without the Caucasian operations as *The Campaign in Bulgaria of 1877–8*.

La Guerre d'Orient en 1877–1878. Etude Stratégique et Tactique des Opérations des Armées Russe et Turque en Europe, en Asie et sur les Côtes de la Mer Noire par un Tacticien Auteur de Plusieurs Ouvrages Militaires: Ouvrage Redigé sur les Documents Officiels, Librairie Baudouin, Paris, 1885.
Excellently documented, this work in four volumes bears all the marks

of having been published under the auspices of the French General Staff. Vol. IV, cinquième partie, covers the Turco-Russian campaign in Armenia.

Ali Fuat Paşa, *Rus-Türk Seferi 1293–1294* (The Russo-Turkish War of 1877–78), Turkish General Staff publication, Istanbul, 1326 (1908); 3 vols. with maps and numerous illustrations.

Hozier, Capt. H. M., *The Russo-Turkish War*, 2 vols., London (undated).
A popular work which has some excellent engravings of battle scenes and contemporary personalities; the second volume contains chapters on the Caucasian front.

An interesting series of original documents from Turkish sources for the period 1875–8 is in course of publication as appendices to *Askeri Mecmua* under the title, *1875–1878 Osmanli Ordusu Savaşlari* (Battles of the Ottoman Army, 1875–8). Gen. Halil Sedes is the editor. The first eight parts (1935–40), *Osmanli-Rus ve Romen Savaşi, 1877–78* (Ottoman-Russian and Rumanian Battles), cover the war in Europe only up to Aug. 1877.

(c) Special Military Studies

Gazi Ahmet Muhtar Paşa, *Sergüzeşti Hayatim, 1294 Anadolu'da Rus muharebesi* (My adventurous life: the Russian campaign in Anatolia in 1878), Istanbul, 1328 (1910).

Gizetti, Maj.-Gen. A. L., *Khronika kavkazskikh voysk* (in 2 parts), Tiflis, 1896.

Gryaznov, *Voyenny obzor perednego teatra v aziatskoy Turtsii*, pts. 1 and 2, published by the Military History Section of the Staff of the Caucasian Military District, Tiflis, 1895, and St Petersburg, 1897.

Zhukov, G. M., *Voyna 1877–1878 gg.*, vol. III, *Voyna v aziatskoy Turtsii*, St Petersburg, 1882.
It contains some crude but interesting illustrations of battle scenes.

Materyaly dlya opisaniya russko-turetskoy voyny 1877–1878 goda na kavkazsko-maloaziatskom teatre, vol. I, published by the Military History Commission of the General Staff, St Petersburg, 1904.

Kishmishev, S., *Voyna v turetskoy Armenii, 1877–78*, St Petersburg, 1884.

Sidel'nikov, N., *Turetskaya kampaniya 1877–1878*, vols. I–II, SPB, 1878–9.

Maslov, A. N., *God voyny v Maloy Azii*, St Petersburg, 1879.

Col. Filippov, *Voyennoye obozreniye aziatskoy Turtsii*, published by the Military Educational Committee, St Petersburg, 1879.

Kolyubakin, Major-Gen. B., *Russko-turetskaya voyna v 1877–1878 na Kavkaze i v Maloy Azii*, published by the Academy of the General Staff, pt. 1 (later parts not issued), St Petersburg, 1906.

Etude Critique des Opérations en Turquie d'Asie pendant la Guerre Turco-Russe en 1877–1878 d'après des Documents Officiels, par un Officier Supérieur Turc, Otto Keil, Constantinople, 1896.

Norman, C. B. (*The Times* correspondent), *Armenia and the Campaign of 1877*, London, 1879.

'Sbornik turetskikh dokumentov o posledney voyne, perevod s turetskago trëkh poslednikh glav sbornika "Zubdetul-Hikiin"', by Ahmet Mi(d)hat Effendi, in *Voyenny Sbornik*, 1879, nos. 4–5.

'Materialy dlya istorii voyny 1877–1878 v aziatskoy Turtsii: Sudebniye prigovory voyennago suda anatoliyskago otdeleniya', *Voyenny Sbornik*, 1880–90.

Williams, Charles, *The Armenian Campaign: A Diary of the Campaign of 1877 in Armenia and Kurdistan*, London, 1878.

'La guerre Russo-Turque par les documents inédits', in *Nouvelle Revue*, 1880, nos. IV–VI.

(*d*) Detailed Accounts of Operations

(i) *Ardahan operation*

Komarov, Gen. V., 'Ardaganski otryad', in *R.S.*, vols. CXXXI, CXXXII (1907), vol. CXXXV (1908).

(ii) *Battle of Zivin*

Heimann, Gen., 'Zivinskoye srazheniye', report in *Inzhenerny Zhurnal*, 1878, no. 12.

Voynov, A. M., 'Boy pod Zivinom', *R.S.*, vol. LIX (July 1888).

Zezemann, Gen., 'Opisaniye ukrepleniya zivinskoy pozitsii', in *Inzhenerny Zhurnal*, 1878, no. 12.

(iii) *Bayazit operation*

Geyns, K. K., 'Bayazitskoye sideniye', *R.S.*, vol. XLV, 1885.

(iv) *Storm of Kars*

Gippius (Hippius), V., *Osada i shturm kreposti Karsa v 1877 godu*, St Petersburg, 1885.

(v) *Black Sea coast*

Kolyubakin, B., 'Vospominaniya ofitsera kobuletskago otryada', in *Voyenny Sbornik*, 1881.

Various writers: *Batum i ego okrestnosti*, Batum, 1906, containing valuable articles by Derzhavin, Janashvili, Stefanov and Gen. Potto.

(vi) *Chechnia and Dağistan*

Baddeley, J. F., 'The Rising of 1877 in Daghestan and Tchetchnia', in *Georgica*, Oct. 1936.

Takho-Godi, A., 'K pyatidesyatiletiyu vosstaniya Chechni i Dagestana (1877)', in *Novy Vostok*, XII, pp. 153–76.

Ivanov, A. I., 'Vosstaniye v Chechne v 1877 g.' in *Istoricheskiye Zapiski*, X, pp. 280–94; *idem*, 'Natsionalno-osvoboditelnoye dvizheniye v Chechne i Dagestane v 60–70-kh gg. XIX veka', *ibid.* XII, pp. 165–99.

(*e*) Biographical

T.M.A., for short notices on Gazi Ahmet Muhtar Paşa Katircioğlu, Ibrahim Derviş Paşa Lovçali and Ismail (Hakki) Paşa Kurt.

R.B.S., for Loris Melikov, Knyaz' Mikhail Tarielovich; Chavchavadze, Knyaz' Yason Ivanovich; Devel, Theodore Danielovich; Heimann (Geiman) Vasili Alexandrovich; Lazarev, Ivan Davidovich; Shelkovnikov, Begbut Martirosovich; Tergukasov, Arzas Artemyevich.

For General Devel see also *Kav. Sbor.* xxv, pp. 1–56, 'Zapiski Develya'; and for an appreciation of Gen. Tergukasov, *Kav. Kal.* xxxvii (1882), and A. Tsezarski's *Opisaniye boyevoy zhizni 3-90 kavkazskago strelkovago batalyona*, Tiflis, 1909, 2 parts (of which only part 1 was available to the authors). For an account in English of Lazarev's campaign in Transcaspia and of his death there in August 1879, see Charles Marvin, *The Russian Campaign against the Turcomans*, London, 1880. For Heimann's 'Vospominaniya' see *Kav. Sbor.* iii (2), pp. 251–376.

D.N.B. (supplementary volumes) contains biographies of General Sir Arnold Kemball and of Hobart Paşa (Hon. Augustus Charles Hobart). For portrait and sympathetic notice of Hobart, see Hozier, *Russo-Turkish War*, i, pp. 310–11.

M. T. Loris Melikov, the chief of the quartette of robust and able Armenians (the others being Lazarev, Tergukasov, Shelkovnikov) who led the Caucasian army to victory in 1877–8, was born in Tiflis in 1825, the son of a merchant, connected, perhaps not very closely, with the princely family of the Meliks of Lori. He was educated at the Lazarevski Institute of Oriental Languages and at a military academy, and served with distinction under the most celebrated commanders in the Caucasus—Freytag, Argutinski, Baryatinski, Bebutov and Muravyev. A corps commander at fifty-one, his career was successful, if not spectacular. But after his Caucasian victories, he was appointed governor-general of the Lower Volga (1879) and the same year minister of the interior and virtually dictator of Russia in the struggle against the Nihilists. The public career of M. T. Loris Melikov did not survive the assassination of Emperor Alexander II on 1 March 1881. He went into retirement abroad and died at Nice in 1888.

By an odd turn of fate, Loris Melikov's antagonist at Zivin and Alaca-dağ, Ahmet Muhtar, became, more than a generation later, grand vizier in the final crisis of the Ottoman empire (1913). And the old gazi, who had been born in the last year of Mahmut II, lived to see the end of both the Russian and Turkish empires (died 1918).

The memoirs of Musa Paşa Kundukov were edited by his grandson, Şevket Kunduk, and published in the *émigré* journal, *Kavkaz*, during the 1930's (French edition, *Le Caucase*, nos. 3–14, Aug. 1937–July 1938; see also a biographical sketch by A. Kantemir in no. 3, and 'Considérations sur les mémoirs de M. P. K.', by Z. Avalishvili in no. 17).

The son of an *aldar* (noble) of the Tagaur Ossetians settled to the north of Daryal, Musa Kundukov entered the Pavlovski Cadet Corps at the age of twelve. In 1837, when still under twenty, he began a brilliant military and administrative career as officer-interpreter to the Emperor Nicholas during the imperial visit to the Caucasus. During the following twenty years Musa Kundukov rose to the rank of major-general in the Russian army, and in 1859, the year of the surrender of Shamyl, was appointed governor of the

Terek province. Kundukov remained the friend and confidant of a succession of high Russian officials, and his contacts with the insurgent leaders enabled him sometimes to play the role of mediator.

His break with the Russian authorities occurred after the fall of Shamyl and, according to his own account, was provoked by the policy of expropriation and deportation directed against the tribes of his region. After numerous protests to the Russian authorities, and following a visit to Istanbul, Musa Kundukov in 1865 migrated with his whole family, and some 3000 Ossetian and Chechen 'hearths' to Turkey where he was provided with lands round the village of Batmantaşi in the district of Tokat. The descendants of these 'Circassians' are still a substantial element in the population round Tokat.

Musa Paşa Kundukov was chief of staff to Ismail Hakki when the latter was ordered by the Porte to surrender Erzurum to the Russians under the terms of the armistice of Adrianople. He seems to have received friendly treatment from the Russian authorities. When they evacuated the city on 7 September 1878, Musa Paşa took charge. He was commended by the Russians (Avalishvili, quoting Esadze, *Istoricheskiya zapiski*) for measures taken to prevent demonstrations against the Christian population during Bayram—a fact which is in curious contrast to Norman's complaints of his fanaticism (*Armenia*, pp. 137–8).

Musa Paşa Kundukov died in Erzurum in 1889. His younger son, Bekir Sami Bey, was the distinguished foreign minister of the first nationalist government in Ankara.

(*f*) The Effect of the War on the Volga Turks

R.S. CXVI (1903), 4, pp. 471–6, 'Turetskiye emissary v Rossii pered voynoy 1877–8 gg.', by P. Yudin.

(6) Book IV

THE CAUCASIAN CAMPAIGNS OF WORLD WAR I

(*a*) General Background

Ahmet Emin (Yalman), *Turkey in the World War* (in Economic and Social History of the World War: Turkish Series), Yale University Press, 1930.

This work by a gifted Turkish writer who visited many fronts as a war correspondent is essential to a perspective of Turkey in the First World War.

Pomiankowski, Lieut. Field-Marshal Joseph, *Der Zusammenbruch des ottomanischen Reiches: Erinnerungen*, Wien, 1928.

Important memoirs of the Austro-Hungarian military attaché in Constantinople.

Djemal Pasha, *Memories of a Turkish Statesman, 1913–1919*, London, 1922.

Ludendorff, Erich von, *Ludendorff's Own Story*, 2 vols., London, no date.

Vol. 2 contains important passages on the Caucasian situation in 1918.

Sanders, Liman von, *Five Years in Turkey*, Annapolis, U.S. Navy Institute, 1937.

Useful for a general view but limited references to Caucasia.

(b) General Military Studies

Larcher, Commandant M.: *La Guerre Turque dans la Guerre Mondiale*, Paris, 1926.

A valuable account of the Turkish military effort in World War I, utilizing Turkish sources. Important bibliography.

**Büyük Harpte Türk Harbi* (Turkey's War in the Great War), 3 parts, Askeri Basimevi, Istanbul.

Mehmet Emin, Col., *Les Evénements sur les Fronts Ottomans pendant la Grande Guerre*, Ecole d'Etat Major, Istanbul, 1922.

Sostav i gruppirovka diviziy i polkov turetskoy armii, publication of the reconnaissance section of the Russian General Staff.

(c) Special Military Studies

Maslovski, E. V. (Quartermaster-General of the staff of the Caucasian front), *Mirovaya voyna na kavkazskom fronte, 1914–1917*, Paris, 1933.

The definitive history of the first three years on the Caucasian front. It contains a valuable list of Russian sources.

Baki, Col., *Büyük Harpte Kafkas Cephesi* (The Caucasian Front in the Great War), 2 vols., Askeri Basimevi, Istanbul, 1933.

Contains original Turkish documents for period up to November 1914 only and a valuable series of maps.

*Baki, Col., *Büyük Harpte Kafkas Cephesi Hatiralari* (Memoirs of the Caucasian Front in the World War), Istanbul.

Guse, Oberst D., 'Die Kämpfe des Weltkrieges an der Kaukasus Front', in *Schweizerische Monatschrift für Offiziere aller Waffen*, 1929–30.

Korsun, N. G., *Strategicheski obzor voyny 1914–1918 gg. na kavkazskom fronte*, pt. 1, 'Ot nachala voyny do Erzerumskoy operatsii', Moscow, 1935.

(d) Detailed Accounts of Operations

(i) *Sarikamiş*

Korsun, N. G., *Sarikamyshskaya operatsiya*, Moscow, 1937.

*Köprülü, Lt.-Col. Şerif, *Sarikamiş Ihata Manevrasi* (The Sarikamiş Encirclement Manoeuvre), Istanbul, 1922.

Akhatkin, K. Z., 'Sarikamish', in *Voyenny Sbornik*, no. 6, Belgrad, 1925.

(ii) *Erzurum*

Korsun, N. G., *Erzerumskaya operatsiya*, Moscow, 1938.

Akhatkin, K. Z., 'Erzerum', in *Chasovoy*, no. 3, Paris, 1929.

(iii) *Malazgirt and Eleşkirt*

Balkan, Col. R., *Büyük Harpte Şark Cephesinde, Sağ Kanad Harekâti. Murat, Van-gölü Havzasinda Türk Rus Süvarilerinin Operatif Hedefleri: 1914 Ekim–20 Ağustos 1915* (The Right Flank Operation on the Eastern Front in the World War. Cavalry operations of the Turks and the Russians in the neighbourhood of the Murat and Lake Van: Oct. 1914–20 Aug. 1915), 1 vol. and atlas, Askeri Mecmua, Tarih Kismi (Military Review, Historical section), Istanbul, 1946.

(iv) *Black Sea*

Novikov, N., *Operatsii flota protiv berega na Chernom more v 1914–1917 gg.*, Moscow, 1937.

(v) *Azerbaijan and Persia*

Baratov, N. N., 'Persidski ekspeditsionny korpus', in *Russki Invalid*, no. 11, Paris, 1930.

**Büyük Harpte Iran Cephesi* (The Persian Front in the Great War), 2 vols., Askeri Basimevi, Istanbul.

**Irak Seferi* (The Iraq Campaign), pts. 1, 3, 4, Askeri Basimevi, Istanbul.

Kiesling, Hans von, *Mit Feldmarschall von der Goltz in Mesopotamien und Persien*, Leipzig, 1922.

Erdmann, Lieut., *Im Heiligen Krieg nach Persien*, Berlin, 1923.

*Nikitin, V. N., *Voyenniye deystviya v Azerbeydzhane 1914–1918: Mosulskoye napravleniye* (in MS.).

Idem, 'Les Assyro-Chaldéens', in *Revue des Sciences Politiques*, Paris, 1920.

Wilson, Sir A. T., *Loyalties: Mesopotamia, 1914–1917*, vol. I, London, 1930.

For the detailed topography and ethnography of the Van-Urmia region, see arts. in *E.I.* by V. Minorsky under heads: 'Maku', 'Sawdj-bulak', 'Shatak', 'Somai', 'Tabriz', 'Urmiya', 'Ushnu', 'Wan'; and by B. Nikitin under 'Shamdinan'.

The region between the Habur and the upper Zab, formerly inhabited by the Nestorians (Assyrians), is now known as Hakâri, after the Kurdish tribe of that name. See art. 'Nestorians' by B. Nikitin in *E.I.* The first modern travellers to describe the geography of the region were W. F. Ainsworth (*Travels in Asia Minor*, London, 1842, vol. II, chaps. XXXVI–XLVI) and A. H. Layard (*Nineveh and its Remains*, London, 1849, vol. I, chaps. VII–VIII; and *Discoveries in Nineveh and Babylon*, London, 1853, chaps. XVII and XIX). The best general description is still *The Cradle of Mankind*, by W. A. and E. T. A. Wigram, London, 1914. (In his lengthy bibliography in *E.I.*, Nikitin ignores this work and cites Layard's *Nineveh* for the right bank of the Zab.)

A German alpine expedition visited Hakâri during the 1930's; see Hans Bobek, 'Zentral Kurdistan' (offprint from *Zeitschrift des deutschen Alpenvereins*, 1937) and *idem*, 'Forschungen in zentral-kurdischen Hochgebirge' in *Petermann's geographische Mitteilungen*, May 1938, with five sheets of the area explored on scale 1:50,000. A further article on glacier formations was printed in *Zeitschrift für Gletscherkunde*, 1939.

For Turkish expeditions to Hakâri during the 1940's, see *Cilo Dağlari* (The Cilo Mountains) by Şinasi Barutçu, Istanbul, 1949 (with brief Turkish-French-German text and some fine photographs). Hakâri is also described in *Dağlarimiz* (Our Mountains) by Muvaffak Uyanik, Istanbul, 1948.

There is a short account of Hakâri by Besim Darkot in *I.A.*

(vi) *Baku and the Turkish invasion of Caucasia*

Dunsterville, Gen., *Adventures of Dunsterforce*, London, 1920.

Karabekir, Gen. Kâzim, *Erzincan'in ve Erzurum'un Kurtuluşu* (The Liberation of Erzincan and Erzurum), Istanbul, 1939.

Dursunoğlu, C., *Milli Mücadelede Erzurum* (Erzurum in the National Struggle), Ankara, 1946.

Korganov, Gen. C., *La Participation des Arméniens à la Guerre Mondiale sur le front du Caucase, 1914–1918*, Paris, 1927.

Rawlinson, Lt.-Col. A., *Adventures in the Near East, 1918–1922*, London, 1923.

Dokumenty i materialy po vneshney politike Zakavkazya i Gruzii, published by the Government of the Georgian Republic, Tiflis, 1919.

Avalishvili, Zourab, *The Independence of Georgia in International Politics*, London, 1935 (English translation of the author's *Nezavisimost' Gruzii*).
Important.

Poidebard, A., *Voyages au carrefour des routes de la Perse*, Paris, 1923.
An original contribution by a gifted French officer.

Idem, 'Rôle militaire des Arméniens sur le front du Caucase après la défection de l'armée russe', in *Revue des Etudes Arméniennes*, II, pp. 143–6, Paris, 1920.

Price, Morgan Phillips, *War and Revolution in Asiatic Russia*, London, 1918.

Donohoe, M. H., *With the Persian Expedition*, London, 1918.

*Baikov, B., *Vospominaniya o revolyutsii v Zakavkazye (1917–20 gg.)*, Berlin, no date.
Memoirs of a leader of the Russian Cadet party in Baku.

*Ratgauzer, Ya., *Revolyutsiya i grazhdanskaya voyna v Baku*, Baku, 1927.

(vii) *Note on Russian naval and air operations against the Zonguldak coalfields*

The coalfields at Zonguldak and Ereğli were an essential source of fuel for Constantinople—particularly in a war when Turkey was cut off from supplies of higher-grade British coal. According to the Soviet naval historian, N. Novikov, this aspect of economic strategy had not been understood by the Russian naval command in the Black Sea prior to the outbreak of war at the end of October 1914.

As early as 1 November 1914, the western Allies, concerned to disrupt coal supplies to the Turkish capital, communicated with *Stavka* who sent a directive to the naval command in the Black Sea to undertake operations against Zonguldak. The fleet left Sevastopol on 4 November. No preparations had been made. Only one copy of the map of the area was available to the Russian naval staff, and only one officer with the fleet (a Lieutenant Tumanov—a Georgian) knew anything about the topography of Zonguldak. On 7 November the battleship *Rostislav* and the cruiser *Kagul* carried out the first bombardment of Zonguldak under conditions of very poor visibility; fires were started but it was not possible to estimate the damage caused. The Russians also sank two colliers and three large troopships, the latter on their way to Trebizond.

The raid did not interrupt production and shipments of coal from Zonguldak. The port was protected by a high breakwater which gave shelter from Russian shell-fire to Turkish ships while loading.

After a delay of nearly two months, the Russian naval command mounted an attempt to seal the port by sinking four blockships at the entrance. Again the expedition was poorly organized. There was little co-ordination between the flag officer (the captain of *Rostislav*) and the commanders of the cruiser *Almaz* and the destroyer flotilla. In the interests of secrecy, the junior commanders had not been properly briefed and during the course of the operation they reported: 'Zonguldak is impossible to find.'

The squadron approached Zonguldak in bad weather during the night of 23–24 December (while Enver's divisions were approaching Sarikamiş). The ships had scattered in the dark and, as dawn broke, the captain of *Rostislav* tried to get them together again. The officer in charge of the blockship *Oleg* reported that during the night he had run into a flotilla of destroyers which he thought were Russian; they turned their searchlights on his superstructure and, thinking that they were trying to identify him, he shouted the name of his ship through a megaphone. The answer came back in Russian, 'Thanks, we have already read that you are *Oleg*. Here comes a present', after which fire was opened and the Russian ship was hit. As a matter of fact the entire episode of destroyers was later proved to have been a figment of the imagination of the Russian officer. The only enemy ship in the area was *Breslau* which, in the early hours of the morning of 24 December had already sunk another blockship. The appearance of *Breslau* in the midst of a strong Russian squadron, which included a battleship, and at a moment when the entire Russian Black Sea fleet was in the neighbourhood, puzzled the Russians. Disorder increased. After the Russian ships had reassembled, the destroyers were sent forward to bombard the port from close quarters, but they were driven off by the fire of new Turkish shore batteries. Thinking that *Goeben*, and not only *Breslau*, was in the neighbourhood, the Russian commander ordered the sinking of the remaining blockships, after which the fleet returned to Sevastopol.

The failure of this attempt forced the Russians to undertake a naval blockade of the coalfield area, which lasted until the outbreak of the revolution in March 1917. It was carried out by pickets of fast destroyers patrolling close inshore, whilst a squadron including battleships and cruisers stood by in case of emergency, some fifty to sixty miles away. Ships were relieved as their fuel ran out, and there were often gaps between the departure of one flotilla and the arrival of another. During the blockade Zonguldak was bombarded more than twenty-five times. Each time the Turks repaired the damage. Their look-outs on the mountains gave adequate warning of the approach of Russian ships so that transport between the mines and the port could be halted or diverted. At the same time, the pitheads, power stations and other installations, hidden in the valleys, were invisible from the sea.

Seeing that naval bombardments did not disrupt coal production, the Russian command decided to raid the mines from the air, using fourteen hydroplanes. On 6 February 1916 (a few days before the storm of Erzurum) the planes were brought to within fifteen miles of Zonguldak on two transports and lowered for a take-off from the sea. Only eleven of them reached their objective, where they found poor visibility and low clouds; further, they were prevented from flying low by accurate Turkish anti-aircraft fire, the existence of which came as a surprise. The results of the raid were insignificant, as most of the bombs were dropped at random. No aircraft were lost during the actual operation and a torpedo attack on the transports undertaken by an enemy submarine was beaten off.

The blockade went on, and the Turks lost an ever-increasing number of steamships and sailing barges, but they always had fresh boats ready to carry the coal. Since July 1915 German U-boats had been in the area, and *Breslau*, Turkish cruisers, and even *Goeben* were used to cover the movements of the colliers. All the same, the gradual destruction of Turkish shipping decreased supplies of coal to the capital, so that at last the German naval command in Constantinople was forced to lay on supplies of German coal by rail. Throughout the war, however, the Turks persisted in their dogged determination to bring supplies of Zonguldak coal by sea; and each time that Russian naval forces were deflected to other operations—such as the landings on the coast of Lazistan (March–April 1916)—Turkish boats would again be carrying coal to Constantinople.

In general, despite the great efforts deployed by the Russian Black Sea fleet, the blockade of Zonguldak was never fully effective. The only way to stop production of coal in the Zonguldak area would have been to carry out landing operations there; this would not have been difficult, as the Turks at first only had two coastal batteries and one battalion of infantry for the defence of the area, whilst lack of communications with the interior would have made it impossible to bring up reinforcements in time to prevent landing parties from blowing up pitheads and other installations. The Russian naval command, however, was opposed to landing operations since it feared that the fleet might become involved to a serious extent in an area far removed from its nearest base at Sevastopol. The naval command further put forward the unpredictability of weather conditions in the Black Sea and held the view that troops would suffer too much from cramped conditions and sea-sickness on board the transports and would be unable to fight with success when landed on a hostile shore.

For details, see N. Novikov, *Operatsii flota protiv berega na Chernom more v 1914–1917 gg.*

(7) Epilogue

'THE GOVERNMENT OF SOUTH-WEST CAUCASIA' AND THE DAĞISTAN REVOLT

(*a*) General

The first detailed history in English of the political history of Transcaucasia during the period of the independent republics was published in New York in 1951: *The Struggle for Transcaucasia (1917–21)*, by Firuz Kazemzadeh. The excellent notes and bibliography cover the Soviet and émigré literature on the subject. Reference may further be made to 'Handbook prepared under the direction of the Historical Section of the Foreign Office', No. 54, *Caucasia*, H.M. Stationery Office, 1920; and *The Baltic and Caucasian States*, edited by John Buchan, London, 1923 (sections on Georgia and Azerbaijan by W. E. D. Allen). Useful bibliographies are included in *B.S.E.*, arts. 'Armyanskaya S.S.R.' and 'Azerbaydzhanskaya S.S.R.', and in *I.A.*, arts. 'Azerbaycan' (by A. Z. V. Togan), 'Azeri' (by M. F. Köprülü) and 'Gence' (Ganja) (by Mirza Bala). See also *Materialy dlya bibliografii Azerbaydzhana*, published by Dom rabotnikov prosveshcheniya, Baku, 1924–5.

The following works on Azerbaijan relate to the Muslim Turkish areas included in the 'Government of South-west Caucasia':

Claims of the Peace Delegation of the Republic of Caucasian Azerbaijan presented to the Peace Conference in Paris, Paris, 1919.

Resulzade, Mehmet Emin (Rassoul-zadé), *L'Azerbaidjan en lutte pour l'indépendance*, Paris, 1930.

Idem, *O Panturanizme v svyazi s kavkazskoy problemoy*, Paris, 1930.

Idem, *Grundzüge der Geschichte Azerbeidschans* (in MS.).

(Mehmet-zade) Mirza Bala, *Milli Azerbaycan Hareketi* (The Azerbaijan People's Movement), Berlin, 1938.

Ziatkhan, Adil Khan, *Aperçu sur l'histoire, la littérature et la politique de l'Azerbeidjan*, Baku, 1919.

Numerous references to Georgia, Azerbaijan and Dağistan are contained in Gen. A. I. Denikin's *Ocherki russkoy smuty* (vols. I and II, Paris, vols. III, IV, V, Berlin, 1924–6).

(*b*) Government of South-west Caucasia

Kars, Istanbul, 1943.

The historical chapter by F. Çelik gives many interesting details of the organization of Turkish resistance to Georgian and Armenian intervention in the Kars province during the years 1919–21.

The *émigré* journal *Kavkaz*, published in Paris (67 numbers, from Jan. 1934 to July 1939), contains numerous articles on events in the south-west Caucasian region. Many of these were reprinted in the French edition of the journal (*Le Caucase*, twenty-six numbers from June 1937 to July 1939) and in the English edition (*The Caucasian Quarterly*, four numbers from Nov. 1937 to Mar. 1939; ref. *C.Q.*).

Kvinitadze, Gen. G., 'The occupation of Ardahan', *C.Q.*, Oct. 1938–Mar. 1939, pp. 23–31.

A picturesque personal account of the occupation of Ardahan by Georgian troops in April 1919.

Idem, 'The lost opportunity', *ibid.* no. 1, Nov. 1937, pp. 46–52.

An account of the abortive Soviet attempt to invade Georgia from Azerbaijan in May 1920. 'The lost opportunity' arose from the refusal of the Georgian government to authorize the occupation of Ganja by Gen. Kvinitadze with a view to supporting a Muslim revolt against the Red Army in Azerbaijan.

Idem, 'The truth about Batum', *ibid.* Jan.–Mar. 1938, pp. 57–63.

A somewhat confused account of Georgian-Turco-Soviet relations in March 1921. See also *ibid.* no. 1, pp. 25–30, *idem*, 'Moscow and Tiflis', for some details on the strength of the Georgian army at the time of the Soviet attack in the spring of 1921.

Avalishvili, Zourab, 'From San Stefano to Batum (1878–1921)', *ibid.* Jan.–Mar. 1938, pp. 64–77.

An acute analysis of Turco-Georgian relations in the crisis of the fall of the Georgian republic.

Idem, 'Regarding certain controversies', *ibid.* Jan.–Mar. 1938.

An account of Batum and Turco-Georgian relations.

Odishelidze, Gen. I. Z., 'O prichinakh porazheniya Gruzii', *Kavkaz*, nos. 18, 19–20, 21–22 (June–Oct. 1935).

An inquiry into the causes of the fall of the Georgian republic by a distinguished Georgian general.

Idem, 'Rol' gruzinskikh menshevikov v voyne 1921 g.', *ibid.* no. 25 (Jan. 1936).

Artsruni, Ashot, 'Fevralskoye vozstaniye v Armenii', *ibid.* no. 26 (Feb. 1936).

Describes the unlucky Dashnak attempt to expel the Red Army from Armenia in Feb. 1921.

Gvaramadze, I., 'Iz nedavney istorii karsskoy oblasti', *ibid.* no. 55 (Oct. 1938).

An account of Georgian and Armenian territorial rivalries in the former Kars province.

For Soviet interpretation of detail of this period, see:

*Khachapuridze, G., 'Adzharistan v period angliyskoy okkupatsii', in *Borba klassov*, no. 12, 1934.

Gukovski, A., 'Pobeda sovetskoy vlasti v Armenii v 1920 godu', in *Istorik Marksist*, no. 11 (87), 1940.

Kuznetsova, S., 'Krakh turetskoy interventsii v Zakavkazye v 1920–1921 godakh', in *Voprosy Istorii*, 1951, no. 9, pp. 143–56.

In *Six Prisons and Two Revolutions* (London, 1925), Oliver Baldwin described conditions in Erevan, Kars and Erzurum during 1919–21.

(*c*) Dağistan. Topographical

Vol. II of J. F. Baddeley's *Rugged Flanks of Caucasus* gives the best description in English of central Dağistan and is of great value for the topography of the area. For historical topography see also Baddeley's *Conquest*.

In *Aus den Hochregionen des Kaukasus*, vol. II, Gottfried Merzbacher describes the eastern section of the main chain of the Caucasus from Tebulosmta to the Addala peaks of the Bogos massif. In chapter XXXIV he gives an account of the descent from the western flanks of Bogos into the Kakhetian plain. In *Kaukasus: Reisen und Forschungen im kaukasischen Hochgebirge*, vol. II, Moriz von Dechy, more briefly, covers the same ground. Both authors reproduce numerous photographs; those of Dechy are superb.

R. von Erckert, apparently a German in the Russian service, spent two years in the Caucasus during the eighties in command of the 21st Infantry Division. In his *Kaukasus und seine Völker*, he has left an account of the topography and tribes of Dağistan (pp. 145–295) which is still useful.

Although published more than fifty years ago, *Pamyatnaya knizhka dagestanskoy oblasti*, edited by E. I. Kozubski (Temir-Khan-Shura, 1895), contains much interesting material and a valuable bibliography. There is a useful tribal map.

Todorski (see below) gives plans of the battle areas in 1921–2. These are based on the Russian maps (1 *verst*: 1 *vershok*) but unfortunately they are badly printed.

(*d*) Dağistan and Northern Caucasia, 1917–21

I.A., art. 'Dağistan', by Mirza Bala contains short bibliographical note. *B.S.E.*, art. 'Dagestanskaya A.S.S.R.', by various authors contains useful maps and some comparatively recent photographs.

Aperçu historique sur les Ciscaucasiens pendant la guerre mondiale and *Compte-rendu des assemblées des peuples de la Ciscaucasie et de leurs travaux legislatifs*, both published by Loeffler, Istanbul, 1918.

Todorski, A., *Krasnaya armiya v gorakh*, published by Voyenny Vestnik, Moscow, 1924, with numerous maps and sketches.

Samurski (Efendiyev), *Dagestan*, Moscow, 1925, with maps; and *Novy Vostok*, III, pp. 230–40, 'Grazhdanskaya voyna v Dagestane'.

Takho-Godi, A., *Revolyutsiya i kontr-revolyutsiya v Dagestane*, published by Dagestanskoye Gos. Izd., Makhach-Kala, 1927, with numerous photographs of personalities and documents.

Allen, W. E. D., 'Military operations in Daghestan, 1917–21', in *Army Quarterly*, 1934–5 (based on Todorski, Samurski and Takho-Godi).

Pavlovich, M., 'Krasny Dagestan', in *Novy Vostok*, III, pp. 220–9; VIII–IX, pp. 12–25.

Apukhtin, V., 'Materialy grazhdanskoy voyny v Chechne', in *Novy Vostok*, VIII–IX, pp. 160–77.

Arsharuni, A., 'Iz istorii grazhdanskoy voyny na Tereke: Kazache-Osetinski blok', in *Novy Vostok*, XXVIII, pp. 165–74.

Takoyev, S., 'K istorii revolyutsionnogo dvizheniya na Tereke', in *Izvestiya Osetinskogo instituta krayevedeniya*, vol. II, Vladikavkaz, 1926, pp. 309–85.

The organization of the Bolshevik movement in northern Caucasia by Buachidze, Ordzhonikidze and Kirov has been described in an anonymous article in *Voprosy Istorii*, 1951, no. 10, pp. 3–21, 'Iz istorii borby za sovetskuyu vlast' na severnom Kavkaze v 1917–1918 godakh'. The article is based on a useful analysis of the contemporary north Caucasian press. See also I. M. Razgon, 'Sergo Ordzhonikidze v gody grazhdanskoy voyny', in *Istoricheskiye Zapiski*, II (1937), pp. 3–48.

(*e*) Languages of Dağistan

For the languages of the Andi and Dido tribes, see A. M. Dirr, *Kratki grammaticheski ocherk andiyskago yazyka* (Tiflis, 1906) and *Materialy dlya izucheniya yazykov i narechiy ando-didoyskoy gruppy* (Tiflis, 1909). For Avar, see art. in *B.S.E.* 'Avarski yazyk' and observations by Baddeley, *Conquest*, pp. xxix ff. Cf. also *B.S.E.*, art. 'Dagestanskiye yazyki', citing Zhirkov, *Grammatika avarskago yazyka* (Moscow, 1924). For Kazi-Kumukh (Lak), see P. K. Uslar, *Lakski yazyk* (Tiflis, 1890). For the language of southern Dağistan, see studies by A. M. Dirr in *Sbor. Mat.* XXXIII (Udin), XXXV (Tabassarani), XXXVII (Agul), XL (Archin), XLII (Rutul), XLIII (Tsakhur).

INDEX

The following abbreviations have been used in the Index: A. (Armenian), Ch. (Cherkess), E. (English), G. (Georgian), R. (Russia, Russian, Russians), T. (Turkey, Turkish, Turks).

In view of the fact that Turkish proper names, etymologically derived from two or more words, can be written—equally correctly—as one word, two words joined by a hyphen, two disjointed words, etc., names consisting of several words are placed in the Index as if they were spelled in one word only. Thus, for example, the entry for Ali Haci occurs before Alikhanov, whilst Ali Paşa comes after Alikhanov.

Ç, ç, Ş, ş, Ö, ö, Ü, ü, ğ, etc., have been treated for purposes of indexing as c, s, o, u, g, etc.

Names and numbers of military units have not been included in the Index.